Photochemistry of Organic Molecules in Isotropic and Anisotropic Media

T0175716

MOLECULAR AND SUPRAMOLECULAR PHOTOCHEMISTRY

Series Editors

V. RAMAMURTHY

Professor
Department of Chemistry
Tulane University
New Orleans, Louisiana

KIRK S. SCHANZE

Professor
Department of Chemistry
University of Florida
Gainesville, Florida

1. Organic Photochemistry, *edited by V. Ramamurthy and Kirk S. Schanze*
2. Organic and Inorganic Photochemistry, *edited by V. Ramamurthy and Kirk S. Schanze*
3. Organic Molecular Photochemistry, *edited by V. Ramamurthy and Kirk S. Schanze*
4. Multimetallic and Macromolecular Inorganic Photochemistry, *edited by V. Ramamurthy and Kirk S. Schanze*
5. Solid State and Surface Photochemistry, *edited by V. Ramamurthy and Kirk S. Schanze*
6. Organic, Physical, and Materials Photochemistry, *edited by V. Ramamurthy and Kirk S. Schanze*
7. Optical Sensors and Switches, *edited by V. Ramamurthy and Kirk S. Schanze*
8. Understanding and Manipulating Excited-State Processes, *edited by V. Ramamurthy and Kirk S. Schanze*
9. Photochemistry of Organic Molecules in Isotropic and Anisotropic Media, *edited by V. Ramamurthy and Kirk S. Schanze*

ADDITIONAL VOLUMES IN PREPARATION

Photochemistry of Organic Molecules in Isotropic and Anisotropic Media

edited by

V. Ramamurthy
Tulane University
New Orleans, Louisiana

Kirk S. Schanze
University of Florida
Gainesville, Florida

CRC Press
Taylor & Francis Group
Boca Raton London New York

CRC Press is an imprint of the
Taylor & Francis Group, an **informa** business

CRC Press
Taylor & Francis Group
6000 Broken Sound Parkway NW, Suite 300
Boca Raton, FL 33487-2742

First issued in paperback 2019

© 2003 by Taylor & Francis Group, LLC
CRC Press is an imprint of Taylor & Francis Group, an Informa business

No claim to original U.S. Government works

ISBN-13: 978-0-367-39551-3

Library of Congress Cataloging-in-Publication Data
A catalog record for this book is available from the Library of Congress.

Visit the Taylor & Francis Web site at
http://www.taylorandfrancis.com

and the CRC Press Web site at
http://www.crcpress.com

Preface

Owing to the dedicated efforts of numerous chemists, photochemistry has reached a stage of maturity—considerable knowledge has been accumulated concerning the excited-state behavior of both organic and inorganic molecules, and basic rules of the discipline have been laid out. To keep this knowledge intact and readily available it is necessary to periodically review the information that has been accumulated. With this in mind, this series continues to provide in-depth coverage of various aspects of photoscience.

Life as we know it cannot exist without light and light has become an integral part of modern life. The research of the type discussed in this volume has led to applications such as photodynamic therapy, LASIK surgery, photolithography (which is central to modern printing and electronic industries), xerography, and solar energy conversion, to name a few. Understanding of phenomena that at the time of initial investigation seemed remotely connected to anything "useful" led to the above major applications. For example, explorations into cinnamic acid photodimerization led to the invention of "photolithography." Investigations of the photofragmentation of organic molecules by an organic photochemist led to the invention of LASIK surgery. Thus, the connection between the basic and applied science is stronger than might be immediately apparent.

One might consider that science progresses in three phase: (1) fundamental observation and exploration to establish the basic rules, (2) manipulation of what has been understood, and (3) exploitation of the fundamental observation toward material benefits. All three are important ingredients for the sustenance and growth of science. In this volume we have attempted to illustrate this process of scientific discovery through eight chapters written by experts who have made significant contributions to the topics discussed.

Chapters 1 through 4 summarize the basic excited-state reactions of organic molecules. Chapters 5 and 6 present the existing knowledge on controlling and manipulating two reactions: electron transfer and oxygenation. Chapters 7 and 8 describe how one exploits the available information in materials applications. These chapters as a group provide a nice illustration of collection, manipulation, and exploitation of knowledge for our benefit.

Chapter 1, by Armesto et al., summarizes the recent results on di-π-methane and related rearrangements. Many of the reactions reported in this chapter were discovered by Armesto's group and expand the potential of this versatile reaction often known as the Zimmerman reaction. This chapter complements many of the reactions discussed previously in this series.

Chapter 2, by Miranda and Galinco, provides a critical survey of the photo-Fries reaction undergone by numerous aromatic esters, amides, and so forth. This chapter is a valuable companion to Chapter 5 by Fleming and Pincock, in Volume 3. Miranda and Galinco's chapter is the sixth chapter devoted to the photochemistry of aromatic compounds in this series.

Chapter 3, by Nicolaides and Tomioka, on the generation and characterization of biscarbenes, bisnitrenes, and carbenonitrenes illustrates how computational methods can serve as a valuable tool in understanding highly reactive intermediates. Given that many of the high-level computations can be performed on desktop computers, computation is likely to become a more common tool in physical organic chemistry laboratories. A future volume in this series will be devoted to computational methods in photochemistry.

Chapters 4 and 5 discuss electron transfer, a fundamental and simple process in chemistry. This topic is not new to this series: at least eight chapters have dealt with electron transfer process in photochemistry. Chapter 4, by Schmoldt et al., focuses on the applications of photoelectron transfer process in organic synthesis. This chapter along with Chapter 7, Volume 1 by Pandey and Chapter 8, Volume 1 by Li is a must for those wishing to use photoelectron transfer (PET) process as a strategy to assemble complex organic molecules. Chapter 5, by Fukuzumi and Imahori, presents information on designing molecules mimicking photosynthetic center. In addition, less appreciated use of metal ions to control the PET process is highlighted in this chapter.

Chapter 6, by Clennan, on oxygenation in zeolites critically summarizes the results on oxidation of organic molecules within zeolites by singlet oxygen and superoxide anion. This chapter complements Chapter 6, Volume 5 by Vasenko and Frei. In addition to Chapter 6, Chapter 7 in this volume and Chapters 3, 4, and 5 in Volume 5 are exceptional resources in the area of excited-state behavior of organic molecules within zeolites.

In Chapter 7, Maas et al., summarize their elegant work on the photophysics of dye molecules included within zeolites. This chapter illustrates how a marriage between photophysics and materials science could lead to "useful" chemistry.

Chapter 8, by Hobley et al., critically evaluates the existing mechanistic results, especially that of time-resolved investigations, on the photochromic process of spiropyrans and related system. The summary provided should be of great value in the development of these systems as photoswitches.

We are grateful to all the chapter authors for their contributions. It was a pleasure to work with them and we were fortunate to be the first to read the well-organized and critically written chapters. We hope that the material herein is a welcome source of information to graduate students as well as researchers in photoscience.

V. Ramamurthy
Kirk S. Schanze

Contents

Preface   *iii*

Contributors   *ix*

Contents of Previous Volumes   *xi*

1. Recent Advances in Di-π-methane Photochemistry: A New Look at a Classical Reaction   1
   Diego Armesto, Maria J. Ortiz, and Antonia R. Agarrabeitia

2. The Photo-Fries Rearrangement   43
   Miguel A. Miranda and Francisco Galindo

3. The Characterization and Reactivity of Photochemically Generated Phenylene Bis(diradical) Species as Revealed by Matrix Isolation Spectroscopy and Computational Chemistry   133
   Athanassios Nicolaides and Hideo Tomioka

4. Photoinduced-Electron-Transfer Initiated Reactions in Organic Chemistry   185
   Philip Schmoldt, Heiko Rinderhagen, and Jochen Mattay

5. Design and Fine Control of Photoinduced Electron Transfer   227
   Shunichi Fukuzumi and Hiroshi Imahori

6. Molecular Oxygenations in Zeolites   275
   Edward L. Clennan

7. Organic–Inorganic Composites as Photonic Antenna   309
    Huub Maas, Stefan Huber, Abderrahim Khatyr, Michel Pfenniger,
    Marc Meyer, and Gion Calzaferri

8. Photo-Switching Spiropyrans and Related Compounds   353
    Jonathan Hobley, Martin J. Lear, and Hiroshi Fukumura

Index   *405*

Contributors

Antonia R. Agarrabeitia   Organic Chemistry Department, Faculty of Chemistry, Complutense University of Madrid, Madrid, Spain

Diego Armesto   Organic Chemistry Department, Faculty of Chemistry, Complutense University of Madrid, Madrid, Spain

Gion Calzaferri   Department of Chemistry and Biochemistry, University of Berne, Berne, Switzerland

Edward L. Clennan   Department of Chemistry, University of Wyoming, Laramie, Wyoming, U.S.A.

Hiroshi Fukumura   Department of Chemistry, Graduate School of Science, Tohoku University, Sendai Miyagi, Japan

Shunichi Fukuzumi   Department of Material and Life Science, Graduate School of Engineering, Osaka University, Osaka, Japan

Francisco Galindo   Department of Inorganic and Organic Chemistry, Universidad Jaume I, Castellón, Spain

Jonathan Hobley   Department of Chemistry, Graduate School of Science, Tohoku University, Sendai Miyagi, Japan

Stefan Huber   Department of Chemistry and Biochemistry, University of Berne, Berne, Switzerland

Hiroshi Imahori   Department of Material and Life Science, Graduate School of Engineering, Osaka University, Osaka, Japan

Abderrahim Khatyr   Department of Chemistry and Biochemistry, University of Berne, Berne, Switzerland

Martin J. Lear   Department of Chemistry, Graduate School of Science, Tohoku University, Sendai Miyagi, Japan

Huub Maas   Department of Chemistry and Biochemistry, University of Berne, Berne, Switzerland

Jochen Mattay   Chemistry Department, Universität Bielefeld, Bielefeld, Germany

Marc Meyer   Department of Chemistry and Biochemistry, University of Berne, Berne, Switzerland

Miguel A. Miranda   Departamento de Química-Instituto de Tecnología Química, Universidad Politécnica de Valencia, Valencia, Spain

Athanassios Nicolaides   Department of Chemistry, University of Cyprus, Nicosia, Cyprus

Maria J. Ortiz   Organic Chemistry Department, Faculty of Chemistry, Complutense University of Madrid, Madrid, Spain

Michel Pfenniger   Department of Chemistry and Biochemistry, University of Berne, Berne, Switzerland

Heiko Rinderhagen   Chemistry Department, Universität Bielefeld, Bielefeld, Germany

Philip Schmoldt   Chemistry Department, Universität Bielefeld, Bielefeld, Germany

Hideo Tomioka   Chemistry Department for Materials, Faculty of Engineering, Mie University, Mie, Japan

Contents of Previous Volumes

Volume 1.   Organic Photochemistry

1. The Photochemistry of Sulfoxides and Related Compounds
    William S. Jenks, Daniel D. Gregory, Yushen Guo, Woojae Lee, and Troy Tetzlaff

2. The Photochemistry of Pyrazoles and Isothiazoles
    James W. Pavlik

3. Photochemistry of (S-Hetero)cyclic Unsaturated Carbonyl Compounds
    Paul Margaretha

4. Photochemistry of Conjugated Polyalkynes
    Sang Chul Shim

5. Photochemistry and Photophysics of Carbocations
    Mary K. Boyd

6. Regio- and Stereoselective 2, + 2 Photocycloadditions
    Cara L. Bradford, Steven A. Fleming, and J. Jerry Gao

7. Photoinduced Redox Reactions in Organic Synthesis
    Ganesh Pandey

8. Photochemical Reactions on Semiconductor Particles for Organic Synthesis
   Yuzhuo Li

9. Photophysics and Photochemistry of Fullerene Materials
   Ya-Ping sun

10. Use of Photophysical Probes to Study Dynamic Processes in Supramolecular Structures
    Mark H. Kleinman and Cornelia Bohne

11. Photophysical and Photochemical Properties of Squaraines in Homogeneous and Heterogeneous Media
    Suresh Das, K. George Thomas, and M. V. George

12. Absorption, Fluorescence Emission, and Photophysics of Squaraines
    Kock-Yee Law

Volume 2.   Organic and Inorganic Photochemistry

1. A Comparison of Experimental and Theoretical Studies of Electron Transfer Within DNA Duplexes
   Thomas L. Netzel

2. Coordination Complexes and Nucleic Acids. Perspectives on Electron Transfer, Binding Mode, and Cooperativity
   Eimer Tuite

3. Photoinduced Electron Transfer in Metal-Organic Dyads
   Kirk S. Schanze and Keith A. Walters

4. Photochemistry and Photophysics of Liquid Crystalline Polymers
   David Creed

5. Photochemical Solid-to-Solid Reactions
   Amy E. Keating and Miguel A. Garcia-Garibay

6. Chemical and Photophysical Processes of Transients Derived from Multiphoton Excitation: Upper Excited States and Excited Radicals
   W. Grant McGimpsey

7. Environmental Photochemistry with Semiconductor Nanoparticles
    Prashant V. Kamat and K. Vinodgopal

Volume 3.   Organic Molecular Photochemistry

1. Solid-State Organic Photochemistry of Mixed Molecular Crystals
    Yoshikatsu Ito

2. Asymmetric Photochemical Reactions in Solution
    Simon R. L. Everitt and Yoshihisa Inoue

3. Photochemical cis-trans Isomerization in the Triplet State
    Tatsuo Arai

4. Photochemical cis-trans Isomerization from the Singlet Excited State
    V. Jayathirtha Rao

5. Photochemical Cleavage Reactions of Benzyl-Heteroatom Sigma Bonds
    Steven A. Fleming and James A. Pincock

6. Photophysical Probes for Organized Assemblies
    Kankan Bhattacharyya

**Volume 4.   Multimetallic and Macromolecular Inorganic
Photochemistry**

1. Metal-Organic Conducting Polymers: Photoactive Switching in
    Molecular Wires
    Wayne E. Jones, Jr., Leoné Hermans, and Biwang Jiang

2. Luminescence Behaviour of Polynuclear Metal Complexes of Copper (I),
    Silver (I), and Gold (I)
    Vivian W.-W. Yam and Kenneth K.-W. Lo

3. Electron Transfer Within Synthetic Polypeptides and De Novo Designed
    Proteins
    Michael Y. Ogawa

4. Tridentate Bridging Ligands in the Construction of Stereochemically
Defined Supramolecular Complexes
Sumner W. Jones, Micheal R. Jordan, and Karen J. Brewer

5. Photophysical and Photochemical Properties of Metallo-1,2-enedithiolates
Robert S. Pilato and Kelly A. Van Houten

6. Molecular and Supramolecular Photochemistry of Porphyrins and
Metalloporphyrins
Shinsuke Takagi and Harou Inoue

Volume 5.   Solid State and Surface Photochemistry

1. Spectroscopy and Photochemical Transformations of Polycyclic Aromatic
Hydrocarbons at Silica – and Alumina – Air Interfaces
Reza Dabestani and Michael E. Sigman

2. Photophysics and Photochemistry in Clay Minerals
Katsuhiko Takagi and Tetsuya Shichi

3. Photochromism of Diarylethenes in Confined Reaction Spaces
Masahiro Irie

4. Charge and Electron Transfer Reactions in Zeolites
Kyung Byung Yoon

5. Time-Resolved Spectroscopic Studies of Aromatic Species Included in
Zeolites
Shuichi Hashimoto

6. Photo-Oxidation in Zeolites
Sergey Vasenko and Heinz Frei

Volume 6.   Organic, Physical, and Materials Photochemistry

1. Photochemistry of Hydroxyaromatic Compounds
Darryl W. Brousmiche, Alexander G. Briggs, and Peter Wan

2. Stereoselectivity of Photocycloadditions and Photocyclizations
Axel G. Griesbeck and Maren Fiege

3. Photocycloadditions with Captodative Alkenes
   Dietrich Döpp

4. Photo- and Electroactive Fulleropyrrolidines
   Michele Maggini and Dirk M. Guldi

5. Applications of Time-Resolved EPR in Studies of Photochemical
   Reactions
   Hans van Willigen

6. Photochemical Generation and Studies of Nitrenium Ions
   Daniel E. Falvey

7. Photophysical Properties of Inorganic Nanoparticle-DNA Assemblies
   Catherine J. Murphy

8. Luminescence Quenching by Oxygen in Polymer Films
   Xin Lu and Mitchell A. Winnik

Volume 7.   Optical Sensors and Switches

1. Buckets of Light
   Christina M. Rudzinski and Daniel G. Nocera

2. Luminescent PET Signalling Systems
   *A. Prasanna de Silva, David B. Fox, Thomas S. Moody, and Sheenagh
   M. Weir*

3. Sensors Based on Electrogenerated Chemiluminescence
   Anne-Margret Andersson and Russell H. Schmehl

4. From Superquenching to Biodetection: Building Sensors Based on
   Fluorescent Polyelectrolytes
   *David Whitten, Liaohai Chen, Robert Jones, Troy Bergstedt, Peter
   Heeger, and Duncan McBranch*

5. Luminescent Metal Complexes as Spectroscopic Probes of Monomer/
   Polymer Environments
   Alistair J. Lees

6. Photorefractive Effect in Polymeric and molecular Materials
    ALiming Wang, Man-Kit Ng, Qing Wang, and Luping Yu

7. Dynamic Holography in Photorefractive liquid Crystals
    Gary P. Wiederrecht

8. Hologram Switching and Erasing Strategies With Liquid Crystals
    Michael B. Sponsler

9. Novel Molecular Photonics Materials and Devices
    Toshihiko Nagamura

10. Molecular Recognition Events Controllable by Photochemical Triggers
    or Readable by Photochemical Outputs
    Seiji Shinkai and Tony D. James

11. Probing Nanoenvironments Using Functional Chromophores
    Mitsuru Ishikawa and Jing Yong Ye

Volume 8.   Understanding and Manipulating Excited-State Processes

1. Ortho Photocycloaddition of Alkenes and Alkynes to the Benzene Ring
    Jan Cornelisse and Rudy de Haan

2. Photocycloaddition and Photoaddition Reactions of Aromatic Compounds
    *Kazuhiko Mizuno, Hajime Maeda, Akira Sugimoto, and Kazuhiko
    Chiyonobu*

3. Singlet-Oxygen Ene-Sensitized Photo-Oxygenations: Stereochemistry and
    Mechanisms
    Michael Orfanopoulos

4. Singlet-Oxygen Reactions: Solvent and Compartmentalization Effects
    Eduardo A. Lissi, Else Lemp, and Antonio L. Zanocco

5. Microreactor-Controlled Product Selectivity in Organic Photochemical
    Reactions
    Chen-Ho Tung, Kai Song, Li-Zhu Wu, Hong-Ru Li, and Li-Ping Zhang

6. Enantioselective Photoreactions in the Solid State
    Fumio Toda, Koichi Tanaka, and Hisakazu Miyamoto

7. Observations on the Photochemical Behavior of Coumarins and Related Systems in the Crystalline State
    Kodumuru Vishnumurthy, Tayur N. Guru Row, and Kailasam Venkatesan

8. Supramolecular Photochemistry of Cyclodextrin Materials
    Akihiko Ueno and Hiroshi Ikeda

9. Photoactive Layered Materials: Assembly of Ions, Molecules, Metal Complexes, and Proteins
    Challa V. Kumar and B. Bangar Raju

10. Fluorescence of Excited Singlet-State Acids in Certain Organized Media: Applications as Molecular Probes
    Ashok Kumar Mishra

11. Photophysics and Photochemistry of Fullerenes and Fullerene Derivatives
    Jochen Mattay, Lars Ulmer, and Andreas Sotzmann

1

Recent Advances in Di-π-methane Photochemistry: A New Look at a Classical Reaction

**Diego Armesto, Maria J. Ortiz, and
Antonia R. Agarrabeitia**
Complutense University of Madrid, Madrid, Spain

I. INTRODUCTION

Studies of di-π-methane photochemical rearrangements have been one of the main areas of research in organic photochemistry for many years (for reviews, see Refs. 1–4). The first example of a reaction of this type was reported by Zimmerman in 1967 in the sensitized irradiation of barrelene 1 that yields semibullvalene 2 [5] (Scheme 1). The reaction has been extended to a large number of acyclic and cyclic 1,4-dienes that yield the corresponding vinylcyclopropanes on irradiation, in the di-π-methane (DPM) version of the rearrangement. This reaction also takes place when a vinyl unit is replaced by an aryl group. A few representative examples of DPM rearrangements are shown in Scheme 1 [6–9].

Studies carried out by different research groups for more than 30 years have shown that this reaction is very general and usually takes place with high chemical and quantum yields affording vinylcyclopropanes that in many instances are difficult to obtain, or not available, by alternative routes. A biradical mechanism has been proposed to account for the rearrangement as

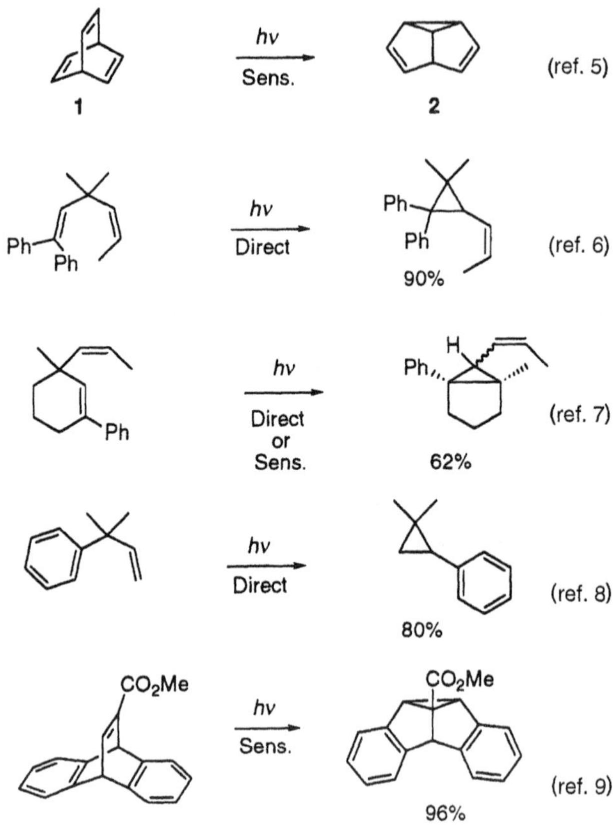

Scheme 1

shown in Scheme 2 for 3,3,5-trimethyl-1,1-diphenyl-1,4-hexadiene [10]. In this pathway, excited 1,4-dienes undergo 2,4-bonding to form a cyclopropyl-1,4-biradical that, by opening of bond **a**, yields the more stable 1,3-biradical, which cyclizes to form the observed product. This mechanism permits one to predict the regioselectivity encountered in the reaction of unsymmetrically substituted 1,4-dienes by simply considering the stability of the two possible 1,3-biradicals resulting from the ring opening of the cyclopropyl-1,4-biradical intermediate. However, alternative concerted mechanisms have been proposed to justify the features of the reaction, particularly the high degree of stereoselectivity observed in some instances [1].

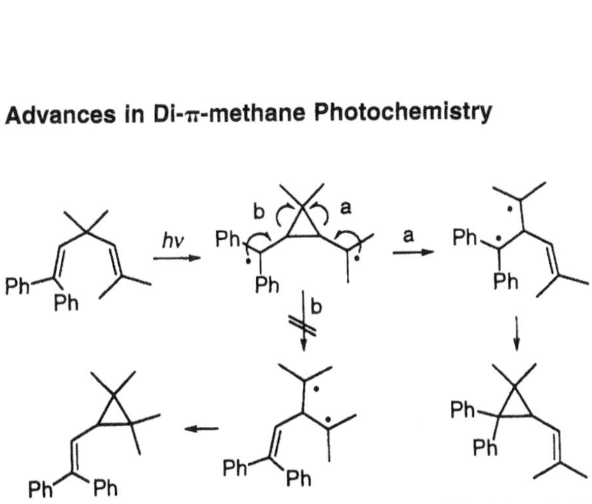

Not obtained

Photoproduct

Scheme 2

The multiplicity of the excited state involved in the rearrangement has been investigated. In general terms, the di-π-methane reaction takes place in the triplet excited state of cyclic 1,4-dienes, whereas the singlet excited state is responsible for the rearrangement of acyclic dienes [4]. However, there are exceptions to this general rule and acyclic 1,4-dienes with electron-withdrawing groups at positions 1 or 3 of the 1,4-diene skeleton rearrange very efficiently in the triplet excited state [11,12]. A large number of studies have been carried out on different aspects of this reaction, such as the influence of substitution, regio selectivity, and stereo selectivity, but is not our intention to discuss these aspects. A detailed summary of these topics can be found in the reviews previously published on the di-π-methane rearrangement [1–4].

Di-π-methane reactions are not restricted to 1,4-dienes. Other 1,4-unsaturated systems, such as β,γ-unsaturated ketones, undergo similar rearrangements yielding cyclopropyl ketones in the oxa-di-π-methane (ODPM) version of the reaction. The first example of the ODPM reaction was reported in 1966, in which irradiation of ketone **3** affords the cyclopropyl ketone **4** [13] (Scheme 3).

Since then, a large number of studies have been carried out on the photochemistry of β,γ-unsaturated ketones and many examples of the ODPM process have been uncovered. Some comprehensive reviews have been published in the last 25 years or so on this subject (for reviews, see Refs. 14–16). The results of these studies on the ODPM rearrangement have demonstrated the following general features. Very often, the reaction takes place with high chemical and quantum yield. The reaction is very general for many cyclic β,γ-unsaturated ketones and it shows a high degree of stereoselectivity, and even enantioselectivity, in some instances. Therefore, it is not surprising that the ODPM reaction has been applied as the key step in the synthesis of some natural products and other highly compli-

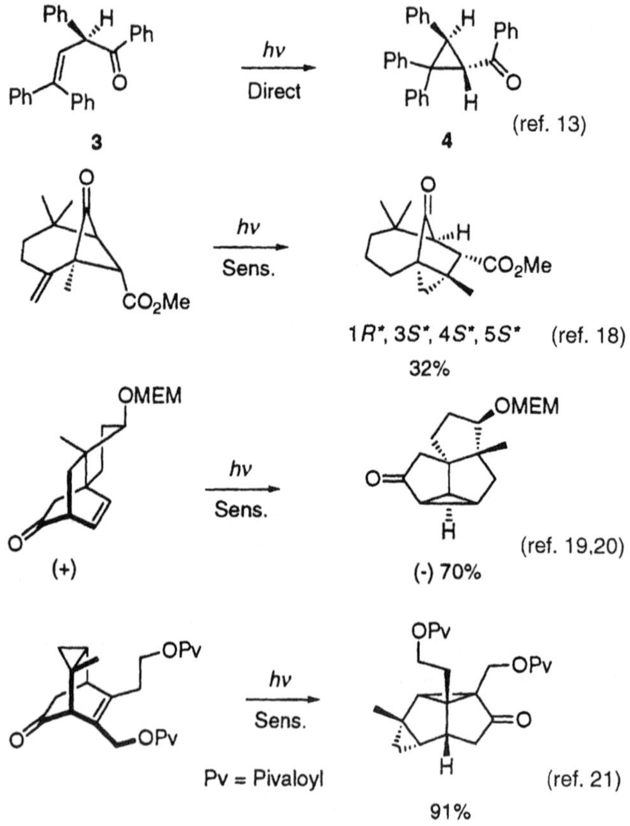

Scheme 3

cated molecules that are difficult to prepare by alternative reaction routes [4,17]. Representative examples of ODPM reactions are shown in Scheme 3 [13,18–21].

A biradical mechanism, analogous to that shown in Scheme 2 for the DPM reaction, is proposed to account for the ODPM rearrangement. This was first postulated by Givens and Oettle to account for the photochemical conversion of benzobicyclo [2.2.2]octadienone into 3,4-benzotricyclo [3.3.0.0^{2,8}]octan-7-one [22]. This mechanism explains the regioselectivity encountered in the rearrangement. In all of the cases studied, the ODPM reaction yields the corresponding cyclopropyl ketones. The alternative ring opening of the cyclopropyl biradical intermediates that would yield oxiranes has never been observed. This regioselectivity can be understood, as shown in Scheme 4 for ketone 5 [23], by considering

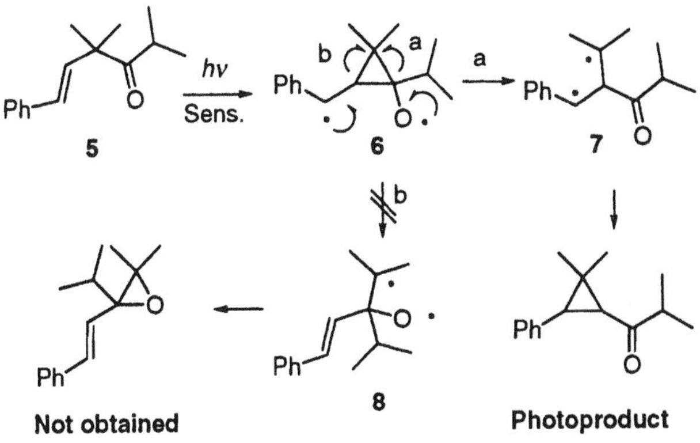

Scheme 4

that the opening of the cyclopropyl biradical intermediate **6** occurs by breaking bond **a** to afford the 1,3-biradical **7**, which is more stable than the alternative 1,3-biradical **8**. As a result, the C—O double bond is always restored.

Concerted mechanisms have also been considered to justify the high degree of stereoselectivity observed in many instances as, for example, in the cases shown in Scheme 3 [13,18–21]. However, the high stereochemical control often observed in many ODPM rearrangements does not necessary imply that the reaction is taking place via concerted mechanisms. A stepwise process is also consistent with the stereochemical outcome of the reaction, where there are conformational or configurational restrictions to rapid C—C rotation. This subject has been extensively discussed and reviewed by Schuster [16].

For more than 15 years, the di-π-methane rearrangements were limited to 1,4-dienes and β,γ-unsaturated ketones. In 1982, we reported the first example of a 1-aza-di-π-methane rearrangement (1-ADPM) in the triplet sensitized photoreaction of the acyclic 1-aza-1,4-diene **9** that yields the corresponding cyclopropyl imine **10** exclusively (see Structures 9–14). Hydrolysis of **10** during isolation affords the corresponding aldehyde **11** [24]. The reaction was extended to other β,γ-unsaturated imines **12** [25,26].

A biradical mechanism similar to those shown in Schemes 2 and 3, for the DPM and ODPM rearrangements, was proposed for the 1-ADPM reaction (Scheme 5). This mechanism explains the regioselectivity observed for the rearrangement, which is analogous to that encountered in the ODPM rearrangement. The reaction always yields cyclopropyl imines, whereas the corresponding aziridines, which could have resulted from the alternative ring opening of the 1,4-cyclopropyl biradical intermediates, have never been observed. However, studies on the influence of substitution on the efficiency of the reaction, carried out by

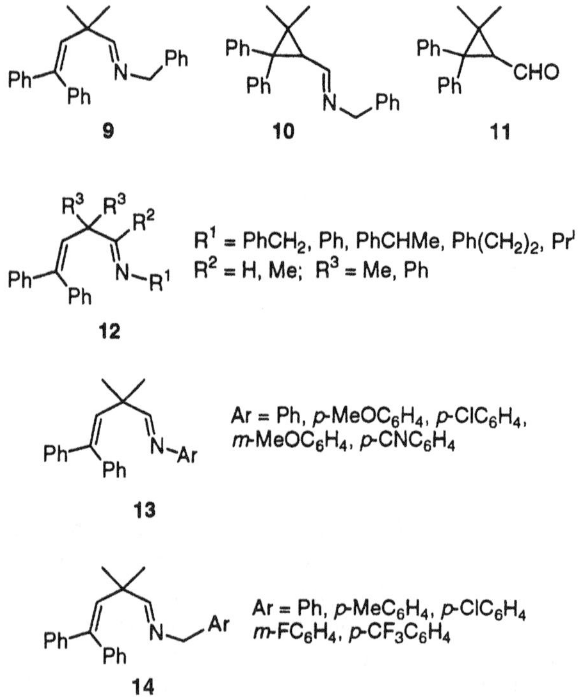

9   **10**   **11**

R^1 = PhCH$_2$, Ph, PhCHMe, Ph(CH$_2$)$_2$, Pr^i
R^2 = H, Me; R^3 = Me, Ph

12

Ar = Ph, p-MeOC$_6$H$_4$, p-ClC$_6$H$_4$,
m-MeOC$_6$H$_4$, p-CNC$_6$H$_4$

13

Ar = Ph, p-MeC$_6$H$_4$, p-ClC$_6$H$_4$
m-FC$_6$H$_4$, p-CF$_3$C$_6$H$_4$

14

Structures 9–14

using a series of N-aryl- (**13**) [27] and N-benzyl- (**14**) [28] β,γ-unsaturated imines, showed that the quantum yield of the cyclization increases with electron-withdrawing groups at the *para* position of the N-aryl or N-benzyl groups. The excellent linear correlation between log Φ and σ· obtained in both cases demonstrates the dependence of reactivity on the conjugative interaction between the aryl group and the nitrogen lone pair in **13** and also on the homoconjugative interaction between the benzyl group and the nitrogen lone pair in **14**. To account for these results, a competitive decay mechanism, involving an electron-transfer process from the imine nitrogen to the alkene group, has been proposed to adversely affect the efficiency of the reaction. This establishes a clear difference between the 1-ADPM mechanism and the DPM and ODPM counterparts in which there is no evidence for a competitive intramolecular single electron transfer (SET) decay pathway (Scheme 5). However, mechanisms involving intramolecular SET have been postulated to account for DPM reactions in some instances [29].

Originally it was considered that one of the main attractive features of the 1-ADPM rearrangement of β,γ-unsaturated imines, from a synthetic point of view, was that it could serve as a surrogate for the conversion of β,γ-unsaturated

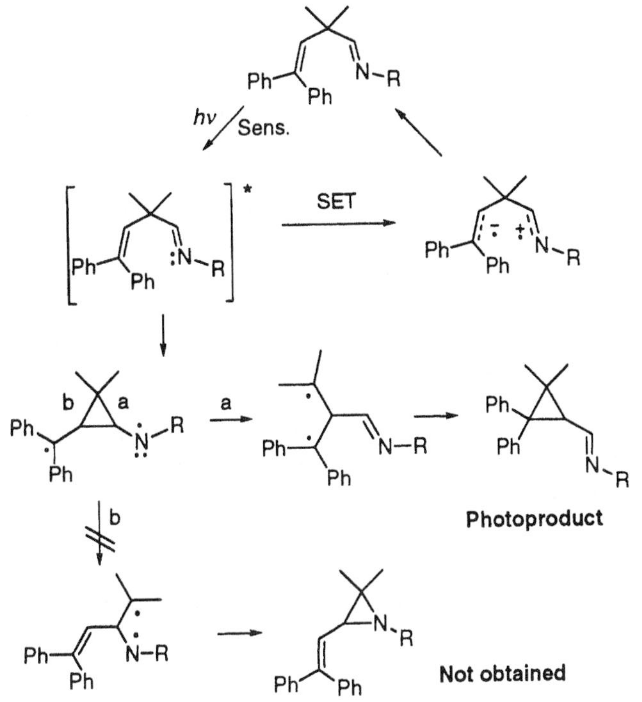

Scheme 5

aldehydes, previously considered to be unreactive in the ODPM mode [17], into the corresponding cyclopropyl aldehydes. The indirect route consists of condensation of the aldehyde with an amine, sensitized irradiation, and hydrolysis. In this way, is possible to obtain cyclopropyl aldehydes that are not accessible by the ODPM rearrangement.

The 1-ADPM reaction was extended to a series of acyclic and cyclic β,γ-unsaturated oxime acetates that, in most cases, yield the corresponding cyclopropyl oxime acetates in high isolated yield after short irradiation times [4]. The extension of the reaction to oxime acetates was an important step in expanding the synthetic potential of the reaction. Thus, although the imines **14** are very difficult to handle, due to their facile hydrolysis to the corresponding aldehydes, the oxime acetates are stable toward hydrolysis and can be isolated by column chromatography on silica gel. Furthermore, these compounds can be easily transformed into nitriles by thermal elimination of acetic acid. The versatility of the cyano group as a precursor to other functional groups, such as carboxylic acids, aldehydes, ketones, and so forth, opens the possibility of using the 1-ADPM reaction for the synthesis of differently substituted cyclopropanes. Representative examples of 1-ADPM rearrangements of oxime acetates are shown in Scheme 6 [30,31]. Other stable C—N double-bond derivatives, such as oximes [32,33], hydrazones [34,35], and semicarbazones [34,35], also undergo 1-ADPM rearrangements. The reaction has been extended to bridged cyclic compounds. Two representative examples of 1-ADPM reactivity in quinoxalinobarrelene derivatives are shown in Scheme 6 [36,37].

From all of the above results, it was clear that the three rearrangements of the di-π-methane type, DPM, ODPM, and 1-ADPM, reported up to 1995, were very general for a large number of β,γ-unsaturated compounds yielding the corresponding cyclopropanes in high chemical yield and quantum efficiency and with high degree of stereochemical control in most cases. In addition, the large number of studies, carried out by different research groups worldwide, have provided considerable information on the influence of substitution, scope, and mechanistic features of these reactions. As a consequence, a general impression developed that there was very little which remained to be uncovered in this area of research.

Some of the main features of these rearrangements can be summarized as follows:

1. Di-π-methane rearrangements are typical examples of reactions that occur in the excited state exclusively. These rearrangements have never been observed in the ground-state chemistry of 1,4-unsaturated compounds.

2. Concerted or biradical mechanisms have been postulated to justify these rearrangements in all the cases studied. There are no data that would suggest the involvement of other intermediates in these processes.

3. Di-π-methane rearrangements always yield cyclopropane derivatives. Three membered ring heterocycles that could be formed in the ODPM and 1-ADPM processes have never been observed.

4. The ODPM rearrangement is restricted to β,γ-unsaturated ketones. The corresponding aldehydes undergo decarbonylation on direct and sensitized irradiation.

However, results from recent efforts in our laboratory have led to a drastic expansion and modification of these ideas, as summarized in the following section.

II. NEW RESULTS FROM STUDIES OF THE DI-π-METHANE REACTIONS

A. The ODPM Rearrangement of β,γ-unsaturated Aldehydes

For many years, the ODPM reactivity of β,γ-unsaturated carbonyl compounds was considered to be restricted to ketones. The photochemical reactivity reported

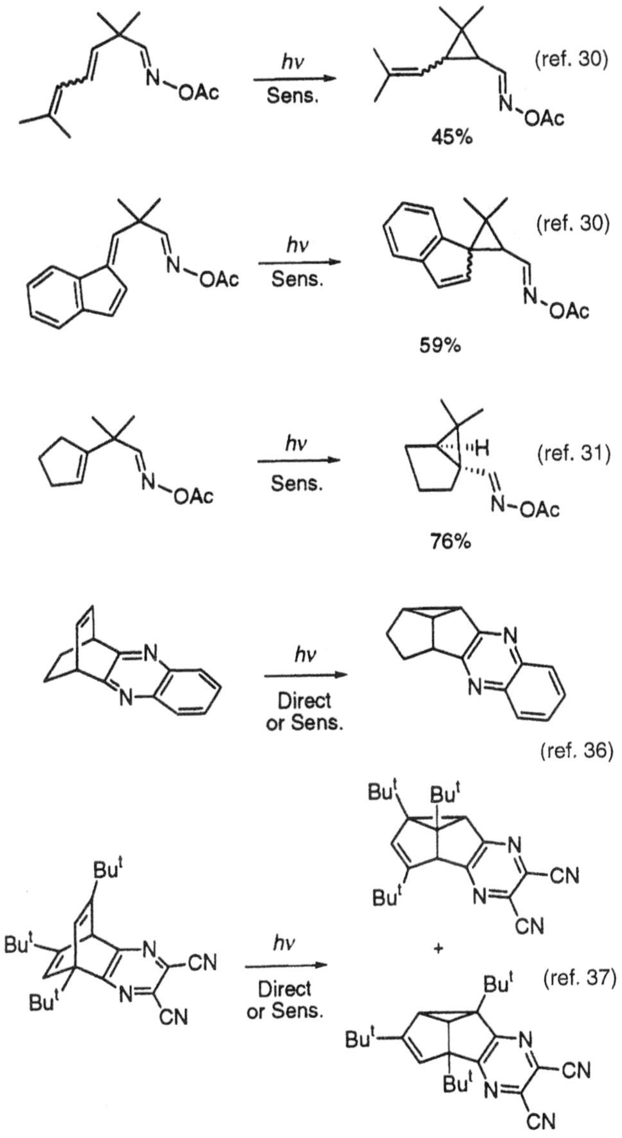
N-OAc
hν
Sens.
(ref. 30)
N-OAc
45%
N-OAc
hν
Sens.
(ref. 30)
N-OAc
59%
N-OAc
hν
Sens.
H
(ref. 31)
N-OAc
76%
N
N
hν
Direct
or Sens.
N
N
(ref. 36)
Bu^t
Bu^t
Bu^t
Bu^t
N
N
CN
CN
hν
Direct
or Sens.
Bu^t
Bu^t
Bu^t
N
N
CN
CN
+
Bu^t
(ref. 37)
Bu^t
Bu^t
N
N
CN
CN

Scheme 6

for β,γ-unsaturated aldehydes, both on direct or sensitized irradiation, was decarbonylation to the corresponding alkenes [4]. Thus, studies carried out by Schaffner et al., more than 20 years ago, demonstrated that β,γ-unsaturated cycloalkenyl aldehydes **15** [38], **16** [39], **17** [39], **18** [40,41], **19** [42] and **20** [43] undergo decarbonylation on direct irradiation (see Structures 15–20). The triplet-sensitized reactivity of aldehyde **17** was also investigated. Again, under these conditions, decarbonylation takes place. In another study by Dürr et al. [44,45], direct or acetophenone-sensitized irradiation of aldehydes **21** was shown to give products

Structures 15–23

resulting from decarbonylation and six-electron cyclization of the *cis*-stilbene type. The photoreactivity of acyclic aldehydes **22** [46] and **23** [47] was also investigated. These aldehydes undergo decarbonylation on direct irradiation. Ambiguous results were reported for the acetophenone sensitization of aldehyde **23**.

From all of these findings, a general conclusion, that is reflected in all of the reviews and monographs on this topic, is that decarbonylation is the normal photoreactivity of β,γ-unsaturated aldehydes. However, two examples of aldehydes that did not follow this general rule were described in early studies. The first example of an ODPM reaction in a β,γ-unsaturated aldehyde was observed by Schaffner et al. [48] in the direct irradiation of the steroidal aldehyde **24** that gives the alkene resulting from decarbonylation and two other compounds, **25** and **26**, derived from the ODPM rearrangement and the 1,3-formyl migration, respectively (Scheme 7). Many years later, Zimmerman and Cassel reported the ODPM triplet reactivity of the aldehyde **27**, which contains two isopropyl groups at C-2, that yields the cyclopropyl aldehyde **28**, quantitatively [49] (Scheme 7). Therefore β,γ-unsaturated aldehydes were considered to be unreactive in the ODPM mode, apart from the two exceptions mentioned here.

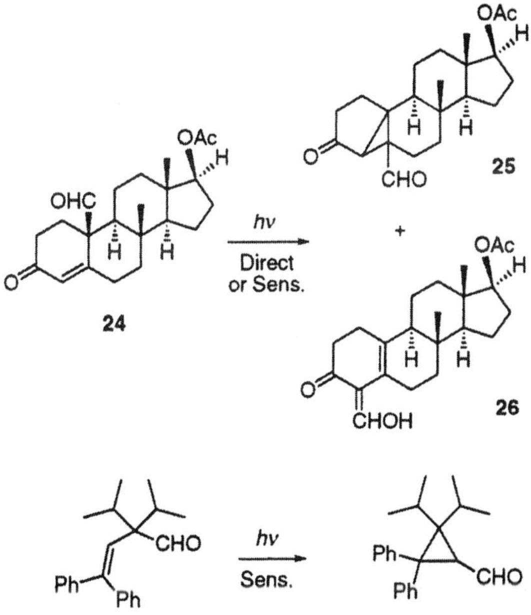

Scheme 7

However, these ideas have changed due to the results emanating from our recent studies [50,51]. Thus, in an investigation of the application of the 1-ADPM rearrangement to the synthesis of bicyclo [*n*. 1.0]alkanes by using ring contraction processes, an imine derivative of the aldehyde **29a** (Scheme 8) was selected as the target molecule. Because the photochemistry of aldehyde **29a** was not described previously, it was important to first establish the photochemical behavior of this compound and to compare its reactivity with that of the corresponding 1-aza-1,4-diene derivatives. We observed that direct irradiation of **29a** affords a mixture of starting material (20%) and a new product identified as the cyclohexene **30** (63%), which results from decarbonylation of aldehyde **29a** (Scheme 8). This result was not surprising because literature precedents have established that decarbonylation is the main photochemical pathway followed by β,γ-unsaturated aldehydes. In contrast, however, acetophenone-sensitized irradiation of **29a** affords the 1,3-migration product **31** (25%) along with the ODPM product **32a** (25%) [51] (Scheme 8).

The photoreactivity of aldehyde **29a** was surprising because of the literature above-mentioned precedents. Therefore, we decided to determine whether this unexpected reactivity could be extended to other related aldehydes, such as **29b** and **29c** (Structures 29–39). Indeed, acetophenone-sensitized irradiation (5 min) of **29b** affords the ODPM product **32b** in 90% isolated yield. Irradiation of **29c** under the same conditions also gives the aldehyde **32c** in 25% yield. In these two instances, the corresponding, 1,3-formyl migration products were not formed [51].

It was clear that there must be structural features in compounds **29a–29c** that are critical for the success of the ODPM process. This is evident not only

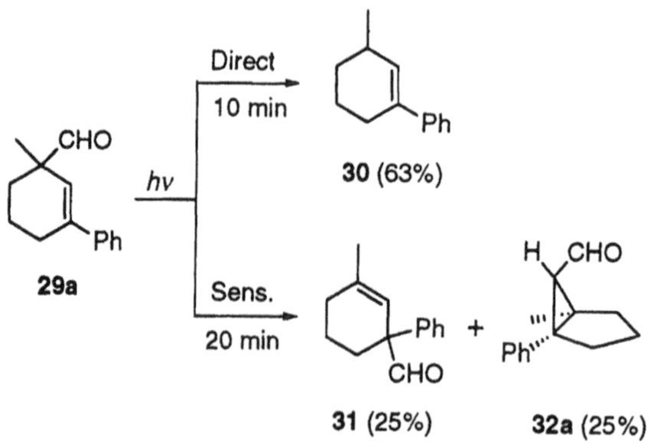

Scheme 8

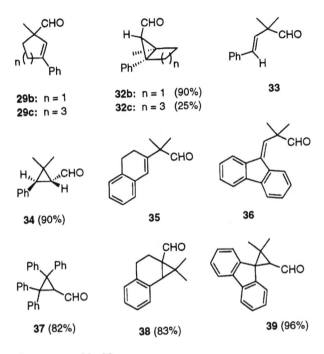

29b: n = 1
29c: n = 3

32b: n = 1 (90%)
32c: n = 3 (25%)

33

34 (90%)

35

36

37 (82%)

38 (83%)

39 (96%)

Structures 29–39

from the fact that compounds **29** undergo rearrangements that have seldom been seen in related compounds but also because of the very high efficiency of the rearrangement reactions. The most likely controlling features are (1) the excitation of the molecule to the $T_1(\pi, \pi^*)$ excited state, (2) the stabilizing influence of the phenyl group on the bridging 1,4-biradical reaction intermediates, and (3) the difficulty of deactivation of the excited state by the free-rotor effect of the vinyl component in cyclic alkenes.

In order to determine whether the partial suppression of the free-rotor effect was required for the success of the ODPM rearrangement process, the study was extended to the aldehyde **33** [51]. When **33** is irradiated (15 min), under similar conditions to those used for **29**, the cyclopropyl aldehyde **34**, resulting from an ODPM rearrangement, was obtained, as the *trans*-diastereoisomer, in 90% isolated yield. This result demonstrated clearly that the ODPM reactivity of β,γ-unsaturated aldehydes is not restricted to cyclic compounds, such as **29**, but can also be extended to acyclic derivatives. Therefore, the suppression of the free-rotor effect is not essential for the success of the rearrangement and the reaction is probably controlled by both the excitation of the molecule to the $T_1(\pi, \pi^*)$

excited state and the stabilizing influence of phenyl groups on the bridging 1,4-biradical reaction intermediates.

To confirm this hypothesis the photoreactivity of aldehyde **23**, previously reported as unreactive in the ODPM mode [47], was reinvestigated. The results obtained show that *m*-methoxyacetophenone-sensitized irradiation of **23** (2 hr) brings about the formation of the cyclopropyl aldehyde **11**, resulting from an ODPM rearrangement, in 57% isolated yield. Further support for the above postulates was obtained from studies of triplet-sensitized irradiation of aldehydes **22**, **35**, and **36**. These substances undergo the ODPM rearrangement to yield the corresponding cyclopropyl aldehydes **37** (82%), **38** (83%) and **39** (96%), respectively [51].

The ODPM reactivity of β,γ-unsaturated aldehydes is not restricted to γ-phenyl-substituted compounds but can also be extended to systems in which the intermediate biradicals are stabilized by conjugation with a vinyl group. Thus, *m*-methoxyacetophenone-sensitized irradiation of **40** (Structures 40–49) for 20 min, affords the cyclopropane derivative **41** (47%) as a 1:8 mixture of *cis:trans* isomers. Similarly, irradiation of **42**, for 15 min, under the same conditions, yields **43** (52%) as the *trans* isomer exclusively [51].

The synthesis of compounds **39**, **41**, and **43** by the ODPM rearrangement opens a novel photochemical route to chrysanthemic acid and other cyclopropane carboxylic acids present in pyrethrins and pyrethroids [52]. In fact, aldehyde **43** can be transformed to *trans*-chrysanthemic acid by simple oxidation. This new synthetic route to ecologically benign insecticides competes with the one previously described by us using the 1-ADPM rearrangement of β,γ-unsaturated oxime acetates [30,53].

The possible competition between ODPM and DPM processes was also studied. Triplet-sensitized irradiation of **44** for 10 min affords the cyclopropyl aldehyde **45** (19%) resulting from an ODPM rearrangement, exclusively. However, under the same conditions, aldehyde **46** yields the cyclopropyl aldehyde **47** (48%), resulting from a DPM rearrangement [51].

In the absence of phenyl or vinyl substituents at the γ-position of the β,γ-unsaturated aldehyde, the rearrangement can also take place, although very inefficiently. Thus, acetone-sensitized irradiation of aldehyde **48**, with diphenyl substitution at C-2, gives the ODPM aldehyde **11** in low yield (8%) and the corresponding alkene resulting from decarbonylation (14%) [51]. However, aldehyde **49**, in which there are no phenyl or vinyl groups to stabilize the biradical intermediates, does not undergo the ODPM rearrangement.

The above-described results have changed the ideas that most photochemists had about the photoreactivity of β,γ-unsaturated aldehydes. From the earlier work carried out mainly by Schaffner, Dürr, and Pratt [4] on a series of compounds of this type, a general consensus surrounding their lack of ODPM reactivity was developed [17]. Most of the compounds studied previously undergo decarbonyla-

Structures 40–49

tion, with only two exceptions. The decarbonylation takes place via either the $S_1(n, \pi^*)$ or $T_2 (n, \pi^*)$ excited states [54,55]. However, based on studies with analogous ketones, ODPM rearrangements of β,γ-unsaturated aldehydes almost surely occur via the $T_1(\pi, \pi^*)$ excited state [50,51]. The lack of ODPM reactivity of the aldehydes previously studied is probably due to the absence of the adequate substitution pattern that would allow efficient transfer of the triplet energy from the sensitizer to the alkene moiety. Therefore, the $T_1 (\pi, \pi^*)$ excited state, necessary for the ODPM rearrangement, would not be generated. However, our studies show that suitably substituted β,γ-unsaturated aldehydes undergo the ODPM rearrangement with high chemical efficiency. The results also indicate that the ODPM rearrangement of β,γ-unsaturated aldehydes occurs when the triplet energy from the sensitizer is efficiently transferred to the alkene moiety generating a $T_1(\pi, \pi^*)$ excited state. In addition, efficient reactions occur when the biradical intermediates are stabilized by phenyl or vinyl substitution [50,51]. β,γ-Unsaturated aldehydes that do not meet these two requirements, as in most of the cases previously probed [4], undergo exclusive photodecarbonylation.

To test the accuracy of the above postulates, our studies were extended to aldehyde **50** (Structures 50–64). In a previous study, Schaffner et al. [39] reported

Structures 50–64

that aldehyde **16** undergoes decarbonylation on irradiation exclusively. If the postulates on the ODPM rearrangement of aldehydes are correct, it would be possible to overcome the lack of ODPM reactivity by modifying the substitution in **16** in such a way so as to ensure efficient triplet energy transfer to the C—C double bond and stabilization of the biradical intermediates. The conjugation of the C—C double bond with a phenyl ring and the suppression of the methyl at C-3 present in **16**, should favor the ODPM rearrangement of **50**. However, all of the attempts to synthesize compound **50** yielded an inseparable mixture of **50** and its isomer **51**. Nevertheless, a 1:1 mixture of **50** and **51** was irradiated using acetophenone as the sensitizer. In spite of the complexity of the photomixture, the ¹H-NMR (nuclear magnetic resonance) spectrum of the crude photolysate suggested the possibility that the bicyclic aldehyde **52**, resulting from an ODPM rearrangement of the aldehyde **50**, is present in the reaction mixture.

In order to overcome the problems regarding the synthesis of **50**, the study was extended to aldehyde **53**. In this instance, the extended conjugation of the alkene moiety in **53** prevented isomerization of the C—C double bond to the endocyclic position, thus allowing the synthesis of this compound as a pure substance. However, contrary to our expectations, sensitized irradiation of **53** gives the diene **54** resulting from decarbonylation, in 39% yield. No ODPM product was formed in this instance. The formation of **54** is reminiscent of a Norrish Type I process. However, in this process, the homolytic bond fission between the carbonyl and the α-carbon is not the result of the excitation of the carbonyl, as is the case in an authentic Norrish Type I reaction [56]. Irradiation of the corresponding methyl ketone **55**, under the same conditions used for **53**, affords **56** resulting from 1,3-acyl migration, in 24% yield. Again, in this instance, no ODPM product was obtained.

As far as we are aware, these observations are the first that show that the well-known Norrish Type I reactions of β,γ-unsaturated carbonyl compounds can take place by excitation of the alkene moiety rather than the carbonyl group. This unusual reactivity may be due to the fact that the $T_1(\pi, \pi^*)$ excited states of **53** and **55** possess sufficient energy to promote the homolytic allylic bond fission to form the stabilized pentadienyl radical **57**. As a result, photodecarbonylation competes favorably with the ODPM rearrangement.

These unexpected results suggest a modification of the conclusions reached in our earlier studies. Specifically, the observations indicate that in order to detect ODPM photoreactivity in β,γ-unsaturated aldehydes, substituents should be present to stabilize intermediate biradicals in the rearrangement pathway, but they should not enhance alternative reactions, such as allylic homolytic cleavage. Further studies will be necessary to confirm this hypothesis and to determine the scope of these new reactions.

Another intriguing observation made in our preliminary studies in this area is that β,γ-unsaturated aldehyde **23** undergoes photoinduced ODPM rearrangement, whereas the analogous methyl ketone **58** does not react in this manner [51]. This result suggests that other aldehydes with substitution patterns similar to those found in ODPM unreactive ketones might participate in this photorearrangement process. Previous studies have shown that methyl ketone **59a** is not an ODPM substrate [57,58]. The corresponding aldehyde **60a**, which has the same substitution pattern as does **59a**, according to our postulates, should undergo the ODPM rearrangement. In order to test this proposal, aldehydes **60a** and **60b** were irradiated using *m*-methoxyacetophenone as the sensitizer. Under these conditions, **60a** leads to generation of the spirocyclic derivative **61a** (36%), as the *trans* diastereoisomer, along with a mixture of starting material and **62** (13%), a compound resulting from 1,3-formyl migration (Scheme 9). Irradiation (2 hr) of the diphenyl analog **60b**, under the same conditions used for **60a**, gives rise to the ODPM product **61b** (24%), the alkene **63** (5%), and recovered starting material (53%) (Scheme 9). The formation of the alkene **63** by irradiation of **60b** might be due to the stabilization of the radical resulting from homolytic fission of the bond between the formyl group and the α-carbon, similar to the reaction observed for compound **53**. In contrast, the methyl ketone derivative **59b** does not undergo the ODPM rearrangement. Sensitized irradiation (17 hr) of this substance provides recovered starting material (71%) and a complex mixture of products. These additional examples add further support to the proposal that β,γ-unsaturated aldehyde are more prone to undergo the ODPM rearrangement than the corresponding methyl ketones.

The large number of studies carried out in the di-π-methane area have established that disubstitution at the central carbon of 1,4-diene substrates is an important structural requirement for efficient rearrangement [1–4,14–16]. In fact, there are only two examples in which β,γ-unsaturated ketones (**3** [13] and **64a** [57,58]), monosubstituted at the α-position, undergo the ODPM rearrangement. The reactivity observed in these two substrates was attributed to the bulk of the substituent at C-2 that compensates for the absence of disubstitution at that α-position.

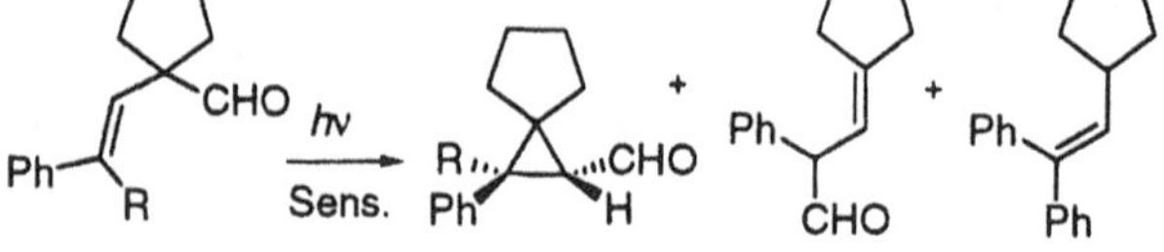

60a: R = H	1 h	**61a** (36%)	**62** (13%)	——
60b: R = Ph	2 h	**61b** (24%)	——	**63** (5%)

Scheme 9

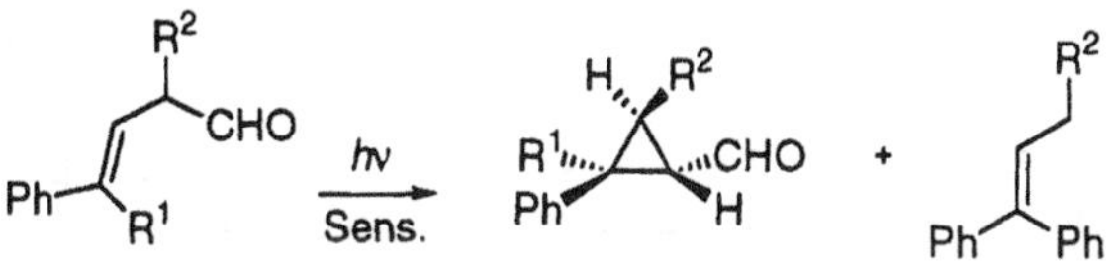

65a: R^1 = H; R^2 = ^{i}Pr   **66a** (21%)
65b: R^1 = Ph; R^2 = ^{i}Pr   **66b** (23%)   **67a** (9%)
65c: R^1 = H; R^2 = Et   **66c** (22%)
65d: R^1 = Ph; R^2 = Et   **66d** (24%)   **67b** (25%)

Scheme 10

In an attempt to establish the limits for ODPM reactivity of β,γ-unsaturated aldehydes, we have extended our studies to a series of aldehydes **65**, (Scheme 10) which are monosubstituted at C-2. Triplet-sensitized irradiation of **65** leads to the formation of the corresponding cyclopropanecarbaldehydes **66** [59] (Scheme 10). The diphenyl-substituted aldehydes **65b** and **65d** yield, in addition to the ODPM products, the corresponding alkenes **67a** and **67b**, resulting from photodecarbonylation. The formation of these alkenes is probably due to stabilization of the radical, formed by allylic cleavage, by diphenyl conjugation. The ODPM rearrangement of aldehydes **65** is diastereoselective, yielding only one diastereoisomer of **66** (Scheme 10).

The ODPM reactivity of aldehydes **65** is surprising because, as mentioned earlier, the only two ODPM reactive acyclic β,γ-unsaturated mono-C-2-substituted ketones are compounds **3** and **64a**. Each of these substrates contains bulky substituents at C-2. However, the observations made with **65c** and **65d**, each having ethyl groups at C-2, clearly demonstrates that the bulk of the C-2 substituent is not an important feature in determining the ODPM reactivity of β,γ-unsaturated aldehydes. However, it should be noted that the isopropyl substituted aldehydes, **65a** and **65b**, do react more efficiently, in qualitative terms, than **65c** and **65d**.

Substrates **64b–64d** were synthesized in order to determine if the aldehydes **65** are more reactive toward the ODPM than the corresponding methyl ketones. Compound (E)-**64c** does not undergo the ODPM rearrangement. Irradiation of this ketone, under the same conditions used for **65**, affords the corresponding diastereoisomer (Z)-**64c** in 50% yield and 25% of recovered starting material. Prolonged irradiation of **64b** and **64d**, under these conditions, results in a complex mixture of highly polar materials (~20%) along with recovered starting material (~80%) only. These results confirm the postulate that β,γ-unsaturated aldehydes are better substrates for ODPM rearrangement. In this regard, there are questions that still remain to be answered. For instance, it is surprising that ketone **64a** is reported to be ODPM reactive, whereas ketones **64b–64d** do not undergo this rearrangement reaction [59].

In summary, the results obtained in this study demonstrate clearly that β,γ-unsaturated aldehydes undergo the ODPM rearrangement very efficiently and that they are more reactive than the corresponding methyl ketones. In addition, our current studies have provided the first examples of reactions, similar to the well-known Norrish Type I process, which take place in the triplet excited state of β,γ-unsaturated carbonyl compounds by excitation of the C—C double bond instead of the carbonyl group. Finally, ODPM rearrangement of mono-C-2-substituted aldehydes is surprising because there are only two precedents of ketones with this type of substitution that undergo this photoreaction. Furthermore, the photoreactions of these substances are stereoselective, yielding only one of the possible diastereoisomeric cyclopropanecarboxaldehyde products. When combined, the results of the current effort broaden the synthetic potential of the ODPM rearrangement and modify some ideas firmly established, but apparently incorrect, about the reactivity of β,γ-unsaturated carbonyl compounds.

B. The 2-Aza-di-π-methane Rearrangement

From the foregoing discussion, it is clear that DPM rearrangements are very general for a variety of 1,4-unsaturated systems, such as, 1,4-dienes, β,γ-unsaturated aldehydes and ketones, and different 1-aza-1,4-diene derivatives. Surprisingly, the literature was devoid of studies describing the photoreactivity of the closely related 2-aza-1,4-diene derivatives. For many years, the only studies in this area were carried out by Mariano and his co-workers [60] on the photochemistry of iminium salts derived from 2-aza-1,4-dienes. The results obtained demonstrated the synthetic utility of the photocyclizations of iminium salts to different heterocycles, in reactions that are initiated by intramolecular single electron transfer [60].

The interesting photoreactivity reported for iminium salts of 2-aza-1,4-dienes and for 1-aza-1,4-dienes prompted us to carry out a study on the photoreactivity of 2-aza-1,4-dienes. Perhaps this is the most interesting type of compound among all the β,γ-unsaturated systems studied so far, because the expected products resulting from rearrangements of the di-π-methane type would be vinylaziridines and/or cyclopropyl imines. The thermal transformations of these compounds make them highly interesting from a synthetic point of view [61,62]. The results obtained from our studies of these systems have gone beyond expectations and unprecedented photorearrangement reactions under SET- and triplet-sensitized conditions have been uncovered.

The first compound selected for this study was the 2-aza-1,4-diene **68** (Scheme 11). Triplet-sensitized irradiation of azadiene **68**, using acetophenone, for 25 min affords two new products, identified as the cyclopropylimine **69** (11%)

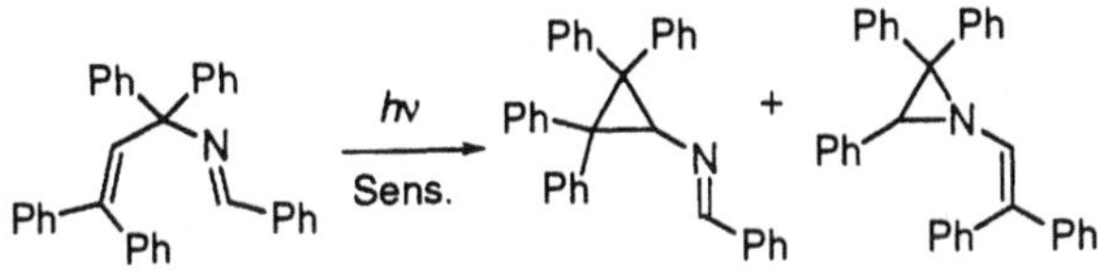

Scheme 11

and the vinylaziridine **70** (3%) (Scheme 11). The former substance very readily hydrolyzes to form the corresponding cyclopropylamine during chromatography [63].

The production of **69** and **70** in this photoreaction is in accord with the operation of mechanistic pathways involving the generation and competitive cleavage of aziridinyl-dicarbinyl biradical **71** (Scheme 12). Thus, the major photoproduct **69** is formed by C—N bond cleavage in **71** (path **a**). The alternative fragmentation of the C—C bond in **71** affords intermediate **72**, which cyclizes yielding aziridine **70** (path **b**). As such, this represents the first example of a 2-aza-di-π-methane (2-ADPM) rearrangement that occurs via a three-membered-

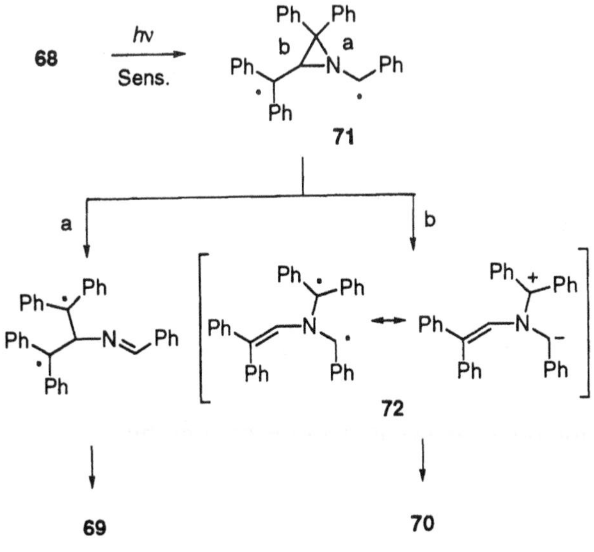

Scheme 12

ring heterocyclic biradical and brings about the formation of a heterocyclic product. Previous attempts by Adam et al. [64] to promote the formation of oxiranes by the DPM arrangement were unsuccessful [64].

The formation of the N-vinylaziridine **70** in the photoreaction of **68** deserves additional comment. Depending on the multiplicity, the intermediate **72** formed by path **b** could be a triplet 1,3-biradical. However, if intersystem crossing occurs along the reaction coordinate, the singlet biradical must be considered as a dipolar azomethine ylide. According to literature precedents, both intermediates, the 1,3-biradical and the ylide, will cyclize to form the observed aziridine. This is the first case in a DPM process where a zwitterion can be postulated as a possible intermediate.

The study was extended to azadiene **73** (Structures **73–87**). Irradiation of this compound under the above conditions yields the cyclopropyl imine **74**, resulting from a 2-ADPM rearrangement, and the cyclopropyl imine **75**, arising by an aryl-di-π-methane reaction. Azadienes **76, 77** and **78** also undergo 2-ADPM rearrangements to afford the corresponding cyclopropyl imines **79, 80,** and **81**, respectively [65]. N-Vinylaziridines were not detected in the product mixtures arising by irradiation of compounds **73, 76,** and **78**. The reason why aziridine formation does not take place in these instances is unclear at this point. It might be due to preferential ring opening of the aziridinyl dicarbinyl biradical by path **a** (Scheme 12) as a result of differences in the stability of the resulting 1,3-biradical or zwitterionic intermediates. Another possible explanation could be that the reaction follows the zwitterionic path in which the resulting intermediate reacts to yield fragmentation products instead of undergoing cyclization to produce the aziridine [65].

The study has been extended to azadienes **82–85**. The substitution patterns present in these compounds has proved to be adequate to promote the ODPM rearrangement in β,γ-unsaturated aldehydes [4] and the 1-ADPM reaction in 1-aza-1,4-dienes [4]. However, triplet-sensitized irradiation of compounds **82–85** yields complex mixtures of products that have not been identified. These results show that a clear difference exists between the DPM reactivity of 2-aza-1,4-dienes and the other DPM processes previously reported [65]. The differences are difficult to explain and further studies are underway to clarify this point.

C. Di-π-methane Rearrangements Via Radical-Cation Intermediates

1. The 2-Aza-di-π-methane Rearrangement Promoted by Electron-Acceptor Sensitizers

Direct or triplet-sensitized irradiations have been used in most of the studies on the photoreactivity of 1,4-unsaturated compounds. However, Zimmerman and Hoffacker have reported the novel cyclization of aryl-substituted 1,4-pentadienes

Structures 73–87

86 to the corresponding benzhydryldihydronaphthalene derivatives **87** using DCA as an electron-acceptor sensitizer [66].

In order to gain a more complete picture of the photoreactivity of azadiene **68**, this compound was irradiated using DCA as a SET sensitizer. Surprisingly, under these conditions, two products are formed that were identified as the vinyla-ziridine **70** (11%) and the cyclopropyl imine **88** (19%) as shown in Scheme 13 [63].

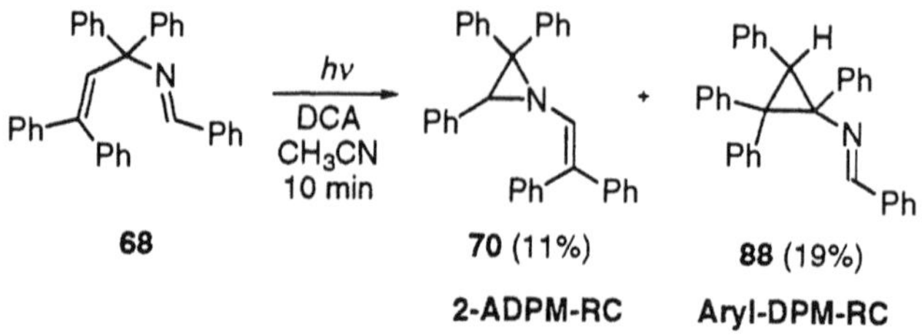

Scheme 13

The formation of **70** and **88** represents the first examples of reactions of the di-π-methane type (a 2-ADPM reaction producing **70** and an aryl-di-π-methane rearrangement generating **88**) that take place via radical-cation intermediates. Of equal interest is the fact that vinylaziridine **70** is obtained in both the triplet- and SET-sensitized photoreactions of **68**. The products generated in this process appear consistent with a pathway in which an initially formed olefin-localized cation-radical intermediate **89** bridges by C—N bond formation to give the aziridinyl intermediate **90**. A ring opening in **90** affords cation-radical **91**, which undergoes back electron transfer and biradical cyclization yielding **70**. A competitive route involving phenyl migration in **89** generates cation-radical **92**, the precursor of **88** (Scheme 14) [63].

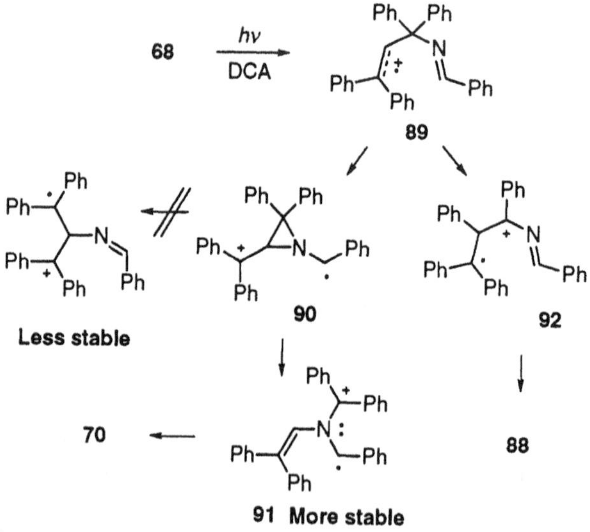

Scheme 14

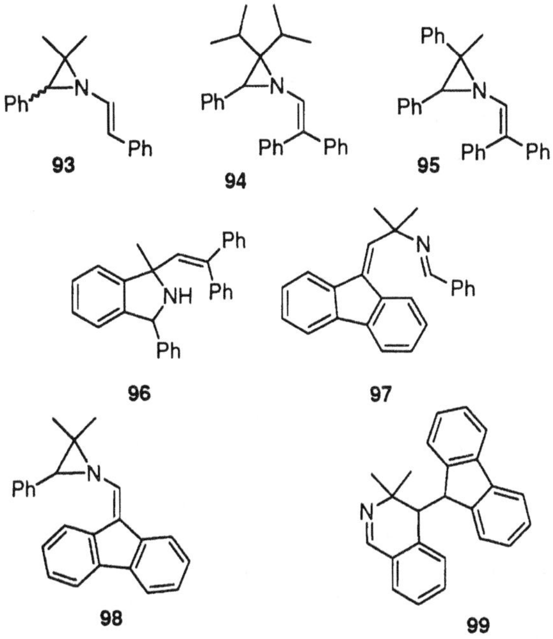

Structures 93–99

Additional examples of these novel SET-promoted photorearrangement reactions were uncovered in our investigations with the 2-azadienes **78** and **82**. These substances yield the corresponding *N*-vinylaziridines **93** and **94**, respectively, upon DCA-sensitized irradiation (Structures 93–99). The study was extended to azadiene **73**. The DCA-sensitized irradiation of **73** yields the aziridine **95** and the dihydroisoindole **96** [65]. However, under these conditions, 2-azadiene **97** affords, in addition to the expected aziridine **98**, the dihydroisoquinoline **99**. The formation of the isoindole **96** can be explain by a pathway in which a radical-cation centered on the C—N double bond attacks the phenyl ring at C-3, as shown in Scheme 15. The formation of the dihydroisoquinoline **99** can be justified by

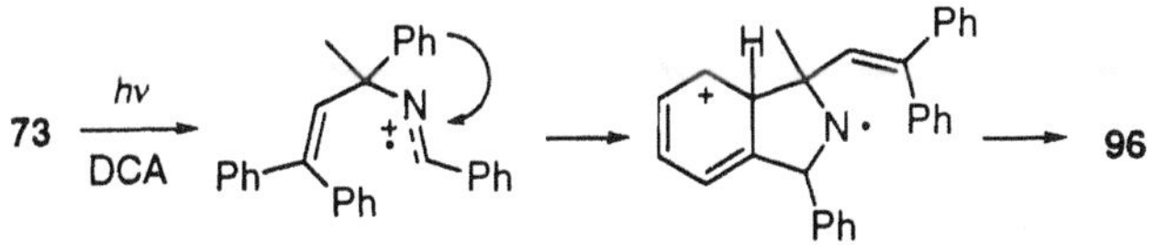

Scheme 15

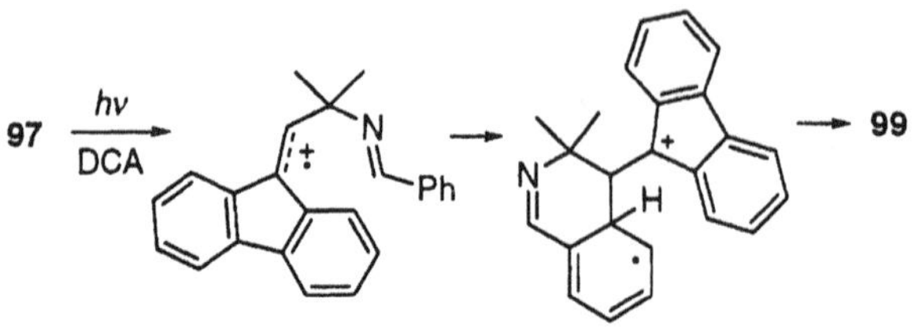

Scheme 16

attack of the radical-cation centered on the C—C double bond onto the phenyl ring at C-1, as shown in Scheme 16. This cyclization is similar to the reaction reported by Zimmerman in the irradiation of 1,4-dienes **86** using electron-acceptor sensitizers [66].

The influence of diphenyl substitution at C-1 of the 2-azadiene skeleton has been investigated. However, when compounds **84** and **85** are irradiated under the above conditions, alternative reactions occur to afford the corresponding dihydrobenzoazepines **100a–100b** in good yields. The formation of **100** can by justified by attack of a radical-cation centered on the C—N double bond onto the phenyl ring present at C-5, as shown in Scheme 17 [65].

Something worth noting about the photoreactivity of 2-aza-1,4-dienes under DCA sensitization is that some of the products obtained result from the rearrangement of radical-cations centered on the C—C double bond, whereas in other

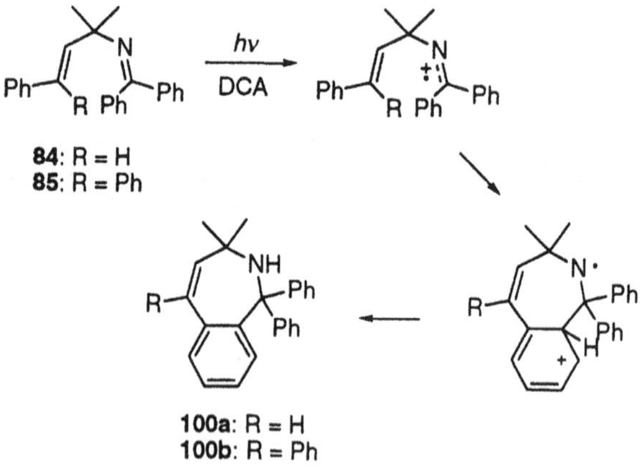

84: R = H
85: R = Ph

100a: R = H
100b: R = Ph

Scheme 17

cases, the observed photoproducts come from radical-cations centered in the C—N double bond. Thus, for example, the formation of compound **88** in the photoreaction of **68** is promoted by 1,2-phenyl migration to a radical-cation centered in the C—C double bond, as shown in Scheme 14. Similarly, the dihydroisoquinoline **99**, formed in the reaction of **97**, results from an electrophilic attack of the radical–cation centered in the C—C double bond into the phenyl ring at C-1 (Scheme 16). However, dihydroisoindol **96**, obtained in the irradiation of azadiene **73** is formed by an electrophilic attack of a radical-cation centered in the C—N double bond into the phenyl ring at C-3 (Scheme 15). The same situation is observed in the reactions of the ketoimines **84** and **85** that yield the corresponding benzoazepines **100a–100b**, respectively (Scheme 17). These results demonstrate that two different radical-cations can be generated under DCA-sensitization photoreaction conditions. One of them results from the oxidation of the functional group with the lowest ionization potential (the C=N group). However, the radical-cation from the group with the highest ionization potential (the C=C moiety) is also formed either directly or by intramolecular SET. These results suggest that the evolution of the radical-cations to the final products is faster that intramolecular oxidation–reduction. As far as we are aware, these are the first examples of generation of two different radical-cations in the DCA-sensitized irradiation of bifunctional molecules with very different ionization potentials.

2. The 1-Aza-di-π-methane Rearrangement Promoted by Electron-Acceptor Sensitizers

The results obtained in the DCA-sensitized irradiation of 2-aza-1,4-dienes suggest the possibility that other β,γ-unsaturated systems might also undergo novel SET-promoted rearrangements via radical-cation intermediates. Therefore, in an extension of this study aimed at probing the generality of this proposal, we have investigated the photochemical properties of the 1-aza-1,4-diene **101**. DCA-sensitized irradiation of **101** for 7 hr leads to formation of the cyclopropylimine **102**, which hydrolyzes to form the corresponding aldehyde **34** during chromatography. (Scheme 18) [67]. This result demonstrates clearly that, like their 2-aza-1,4-dienes

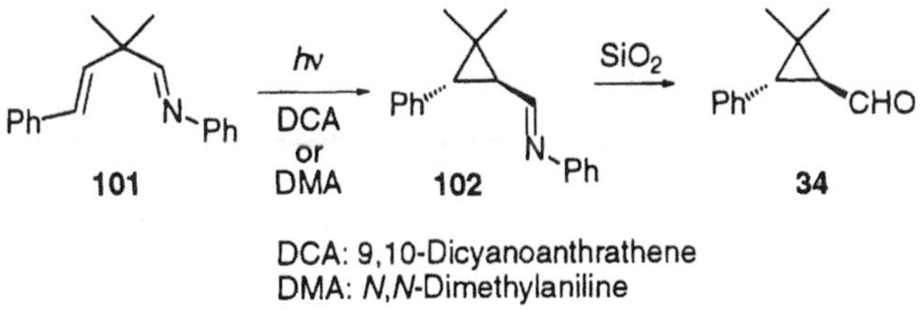

DCA: 9,10-Dicyanoanthrathene
DMA: N,N-Dimethylaniline

Scheme 18

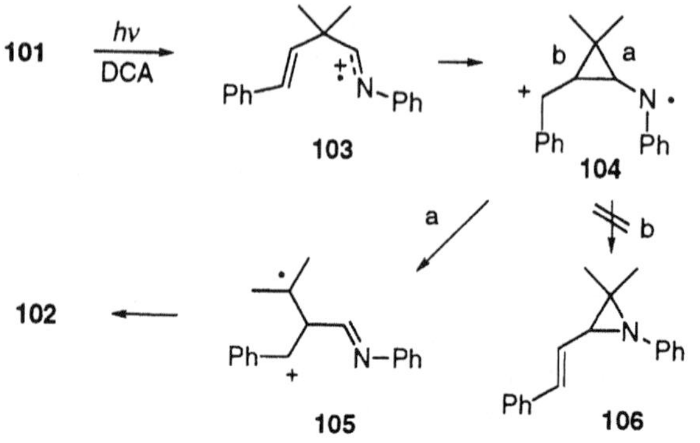

Scheme 19

analogs, 1-aza-1,4-dienes such as **101** also undergo di-π-methane-type radical-cation rearrangement reactions.

The mechanism shown in Scheme 19 accounts for the formation of **102**. The pathway involves the generation of the imine localized radical-cation intermediate **103** that bridges to give the cyclopropyl radical-cation **104**. A ring opening of bond **a** in **104** generates **105**, which yields **102** by back electron transfer and biradical cyclization. The alternative cleavage of bond **b** that would have yielded the vinylaziridine **106** is not observed probably because the resulting radical-cation would be considerably less stable than intermediate **105** (Scheme 19).

Our attention is next directed at determining whether or not the reaction could be extended to other 1-aza-1,4-dienes in which the C═N bond would be less readily hydrolyzed, thus facilitating handling and isolation of starting materials and photoproducts. Our previous studies of the 1-ADPM rearrangement have demonstrated that β,γ-unsaturated oxime ethers [33] and oxime esters [4] rearrange efficiently on triplet-sensitized irradiation and, furthermore, that the reactants and products of these processes are sufficiently stable to allow for their isolation by silica gel chromatography. Therefore, the study was extended to the oxime ether **107a** and the oxime acetate **107b** (Structures 107–116).

The DCA-sensitized irradiation of **107a** for 13 hr affords, after column chromatography on silica gel, the *trans*-cyclopropane derivative **108a** (10%) as a 1:1 mixture of C═N bond *E:Z*-isomers. Similarly, irradiation of the oxime acetate **107b** under these conditions for 2.5 hr affords, after chromatography, the *trans*-cyclopropane derivative **108b** (12%). These results show that the novel 1-ADPM rearrangement promoted by electron-transfer sensitization can be extended to other C—N double-bond derivatives.

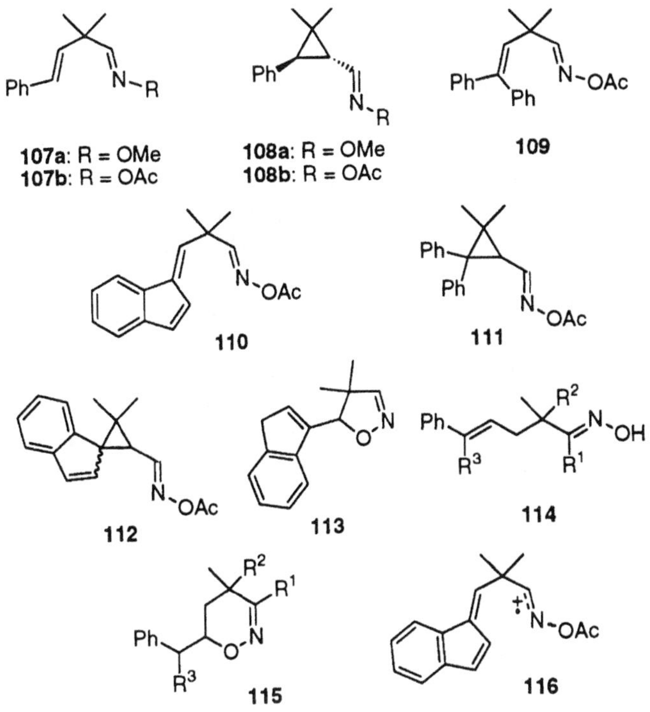

107a: R = OMe
107b: R = OAc

108a: R = OMe
108b: R = OAc

109

110

111

112

113

114

115

116

Structures 107–116

In qualitative terms, the rearrangement reaction is considerably more efficient for the oxime acetate **107b** than for the oxime ether **107a**. As a result, the photochemical reactivity of the oxime acetates **109** and **110** was probed. Irradiation of **109** for 3 hr, under the same conditions used for **107**, affords the cyclopropane **111** (25%) as a 1:2 mixture of Z:E isomers. Likewise, DCA-sensitized irradiation of **110** for 1 hr yields the cyclopropane derivative **112** (16%) and the dihydroisoxazole **113** (18%). It is unclear at this point how **113** arises in the SET-sensitized reaction of **110**. However, this cyclization process is similar to that observed in our studies of the DCA-sensitized reaction of the γ,δ-unsaturated oximes **114**, which affords the 5,6-dihydro-4*H*-1,2-oxazines **115** [68]. A possible mechanism to justify the formation of **113** could involve intramolecular electrophilic addition to the alkene unit in **116** of the oxygen from the oxime localized radical-cation, followed by transfer of an acyl cation to any of the radical-anions present in the reaction medium.

The results obtained in this study show that the di-π-methane rearrangement of 1,4-unsaturated systems via radical-cations is not restricted to the 2-aza-1,4-

dienes and that it can be extended to differently N-substituted 1-aza-1,4-dienes. However, the yield of products in these reactions are considerably lower that in the triplet-sensitized photoreactions of 1-aza-1,4-dienes. Therefore, the synthetic interest of these new reactions is limited. All attempts to increase the yields of products, using DCA as the sensitizer, resulted in destruction of the starting material and the photoproducts. However, recent results show that replacing the DCA by dicyanodurene (DCD) as an electron-acceptor sensitizer and biphenyl as the cosensitizer results in considerably higher yields. This opens the possibility of using these reactions for synthetic purposes [69].

We believe that the rearrangements of the di-π-methane type observed in the DCA-sensitized irradiations of 1-aza- and 2-aza-1,4-dienes are important because the di-π-methane process has been considered until now a paradigm of reactions that take place in the excited-state manifold only. Our results show that rearrangements of this type can also occur in the ground states of radical-cation intermediates. This opens the possibility of promoting di-π-methane-type rearrangements by alternative thermal means.

D.   Di-π-methane Rearrangements Via Radical-Anion Intermediates

The aza-di-π-methane (ADPM) rearrangement of aza-1,4-dienes via radical-cations suggests the possibility that other radical-ion intermediates (e.g., radical-anions) could also be responsible for this rearrangement reaction. In order to test this proposal, the azadiene 101 was irradiated for 20 min in acetonitrile using N,N-dimethylaniline (DMA) as an electron-donor sensitizer. The reaction leads to formation of the cyclopropylimine 102. Separation of product mixture by column chromatography on silica gel affords the aldehyde 34 (21%) resulting from hydrolysis of the imine 102, (Scheme 18) [70].

The study was extended to 1-azadienes 117, 118, and 119. The DMA-sensitized irradiation of these compounds under the above-described conditions yields cyclopropylimine photoproducts 120, 121, and 122 that, after silica gel column chromatography, afford the respective cyclopropanecarbaldehydes 11 (15%), 38 (11%), and 123 (15%). The results clearly show that 1-ADPM rearrangements of azadienes 101, 117, 118, and 119 can be photosensitized by DMA, a well-known electron donor [71]. However, the possible involvement of radical-anions in these reactions is tentative because we previously observed triplet 1-ADPM reactivity of azadiene 117 by using acetophenone as the sensitizer [25]. Because the DMA has a triplet energy of 68.4 kcal/mol [71], efficient energy transfer to the diphenylvinyl unit (approximate triplet energy of 53–62 kcal/mol [11]) is possible in these processes. The question that arises is whether or not the DMA-sensitized reactions take place via SET or triplet energy-transfer mechanisms. In order to clarify this issue, azadienes 101, 117, 118, and 119 were irradiated for the same period of time, by using DMA in the nonpolar solvent hexane. Under this condition, radical-anion formation should not be favored. We observed that DMA-sensitized reactions of **101**, **117**, and **118** takes place in hexane to yield the corresponding cyclopropanecarbaldehydes **34** (2%), **11** (2%), and **123** (4%), respectively. The yields of these processes are low, which supports the proposal that radical-anions are involved as intermediates in photoreactions in acetonitrile. However, in the case of compound **119**, the yields of aldehyde **38** from the acetonitrile and hexane reactions were similar (∼11%), casting doubts again on whether or not this process occurs via SET or triplet energy sensitization.

In an attempt to obtain a more definite answer to this question, we searched for a substrate that would not undergo the di-π-methane rearrangement in the triplet excited-state manifold. With this goal in mind, the study was extended to the 1,4-diene **124** (Scheme 20) Numerous studies have demonstrated that triplets of acyclic 1,4-dienes do not undergo di-π-methane rearrangements [1–4]. The photochemical reactivity of **124** was studied earlier by Zimmerman and Pratt [10] who showed that it rearranges efficiently to yield the cyclopropane **125** on direct irradiation but that the process is very inefficient when benzophenone triplet sensitization is used. Therefore, compound **124** was considered to be a good candidate to test the possibility of promoting di-π-methane reactions by using electron-donor sensitizers. Irradiation of **124** in acetonitrile, for 3 hr, by using

Structures 117–123

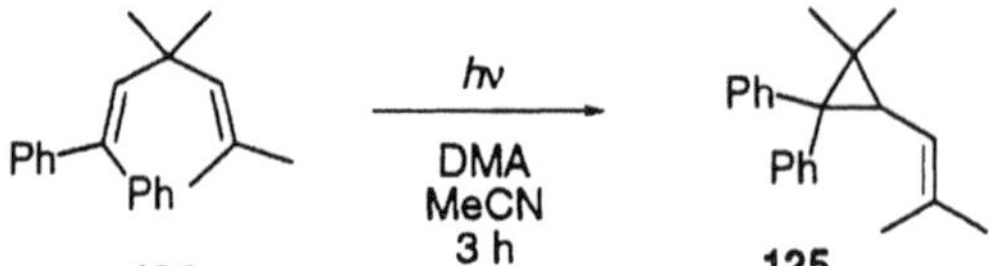

DMA/Acetonitrile	34%
DMA/Hexane	22%
Benzophenone/Hexane	5%

Scheme 20

DMA as a sensitizer, yielded cyclopropane **125** in 34% yield (Scheme 20). Changing the solvent from acetonitrile to hexane does not modify significantly the efficiency (22%) of the process. However, when diene **124** was irradiated in hexane for 3 hr, with benzophenone as the sensitizer, the cyclopropane **125** was obtained in only 5% yield, in agreement with the earlier results [10]. Because the triplet energy of benzophenone (69.2 kcal/mol) [71] is comparable to the value reported for the DMA (68.4 kcal/mol) [71], both sensitizers should be able to transfer their triplet energy to the diphenylvinyl unit with the same efficiency. Therefore, the decrease in the yield of product observed in the reactions run in hexane, caused by changing the sensitizer from DMA to benzophenone, cannot be explained by less efficient triplet energy transfer from the benzophenone to the diene **124**. An alternative explanation could invoke the possible involvement of singlet energy transfer from the DMA to the diene **124**. However, this possibility seems unlikely because the energy of the diphenylvinyl singlet excited state (97.9 kcal/mol) [71] is considerably higher than the singlet excited state of DMA (88.8 kcal/mol) [71].

The above results lead to the conclusion that rearrangement of **124** to the cyclopropane **125**, occurring in the reactions sensitized by DMA, must take place via radical-anion intermediates. Considering that the C—N double bond in compounds **101**, **117**, **118**, and **119** should be a better electron acceptor than the diphenylvinyl unit in **124**, it is logical to assume that the rearrangement of the 1-azadienes also takes place via the same types of radical-ion intermediates.

A possible mechanism for these reactions is shown in Scheme 21 for compound **101**. The absence of a solvent polarity effect on the efficiency of photoreactions of **119** and **124** might be due to a very fast rearrangement of the radical-anion **126** within solvent cages (Scheme 21, path **a**). In cases in which this intermediate escapes from the cage before rearrangement occurs, a significant influence of the polarity of the solvent would have been observed. This is the situation in DMA-sensitized reactions of **101**, **117**, and **118** (Scheme 21, path **b**).

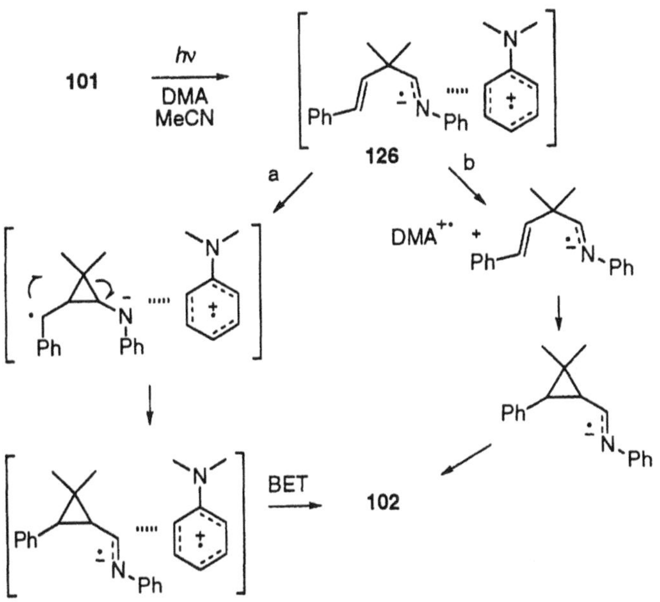

Scheme 21

These results provide strong evidence in favor of the involvement of radical-anions in electron-donor-sensitized 1-aza-di-π-methane rearrangements of aza-dienes **101**, **117**, **118**, and **119** and also in the di-π-methane reaction of diene **124**. These observations open new lines of research in an area in which, due to the large number of studies carried out for more than 30 years, apparently there were very few things that remained to be uncovered. Further studies are in progress to determine the scope and synthetic applications of these reactions.

E. Other New Results on Di-π-methane Rearrangements

1. Mechanistic Aspects

A study on mechanistic aspects of di-π-methane rearrangements has been published recently [72]. The kinetic modeling of temperature-dependent datasets from photoreactions of 1,3-diphenylpropene and several of its 3-substituted derivatives **127a–127d** (structures **127** and **128**) show that the singlet excited state decays via two inactivated processes, fluorescence and intersystem crossing, and two activated processes, *trans–cis* isomerization and phenyl–vinyl bridging. The latter activated process yields a biradical intermediate that partitions between forma-

127   128

127a, 128a: $R^1 = R^2 = H$
127b, 128b: $R^1 = Me, R^2 = H$
127c, 128c: $R^1 = R^2 = Me$
127d, 128d: $R^1 = Ph, R^2 = H$

Structures 127 and 128

tion of ground-state reactant and the corresponding di-π-methane product **128a–128d**. These results are in agreement with the conventional mechanism via biradical intermediates, proposed by Zimmerman, rather than the Bernardi–Robb concerted 1,2-vinyl migration [73]. The conclusion from these results is that Bernardi–Robb energy surface for 1,4-pentadiene is not applicable to more complex molecules such as the 1,3-diphenylpropenes.

2. The Tri-π-methane Rearrangement

The photochemistry of tri-π-methane systems has been investigated. According to the conventional biradical mechanism normally accepted for the di-π-methane rearrangement, the presence of a vinyl group at the methane carbon would allow for observation of alternative cyclizations of the intermediate biradicals that could yield cyclopentenes in addition to the cyclopropanes normally observed in these reactions (Scheme 22). However, studies carried out by Zimmerman [74] on the solution photochemistry of triene **129** (Structures 129–134) and by us on the sensitized irradiation of β,-γ-unsaturated aldehydes **44** and **46** [51] and 1-aza-dienes **130** and **131** [33] with vinyl substituents at the methane carbon showed that these compounds rearrange to the corresponding vinylcyclopropanes **132, 45, 47, 133,** and **134,** exclusively. The cyclopentene derivatives that could be formed by the alternative cyclization path were not observed. The preferred cyclization to cyclopropanes was interpreted by Zimmerman as the result of the allylic moiety being s-trans and therefore incapable of closing to a cis-cyclopentene [74].

The first example of a tri-π-methane reaction was reported by Zimmerman et al. in their study of the irradiation of triene **135** in the solid state that affords cyclopentene **136** exclusively (for reviews, see Refs. 75–78). In this instance the cisoid conformation of the biradical, necessary for cyclopentene formation, is

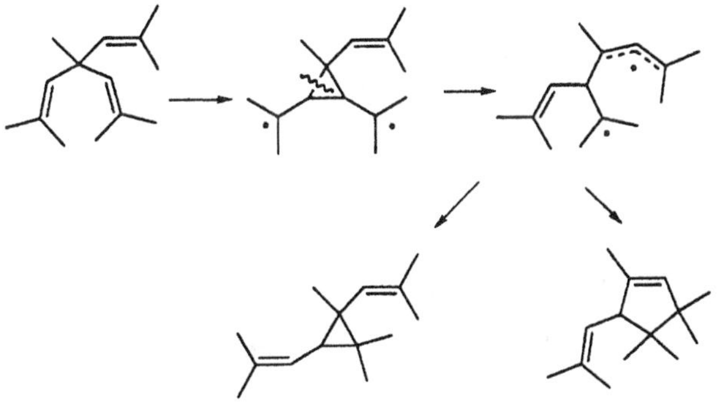

Di-π-methane   Tri-π-methane

Scheme 22

enforced by the confines of the surrounding crystal lattice. Irradiation of **135** in solution affords the corresponding cyclopropanes **137a–137c**, exclusively (Scheme 23).

In this regard, it is worth mentioning that there are many reports of differences in di-π-methane reactivity in the solid state versus solution. In some instances, differences in regio selectivity and stereoselectivity have been observed as well as alternative reactions [75–78]. However, these results are difficult to

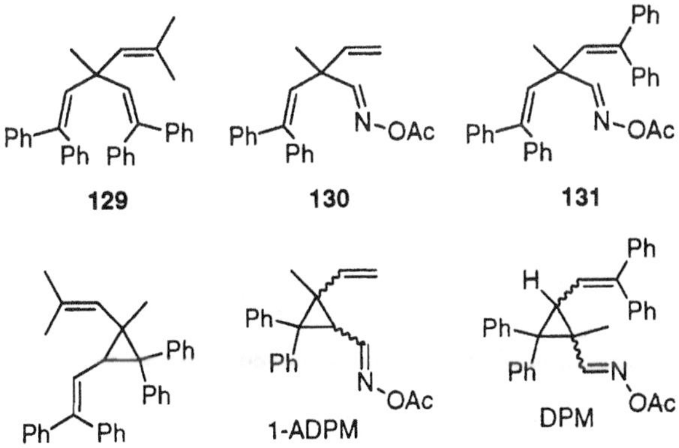

129   130   131

132   1-ADPM   134

133   DPM

Structures 129–134

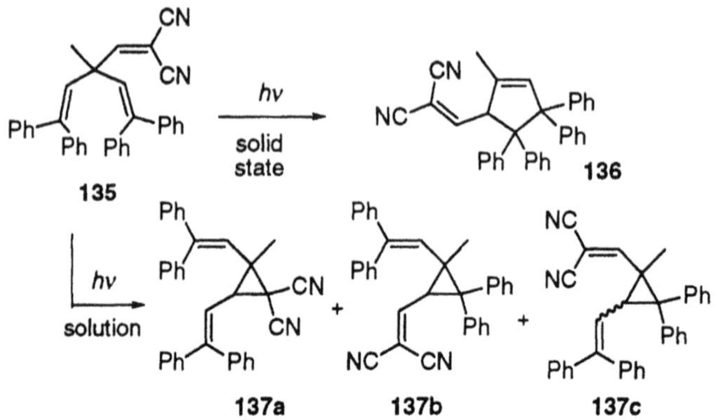

Scheme 23

rationalize because the outcome of the reaction depends mainly on the confines of the crystal lattice in each particular case.

However, in a recent publication, Zimmerman and Cirkva have reported the first examples of the elusive tri-π-methane rearrangement in solution [79]. Thus, direct irradiation of trienes **138a** and **138b** yields a mixture of cyclopropanes **139a** and **139b** and cyclopentenes **140a** and **140b**, respectively (Scheme 24). However, the direct irradiation of the di-π-methane photoproducts **139a** and **139b** led to the tri-π-methane products **140a** and **140b**, respectively, giving rise to the question of whether the cyclopentene products are formed directly via a tri-π-methane rearrangement or as a consequence of a secondary reaction of the vinyl-cyclopropane products (Scheme 24). To clarify this issue, a kinetic study of the

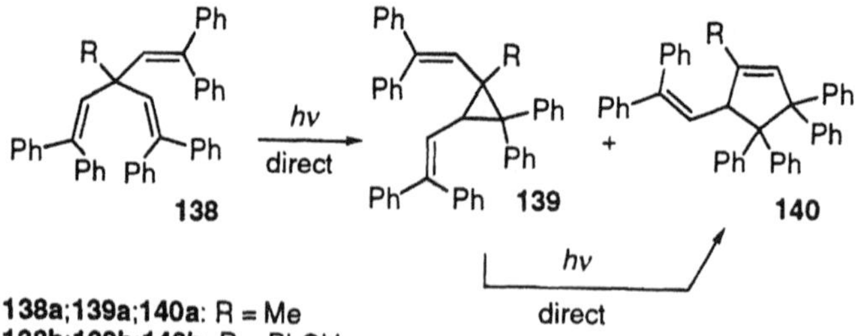

138a;139a;140a: R = Me
138b;139b;140b: R = PhCH$_2$

Scheme 24

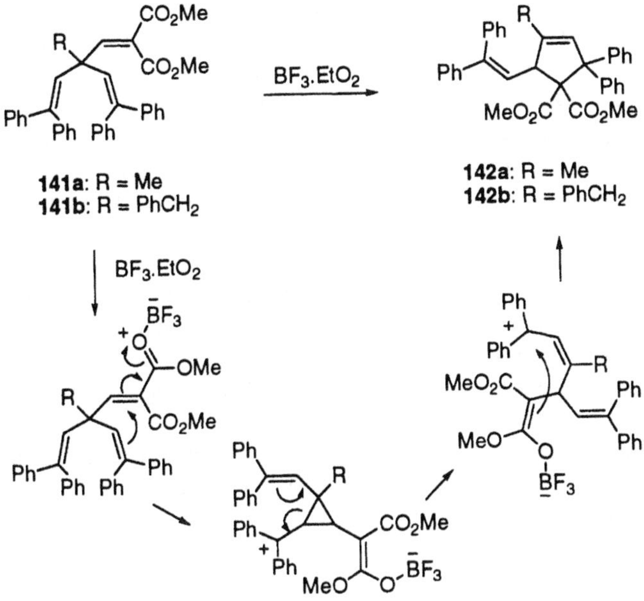

Scheme 25

reaction was carried out demonstrating that most of the five-membered-ring product truly comes from a direct tri-π-methane reaction.

Another interesting observation in this study is the boron trifluoride etherate-catalyzed rearrangement of tri-π-methane systems **141** that afford the corresponding cyclopentenes **142**. These reactions can be considered as the first examples of tri-π-methane rearrangements in the ground state. Interestingly, compounds **141** only undergo conventional di-π-methane reactions on irradiation. The mechanism shown in Scheme 25 is proposed to account for this novel reaction [79].

III. CONCLUSIONS

Di-π-methane rearrangements have been the subject of a large number of studies by different research groups worldwide for more than 30 years. As a consequence, a general impression developed that there were very few things that remained to be uncovered in this area of research. However, the results obtained in the last 5 years or so, in our laboratory and from other research groups, have modified drastically some ideas that apparently were firmly established. Thus, for example, di-π-methane processes were considered the paradigm of reactions that occur in the excited-state manifold only. However, our results in the irradiation of different 1,4-unsaturated systems using electron-donor and electron-acceptor sensitizers demonstrate that these reactions take place via radical-anion and radical-cation intermediates in the ground state, respectively. In addition, the reactivity previously reported for β, γ-unsaturated aldehydes was decarbonylation. In spite of this, our studies demonstrate that these compounds undergo ODPM rearrangements more efficiently than the corresponding methyl ketones, contrary to the common belief. New reactions, similar to the well-known Norrish Type I processes, that take place via the T_1 (π, π*) excited state, instead of by excitation of the carbonyl group, have been uncovered. Finally, N-vinylaziridines are obtained in the triplet- and DCA-sensitized irradiations of 2-aza-1,4-dienes. These are the first examples of formation of three-membered-ring heterocycles in di-π-methane reactions.

When combined, these results open new lines of research in an area in which apparently there was very little to uncover and broaden the synthetic potential of the DPM rearrangements.

REFERENCES

1. Hixson, S. S.; Mariano, P. S.; Zimmerman, H. E. *Chem. Rev.* **1973**, *73*, 531–551.
2. Zimmerman, H.E. In *Rearrangements in Ground and Excited States*; DeMayo, P., ed.; Academic Press: New York, 1980; Vol. 3, pp. 131–166.
3. Zimmerman, H.E. In *Organic Photochemistry*; Padwa, A., ed.; Marcel Dekker: New York, 1991; Vol. 11, pp. 1–36.
4. Zimmerman, H. E.; Armesto, D. *Chem. Rev.* **1996**, *96*, 3065–3112.
5. Zimmerman, H. E.; Binkley, R. W.; Givens, R. S.; Sherwin, M. A. *J. Am. Chem. Soc.* **1967**, *89*, 3932–3933.
6. Zimmerman, H. E.; Pratt, A. C. *J. Am. Chem. Soc.* **1970**, *92*, 6267–6271.
7. Mariano, P. S.; Ko, J-K. *J. Am. Chem. Soc.* **1972**, *94*, 1766–1767.
8. Fasel, J-P.; Hansen, H-J. *Chimia*, **1981**, *35*, 9–12; Zimmerman, H. E.; Swafford, R. L. *J. Org. Chem.* **1984**, *48*, 3069–3083.
9. Ciganek, E. *J. Am. Chem. Soc.* **1966**, *88*, 2882–2883.
10. Zimmerman, H. E.; Pratt, A. C. *J. Am. Chem. Soc.* **1970**, *92*, 6259–6267.
11. Zimmerman, H. E.; Armesto, D.; Amezua, M. G.; Gannett, T. P.; Johnson, R. P. *J. Am. Chem. Soc.* **1979**, *101*, 6367–6383.
12. Zimmerman, H. E.; Factor, R. E. *Tetrahedron* **1981**, *37* (Suppl. *1*), 125–141.
13. Tenney, L. P.; Boykin, D. W., Jr.; Lutz, R. E. *J. Am. Chem. Soc.* **1966**, *88*, 1835–1836.
14. Dauben, G.W.; Lodder, G.; Ipaktschi, J. *Topics Curr. Chem.* **1975**, *54*, 73–114.
15. Houk, K. N. *Chem. Rev.* **1976**, *76*, 1–74.
16. Schuster D. I. In *Rearrangements in Ground and Excited States*; DeMayo, P., ed.; Academic Press: New York, 1980; Vol. 3, pp. 167–279.

17. Demuth, M. In *Organic Photochemistry*; Padwa, A., ed.; Marcel Dekker: New York, 1991; Vol. 11, pp. 37–109.
18. Murato, K.; Frei, B.; Scheweizer, W. B.; Wolf, H. R.; Jeger, O. *Helv. Chim. Acta* **1980**, *63*, 1856–1866.
19. Demuth, M.; Hinsken, W. *Angew. Chem., Int. Ed. Engl.* **1985**, *24*, 973–975.
20. Demuth, M.; Hinsken; W. *Helv. Chim. Acta* **1988**, *71*, 569–576.
21. Paquette, L. A.; Kang H. -J.; Ra, C. S. *J. Am. Chem. Soc.*, **1992**, *114*, 7387–7395.
22. Givens, R. S.; Oettle, W. F. *Chem. Commun.* **1969**, 1164–1165.
23. Dauben, W. G.; Kellogg, M. S.; Seeman, J. I.; Spitzer, W. A. *J. Am. Chem. Soc.* **1970**, *92*, 1786–1787.
24. Armesto, D.; Martin, J. F.; Perez-Ossorio R.; Horspool, W. M. *Tetrahedron Lett.* **1982**, *23*, 2149–2152.
25. Armesto, D.; Horspool, W. M.; Martin, J-A. F.; Perez-Ossorio R. *J. Chem. Res.* **1986**, (S), 46–47; (M), 0631–0648.
26. Armesto, D.; Langa, F.; Martin, J-A. F.; Perez-Ossorio R.; Horspool, W. M. *J. Chem. Soc., Perkin Trans. I* **1987**, 743–746.
27. Armesto, D.; Horspool, W. M.; Langa, F.; Perez-Ossorio R. *J. Chem. Soc., Perkin Trans. II* **1987**, 1039–1042.
28. Armesto, D.; Horspool, W. M.; Langa, F. *J. Chem. Soc., Perkin Trans. II* **1989**, 903–906.
29. Mariano, P. S.; Stavinoha, J. L. In *Synthetic Organic Photochemistry*; Horspool, W. M., ed.; Plenum Press: New York, 1984; pp. 145–257.
30. Armesto, D.; Gallego, M. G.; Horspool, W. M.; Agarrabeitia, A. R. *Tetrahedron* **1995**, *51*, 9223–9240.
31. Armesto, D.; Ramos, A. *Tetrahedron* **1993**, *49*, 7159–7168.
32. Armesto, D.; Ramos, A.; Mayoral, E. P. *Tetrahedron Lett.* **1994**, *35*, 3785–3788.
33. Armesto, D.; Ortiz, M. J.; Ramos, A.; Horspool, W. M.; Mayoral, E. P. *J. Org. Chem.* **1994**, *59*, 8115–8124.
34. Armesto, D.; Horspool, W. M.; Mancheño, M. J.; Ortiz, M. J. *J. Chem. Soc., Perkin Trans. I* **1990**, 2348–2349.
35. Armesto, D.; Horspool, W. M.; Mancheño, M. J.; Ortiz, M. J. *J. Chem. Soc., Perkin Trans. I* **1992**, 2325–2329.
36. Behr, J.; Braun, R.; Martin, H. -D.; Rubin, M. B.; Steigel, A. *Chem. Ber.* **1991**, *124*, 815–820.
37. Liao, C. -C.; Hsieh, H. -P.; Lin, S. -Y. *J. Chem. Soc., Chem. Commun.* **1990**, 545–547.
38. Baggiolini, E.; Hamlow, H. P.; Schaffner, K.; Jeger, O. *Chimia* **1969**, *23*, 181–182.
39. Baggiolini, E.; Hamlow, H. P.; Schaffner, K. *J. Am. Chem. Soc.* **1970**, *92*, 4906–4921.
40. Bozzato, G.; Schaffner, K.; Jeger, O. *Chimia* **1966**, *20*, 114–116.
41. Baggiolini, E.; Berscheid, H. G.; Bozzato, G.; Cavalieri, E.; Schaffner, K.; Jeger, O. *Helv. Chim. Acta* **1971**, *54*, 429–449.
42. Saboz, J. A.; Iizuka, T.; Wehrli, H.; Schaffner, K.; Jeger, O. *Helv. Chim. Acta* **1968**, *51*, 1362–1371.
43. Hill, J.; Iriarte, J.; Schaffner, K.; Jeger, O. *Helv. Chim. Acta* **1966**, *49*, 292–311.
44. Dürr, H.; Heitkämper, P.; Herbst, P. *Tetrahedron Lett.* **1970**, 1599–1600.

45. Dürr, H.; Herbst, P.; Heitkämper, P.; Leismann, H. *Chem. Ber.* **1974**, *107*, 1835–1855.
46. Adam, W.; Berkessel, A.; Hildenbrand, K.; Peters, E. -M.; Peters, K.; von Schnering, H. G. *J. Org. Chem.* **1985**, *50*, 4899–4909.
47. Pratt, A. C. *J. Chem. Soc., Perkin Trans. 1* **1973**, 2496–2499.
48. Pfenninger, E.; Poel, D. E.; Berse, C.; Wehrli, H.; Schaffner, K.; Jeger, O. *Helv. Chim. Acta* **1968**, *51*, 772–803.
49. Zimmerman, H. E.; Cassel, J. M. *J. Org. Chem.* **1989**, *54*, 3800–3816.
50. Armesto, D.; Ortiz, M. J.; Romano, S. *Tetrahedron Lett.* **1995**, *36*, 965–968.
51. Armesto, D.; Ortiz, M. J.; Romano, S.; Agarrabeitia, A. R.; Gallego, M. G.; Ramos, A. *J. Org. Chem.* **1996**, *61*, 1459–1466.
52. Naumann, K. *Synthetic Pyrethroid Insecticides*; Springer-Verlag: Berlin, 1990; Vols. 4 and 5.
53. Armesto, D.; Gallego, M. G.; Horspool, W. M.; Bermejo, F. Patent ES 9100648, 13th March **1991**; WO 92/16499, Patent PCT/ES 92/00017, 13th February **1992**.
54. Dalton, J. C.; Shen, M.; Snyder, J. J. *J. Am. Chem. Soc.* **1976**, *98*, 5023.
55. Schaffner, K. *Tetrahedron* **1976**, *32*, 641; Schuster, D. I.; Calcaterra, L. T. *J. Am. Chem. Soc.* **1982**, *104*, 6397.
56. Braslavsky, S. E.; Houk, K. N. *Pure Appl Chem* **1988**, *60*, 1055–1106.
57. van der Weerdt, A. J. A.; Cerfontain, H. *Recl. Trav. Chim. Pays-Bas* **1977**, *96*, 247–248.
58. van der Weerdt, A. J. A.; Cerfontain, H. *Tetrahedron*, **181**, *37*, 2121–2130.
59. Armesto, D.; Ortiz, M. J.; Agarrabeitia, A. R.; Aparicio-Lara, S. *Synthesis* **2001**, 1149–1158.
60. Mariano, P. S. In *CRC Handbook of Organic Photochemistry and Photobiology*; Horspool, W. M.; Soon, P. -S., eds.; CRC Press: New York, 1995; pp. 867–878.
61. Rao, M. V. B.; Suresh, J. R.; Kumar, A.; Ila, H.; Junjappa, H. *J. Indian Chem. Soc.* **1997**, *74*, 955–960.
62. Goldschmidt, Z.; Crammer, B. *Chem. Soc. Rev.* **1988**, *17*, 229.
63. Armesto, D.; Caballero, O.; Amador, U. *J. Am. Chem. Soc.* **1997**, *119*, 12659–12660.
64. Adam, W.; Berkessel, A.; Krimm, S. *J. Am. Chem. Soc.* **1986**, *108*, 4556–4561.
65. Caballero, O. Ph.D. thesis, Universidad Complutense, 2001.
66. Zimmerman, H. E.; Hoffacker, K. D. *J. Org. Chem.* **1996**, *61*, 6526–6534.
67. Ortiz, M. J.; Agarrabeitia, A. R.; Aparicio-Lara, S.; Armesto, D. *Tetrahedron Lett.* **1999**, *40*, 1759–1762.
68. Armesto, D.; Austin, M. A.; Griffiths, O. J.; Horspool, W. M.; Carpintero, M. *J. Chem. Soc., Chem. Commun.* **1996**, 2715–2716.
69. Armesto, D.; Ortiz, M. J.; Agarrabeitia, A. R.; Aparicio-Lara, S.; Liras, M.; Martin-Fontecha, M. Unpublished results.
70. Armesto, D.; Ortiz, M. J.; Agarrabeitia, A. R.; Martin-Fontecha, M. *J. Am. Chem. Soc.* **2001**, *123*, 9920–9921.
71. Kavarnos, G. J.; Turro, N. J. *Chem. Rev.* **1986**, *86*, 401–449.
72. Lewis, F. D.; Zuo, X.; Kalgutkar, R. S.; Wagner-Brennan, J. M.; Miranda, M. A.; Font-Sanchis, E.; Perez-Prieto, J. *J. Am. Chem. Soc.* **2001**, *123*, 11883–11889.
73. Reguero, M.; Bernardi, F.; Jones, H.; Olivucci, M.; Ragazos, I. N.; Robb, M. A. *J. Am. Chem. Soc.* **1993**, *115*, 2073–2074.

74. Zimmerman, H. E.; Zuraw, M. J. *J. Am. Chem. Soc.* **1989**, *111*, 7974–7989.
75. Ramamurthy, V., ed. *Photochemistry in Organized and Constrained Media*; VCH: New York, 1991.
76. Ohashi, I. Ed. *Reactivity in Molecular Crystals;* VCH: Weinheim, 1994.
77. Sakamoto, M. *Chem. Eur. J.* **1997**, *3*, 684.
78. Ito, Y. *Synthesis* **1998**, 1.
79. Zimmerman, H. E.; Cirkva, V. *J. Org. Chem.* **2001**, *66*, 1839–1851.

2

The Photo-Fries Rearrangement

Miguel A. Miranda
Universidad Politécnica de Valencia, Valencia, Spain

Francisco Galindo
Universidad Jaume I, Castellón, Spain

I. INTRODUCTION

In 1960, Anderson and Reese [1] reported that cathecol monoacetate (**1**) under-goes rearrangement to 2,3- and 3,4-dihydroxyacetophenone (**2** and **3**, respec-tively) upon illumination with ultraviolet (UV) light (Scheme 1). This report was the first one of a long list of articles and reviews [2–16] dealing with the photorrerarrangement of aromatic esters, which was extended to amides, carbon-ates, carbamates, thioesters, sulfonates, and related compounds.

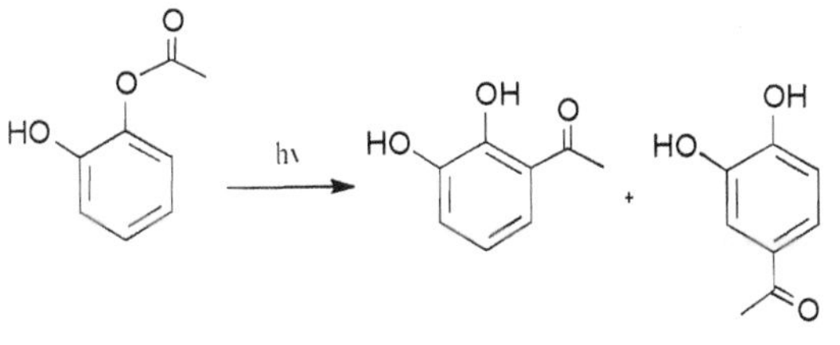

1	**2** (22%)	**3** (18%)

Scheme 1

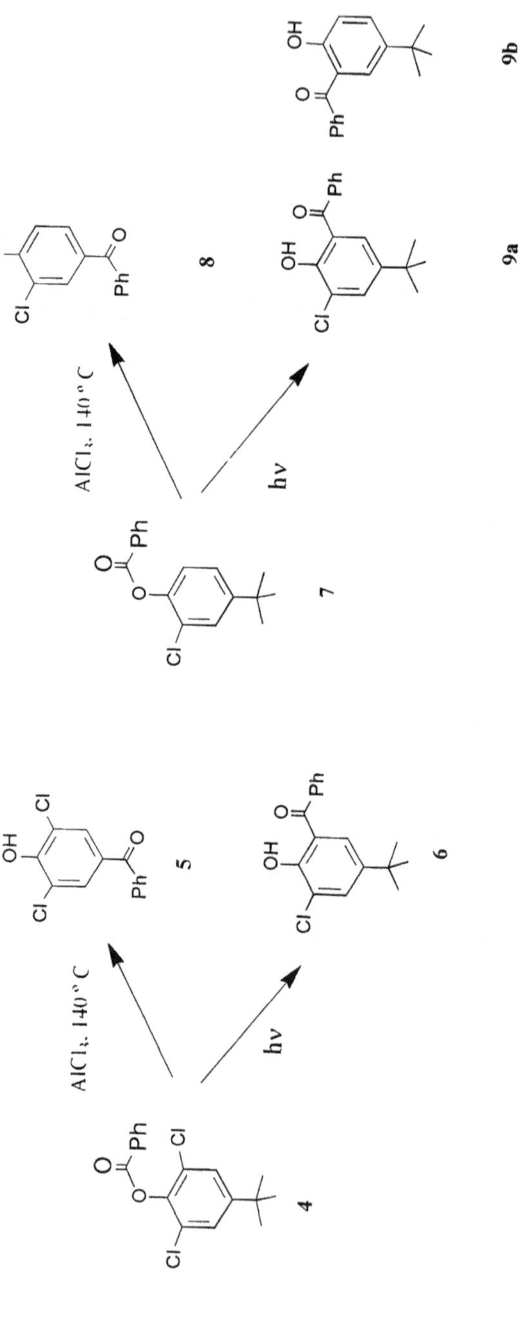

Scheme 2

Due to its clear resemblance with the classical Lewis acid-catalyzed Fries rearrangement, [7–9], the term *photo-Fries* has been coined to designate this kind of photochemical processes. Although leading in most cases to identical products, the thermal and photochemical reactions are completely different from the mechanistic point of view. In 1962, Kobsa reported [10] the illustrative results of a comparative investigation between the thermal and the photochemical rearrangement of several aromatic esters. Although 2,6-dichloro-4-*tert*-butylphenyl benzoate (**4**) undergoes substitution of *tert*-butyl by benzoyl upon acid catalysis with aluminum trichloride (product **5**), irradiation results in substitution of the chlorine atom to give 3-chloro-2-hydroxy-5-*tert*-butylbenzophenone (**6**) (Scheme 2). With 2-chloro-4-*tert*-butylphenyl benzoate (**7**), striking differences are also observed between the reactions carried out under both conditions. These results clearly suggested a fundamental difference in reaction mechanism between the classical and the light-induced rearrangements.

The present work deals with the photo-Fries rearrangement (PFR) and covers the literature appearing since the first report of Anderson and Reese to the beginning of 2002. It includes both the mechanistic aspects and the synthetic applications of this reaction. The topic of the PFR in polymers is also treated. In addition, a section has been devoted to the photo-Fries rearrangement in organized media, a field of recent interest.

II. REACTION MECHANISM

A. The Model

The mechanistic origin of the PFR products was a matter of discussion from the beginning. The archetypal rearrangement is that of phenyl acetate (**10**), which has been widely employed as a model to understand the photoprocess. Its irradiation in solution affords *ortho*-hydroxyacetophenone (**11**), *para*-hydroxyacetophenone (**12**), and phenol (**13**) (Scheme 3). Two alternative mechanistic pathways were

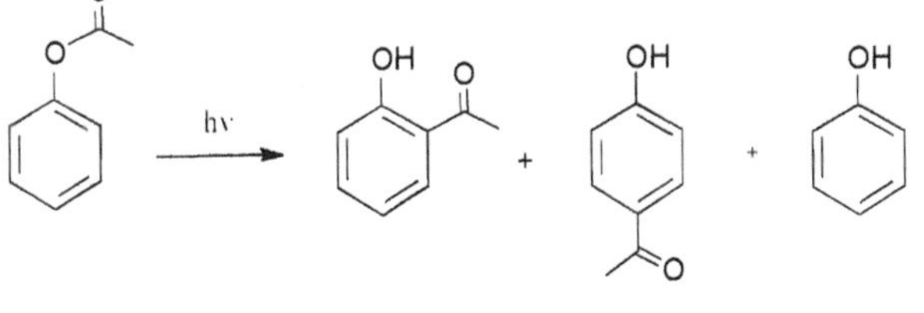

| 10 | 11 | 12 | 13 |

Scheme 3

initially proposed to account for the observed photoproducts; however, one of them was soon rejected based on the experimental evidences.

A concerted, symmetry-allowed process involving a four-membered-ring intermediate like that depicted in the Scheme 4 (intermediate **A**) was invoked, to explain the formation of an ortho product. In analogy with this 1,3-sigmatropic shift, another bridged intermediate would explain formation of the para-rearranged phenol (intermediate **B** in Scheme 4) [1,10–13]. To date, none of those bridged intermediates has been observed by any of the techniques used for the detection of transient species. Likewise, attempts to isolate trapping products derived therefrom have been unsuccessful.

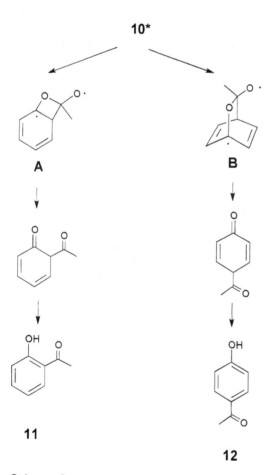

Scheme 4

The alternative mechanism consists in the homolytic scission of the ester carbonyl-oxygen to give a radical pair enclosed in a solvent cage. Due to delocalization, with significant spin densities in the ortho and para positions of the phenoxyl radical, as well as in the oxygen atom (but not in the meta position) [14], in cage recombination yields either the starting ester or 2,4- and 2,5-cyclohexadienones. The latter tautomerize to the final products (Scheme 5) [10,15–20]. The origin of the phenol would be hydrogen abstraction by the phenoxyl radical from the solvent.

Thus, both models postulate cyclohexadienones as necessary reaction steps, although arising by very different paths.

The modern techniques for detection of short-lived species provided the necessary evidences to establish the three main mechanistic features of the photo-Fries rearrangement:

1. It occurs from the excited singlet state.
2. It involves homolytic scission of the O—CO bond.
3. It is an intramolecular process.

These ideas will be discussed in the following subsections, where most of the attention will be devoted to the mechanistic studies with aromatic esters, which have been the subject of an overwhelming majority of the research efforts. Nevertheless, the same reaction mechanism has been shown to be valid for the PFR of anilides, thioesters, sulfonates, and so forth. Furthermore, it is also applicable to the photo-Claisen rearrangement [i.e. the migration of alkyl (or allyl, benzyl, aryl,)] groups of aromatic ethers to the ortho and para positions of the aromatic ring [21,22].

B. Sensitization and Quenching

Attempts to sensitize the rearrangement of aryl esters with triplet energy donors have resulted unsuccessful. On the other hand, the rearrangements of 4-methylphenyl acetate [12,13] and 4-methylphenyl-N-methyl carbamate [23] are not afected by naphthalene (E_T = 60.9 kcal/mol). Another example of inefficient quenching is the case of 4-methylphenyl benzoate, which is insensitive to the presence of biacetyl (E_T = 54.9 kcal/mol) [5,6]. Likewise, sensitization of 4-methylphenyl acetate with compounds of high triplet energy (i.e., acetophenone, E_T = 73.6 kcal/mol) does not yield the observed rearrangement products of direct irradiations (2-hydroxy-5-methylacetophenone and 4-methylphenol) [12]. Not even oxygen is able to quench the photo-Fries rearrangement. Shizuka and co-workers have measured the quantum yields for formation of *ortho*-hydroxyacetophenone (**11**), *para*-hydroxyacetophenone (**12**), and phenol (**13**) from phenyl acetate (**10**) in cyclohexane; they were 0.17, 0.15, and 0.06, respectively, both in aerated and degassed solutions [24].

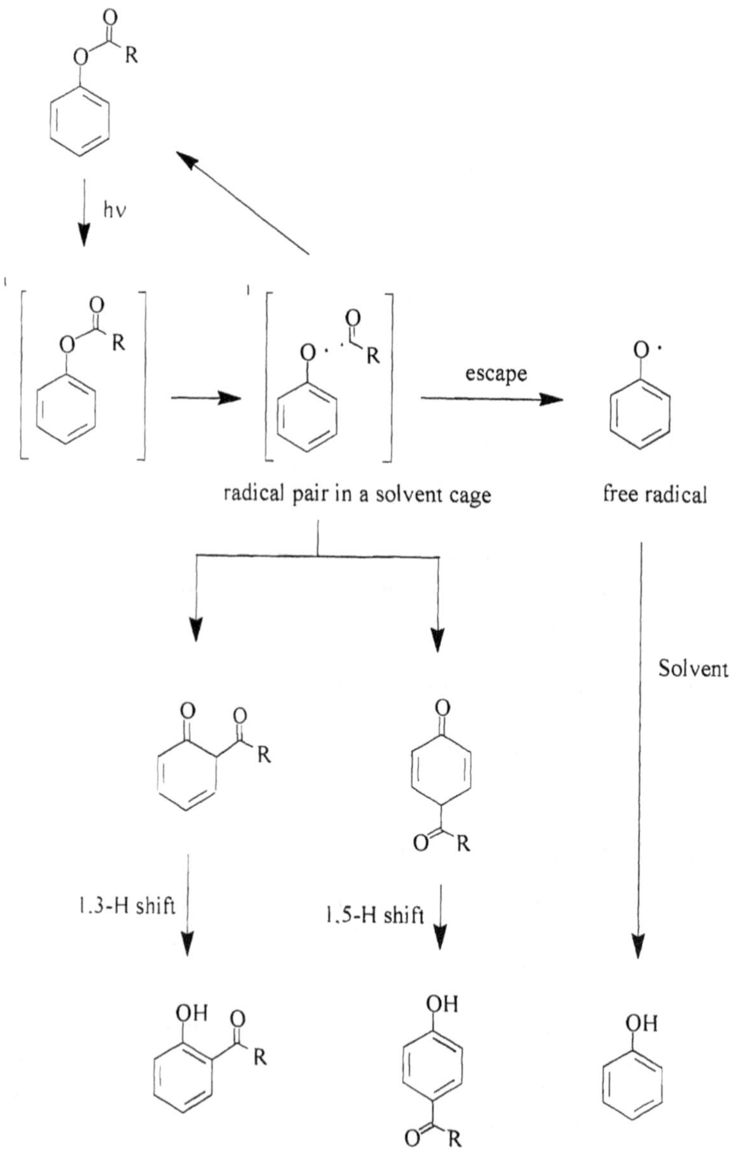

Scheme 5

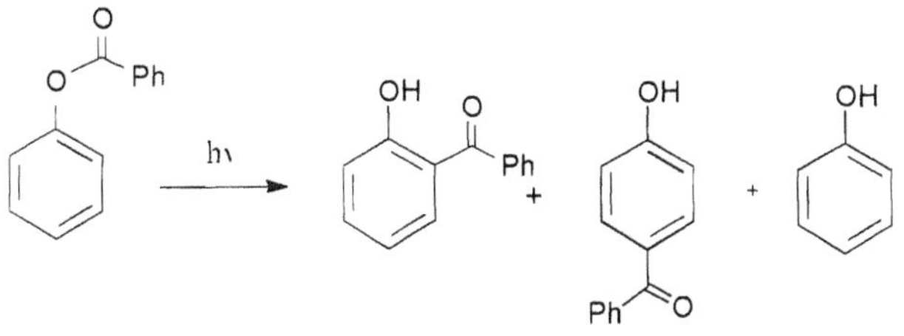

14   15   16   13

Scheme 6

However, benzene (a well-known singlet sensitizer) effectively transfers the photochemically acquired energy to phenyl benzoate (**14**) to give *ortho*-hydroxybenzophenone (**15**), *para*- hydroxybenzophenone (**16**), and phenol (**13**) (Scheme 6) [25]. Additional evidence for singlet energy transfer from the aromatic solvent is provided by the fact that some aromatic esters undergoing rearrangement in dioxane solutions containing benzene or toluene as sensitizers do not react if *para*-terphenyl or biacetyl are present in the medium [5,6]. This has been interpreted as quenching of the photochemically excited singlets of benzene and toluene by *para*-terphenyl and biacetyl.

The implication of a *singlet excited state* in the PFR seems now well established if considering the overwhelming evidences; however, some recent studies point that in certain cases the rearranged products could arise from upper triplet states [26,27]. This contribution is usually minor. For the case of 1-naphthyl acetate, it has been estimated that less than 10% of the rearranged photoproducts have a triplet origin. [28].

C. Solvent Effects

If the photo-Fries reaction would occur via a concerted mechanism, the absence of solvent should be of minor importance for the formation of rearranged products. However, conclusive evidence supporting the radical pair mechanism arises from the experiments carried out with phenyl acetate (**10**) in the vapor phase. The major product in the irradiations of **10** is phenol (**13**), which accounts for ~ 65% of the photoproducts. Under these conditions, less than 1% of *ortho* -hydroxyacetophenone (**11**) appears to be formed [19,20]. Conversely, when a high cage effect is expected, as in rigid matrixes (i.e., polyethylene), the result is completely different, and phenol is practically absent from the reaction mixtures [29]. In the intermediate situation (liquid solution), both rearranged products and phenol are formed in variable amounts depending on solvent properties. These observations

Table 1 Product Yield in the PFR of **14**

Solvent	15 + 16 (%)
Cyclohexane	40
Ether	33
Dioxane	36
Benzene	33
Isopropanol	77
Methanol	76
Ethanol	78
tert-Butanol	84

Source: Ref. 31.

fit well in the model which explains phenol formation as occurring via escape from the cage and rearrangement products through in-cage recombination of the radical fragments arising from the initial homolysis. In the case of a rigid matrix, the rearrangement should be favored, whereas in the vapor phase, the absence of the cage would result in a rapid separation of the radicals, not allowing recombination. The absence of coupling products in the gas phase has also been observed for acetanilide [30].

Solvent polarity has a strong influence on the yield of rearranged products as indicated in Table 1 for the PFR of phenyl benzoate (**14**) in several solvents [31]. Polar solvents favor the rearrangement, whereas nonpolar solvents favor phenol formation. An experiment carried out with methanol–ether mixtures

Table 2 Product Distribution in the Irradiation of **14** in Methanol–Ether Mixtures

Methanol (vol %)	Conversion (%)	Photoproducts		
		15 (%)	16 (%)	13 (%)
0	92	18	18	44
30	93	26	22	32
50	96	38	20	23
70	98	46	21	17
100	99	59	20	13

Source: Ref. 31.

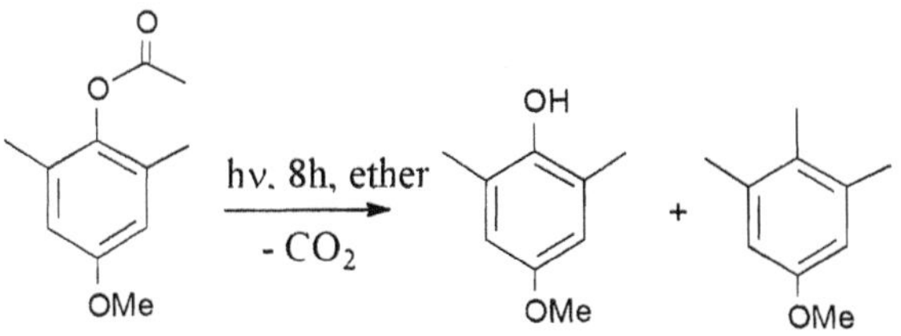

17   18 (3.4%)   19 (31.4%)

Conversion = 93.8%

Scheme 7

showed a drastic decrease in the yield of phenol (13), with increasing methanol concentration (see Table 2).

Diethyl ether has been shown to enhance decarboxylation [32] as a secondary reaction after C—O bond cleavage when the ortho and para positions are blocked. Scheme 7 shows this effect for compound 17. Some esters with the ortho and para positions free to react but with bulky substituents at the meta position also undergo decarboxylation in ether, tetrahydrofuran, and dioxane [33,34].

However, solvent viscosity, rather than polarity, has been a useful tool for mechanistic purposes. Although the quantum yield of the ortho-rearranged product of 4-methylphenyl acetate (20) does not change with viscosity of the medium, the formation of 4-methylphenol (22) is highly sensitive to this factor. Thus, its quantum yield is 0.45 in ethanol (1.00 cP) but only 0.02 in Carbowax 600 (109 cP) (Scheme 8; Table 3) [13]. This clearly supports the mechanism involving caged radical pairs. A related aspect is the *intramolecular nature of the process* confirmed by the lack of cross-coupling products in crossover experiments with mixtures of different esters [10].

D. Substituent Effects

As a general rule, electron-donating substituents attached to the phenolic ring enhance photoreactivity, whereas electron-withdrawing substituents produce the opposite effect [5,6,29]. For instance, the relative photo-Fries quantum yield is more than two times higher for *para*-hydroxyphenyl benzoate than for *para*-acetylphenyl benzoate [5,6]. Esters with susbstituents such as Cl or CH_3 at the para position, as well as the unsubstituted compounds, have intermediate

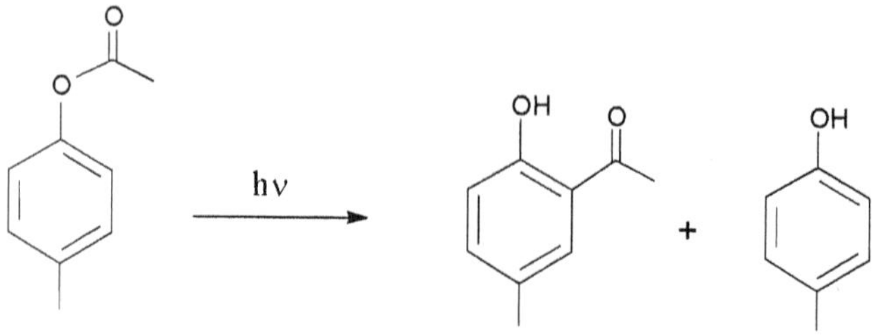

20   **21**   **22**

Scheme 8

reactivities. The relative quantum yields of PFR are given in Table 4. The nitro-substituted esters undergo alcoholysis and reduction rather than rearrangement [35].

The opposite trend is observed upon the introduction of changes in the acyl moiety (i.e., the higher the electron density, the lower the reactivity). Thus, upon irradiation of several para-substituted benzoates, the following order of reactivity is observed (relative PFR quantum yield in parentheses): CN (1.00) > Cl (0.87) > CH$_3$ (0.65) > H (0.49) [5,6]. In this connection, it is worth mentioning that the PFR of phenyl salicylate (**23**) is intramolecularly quenched in nonpolar solvents such as hexane [36]. As Scheme 9 shows, the intramolecular hydrogen bond between the ortho hydroxyl and the carbonyl group through a six-membered ring is reflected in the low photoreactivity of **23** [9% conversion, 4% ortho product

Table 3 Product Quantum Yield in the PFR of **20** in Solvents of Different Viscosities

Solvent	Viscosity (cP, 30°C)	Product quantum yield	
		21	**22**
Ethanol	1.00	0.17	0.45
Isopropanol	1.73	0.17	0.09
tert-Butanol	3.00	0.18	0.07
Carbowax 400	76	0.17	0.03
Carbowax 600	109	0.16	0.02

Source: Ref. 13.

Table 4 Relative Quantum Yield in the PFR of Several Substituted Phenyl Benzoates

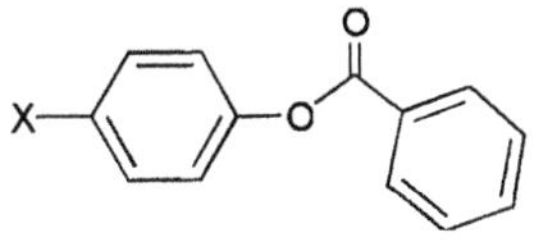

X	Relative quantum yield of PFR
OH	1.00
CH$_3$	0.84
H	0.82
Cl	0.68
COCH$_3$	0.37
NO$_2$	0.01

Source: Ref. 5.

(**24**), and 1% of para product (**25**)]. The same reaction in methanol, where the intramolecular hydrogen bond is disrupted to a great extent, occurs with better efficiency (68% conversion, 28% of **24**, and 32% of **25**). Analogous observations have been reported for salicyl anilide [36].

When Cl and MeO substituents are attached to the phenolic ring, the PFR may result in displacement of those substituents by the acyl moiety. An example of chlorine displacement has been given in Section I, whereas Scheme 10 shows the case of 2-methoxy-4-methylphenyl benzoate (**26**), where the reaction mixture contains 2-hydroxy-5-methylbenzophenone (**27**), a product of methoxy substitution [37]. Hageman also reported the substitution of MeO by acetyl in the photolysis of 4-methoxyphenyl acetate and related esters [32].

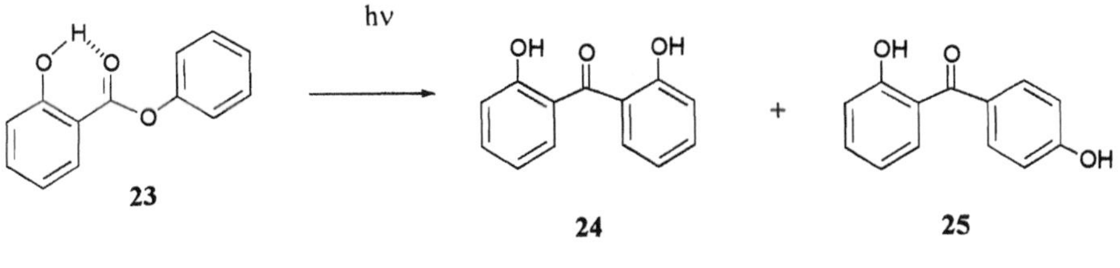

Scheme 9

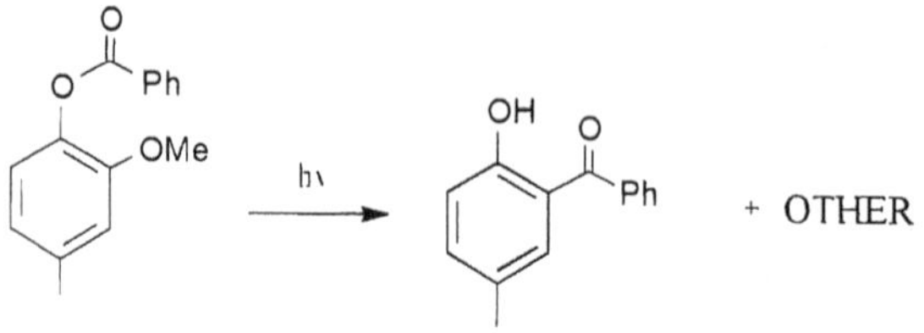

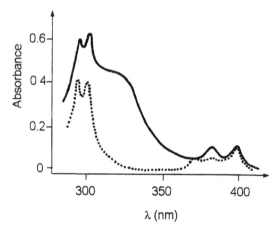

26   **27** (6% in benzene. 7% in ethanol)

Scheme 10

E. Flash Photolysis Studies

1. Aryloxy Radicals

Kalmus and Hercules pioneered the studies using conventional lamp-flash photolysis. They clearly demonstrated the intermediacy of phenoxy radicals and cyclohexadienones in the PFR of phenyl acetate (**10**) [17,18]. Independently, Humphrey and Roller reached similar conclusions from flash-photolysis studies on bis(4-*tert*-butylphenyl) carbonate [38].

  When phenyl acetate (**10**) is photolyzed in solution with a conventional flash lamp, the transient absorption of the phenoxy radical can be readily observed, with a lifetime of 0.2–0.3 ms. The same intermediate has been detected in different solvents such as ethanol and Freon. Figure 1 presents two transient absorption

Figure 1 Flash spectra of phenyl acetate(——) and phenol (. . .) in hexane. Delay time: 15 μsec. (Adapted from Ref. 18.)

spectra for comparison: the first one results from the photolysis of phenyl acetate and the second from the photolysis of phenol, which is also known to give phenoxy radicals. As can be seen, the positions of the bands are coincident. The photolysis of acetanilide and 2,6-dimethylphenyl acetate produces the typical absorptions of anilino and 2,6-dimethylphenoxy radicals. These observations are direct evidences for homolytic cleavage.

Likewise, with 1- and 2-naphthyl esters, the typical absorption of 1- and 2-naphthoxy radicals has been observed [28,39]. In the absence of quenchers, the superposition of the bands corresponding to the radicals and the triplet of the esters, has been registered but in the presence of piperylene, an effective triplet quencher, the absorption of the triplet is quenched but not that corresponding to the naphthoxy radical. The measured quantum yield for the formation of the 1-naphthyl acetate triplet state is 0.4 ± 0.2 in acetonitrile and 0.35 ± 0.15 in methanol. This would explain the low reactivity of 1-naphthyl acetate to give photo-Fries rearrangement, because after absorption of light the ester would give intersystem crossing very efficiently to the unreactive triplet state.

2. Cyclohexadienones

In addition to the absorptions attributable to aryloxy radicals, Fig. 1 displays a broad shoulder around 315 nm, much longer-lived, which is assigned to a 2,4-cyclohexadienone. This intermediate decays with a rate of ~ 1.25 ± 0.1 sec^{-1} in hexane solution, to give 2-hydroxyacetophenone (**11**) via 1,3-hydrogen shift. The rate of appearance of **11** is coincident with the decay rate of the dienone.

More recently, Shizuka and co-workers have re-examined the photolysis of phenyl acetate (**10**) using a laser as the excitation source instead of a conventional lamp, but focusing their attention on the sigmatropic hydrogen shift leading from the cyclohexadienones to the final products [40,41]. In methylcyclohexane as the solvent, it has been possible to follow the evolution with time of the absorption bands corresponding to *para*-hydroxyacetophenone (~ 270 nm, rising), 6-acetyl-2,4-cyclohexadienone) (~ 305 nm, decaying), and *ortho*-hydroxyacetophenone (~ 325 nm, rising). A biexponential fit of the signal at 325 nm gives nearly coincident rates of decay (6-acetyl-2,4-cyclohexadienone) and formation (*ortho*-hydroxyacetophenone), which are equal to the rate of 1,3-hydrogen shift (Fig. 2). The process of H migration has been found to be concentration dependent in nonpolar systems; thus, the actual constant has been determined by means of Eq. (1) by extrapolation at [ester] = 0 M, where $k_{1,3}$ is the rate constant of 1,3-hydrogen shift at a given ester concentration, $k_{1,3}^0$ is the rate constant at 0 M of ester and $k_{1,3}^{se}$ is the constant for the self-enhanced 1,3-shift:

$$k_{1,3} = k_{1,3}^0 + k_{1,3}^{se} \left[\text{ester}\right] \qquad [1]$$

This analysis gives the following values in methylcyclohexane: $k_{1,3}^0 = 3.6 \pm 0.3$ sec^{-1}, $k_{1,3}^{se} = 0.033 \pm 0.006$ M^{-1}/sec. As can be seen, the value of

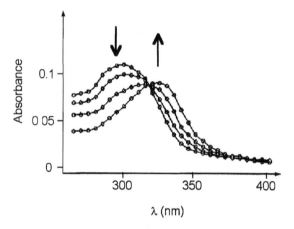

Figure 2 Time-resolved transient absorption spectra obtained by flash photolysis of phenyl acetate in methylcyclohexane (266 nm excitation). Delay times: 50, 100, 200, and 400 msec. (Adapted from Ref. 41.)

$k_{1.3}^0$ obtained by Shizuka and co-workers is very similar to that obtained by Kalmus and Hercules 20 years earlier ($1.25 \pm 0.1 \ \text{sec}^{-1}$). In the case of phenyl acetate-d_5, there is also a 1,3-deuterium shift; however, according to expectations, the process occurs more slowly than in the nondeuterated compound. The observed kinetic isotope effect is $k_{1.3}^0 / k_{1.3}^0$ (deuterated) $= 3.8$.

The rate for the 1,5-hydrogen shift in methylcyclohexane was found to be much slower than the corresponding 1,3-shift ($k_{1.5}^0 = 0.065 \pm 0.006 \ \text{sec}^{-1}$). This has been rationalized taking into account the ionic interactions in the cyclohexadienones rather than the orbital symmetry of the system. According to the Woodward–Hoffmann rules [42], the 1,3-shift should be symmetry forbidden and the 1,5-shift should be symmetry allowed. This contrasts with the experimental results. PM3 calculations show that in the *ortho*-cyclohexadienone, the migrating hydrogen bears a positive charge ($+0.1079$), whereas the oxygen atom has a negative character (-0.3215). The same is true for the *para*-cyclohexadienone. To confirm the role of ionic factors, solvents of different polarity have been used and the $k_{1.3}^0$ and $k_{1.5}^0$ values were measured. As shown in Table 5, the higher the solvent polarity, the higher the rate constant for the shift process (1,3 and 1,5), although with some exceptions. In the case of protic solvents like methanol or ethanol, it has been proposed that basic catalysis (proton exchange) is involved. Triethyl amine has been found to enhance the shift process, whereas trifluoroethanol (an acidic solvent) does not have a clear influence on the values of $k_{1.3}^0$ and $k_{1.5}^0$ with respect to the situation in methylcyclohexane. This agrees with the

Table 5 Rates of 1,3- and 1,5-H Shifts in Different Solvents

Solvent	$k_{1,3}^0$ (sec^{-1})	$k_{1,5}^0$ (sec^{-1})
Methylcyclohexane	3.6 ± 0.6	$(6.5 \pm 0.6) \times 10^{-2}$
Acetonitrile	$(5.0 \pm 1.0) \times 10$	$(1.2 \pm 0.4) \times 10^2$
Diethyl ether	$(2.9 \pm 0.2) \times 10^2$	$(8.8 \pm 0.6) \times 10$
Tetrahydrofuran	$(3.0 \pm 0.1) \times 10^2$	$(1.5 \pm 0.1) \times 10$
Methanol	$(3.0 \pm 0.4) \times 10^5$	$(2.5 \pm 0.2) \times 10^5$
Ethanol	$(3.3 \pm 0.5) \times 10^5$	$(4.4 \pm 0.6) \times 10^5$
n-Propanol	$(2.7 \pm 0.1) \times 10^5$	$(2.7 \pm 0.2) \times 10^5$
i-Propanol	$(2.8 \pm 0.2) \times 10^5$	$(2.8 \pm 0.1) \times 10^5$
Trifluoroethanol	$(3.0 \pm 1.2) \times 10$	$(8.2 \pm 1.1) \times 10$

Source: Ref. 41.

proposed basic catalysis. Finally, it is worth mentioning that in alcohols, there is no self-enhancement, probably due to the predominating effect of the solvent on the shift process.

Based on temperature effects, it has been concluded that the 1,3-hydrogen shift occurs via quantum mechanical tunneling in the electronically ground state of the *ortho*-cyclohexadienone

As can be seen, the pathway from the cyclohexadienone to the ortho-rearranged phenol does not imply a jump over an energy barrier (34 kcal/mol), but a tunneling from the vibrational energy levels located at $E(\nu_0) = 0$ kcal/mol and $E(\nu_1) = 3.9$ kcal/mol (4.4 kcal/mol for the deuterated compound). Analogous results have been reported for the photo-Fries rearrangement of 2,4-dimethoxy-6-(*para*-tolyloxy)-s-triazine (**28**) to give 2,4-dimethoxy-6-(2-hydroxy-5-methylphenyl)-s-triazine (**29**) (Scheme 11) [43].

A further step in the study of cyclohexadienones as PFR reaction intermediates has consisted in the secondary photolysis of these relatively long-lived species [44]. By means of the two-laser two-colour technique, it has been observed that phenyl acetate is first converted into an *ortho*-cyclohexadienone by excitation with a Nd-YAG laser at 266 nm. If a second laser pulse (excimer laser operating at 308 nm) is fired with a convenient delay with respect to the first laser, the signal of the dienone bleaches concomitantly with the growth of a new signal centered around 330 nm. According to the literature data on the photolysis of stable 6,6-disubstituted cyclohexa-2,4-dienones [45], this new signal can be safely assigned to the dienic ketene depicted in Fig. 3.

F. Chemically Induced Dynamic Nuclear Polarization

The PFR of *para*-cresyl *para*-chlorobenzoate (**30**) (Scheme 12) has been studied by means of chemically induced dynamic nuclear polarization (CIDNP) [46]. The

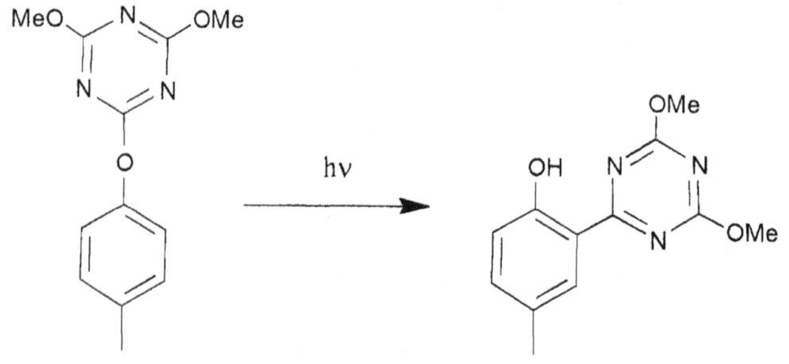

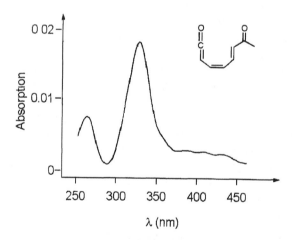

28   **29**

Scheme 11

appearance of strong polarized signals in the NMR (nuclear magnetic resonance) spectrum recorded during steady-state irradiation of **30** supports the radical mechanism.

The polarized signals correspond to the starting ester (**30**), the ortho-rearranged product (**31**) and *para*-cresol (**32**). The variation in intensity of the signals

Figure 3 Absorption spectrum obtained from the difference between the spectrum of the photolysis of phenyl acetate (first pulse at 266 nm) and the spectrum after a second laser pulse at 308 nm. (Adapted from Ref. 44.)

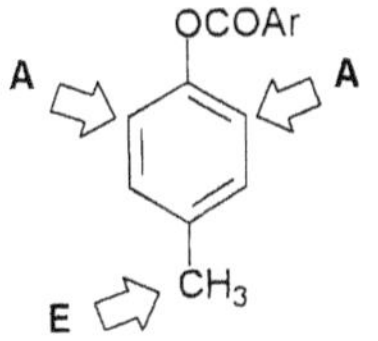

OCOAr
A   A
E   CH₃

30

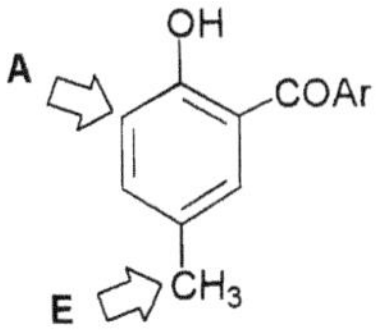

OH
A   COAr
E   CH₃

31

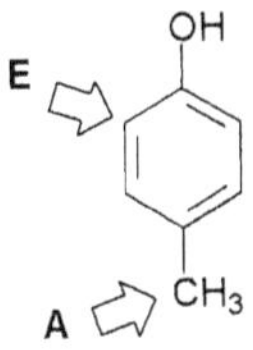

OH
E
A   CH₃

32

Ar = p-Cl-phenyl

A = H polarized in **absorption**
E = H polarized in **emission**

Scheme 12

of **30** indicates that there is some coupling of the primary radical pair to give the starting material. On the other hand, **31** has also polarized protons in spite of not being a direct radical-coupling product but a secondary product (formed by hydrogen migration in the cyclohexadienone precursor). This clearly means that the lifetime of the cyclohexadienone (τ) is considerably shorter than the relaxation times (T) of its polarized protons ($\tau \ll T$). Moreover, in addition to confirming the radical nature of the PFR, the CIDNP technique has allowed one to distinguish between intramolecular (in-cage) and intermolecular (out-of-cage) processes. Assuming the involvement of a singlet excited state and according to Kaptein's rules [47] and electron spin resonance (ESR) parameters for the involved radicals, it has been established that products **30** and **31** are originated by recombination inside the solvent cage, whereas product **32** is formed after escape of the aryloxy radical from the cage. This is supported by the opposite polarization of the same protons observed in **30** and **31** with respect to **32** (see Scheme 12)

More recently, the PFR of 1- and 2-naphthyl acetates has been studied by laser-induced CIDNP [28,39]. It has been found that C—O homolysis takes place not only from the first excited singlet state (S_1), but also from an upper triplet state (T_2). This is due to a very efficient intersystem crossing from S_1 to T_2, which is reactive contrarily to T_1 which gives only T–T annihilation. The final

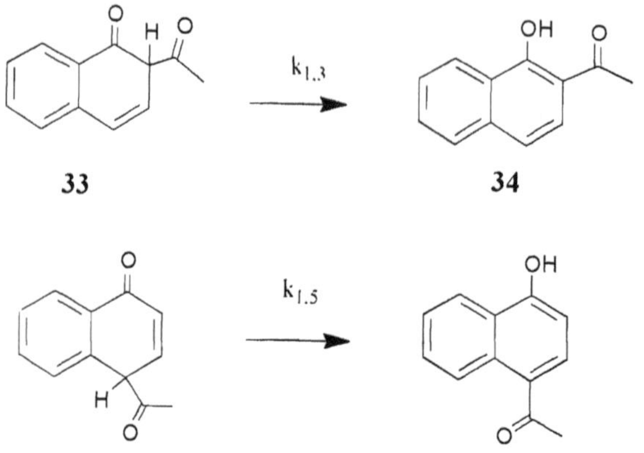

Scheme 13

rearrangement products arise mainly from the S_1 (>90%). The rate constants for hydrogen migration in the corresponding cyclohexadienones have been measured by means of time-resolved studies, asuming that these rates are equal to the rate of signal increase in flash-CIDNP experiments. The resulting values for these constants are $(1.1 \pm 0.1) \times 10^6$ and $(4.5 \pm 0.5) \times 10^5$ sec^{-1} for the 1,3- and 1,5-H shifts, respectively (Scheme 13).

G. Other Mechanistic Evidences

1. Kinetic Isotope Effects

The kinetic isotope effect (KIE) produced in the photorearrangement of 4-methoxyphenyl acetate to 2-hydroxy-5-methoxyacetophenone has been measured by determining the isotope ratios before and after irradiation (k_b and k_a, respectively) in the starting material and the rearranged products. A value of KIE = k_a/k_b different from unity would indicate that the rearrangement proceeds along an activation energy barrier. This is neither the case for the ^{14}C [48,49] nor for the case of ^{18}O isotope, both included in the carbonyl moiety [49]. In fact, the obtained values are KIE (^{14}C) = 0.9988 ± 0.0051 and KIE (^{18}O) = 1.0000 ± 0.0023.

2. Internal Isotope and External Magnetic Field Effects

Magnetic field effects have been used to study the photo-Fries rearrangement of 4-methoxyphenyl acetate [49] and 1-naphthyl acetate [50]. These effects can be subdivided into internal and external.

To determine the internal effects, it is necessary to label the ester with a magnetically active isotope like ^{13}C. This type of effect has been widely studied by Turro [51] and consists in the enrichment of the singlet radical pair in ^{12}C, due to enhanced intersystem crossing of the ^{13}C-labeled pairs. In the case of 4-methoxyphenyl acetate, a value of 0.9511 ± 0.0042 for k_1/k_2 was found, where k_1 represents the ratio of the isotope ^{13}C in the ester before irradiation and k_2 is the value after the irradiation [49]. This confirms that recombination to give the starting product occurs in the singlet radical pair. For 1-naphthyl acetate, similar results can be reached and a value of $k_1/k_2 = 0.96 \pm 0.02$ has been reported [50]. The rates of intersystem crossing (ISC) to the triplet radical pair from the singlet radical pair have been calculated in terms of the hyperfine coupling (hfc) mechanism The values for the radical pairs with ^{12}C and ^{13}C are 1.83×10^8 and 1.73×10^9 sec^{-1}, respectively [50].

The external magnefic field effect on the PFR of 1-naphthyl acetate has also been reported to be > 1 for a labeled sample (^{13}C) and equal to 1 for a nonlabeled sample (^{12}C) of ester; this means that the precursor of the reaction product (2-acetylnaphthol) is a singlet radical pair.

3. Spontaneous Raman Spectroscopy

Beck and Brus [52] have recorded the spontaneous Raman spectrum immediately after laser photolysis (266 nm) of phenyl acetate in water, observing three different species, with different lifetimes. One of the signals has been safely assigned to the phenoxy radical by comparison with the known Raman spectrum of this radical generated by irradiation of phenol. The other two signals were much longer lived and were tentatively assigned to 6-acetyl- and 4-acetylcyclohexadie-nones. The same spectroscopic study was performed for phenyl propionate, yielding identical results.

H. Detection and Fate of the Acyl Radical

So far, the work dealing with the PFR mechanism has paid more attention to the detection and fate of aryloxy radicals than to acyl radicals. It is generally accepted that aryloxy radicals afford phenols (or naphthols) through H abstraction from the solvent, but little has been said about the other partners of the radical pair, namely the acyl radicals. Their detection has been accomplished both directly and indirectly.

Indirect detection of the acyl radicals in the PFR has been achieved by means of *spin trapping* with 2-methyl-2-nitrosopropane and subsequent ESR detection of the resulting stable radicals [53]. The ESR technique has also been used for the study of the PFR of diphenyl carbonate but only a line ($g = 2.005$) corresponding to the phenoxy radical could be detected [54]. The recorded signal in the spin-trapping experiment consists of a typical sharp triplet with small

Table 6 Hyperfine Coupling Constants of the Acyl Radicals Detected in the Photolysis of Phenyl Acetate, Phenyl Benzoate and Acetanilide

$$Bu^t\text{-}N\text{=}O \ + \ RCO \cdot \ \longrightarrow \ Bu^t\text{--}\overset{\overset{\displaystyle O \ \cdot}{\displaystyle \|}}{N}\text{--}COR \quad (\text{ESR detection})$$

		a_N (G)			
Compound	Radical	Methanol	Acetonirile	Benzene	Dioxane
Phenyl acetate	$CH_3CO\cdot$	7.8	7.9	7.8	8.0
Phenyl benzoate	$PhCO\cdot$	8.1	8.1	8.0	8.0
Acetanilide	$CH_3CO\cdot$	7.8	7.9	7.8	8.0

Source: Ref. 53.

hyperfine splittings of the nitrogen of the nitroxy group. This corresponds to the trapped acetyl radical, which is an unambiguous proof of its formation. Table 6 shows the hyperfine coupling constants (a_N) for the photolysis of phenyl acetate, phenyl benzoate, and acetanilide in different solvents.

Time-resolved Fourier-transform infrared spectroscopy (TR-FTIR) has served as an efficient tool for the *direct detection* of C=O stretching frequencies of transient species [55]. Recent work by Frei and Vasenkov [56] describes TR-FTIR investigations on the PFR of 1-naphthyl acetate included in NaY zeolite. A transient absorption (2127 cm^{-1}) with a long lifetime (75 $\pm$ 11 μsec) has been reported (Fig. 4). Irradiation of a ^{13}C-labeled sample caused a 34-cm^{-1} red shift in the vibration frequency. The transient has been attributed to the acetyl radical interacting with positive Na$^+$ cations of the zeolite despite the significantly different wave numbers reported for the acetyl radical in other media: 1864 cm^{-1} in solution [57] and 1842 cm^{-1} in the argon matrix [58]. The authors rationalize this shift by considering that in the strongly ionic environment of the zeolite cavities, the radical would be interacting with sodium ions, acquiring in this way a positive character, which would make it more similar to the acetylium cation (CH_3CO^+). This agrees well with the fact that irradiation of pinacolone/NaY, which can give CH_3CO — but not CH_3CO^+, gives rise to the same 2127-cm^{-1} absorption than that found for the 1-naphthyl acetate/NaY system.

The reactivity of acyl radicals inside and outside the solvent cage has been a matter of discussion. It has been postulated that aryloxy and acyl radicals could disproportionate within the cage to give phenol (or naphthol) and ketene (**37**) but the results are not conclusive (Scheme 14). On the one hand, photolysis of 1-naphthyl acetate in a solvent without abstractable H-like Freon 113 (1,1,2-trichloro-1,2,2-trifluoroethane) yields low amounts of 1-naphthol (< 0.1%) in comparison with the same reaction in acetonitrile (7%) [50]. This reveals that dispropor-

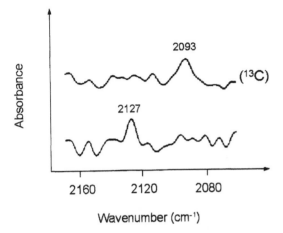

Figure 4 Step-scan FTIR spectra of 1-naphthyl acetate photodissociation (^{12}C and ^{13}C carbonyl species). (Adapted from Ref. 56.)

tionation can actually occur, albeit to a small extent. On the other hand, strong polarized signals at δ 2.08 ppm have been observed in steady-state CIDNP measurements with 1-naphthyl acetate in methanol-d_4; they have been assigned to the protons of ketene (**37**) [28]. For 2-naphthyl acetate, the signal at δ 2.14 ppm has also been assigned to ketene, although no traces of ketene are observed in the NMR spectrum after prolonged irradiation. In methanol-d_4, the observed signals at δ 2.01 ppm and δ 1.98 ppm have been assigned to methyl acetate and acetic acid, respectively, the addition products of methanol and residual water to **37** [39]. Other reported addition products to ketene intermediates are those resulting from

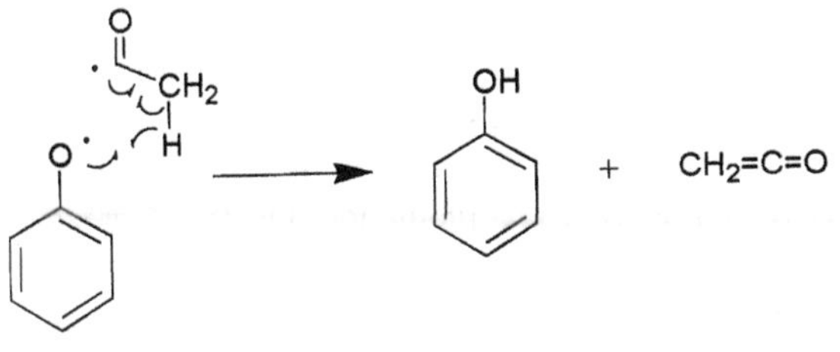

O
‖
C–CH₂
O⸴ ⸢H   →   ⬡–OH   +   CH₂=C=O

13   37

Scheme 14

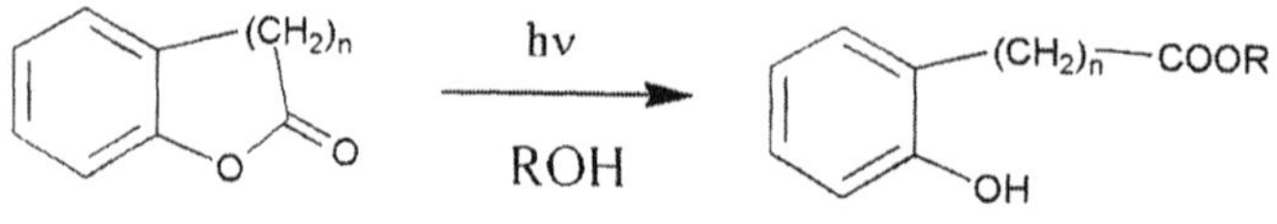

38   39

Scheme 15

lactones as depicted in Scheme 15 [11,59]. With *CIDNP*, it has also been possible to detect signals corresponding to acetaldehyde (δ 9.89 ppm) what would be explained by hydrogen abstraction out of the solvent cage [28]. In this context, it must be mentioned that high yields of aldehydes are obtained in the photolysis of 2-naphthyl thioesters in the presence of a hydrogen source like 1,4-cyclohexadiene [60].

Colussi and co-workers have studied reactions of phenoxy and acyl radicals resulting from homolysis of the C—O bond of phenyl acetate [61] and phenyl benzoate [62]. They concluded that for phenyl acetate, the main source of H to form phenol is the acetyl radical because the reaction PhO · + CH$_3$CO · → PhOH + CH$_2$=C=O is highly exothermic ($\Delta H = -42$ kcal/mol). For phenyl benzoate, which has no α-carbonyl protons and cannot give rise to ketene formation, it was proposed that benzoyl radicals would add to a second molecule of starting ester, initiating a chain polymerization reaction, this would account for the ester disappearance quantum yields higher than 1. The polymeric material would be made up by units containing the cyclohexadiene moiety, which would be the source of hydrogen for phenol formation, instead of the solvent.

I. An Intermolecular Approach to the PFR

All of the evidence discussed so far supporting the back reaction to regenerate the starting ester from the primary radical pair are based on indirect measurements (CIDNP, magnetic isotope effects), due to the impossibility of distinguishing between the "reacted" ester resulting from recombination and the "unreacted" noncleaved ester. However, the independent generation of the radicals implied in the PFR has provided a direct proof of the occurrence of recombination and, on the other hand, allows one to make a rough estimation about the extent of such C—O coupling [63]. This methodology has also been applied to the radicals (phenoxy and alkyl) participating in the photo-Claisen rearrangement [64]. For

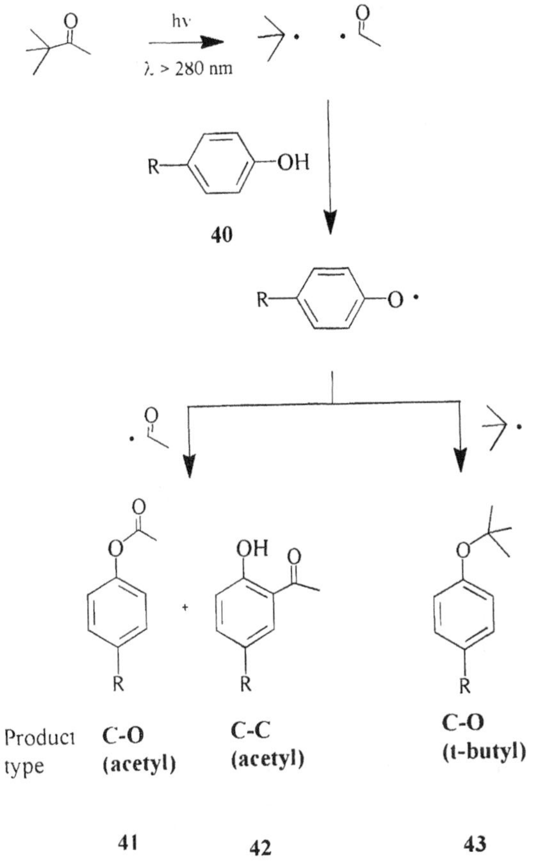

Scheme 16

this purpose, pinacolone was photolyzed in the presence of phenols. Scheme 16 summarizes the mechanism and products of the photolysis of pinacolone/phenols pairs and Table 7 provides quantitative data. Pinacolone becomes excited, because it is the only light-absorbing species at the irradiation wavelengh (>280 nm). The primary photoprocess is the well-known Norrish Type I α-cleavage, to yield acetyl and *tert*-butyl radicals. Interaction of these radicals with the phenol present in the medium brings about H abstraction to afford the corresponding phenoxy radical, which, in turn, recombines with other acetyl/*tert*-butyl radicals present in solution. Hence, the final products are phenyl esters and hydroxyacetophe-nones, as in a *normal* intramolecular PFR, but obtained intermolecularly. The

Table 7 Product Distribution in the Irradiation of Pinacolone/Phenol Pairs

R	Conversion (%)	Mass balance (%)	Product distribution (%)			
			C—O (Acetyl)	C—C (Acetyl)	C—O (tert-Butyl)	Other
tert-Bu	49	92	22	36	42	—
Ph	50	89	27	31	42	—
CF₃	25	83	—	37	63	—
OMe	100	81	11	15	30	44ᵃ

[a] Includes two products: *para-tert*-butylphenol (36%) and *para*-acetylphenol (8%).
Source: Ref. 63.

detected *tert*-butyl ethers are the corresponding intermolecular photo-Claisen C—O recombination products.

When the reaction is run with *para*-methoxyphenol up to 44% of photoproducts correspond to substitution of the methoxy group by *tert*-butyl and acetyl (Scheme 17). This is in agreement with the observation that photolysis of 4-methoxyphenyl acetate yields, among other products, *para*-acetyl phenol [32].

J. Theoretical Studies

A unified theoretical explanation using molecular orbital theory has been proposed. Grimme [65] investigated the PFR of phenyl acetate as well as the photo-Claisen rearrangement of allyl phenyl ether and the β-cleavage of *para*-substituted phenoxyacetones. A unified description of the three reactions has been invoked according to MNDOC-CI and AM1/AM1-HE calculations. No matter what ex-

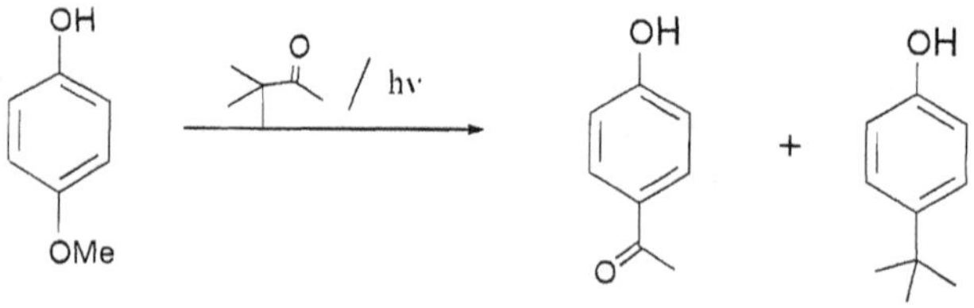

40a   12   40b

Scheme 17

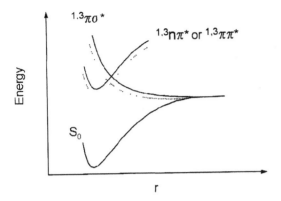

Figure 5 Energy diagram of ground and excited states of phenyl esters (r = reaction coordinate). (Adapted from Ref. 65.)

cited state is reached upon photochemical excitation ($n\pi^*$ or $\pi\pi^*$, both nondissociative), surface crossing to a dissociative state occurs ($\pi\sigma^*$), where the homolysis, and hence the rearrangement, takes place. In order to jump from one potential energy surface to another, an energy barrier must be surpassed; for phenyl acetate, this barrier has been estimated to be in 20.7 kcal/mol from S_1 and 30.9 kcal/mol from T_1. This is illustrated in Fig. 5.

III. SUBSTRATES UNDERGOING PFR/SCOPE OF THE PFR

A great variety of substrates undergo photochemical rearrangements analogous to the typical acyl migrations of phenyl acetate or phenyl benzoate. Structurally, all of the substrates have a heteroatom (oxygen or nitrogen) bound to an aromatic ring and to an acyl moiety. In some cases, the heteroatom can be sulfur or, less frequently, selenium or tellurium. Scheme 18 depicts a general model for a photo-Fries reaction: the substrate (Ar—X—Y) undergoes homolysis to give Ar—X · and Ar—Y · Radical Y · is the migrating species, and the general structure of the final products would be Y—Ar—XH. The purpose of this section is to illustrate the wide variety of reactions that can be classified as photo-Fries rearrangements. It is not intended to give an exhaustive compilation of PFR processes, but rather a number of representative examples reported in the chemical literature in order to show the scope and limitations of this photochemical reaction.

A. Esters (X=O, Y=COR)

The PFR of phenolic esters is by far the best known of this group of rearrangements. The PFR of phenyl acetate (Scheme 3) is the prototype of all of them.

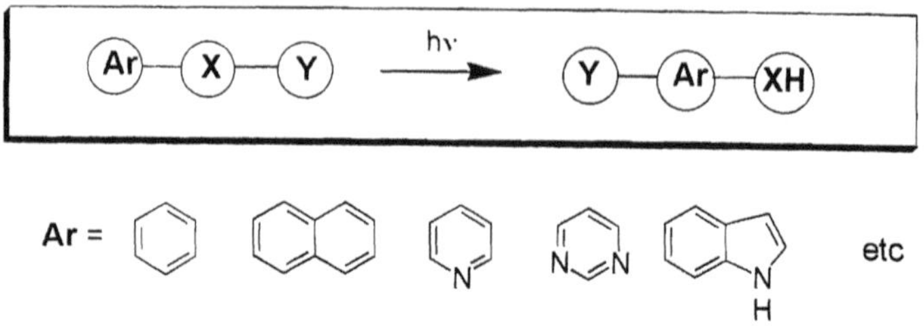

Ar = (benzene) (naphthalene) (pyridine) (pyrimidine) (indole)   etc

X = O. N. S. Se, Te

Y = COR, COOR. COCO$_2$R, CSR, SO$_2$R

Scheme 18

The acyl moiety can be aliphatic or aromatic, as has been shown with several acetates and benzoates.

Phenol esters of α,β-unsaturated carboxylic acids have an interesting reactivity due to the synthetic utility of the resulting hydroxychalcones (Scheme 19). This aspect will be illustrated in Section IV. However, from the basic point of view, it is worth mentioning that the cis or trans configuration of the olefinic part of the acyl moiety can have a marked influence on the photochemical reactivity of the ester. When *para*-methoxyphenyl fumarates are irradiated, the "normal" ortho-rearranged products are obtained. By contrast, irradiation of *para*-methoxyphenyl maleates does not lead to rearrangement. Instead, cyclization products are obtained (Scheme 20).

Thus, the cis configuration of the acyl moiety favors intramolecular cyclization, competing effectively with migration to the ortho position [66]. The same type of side reaction has been observed with phthalates due to the proximity of

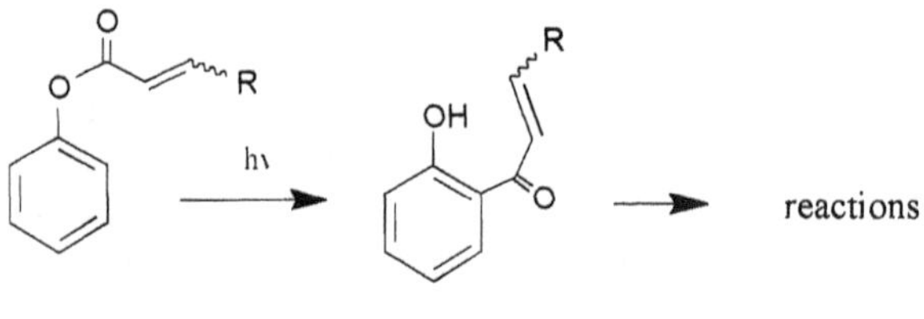

44   45

Scheme 19

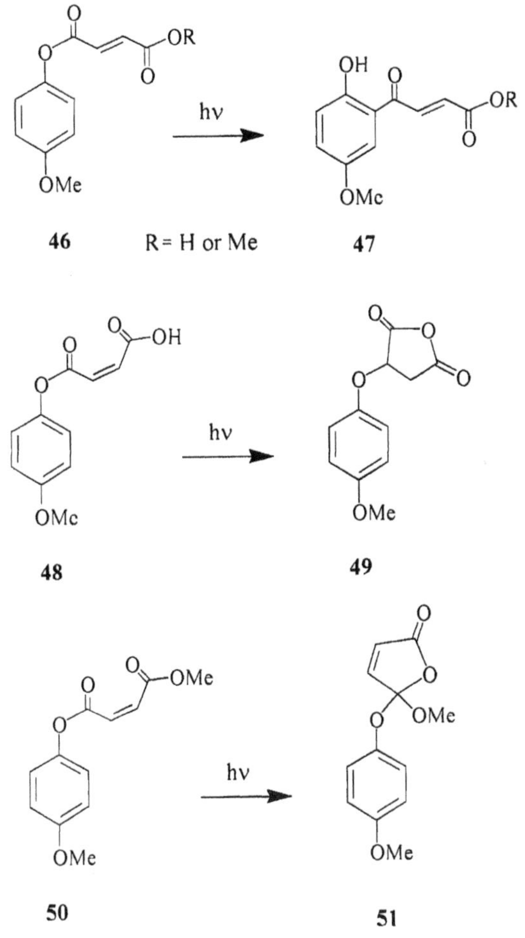

46   R = H or Me   **47**

48   **49**

50   **51**

Scheme 20

the two carboxylate groups [67]. Thus, methyl *para*-tolyl phthalate and di-*para*-tolyl phthalate yield the benzofuranones depicted in Scheme 21; however, if one carboxylate is replaced by a cyano group, the normal PFR takes place.

Other unsaturated esters include cinnamates [68,69] and phenylpropiolates (triple bond) [70]. Esters with allenic functionality like **58** have been reported to afford the ortho product **59** in 4% yield (Scheme 22) [71].

The esters of naphthol have also been reported to undergo photochemical rearrangement. 1-Naphthyl acetate (**60**) [28,39,50] gives 2- and 4-acetyl-1-naph-

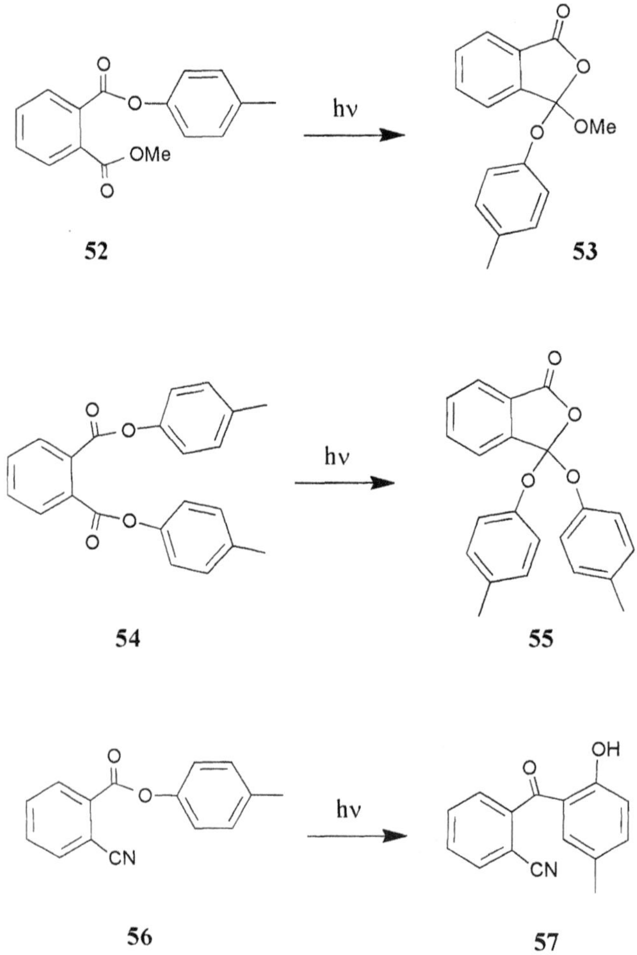

52   hv   53

54   hv   55

56   hv   57

Scheme 21

thols, whereas irradiation of 2-naphthyl acetate (**62**) [72–75] results in acyl substitution at positions 1 (major), 3, and 6 (Scheme 23). Weiss and co-workers have reported [76,77] products substituted at position 8 but in minimal amounts. 2-Naphthyl benzoate has also been reported to undergo photochemical Fries rearrangement [78].

Photochemical rearrangement has been reported for heterocyclic compounds, as shown in the esters of hydroxypyridine [79–81]. An example is shown

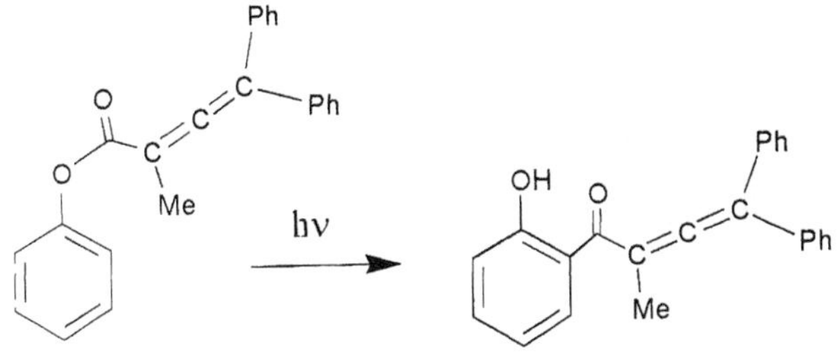

58

59 (4 %)

Scheme 22

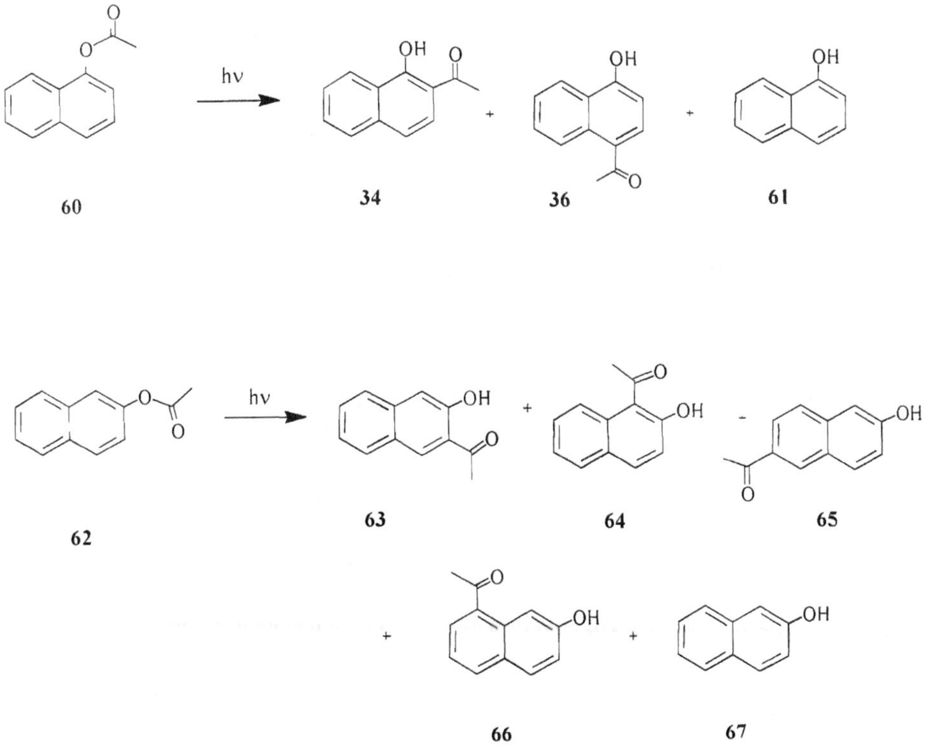

60

34

36

61

62

63

64

65

66

67

Scheme 23

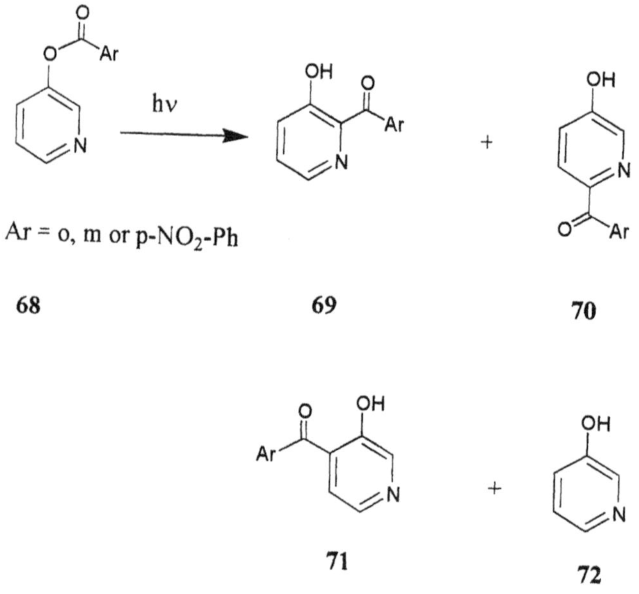

Ar = o, m or p-NO$_2$-Ph

68   69   70

71   72

Scheme 24

in Scheme 24, which illustrates the PFR of several 3-hydroxypyrydine nitrobenzo-ates. The heteroaromatic ring can be found in the acyl part as well. Phenolic esters of nicotinic and isonicotinic acids (**73** and **76**) rearrange to the para position of the phenolic ring (Scheme 25) [82].

The PFR aso can take place with esters of indole [83]. 5-Acetoxyindole rearranges mainly to position 4 (40% yield), whereas 6-acetoxyindol rearranges largely to position 7 (66% yield). For 5,6-diacetoxyindoles, the rearrangement is regioselective to position 7 as shown in Scheme 26 for **78** and **79** [84].

With phenylcumarins [85], preferential formation of one regioisomer has been observed. Interestingly, the Lewis-acid-catalyzed reaction of the same esters leads to the other possible regioisomer, as indicated Scheme 27.

In the above cases, the photoreactivity of indole, coumarin, pyridine, and so forth, does not compete effectively with PFR, which becomes the major reaction. However, in the case of the chromene depicted in Scheme 28, the electrocyclic ring opening prevails over PFR. By contrast, the analogous chromane undergoes clearly acyl migration [86].

It has been reported [25] that PFR of ferrocene **96** of Scheme 29 produces the phenol acylated at the para position (**97**). The same authors have found that

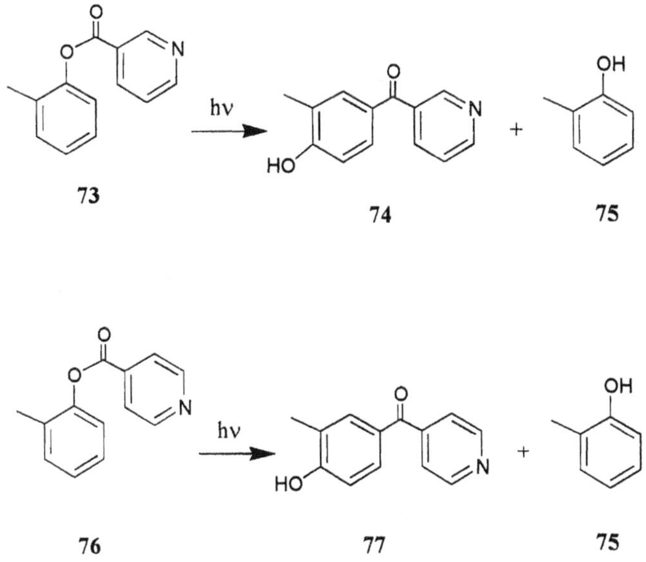

Scheme 25

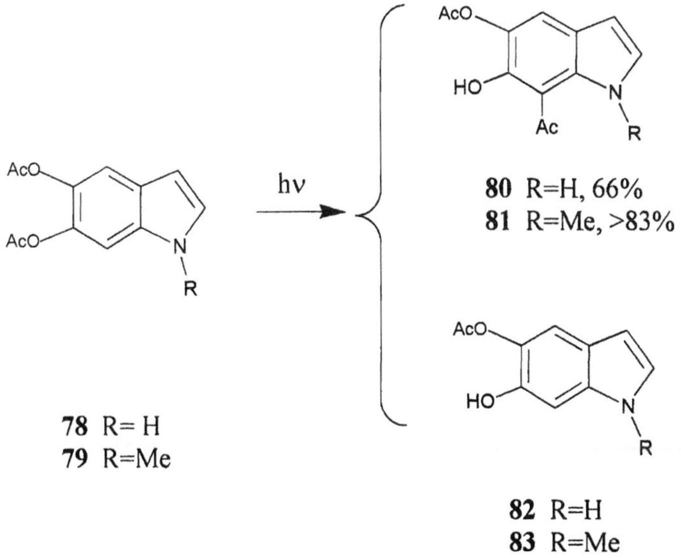

78 R= H
79 R=Me

80 R=H, 66%
81 R=Me, >83%

82 R=H
83 R=Me

Scheme 26

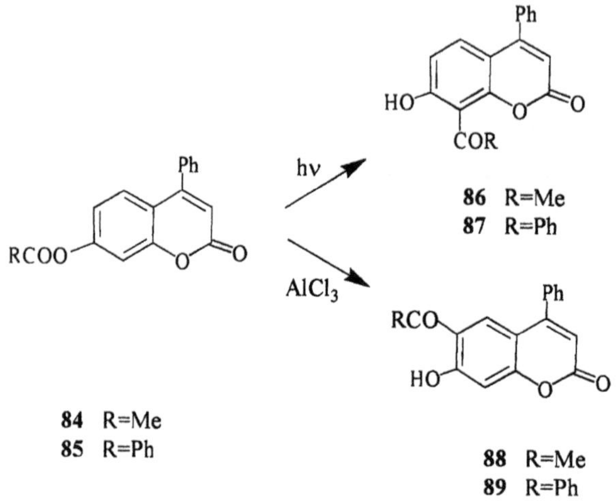

86 R=Me
87 R=Ph

84 R=Me
85 R=Ph

88 R=Me
89 R=Ph

Scheme 27

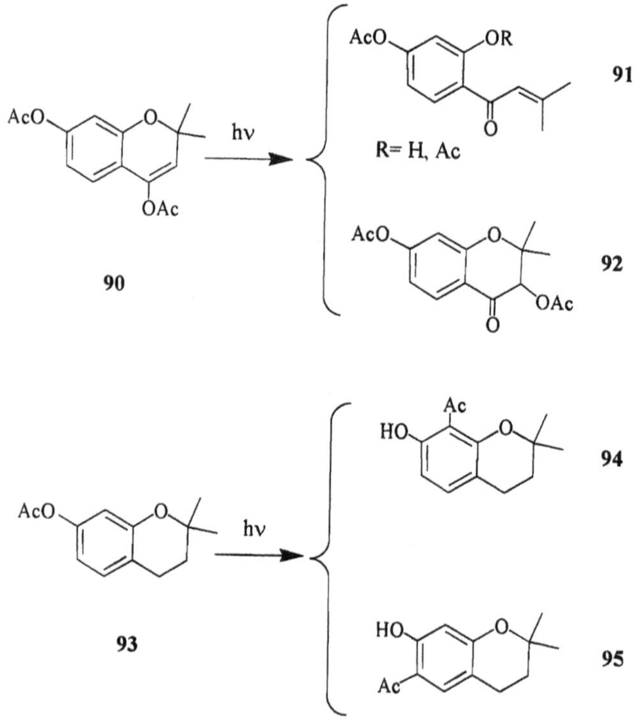

R= H, Ac

91

92

90

94

93

95

Scheme 28

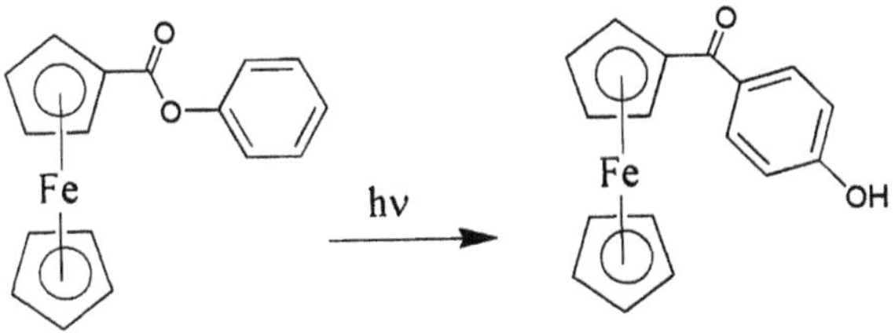

96   97

Scheme 29

3,4-benzocoumarin (**98**), where coupling to the ortho or para positions is not possible, gives the meta-rearranged product (**99**), in very low yields (0.08%) (Scheme 30).

B. Amides (X=N, Y=COR)

The aromatic amides have attracted considerable interest from the photochemical point of view. However, anilides are less prone to photochemical rearrangements than the analogous phenyl esters. For this reason, the side reactions involving other parts of the molecule may compete favorably with PFR. Shizuka started the study of anilides in the 1960, and thereafter a number of papers have appeared dealing with the PFR of N-acyl anilines, carbazoles, indoles, and so forth.

Upon excitation to the singlet state [87], acetanilide (**100**) reacts to 2-amino-acetophenone (**101**), 4-aminoacetophenone (**102**), and aniline (Scheme 31). The quantum yield of product formation is considerably lower than that of phenyl

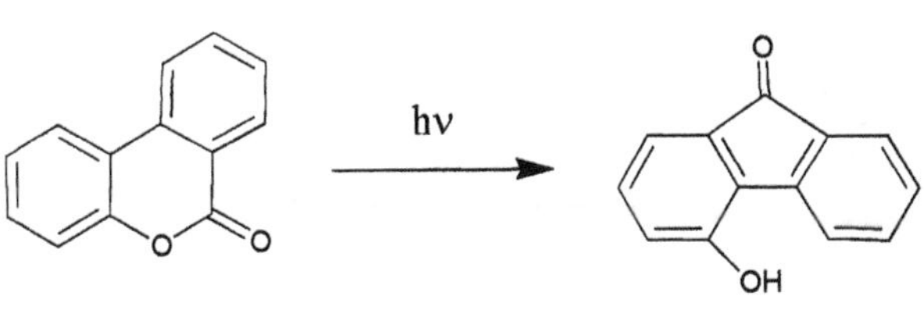

98   99   0.08 %

Scheme 30

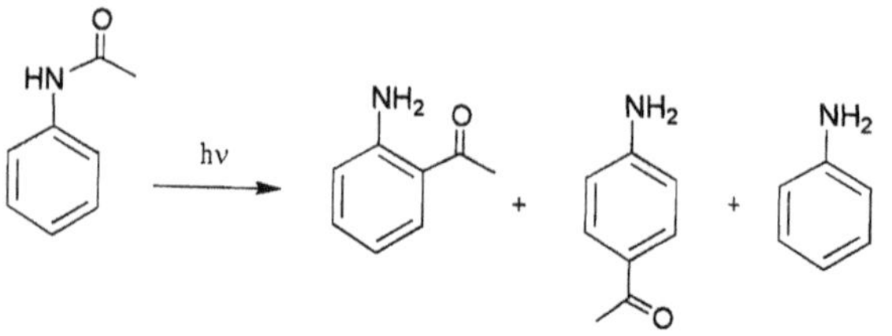

Scheme 31

acetate [88,89]. In cyclohexane as solvent, 2-aminoacetophenone is produced with a quantum yield of 0.07 (as compared with 0.17 for *ortho*-hydroxyacetophenone from phenyl acetate [24]) and 4-aminoacetophenone with a quantum yield of 0.06 (0.15 for *para*-hydroxyacetophenone from phenyl acetate [24]). The quantum yields of anilide rearrangement are collected in Table 8 for several solvents. As it can be seen, they decrease with solvent polarity. This has been attributed to quenching of the singlet of anilides by hydrogen-bonding with the solvent.

Likewise, benzanilide affords 2- and 4-aminobenzophenones in comparable yields (14% and 12%, respectively) but only traces of aniline (see Table 9) [90].

It has been reported that 2,4-dimethoxyacetanilide (**104**) rearranges to 2-amino-3,5-dimethoxyacetophenone (**105**) (63%) and 2,4-dimethoxyaniline (**106**) (8%), the "normal" PFR products. In addition, there is 16% of photoproducts arising from substitution of one methoxy group (2-amino-5-methoxyacetophenone [**107**] and 3-methoxy-4-aminoacetophenone [**108**]; Scheme 32) [91]. This behavior is similar to that of 2,4-dimethoxyphenyl acetate.[33]

Table 8   Product Quantum Yield in the PFR of Acetanilide

	Quantum yields	
Solvent	ϕ (**101**)	ϕ (**102**)
Cyclohexane	0.07	0.06
Ether	0.043	0.037
Ethanol	0.017	0.014
Water	0.0043	0.0018

Source: Ref. 88.

Table 9 Photoproduct Yield in the PFR of Several Anilides

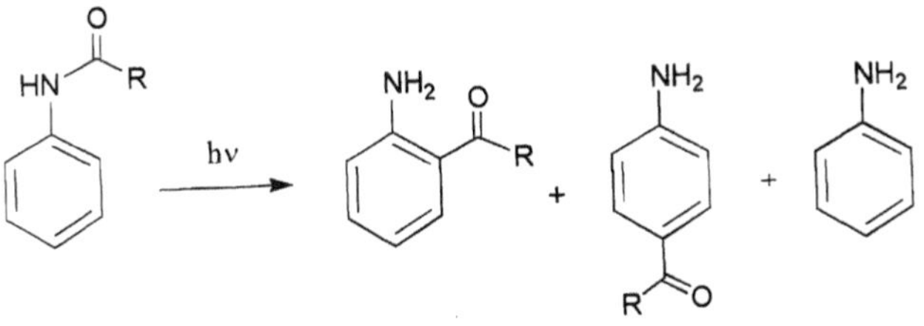

	Yield (%)		
R	Ortho product	Para product	Aniline
Me	20	25	18
Et	22	25	17
Pr	17	23	26
Ph	14	12	Trace

Source: Ref. 90.

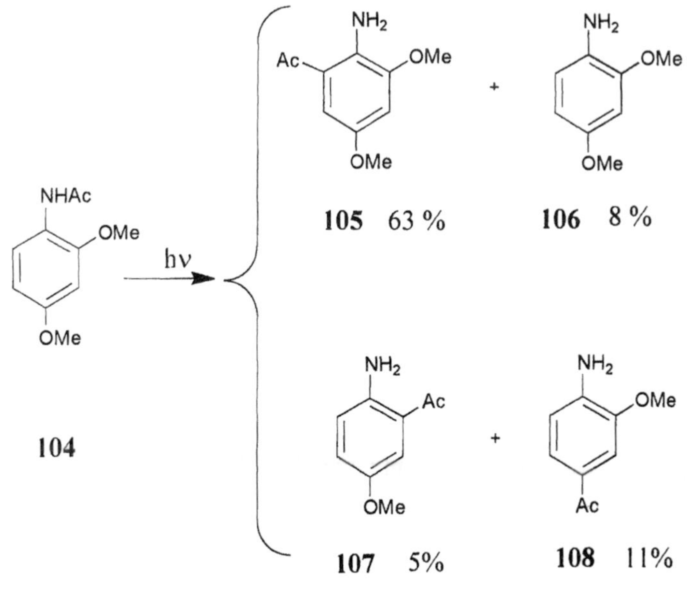

Scheme 32

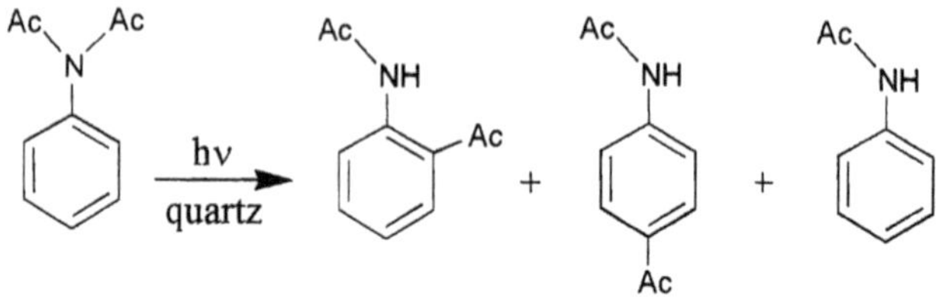

109 110 30% 111 25% 100 45%

Scheme 33

In the case of imides, only one of the acyl groups migrates, as shown for *N,N*-diacetylaniline (**109**) in Scheme 33. The products are 2-acetamidoacetophenone (**110**) (30%), 4-acetamidoacetophenone (**111**) (25%), and acetanilide (**100**) (45%) [92,93].

The remaining *N*-acetyl group of **110** and **111** does not rearrange probably because intersystem crossing to the triplet state is very efficient in the acetophenones and, consequently, the prospects of PFR (from the singlet) are much lower. Likewise, *N,N*-dibenzoylaniline rearranges only once. The disubstituted imide **112** gives a complex mixture. The absence of acetophenones indicates the predominance of benzoyl migration (instead of acetyl) (Scheme 34).

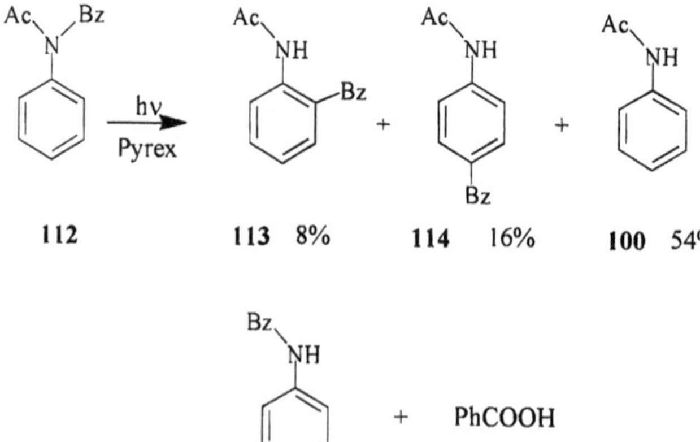

112 113 8% 114 16% 100 54%

115 18% 116 15%

Scheme 34

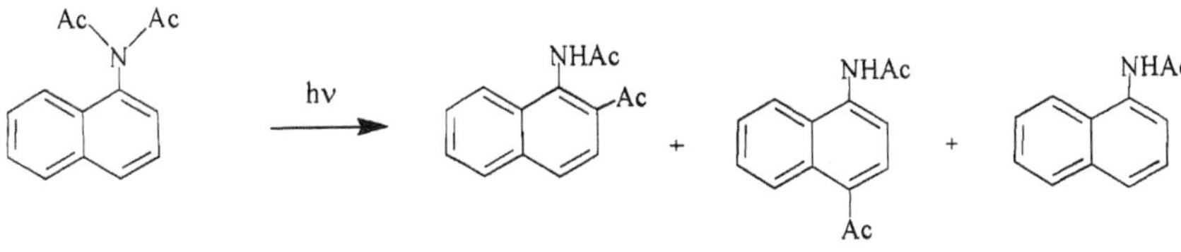

117 **118** 17% **119** 19% **120** 31%

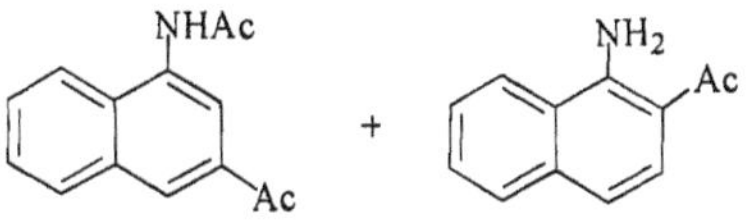

121 1% **122** 1%

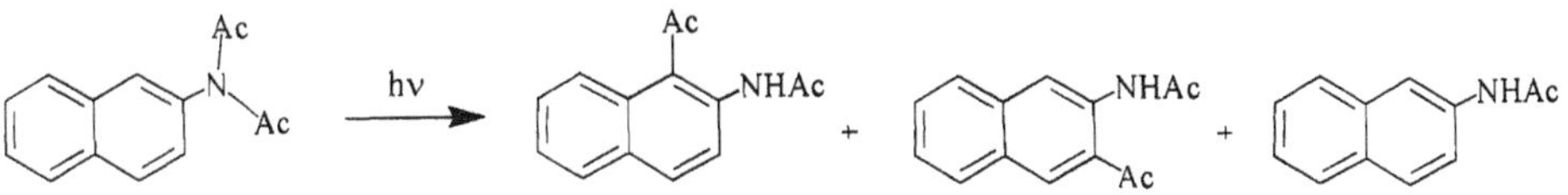

123 **124** 20% **125** 2% **126** 70%

Scheme 35

Similar results have been obtained upon irradiation of naphthylimides [92,93] (Scheme 35). However, 12- and 14-membered macrocyclic imides give rise to photoproducts diacylated at the aromatic ring [94].

Irradiation of *N*-phenyl lactams affords cyclic products selectively rearranged to the ortho position. Studies on the influence of the chain legth on the regioselectivity of the PFR of several lactams have found that, even with 11 methylenes between the nitrogen atom and the carbonyl group, the coupling product was always the ortho ketone (Scheme 36) [95,96].

In the *N*-acyl-2,3-dihydrobenzoxazol-2-ones (**129**) (Scheme 37), there are three cleavable X—CO bonds, but only the free acyl fragment is capable of migrating to the ortho and para positions [97,98].

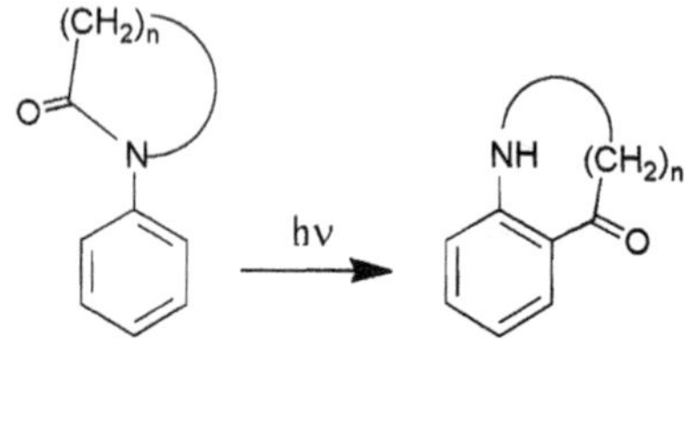

n	yield (%)
5	60
6	83
11	80

127   128

Scheme 36

129

hv

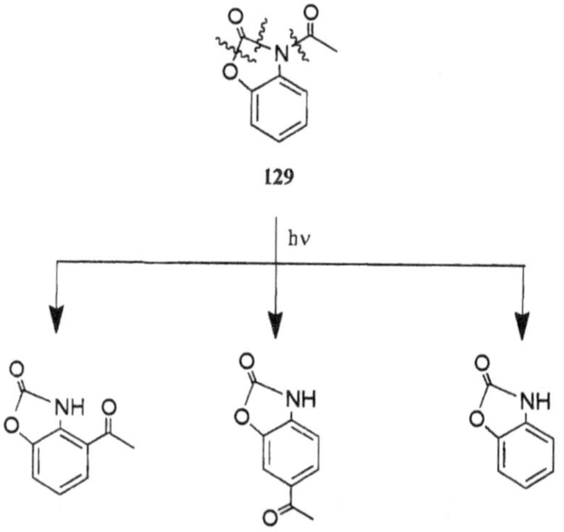

130

131   132

Scheme 37

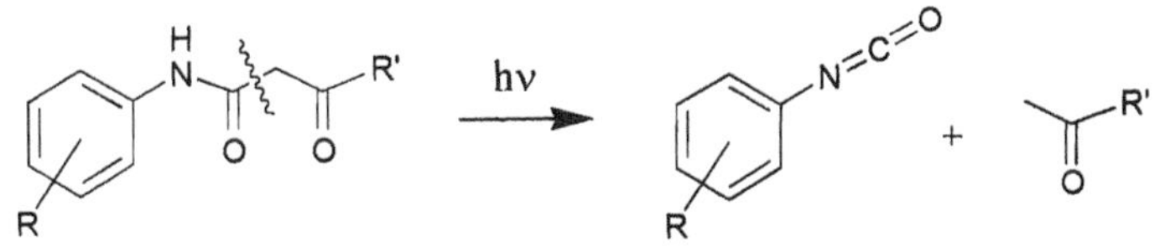

133   134

Scheme 38

Some phenyl esters give rise to photodecarbonylation as a side reaction [37]. This is also the case for certain phenylpropionamides, which implies cleavage of the CO—C bond in the acyl radical, after the breaking of the N—CO bond [99]. If CO—C cleavage occurs prior to N—CO cleavage, then an isocyanate is formed (Scheme 38) and PFR does not occur.

Neighboring groups at the ortho position of the aniline moiety can bring about side reactions that would compete with the photo-Fries rearrangement [100,101]. (Some examples are shown in Scheme 39.)

A number of α,β-unsaturated anilides undergo cyclization instead of rearrangement up to 80% yield. [102–110]. For instance, N-acylanilides 145 and 147 of Scheme 40 have been reported to afford octahydrophenanthridones (146) and dihydroquinolones (148) [111].

Photo-Fries products are also formed upon irradiation of some pyridine [112,113] and indole derivatives [114,115]. This is shown in Schemes 41 and 42.

For instance, when N-acetylindole (153) is irradiated, the three rearranged products are 3-acetylindole (154) (65%), 6-acetylindole (155) (20%), and 4-acetylindole (156) (8%). From this and other experiments, it has been claimed that the order of reactivity toward photo-Fries rearrangement in the indolyl system is 3 > 6 > 4.

Likewise, the N derivatives of carbazoles undergo PFR [116–118]. Upon irradiation in cyclohexane N-acetyl-carbazole (157) gives the following products (quantum yields in parentheses) [119]: 1-acetylcarbazole (158) (0.10), 3-acetyl-carbozole (159) (0.09), and carbazole (160) (0.06) (Scheme 43). The analogous N-acetyldiphenylamine gives the ortho- and para-acetylated products along with diphenylaniline with quantum yields of 0.20, 0.096, and 0.048, respectively [119]. Table 10 shows that these values decrease from cyclohexane to ethanol, but less markedly than in the case of acetanilide (see Table 8).

In chlorinated solvents, the conversion is much lower and chlorinated products are formed [120] (Scheme 44). Moreover, fluorescence quenching and/or

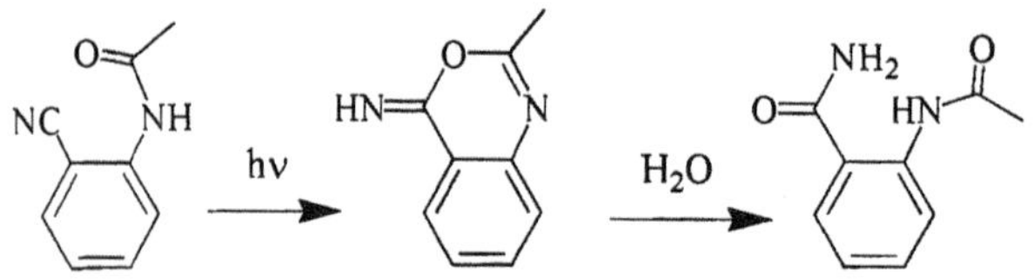

135   136   137

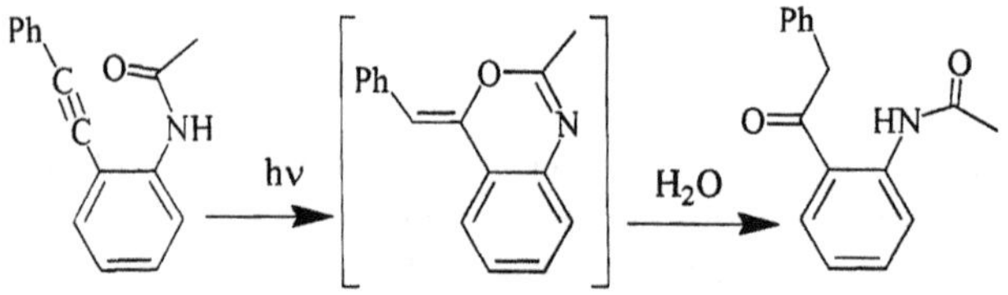

138   139   140

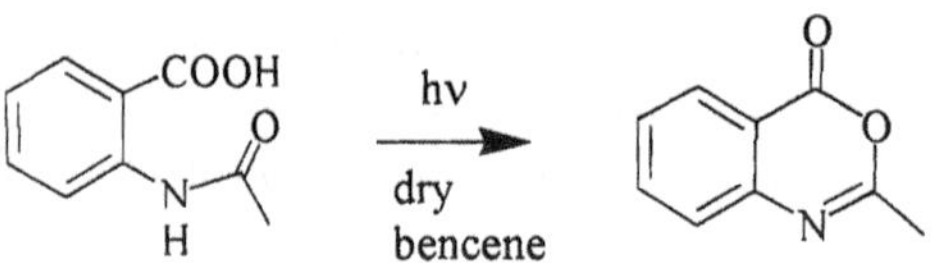

141   142

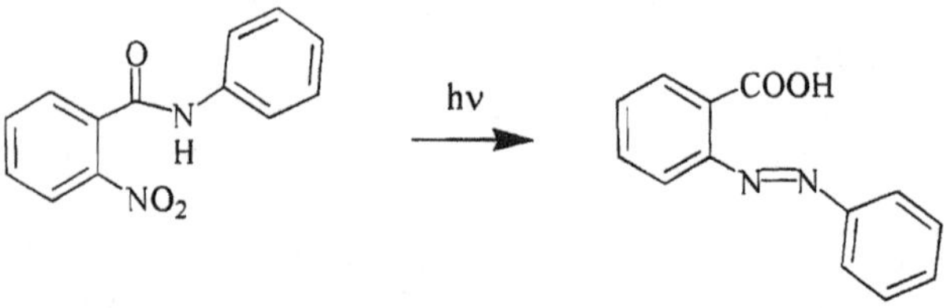

143   144

Scheme 39

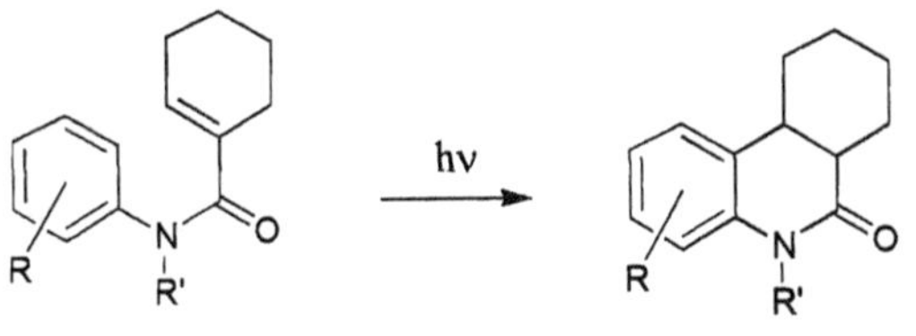

Scheme 40

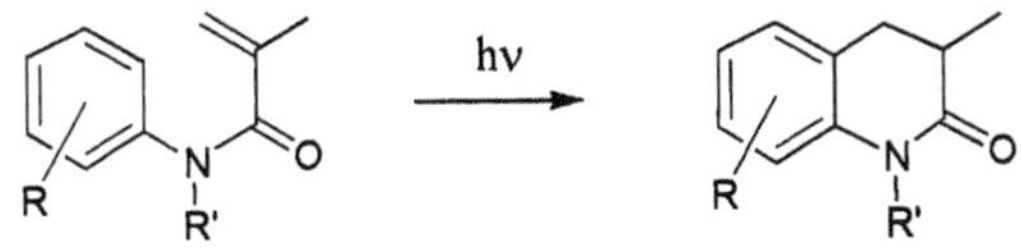

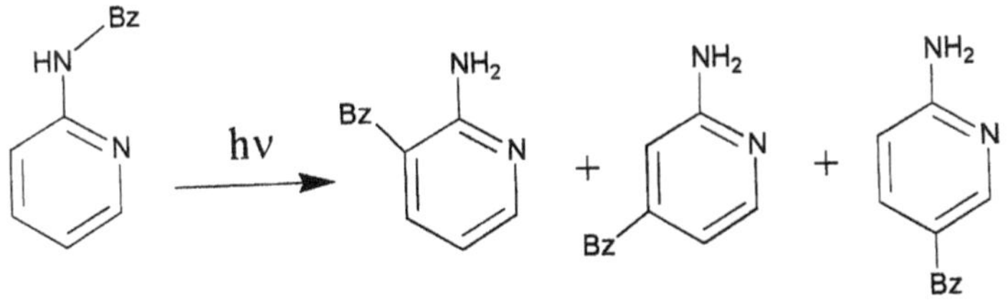

Scheme 41

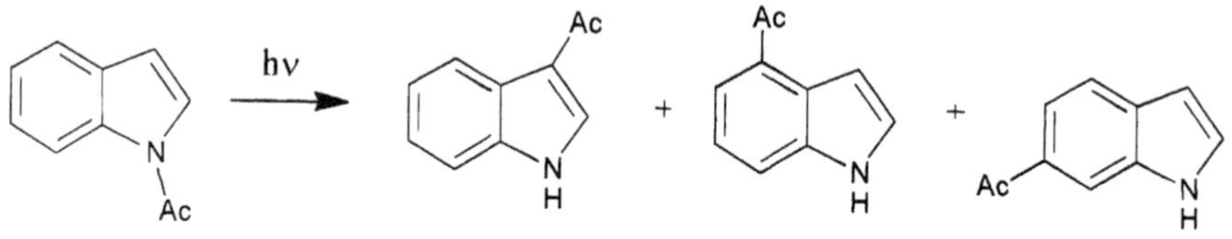

153   154   65%   155   8%   156   20%

Scheme 42

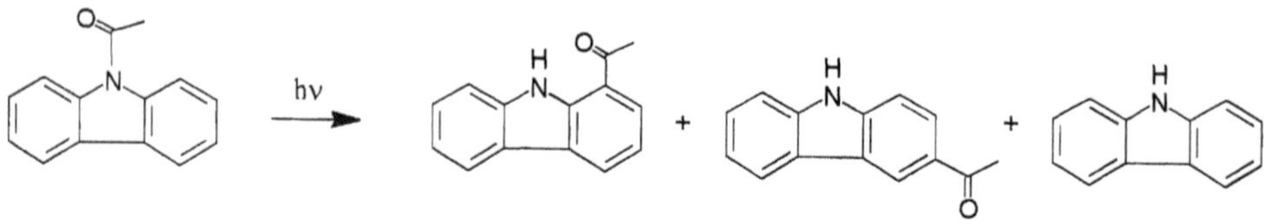

157   158   159   160

Scheme 43

Table 10   Photoproduct Quantum Yields in the PFR
of N-Acetyl Diphenylamine and N-Acetylcarbazole

		Quantum yield	
Substrate	Solvent	ϕ (ortho)	ϕ (para)
a	Cyclohexane	0.20	0.096
	Ethanol	0.15	0.064
b	Cyclohexane	0.10	0.09
	Ethanol	0.07	0.05

Note: a: N-acetyl diphenylamine; b: N-acetylcarbazole.
Source: Ref. 119.

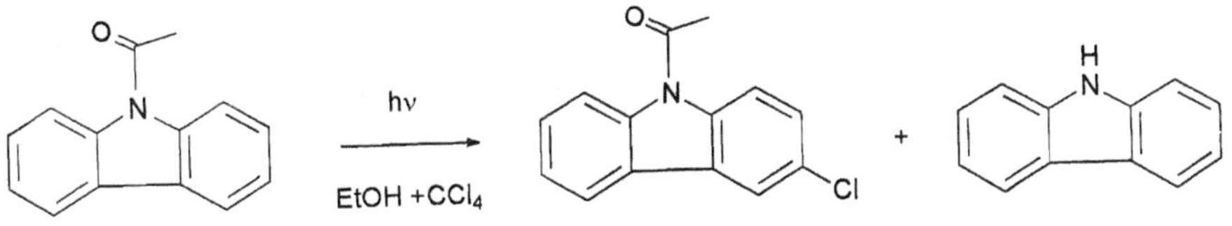

157   161   160

Scheme 44

exciplex emission may be observed. Aromatic chlorination is expected as occurring via electron transfer from the N-acyl carbazole to the solvent: this would compete with PFR.

Related substrates undergoing PFR include N-acetyl and N-(2-naphthoyl) esters of 5H- benzo [b] carbazole (Scheme 45) [121]. Interestingly, the N-(2-naphthoyl) ester of carbazole **164** does not undergo PFR [121].

C. Carbonates (X=O, Y=COOR) and Carbamates (X=N, Y=COOR)

The photochemistry of aryl carbonates and aryl carbamates is interesting, as these compounds are building blocks of polycarbonates and polyurethanes. Diphenyl carbonate (**165**) has been irradiated with ultraviolet (UV) light [122,123] and

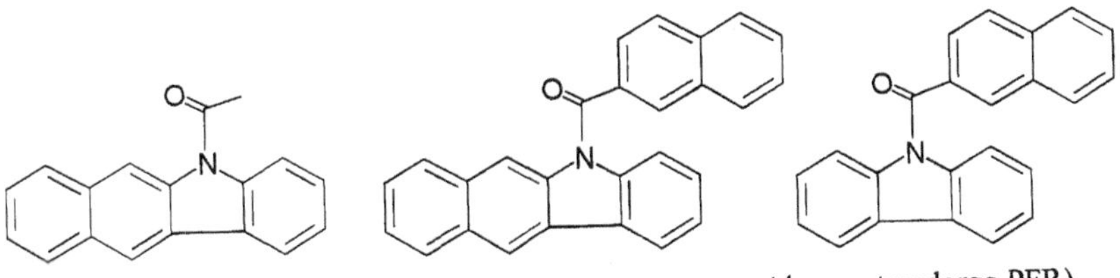

(does not undergo PFR)

162   163   164

Scheme 45

with γ-radiation. [124,125]. In both cases, the phenoxy radical is produced and detected by ESR. The photochemical reaction consists in a two-step migration: The first step would lead to phenyl salicylate (not detected) and the second would be rearrangement to 2,2′-dihydroxybenzophenone (**166**) and 2,4′-dihydroxybenzophenone (**167**) along with phenol (**13**) (Scheme 46). The ortho/para ratio depends on the reaction conditions (from 0.4 to 1.9). In the photolysis of phenyl carbonate, a volatile fraction formed by carbon dioxide (22%) and carbon monoxide (78%) is also formed.

The PFR also takes place with aryl alkyl carbonates [125]. Methoxy-substituted dervatives can undergo substitution of MeO by the acyl moiety, as has been reported for esters and amides. This is shown in Scheme 47 for 2-methoxyphenyl ethyl carbonate (**168**) [126]. On the other hand, the overall quantum yield of photoproducts is 10-fold lower in the para-than in the ortho- or (meta- methoxyphenyl. Chlorophenyl ethyl carbonates do not rearrange, but undergo C—Cl homolysis. The efficency of photodechlorination follows the order para < meta < ortho [127].

Photolysis of carbamates can cleave either the C—O or the C—N bond. Although it is possible to find examples of the first type [128], the second one has received much more attention [129–140]. Thus, the N-phenylethyl carbamate (**173**) would be the equivalent to phenyl acetate in the PFR of esters or to acetani-

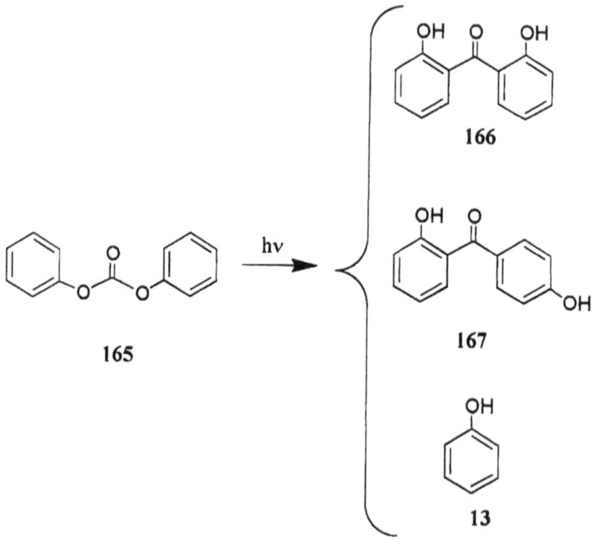

Scheme 46

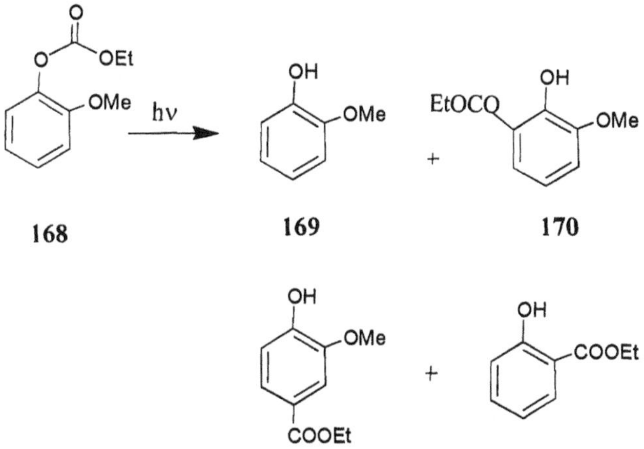

Scheme 47

lide in the PFR of amides. The use of ESR has allowed to detect the anilinyl radical in the PFR of **173** [141]. The photoproducts are ethyl 2-aminobenzoate (**174**), ethyl 4-aminobenzoate (**175**), and aniline. Urethane (H_2NCO_2Et), formanilide, diphenylurea, N-ethylamine, ethylene, methane, and polymer have also been detected [142]. The quantum yields of formation of ortho and para products are strongly solvent dependent, but, in general, they are quite low (<0.01). Table 11 shows the product quantum yields in the photochemical reaction of N-phenylethyl carbamate in several solvents [143].

The photo-Fries rearrangement of heterocyclic carbamates has also been observed [144].

D. Other

Apart from aryl esters and amides of carboxylic acids, aryl carbonates, and carbamates, other types of organic compound have been reported to undergo PFR. They include oxalates, formiates, sulfonates, sulfonamides, thioesters, selenoesters, and telluroesters.

1. Oxalates (X=O, Y=COCOOR) and Formiates (X=O, Y=CHO)

In spite of being carboxylic acid derivatives, these two families of compounds have been put together in this section because their photochemical rectivity is

Table 11 Product Quantum Yield in the PFR of **173**

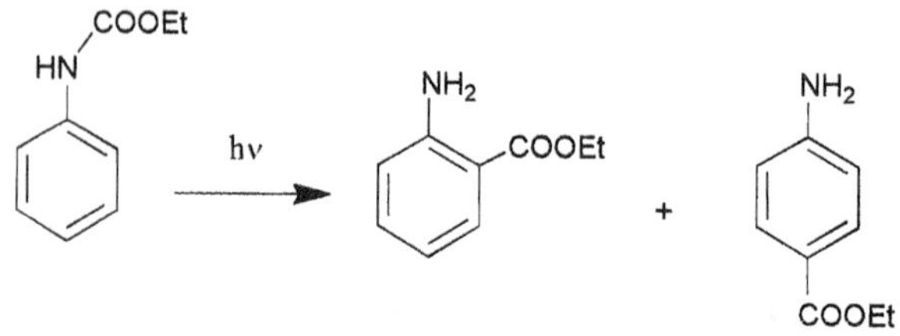

173	**174**	**175**

Solvent	Quantum yield	
	ϕ (**174**) $\times$ 10³	ϕ (**175**) $\times$ 10³
Cyclohexane	8.1	5.7
n-Butyl ether	4.2	—
Ethyl ether	4.0	2.8
Acetonitrile	3.4	—
THF	2.4	—
tert-BuOH	1.8	—
i-PrOH	1.4	—
EtOH	1.5	—
MeOH	1.6	1.1
Water	0.8	0.4

Source: Ref. 143.

commonly reduced to decomposition, to give a mixture of phenols and polymer [123]. In fact, oxalates are good generators of aryloxy radicals if the appropriate substituents are chosen [145]. However, the typical photo-Fries migration occurs in methyl phenyl oxalate, together with formation of hydroxybenzoates and a carbonate as decarbonylation products (Scheme 48) [146].

The photochemistry of formiates is solvent dependent. Thus, the only product obtained upon iradiation of 4-*tert*-butylphenyl formiate in ethanol is 4-*tert*-butylphenol (together with polymer), whereas in benzene, 7% of 2-hydroxy-5-*tert*-butylbenzaldehyde is formed [123].

2. Sulfonates (X=O, Y=SO₂R), Sulfonamides (X=NH, Y=SO₂R), and Sulfamates (X=O, Y=SO₃R)

Esters of sulfonic acids can rearrange photochemically to give sulfones in moderate yields [147]. Thus, phenyl benzenesulfonate (**182**) cleaves photochemically producing 15% of *ortho*-hydroxyphenyl phenyl sulfone (**183**) and 31.1% of *para*-hydroxyphenyl phenyl sulfone (**184**). Besides, 17.4% of phenol (**13**) is formed

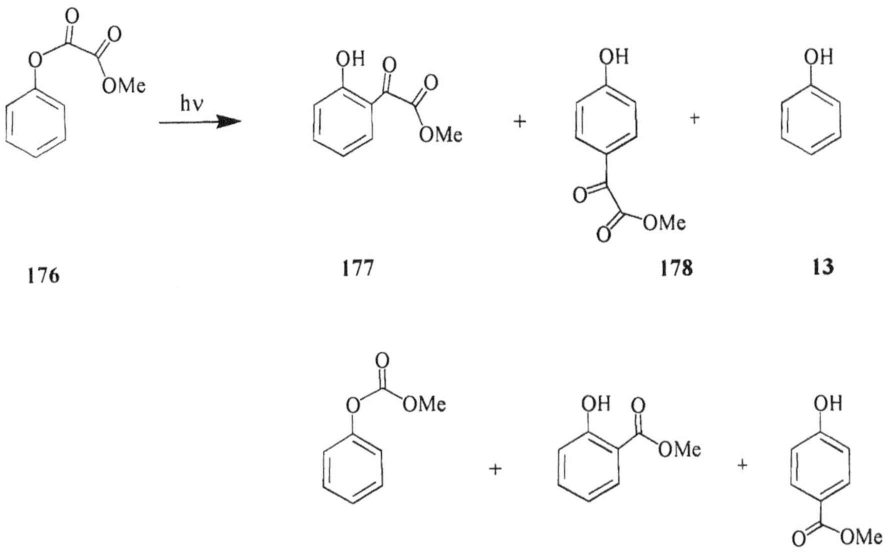

176 177 178 13

179 180 181

Scheme 48

(Scheme 49) [148]. Recently, it has been demonstrated that phenolic sulfonate esters used as photoacid generators cleave into radical fragments; the resulting phenoxyl radical has been detected by transient optical absorption [149]. Similar photochemical rearrangements have been reported in pyrymidinic systems [150]. An example is shown in Scheme 50 [151].

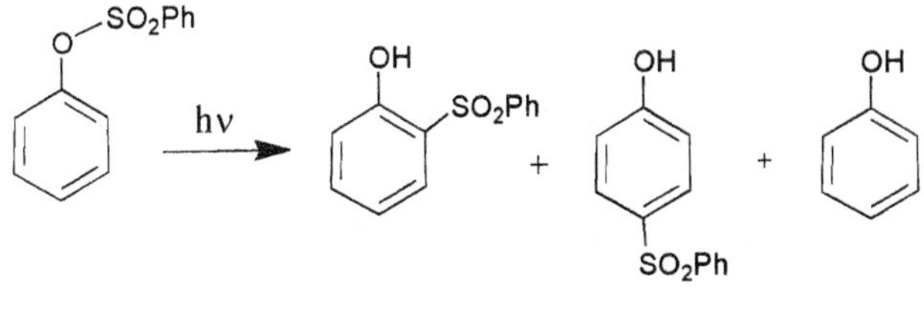

182 183 184 13

Scheme 49

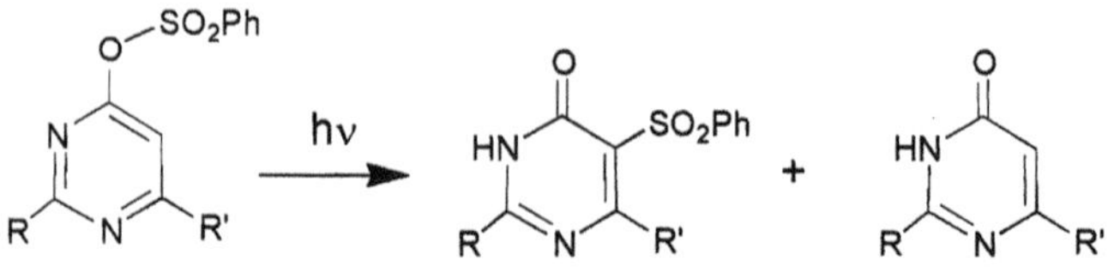

185   186a   186b

Scheme 50

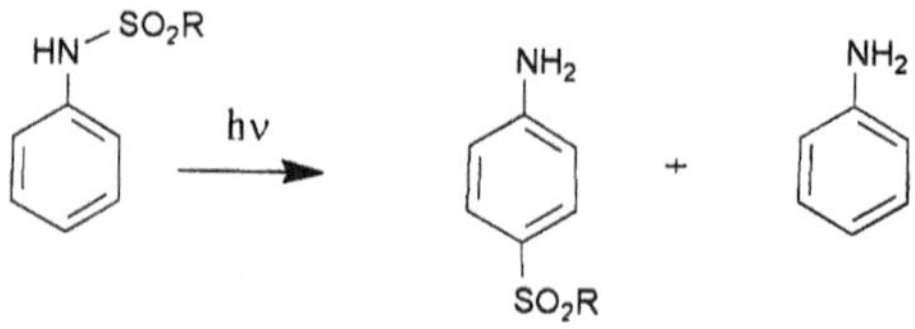

187   188   103

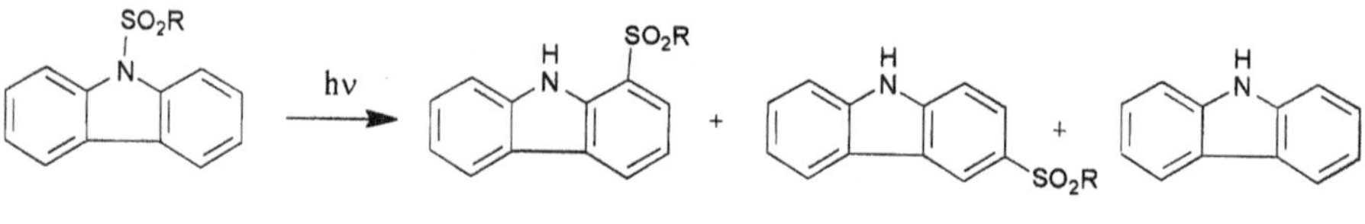

189   190   191   160

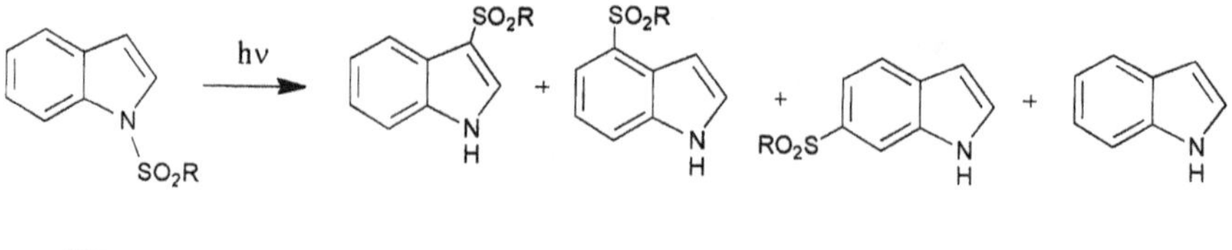

192   193   194   195   196

Scheme 51

Sulfonamides undergo rearrangement in an analogous way. The aromatic ring can be aniline [152], carbazole [153], or indole [114]. Some examples are outlined in Scheme 51.

N-Acyl derivatives of azepine like the iminostilbene **197** (Scheme 52) react in different ways depending on the nature of the N substituent: if it is acyl, aroyl, or ethoxycarbonyl, the preferred reaction is dimerization; on the other hand, N-tosyl compounds give PFR [154].

Finally, the sodium salt of phenylsulphamic acid (**200**) and its *para*-methyl derivative rearrange to give the corresponding *ortho*- and *para*-anilinesulfonic acids (**201** and **202**) and aniline. In the case of phenylsulphamic acid, ~40% of the *para*-sulfonic acid is obtained (along with ~5% of the meta and 22% of the ortho product (**201**)) (Scheme 53).

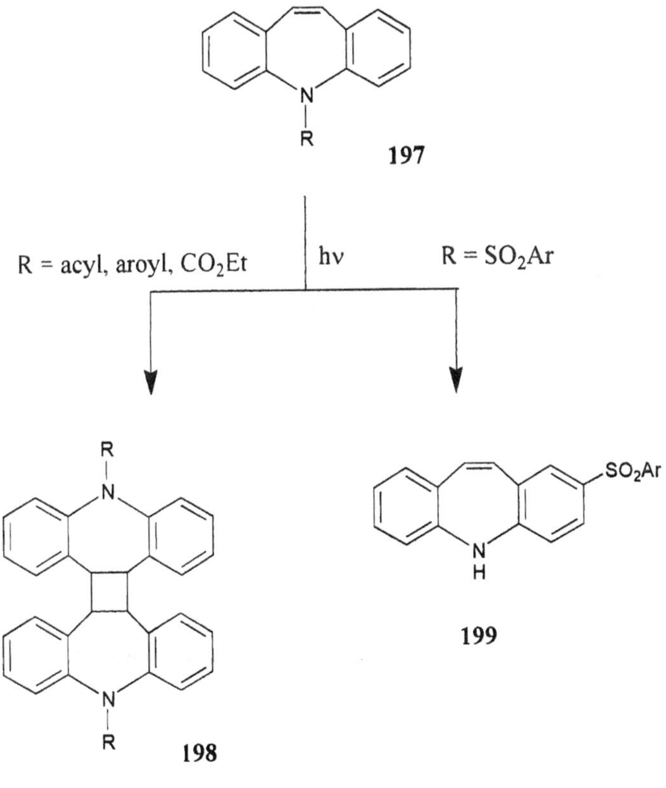

Scheme 52

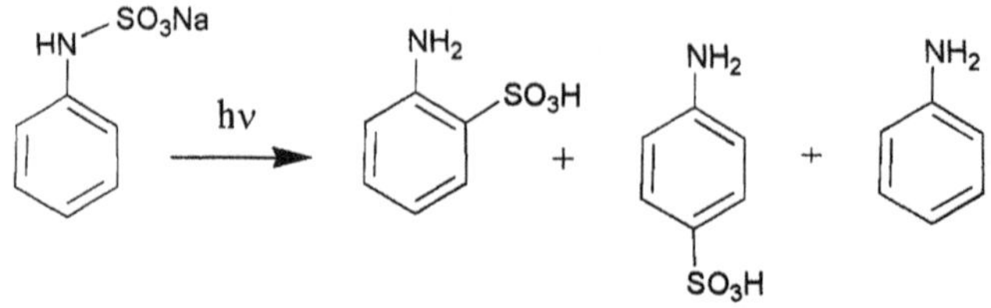

200 **201** **202** **103**

Scheme 53

3. Thioesters (X=S, Y=COR or X=O, Y=CSR)

Irradiation of S-phenyl thiolacetate (**203**) in benzene gives diphenyl disulfide (**206**) as major product (52%). However, minor amounts of ortho-rearranged (**204**) and para-rearranged (**205**) products are also formed, along with thiophenol (17%) and methyl phenyl thioether (**208**) (19%) (Scheme 54) [155].

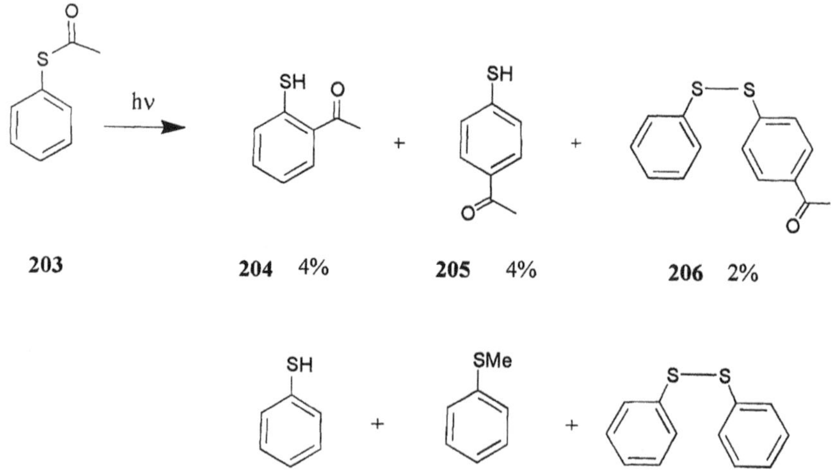

203 **204** 4% **205** 4% **206** 2%

207 17% **208** 19% **209** 52%

Scheme 54

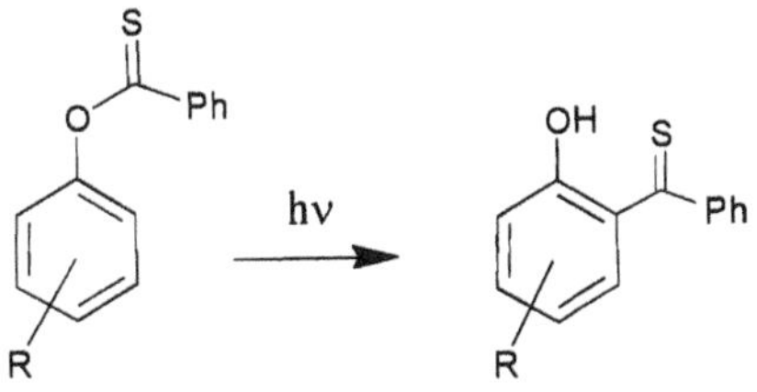

210   211

Scheme 55

Photolysis of 2-naphthyl thioesters in the presence of a hydrogen donor such as 1,4-cyclohexadiene yields aldehydes (80–100%); no photo-Fries reaction products have been reported [60].

If the sulfur atom belongs to the acyl moiety, the rearrangement is also possible (Scheme 55) [156].

4. Selenoesters (X=Se, Y=CO) and Telluroesters (X=Te, Y=CO)

Although the photochemistry of aryl selenoesters and aryl telluroesters is not fully developed, there are some photoreactions of these compounds that resemble a photo-Fries process. Se-*para*-tolyl selenobenzoate (**212**) gives, upon irradiation, selenocresol (**214**), benzaldehyde, and the benzophenone **213** (Scheme 56), which is clearly a photo-Fries product [157,158]. Starting from Se-phenyl 2-chlorosele-

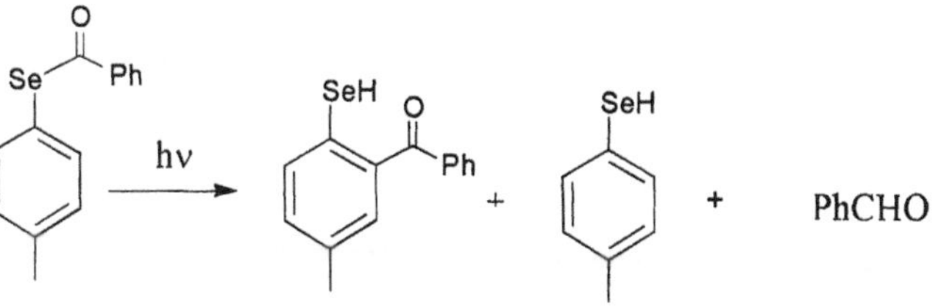

212   213   214

Scheme 56

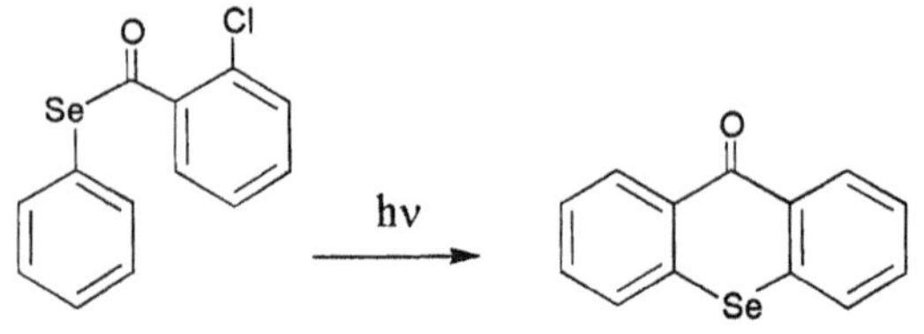

215   216

Scheme 57

nobenzoate (**215**), selenoxantone (**216**) is formed with a 19% yield (Scheme 57) [159,160].

Telluroesters give, on irradiation, a mixture of products that might arise from the breaking of the Te—CO bond. An example is the irradiation of Te-4-methylphenyl 2-methylthiotellurobenzoate ester (**217**), which gives Te (25%), di-(4-methylphenyl)ditelluride (**218**) (33%), 2-methylthiobenzaldehyde (**219**) (13%), 2-methyltelluroxantone (**220**) (22%), bis-[4-methyl-2- [2-(methylthio)-benzoyl]-phenyl]-ditelluride (**221**) (11%), and 5-methyl-2-(4- methylphenyltel-luro)-2′-methylthiobenzophenone (**222**) (2%) (Scheme 58) [159,160].

5. Enol Esters and Enol Amides

Aliphatic systems, like enol esters [161–165] and enol amides [166–170] with a C=C—X—Y moiety, can also give rise to photochemical rearrangement to compounds of the type Y—C=C—XH, which is the enol form of Y—CH—C=X. By extension, these reactions have been called photo-Fries rearrangements. The quantum yield of product formation decreases with increasing conjugation of the migrating RCO group. This can be seen in Table 12 for the rearrangement of several isopropenyl esters [171]. A secondary photoreaction of the β-diketone **225** can result in the formation of another enol ester (Scheme 59) [172,173]

6. 1,3-Benzodioxan-4-ones

Irradiation of 1,3-benzodioxan-4-ones appropriately substituted at position 2 with an aryloxy or thioaryloxy group gives rise to the formation of the same photoproducts as in the PFR of the open ring isomers. Scheme 60 depicts the structures of *para*-methoxyphenyl *ortho*- acetoxybenzoate **229** and its cyclic isomer, 2-(*para*-methoxyphenoxy)-2-methyl-1,3-benzodioxan-4-one **230**. Irradiation of **229** or

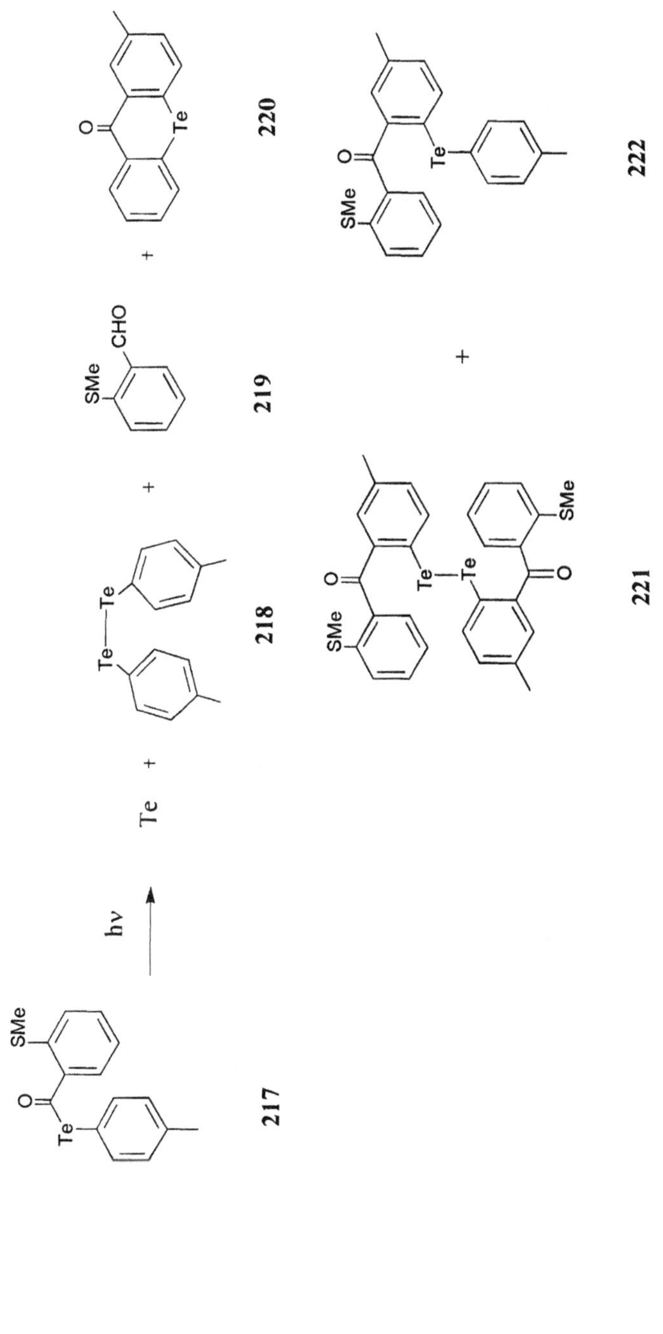

Scheme 58

Table 12 Product Quantum Yield in the Irradiation of **223**

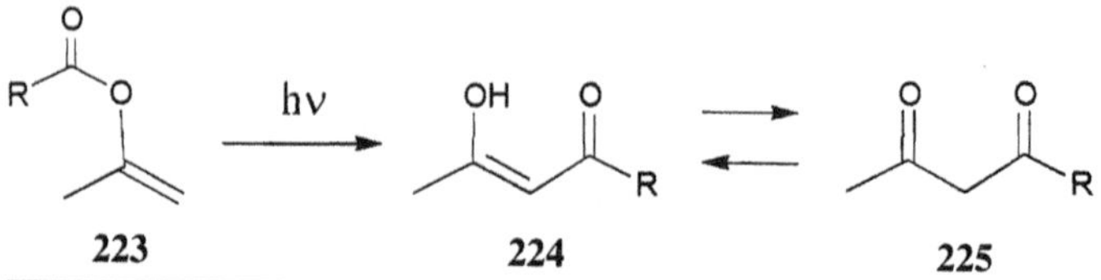

R	φ (hexane)
Me	0.18
Ph	0.12
2-Naphthyl	0.08
2-Anthryl	0.02

Source: Ref. 171.

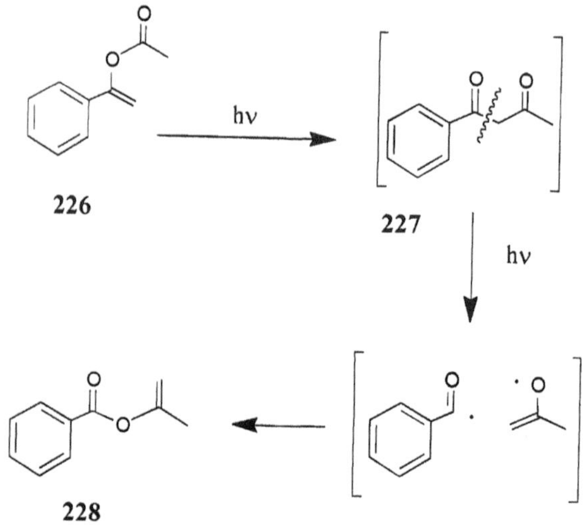

Scheme 59

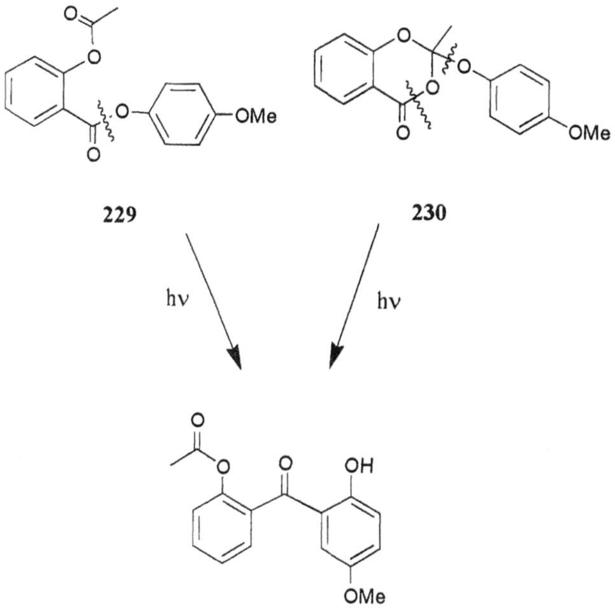

229   230

hv   hv

231

Scheme 60

230 in benzene leads to 2'-acetoxy-2-hydroxy-5-methoxybenzophenone (**231**) (Scheme 60) but with higher efficency in the case of the benzodioxanone (**230**) [174].

IV. SYNTHETIC APPLICATIONS

Acylation of aromatic rings is a common reaction step in chemical synthesis. However, sometimes the use of a Friedel–Crafts reaction or a Lewis-acid-catalyzed Fries rearrangement requires protection of functional groups, in view of the strong acidic conditions employed. To circumvent these difficulties, some authors have exploited the mild conditions of the photochemical Fries rearrangement with good results. In the following subsections, some examples will be used to illustrate the synthetic utility of the PFR of phenyl esters.

A. Some Selected Examples

If the acyl moiety bears the appropriate functional groups, photo-Fries rearrangement may be followed by reaction between the phenolic hydroxyl and the reactive part of the migrated acyl. Thus, making use of unsaturated acyl derivatives, a number of synthetic pathways have been developed for the preparation of *flavonoids*, an important class of natural products [175,176].

A photochemical approach to the synthesis of *precocenes I, II,* and *III*, interesting as juvenile hormome inhibitors [177,178], has been deviced starting from the phenolic esters of 3-methylcrotonic acid. This is shown in Scheme 61 for the synthesis of precocene I (**235**) [179] and in Scheme 62 for the synthesis of precocenes II (**238**) and III (**239**) [180]. The rate of the cyclization process leading to chromanones, like **234** or **237**, has been found to depend on the stereochemistry of the double bond. Thus, ring closure is more rapid for the trans than for the cis isomer, although no clear-cut trend can be established [181].

Whereas the cyclization of chalcones or analogous compounds leads to six-membered rings, in accordance with the Baldwin rules [182], the same reaction with a triple bond can afford compounds with six-membered (6-endo-dig) or five-membered (5-exo-dig) rings, depending on the reaction conditions. This property

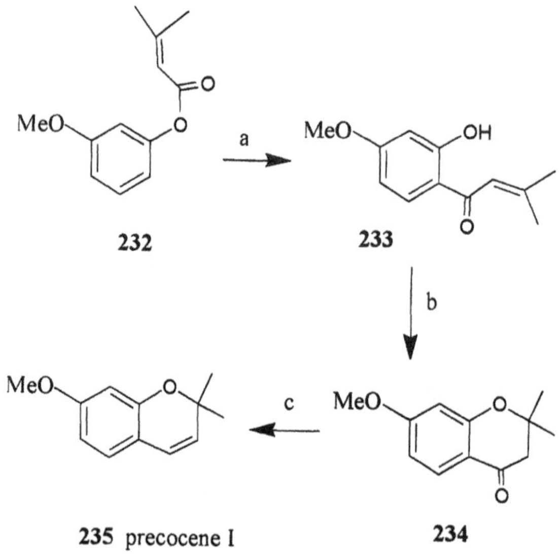

232

233

235 precocene I

234

a: hν
b: base
c: reduction and dehydration

Scheme 61

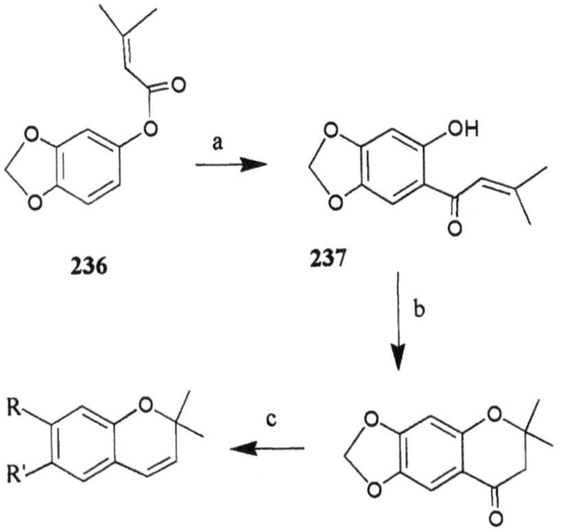

236   a   **237**

b

238 precocene II,   R = OMe, R' = OMe

239 precocene III, R,R' = OCH$_2$O

a. hv
b base
c. methoxide or ethoxide, reduction and dehydration

Scheme 62

has been used to explore application of PFR to the synthesis of *flavones* and *aurones*. Irradiation of aryl phenylpropinoates gives 2-hydroxyaryl phenylethynyl ketones. Subsequent basic treatment leads to flavones **242** or aurones **243**, as indicated in Scheme 63. Table 13 shows the percentages of each product depending on the base and solvent used [183].

A carbonyl group can also be effective in the cyclization step following the rearrangement. Thus, *para*-methoxyphenyl 3-oxobutanoate (**244**) is rearranged to the β-diketone **245**, which is, in turn, cyclized to chromone **246** in high yield (Scheme 64) [184].

The oxygens of methoxynaphthazarin **247** (Scheme 65) (white arrows) can be selectively ortho acetylated/methylated. Starting from the acetates or polyacetates of Scheme 65, it is possible to prepare a variey of *polycyclic hydroxyquinone*

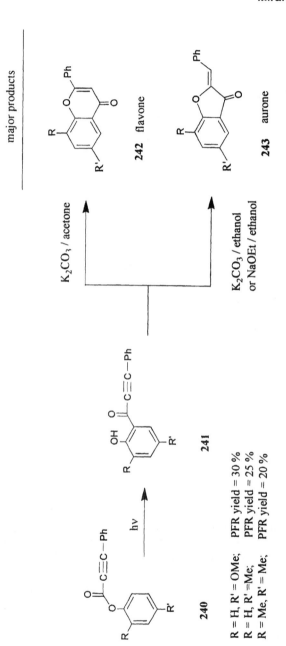

major products

242 flavone

243 aurone

K$_2$CO$_3$ / acetone

K$_2$CO$_3$ / ethanol
or NaOEt / ethanol

hv

240

241

R = H, R' = OMe;   PFR yield = 30 %
R = H, R' = Me;   PFR yield = 25 %
R = Me, R' = Me;   PFR yield = 20 %

Scheme 63

Table 13 Product Yield of the Irradiation of **240**

Substrate			Yield (%)		
R	R'	Reagent	Flavone	Aurone	Flavone/Aurone
H	OMe	A	45	7	0.15
		B	15	65	4.33
		C	11	54	4.91
H	Me	A	85	8	0.09
		B	12	77	6.42
		C	15	79	5.27
Me	Me	A	57	7	0.12
		B	4	66	16.50
		C	18	62	3.44

Note: A: K$_2$CO$_3$/acetone: B: NaOEt/etanol; C: K$_2$CO$_3$/etanol.
Source: Ref. 183.

derivatives C acetylated at the indicated positions (black arrows), using the PFR as the key step [185].

Salicylic [186] and thiosalicylic [187] acid derivatives can also be subjected to PFR, providing straightforward entries to the synthesis of *xanthones* and *thiox-anthones* (Scheme 66).

Phenyl esters of anthranylic acid give upon irradiation *ortho*-amino-benzophenones, and their N-acetyl derivatives **264** give *ortho*-N-acetamido-benzophenones **265**, which can be converted into *quinolones* (**266**) with basic treatment (Scheme 67) [188].

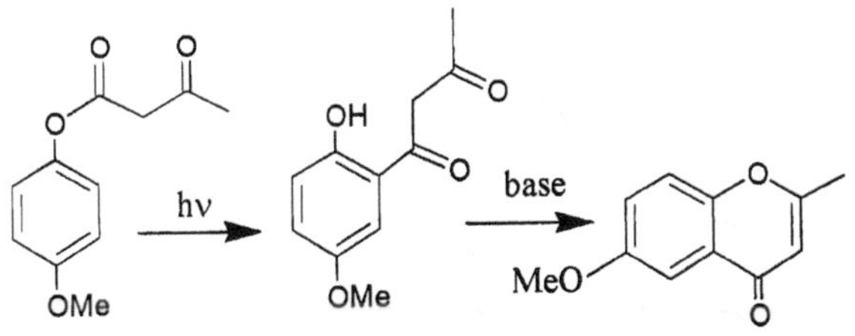

| 244 | 245 | 246 |

Scheme 64

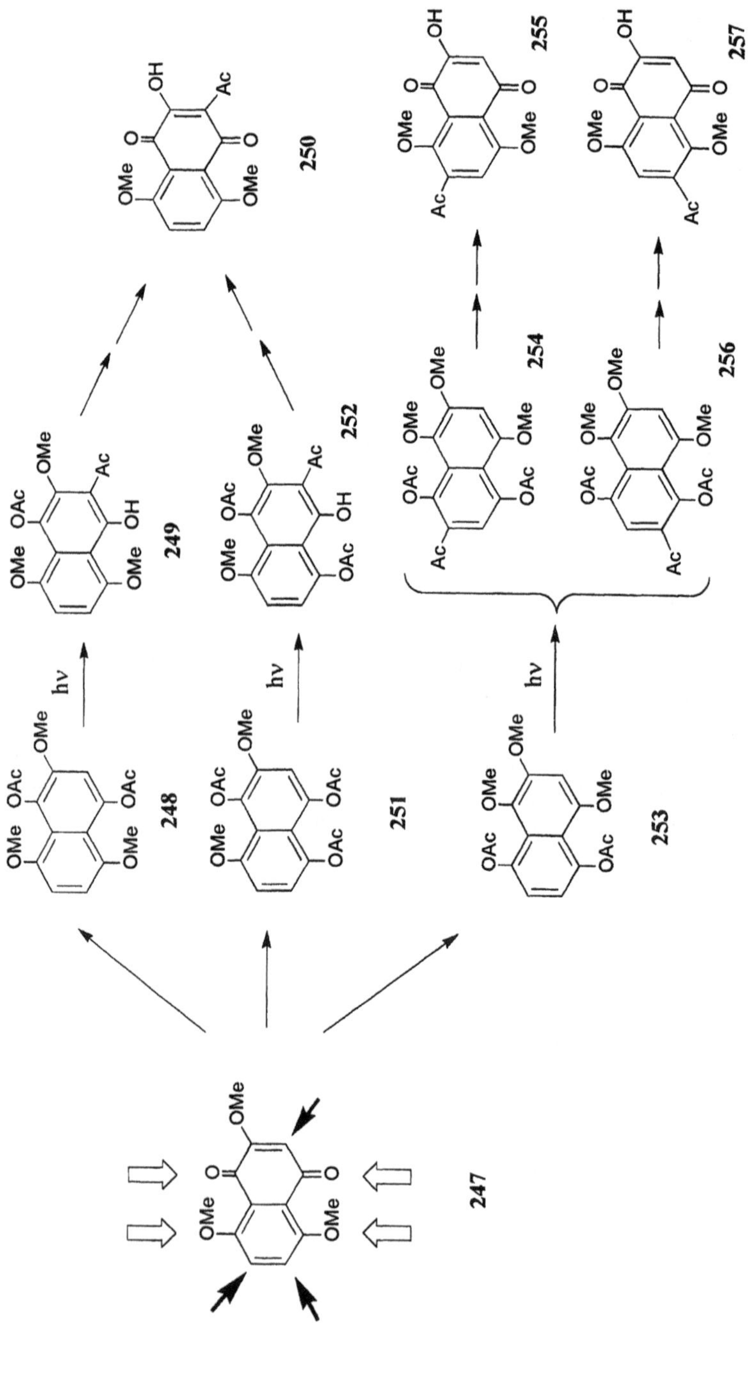

Scheme 65

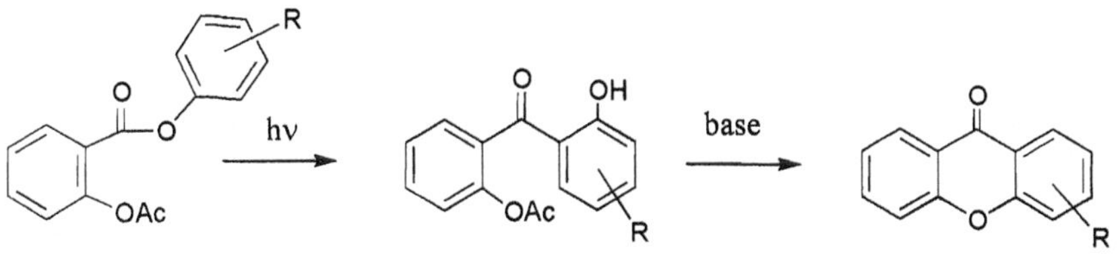

258   259   260

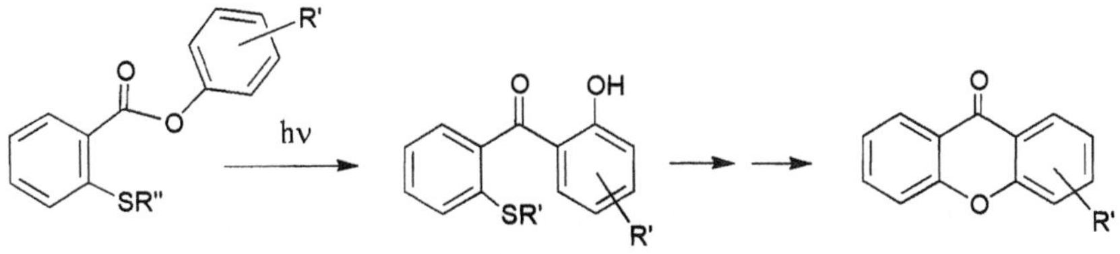

261   262   263

Scheme 66

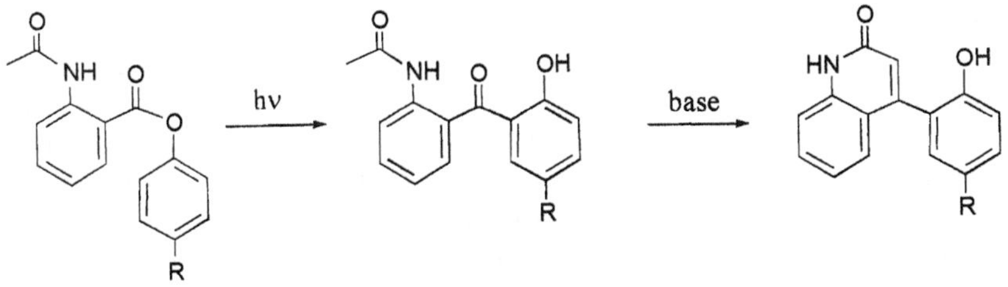

264   265   266

Scheme 67

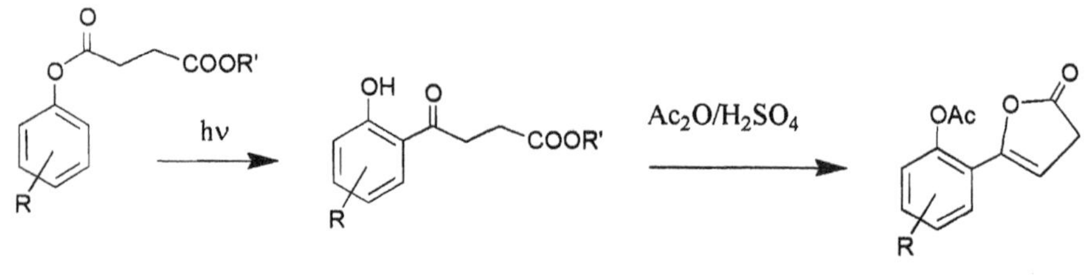

267   268

269

Scheme 68

The photo-Fries rearrangement of aryl hydrogen (or methyl) succinates **267** leads to 4-(2-hydroxyaryl)-4-oxobutanoic acids (or methyl esters **268**), which are readily cyclized to 5-(2-acetoxyaryl)-2(3*H*)-furanones (**269**) (Scheme 68). [189–191] Photolysis of **269** [191] or the analogous open-chain enol acetates [192,193] leads to chromones.

Bikaverin (**272**) is a red pigment having antiprotozoal and antitumor activities. Several approaches have been made to its synthesis; some of them include the PFR as a key step. One strategy uses diester **270** to obtain **271** in 27% yield (Scheme 69) [194]. In another approach, the 1,4-dihydroxy-5,8-dihydronaphthalene ester **273** is irradiated in benzene to obtain ketone **274** whose cyclization gives a xanthone (**275**) that is structurally related to Bikaverin (Scheme 69) [195]. Similar approaches have been studied for the synthesis of *adriamycin* [196] and 9- *deoxydaunomycinnone* [197]. Also, PFR has been recently used as the key step in the synthesis of a 2,3-dimethoxyrotenoid, using sunlight as the excitation source [198].

Diacetylhexahydromarmesin **276** has been converted into the ortho-rearranged product **277** in 30.6% yield among other photoproducts. Transformation of **277** into methyl hexahydrorutaretin methyl ether **278** served to determine the absolute configuration of *rutaretin methyl ether* (**279**) (Scheme 70) [199].

Arizonine (**282**), a tetrahydroisoquinoline alcaloid, has been recently synthesized in 35% yield from isovanilline, using a photochemical rearrangement like that depicted in Scheme 71. The yield of the photochemical step is 55%, and other byproducts (not shown in Scheme 71) are also formed [200]. The same authors reported the synthesis of *caseadine* following an analogous procedure.

The synthesis of some *alcaloids* has been achieved by means of the amide version of PFR [201]. Another application is the recent synthesis of *diazonamides*, an unusual group of alkaloids [202]. In this case, a macrolactam was irradiated,

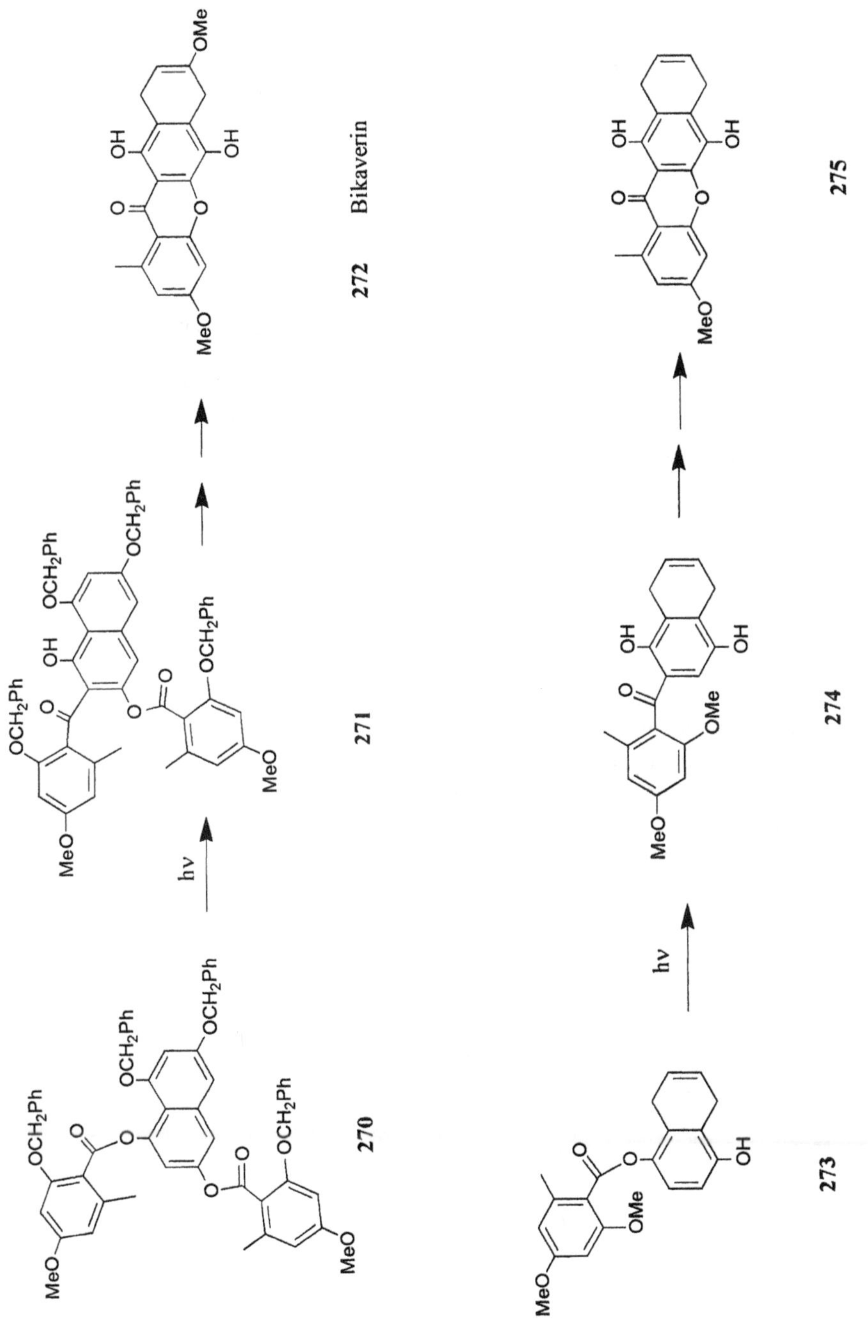

Scheme 69

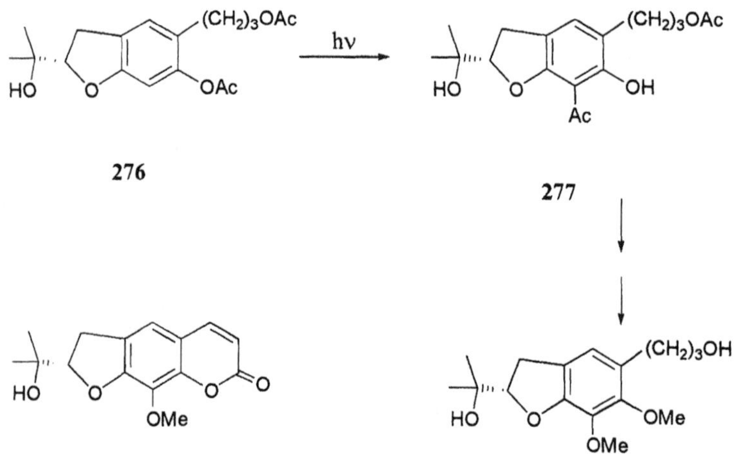

276

277

Rutaretin methyl ether· 279

278

Scheme 70

yielding the ortho-rearranged product very efficiently (76%). This is remarkable, because rearrangement requires homolysis of the macrocycle, but at the same time, very efficiently reformed by radical coupling, probably as a consequence of the singlet nature of the PFR. Finally, the root growth-promoting factor, *capillarol*, and its derivatives, have been synthesized by PFR, using *para*-hydroxy-benzaldehyde as the starting material [203].

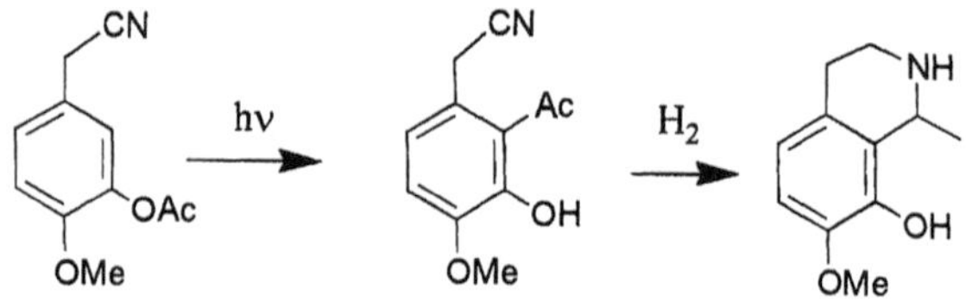

280   281   282

Scheme 71

B. Improvements to the PFR

Usual problems for the application of the PFR to organic synthesis are the occurrence of competing photochemical reactions (like trans–cis isomerization) or the presence of deactivating groups in the phenolic ring. In this subsection, it will be shown how these problems can be circumvented.

Some time ago, it was reported that the use of powdered *potassium carbonate* suspended in hexane enhances the yields of the PFR of phenyl esters (see Table 14) [204].

Polyacylation of hydrocarbons is usually difficult to achieve by means of Friedel–Crafts reactions because the first acyl group deactivates the ring toward introduction of further acyls. On the other hand, polyacylation of phenols by Lewis-acid-catalyzed Fries reaction or by photochemical–Fries rearrangement, is disfavored for the same reason. In the case of the photochemical reaction, this problem has been avoided by using *cyclic acetals as carbonyl blocking groups*. This procedure is limited to the photochemical reaction because acetals do not survive under the acidic conditions of the classical acylation methods. Thus, irradiation of the acetals of *ortho*- and *para*-acetoxyacetophenone give the corre-

Table 14 PFR of Several Aryl Acetates in the Presence and Absence of Potassium Carbonate

(for R' = H)

		Ortho product yield (%)	
R	R'	With K_2CO_3	Without K_2CO_3
H	H	78	13
Me	H	74	32
H	Me	86	35
H	OMe	89	75
H	Cl	88	49
Me	Me	90	34

Source: Ref. 204.

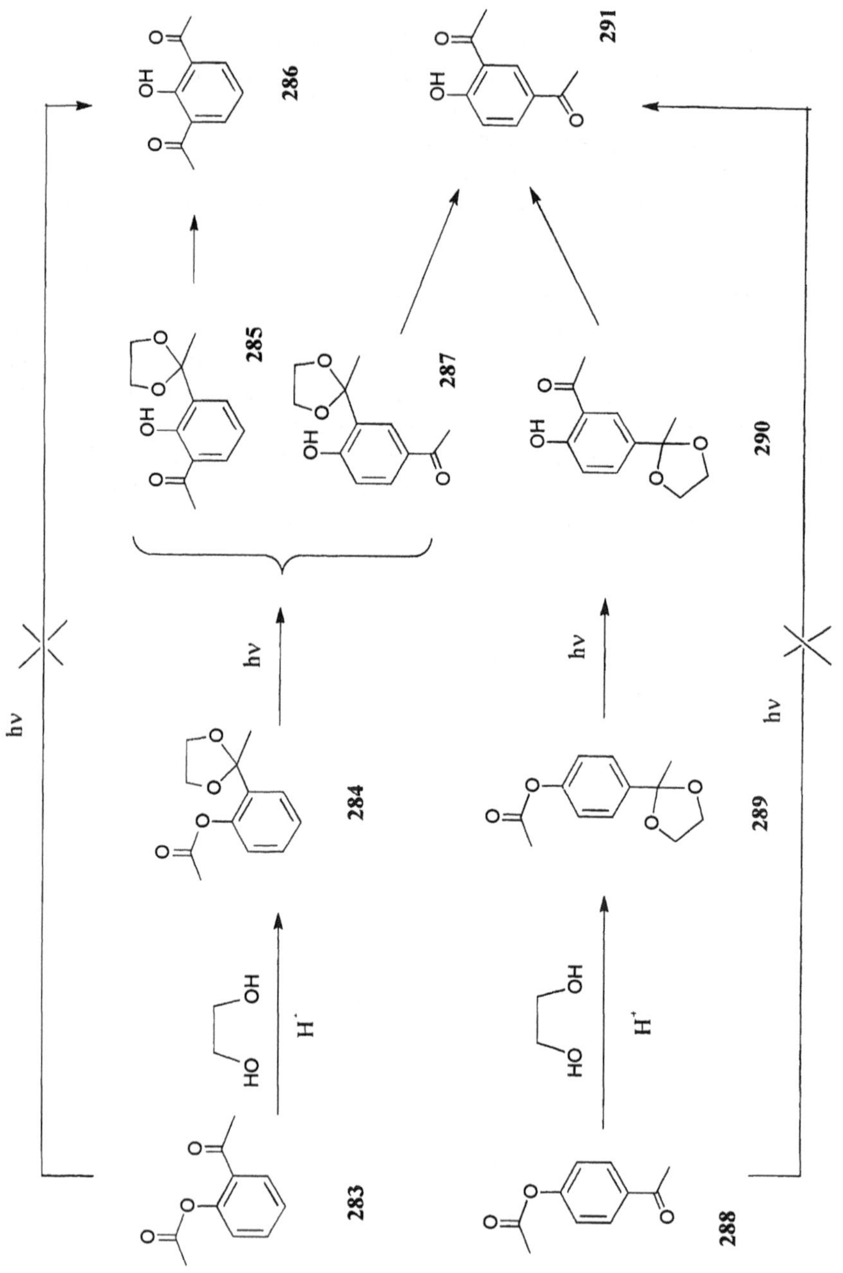

Scheme 72

sponding migration products along with a fraction of deacetalyzed ketones (Scheme 72) [205]. The efficiency of the diacylation is moderate (~30%), but the nonprotected acetophenones do not undergo photochemical rearrangement. This could be due to efficient intersystem crossing to the triplet state, where no C—O cleavage takes place.

Protection of the carbonyl group of an *ortho*-acyolxyacetophenone can also be achieved by synthesizing its *enol acetate*; again, the PFR is more efficient than in the unprotected molecule. After photolysis of the phenyl ester, intramolecular transacetylation takes place, leading to the corresponding acetoxyphenones (Scheme 73) [206].

The *combination of the carbonyl protection method and the addition of the potassium carbonate method* is advantageous: 2,4-Diacetylphenol can be obtained from the acetal of 4-acetylphenyl acetate in 80% yield by irradiation in hexane in the presence of anhydrous K_2CO_3 [207]. In this experiment, the base avoids deprotection of the starting material.

The protective method has also been employed with β-ketoesters. In this case, the goal is to avoid keto-enol photoisomerization that is an efficient energy-wasting channel. Scheme 74 shows that direct photorearrangement of aryl benzoyl acetates (**298**) to the *ortho*-hydroxydibenzoylmethanes (**299**) is poor, whereas irradiation of the related acetal derivatives gives higher yields [208]. The resulting *ortho*-hydroxydibenzoylmethanes are precursors for the synthesis of flavones. Related flavonoids can be obtained in similar yields by PFR of aryl dihydrocinnamates [209].

Another problem for the synthetic use of PFR is that the ortho-rearranged products may act as internal light filters, stopping the reaction. In synthetic routes leading to chromanones, chromones, and related compounds, this is of vital importance because the overall yield is limited by the photochemical step. Improved yields can be obtained if the α,β-unsaturated *ortho*-hydroxyphenones resulting from PFR are removed and cyclized to chromanones. This can be acomplished in one pot by irradiation in a *two-phase system* benzene/10% aqueous NaOH, whereby chromanones are directly obtained from phenyl crotonates in 80–90% yields [210].

V. THE PFR OF POLYMERS

Due to the importance of polymer chemistry for the chemical industry, considerable effort has been devoted to the study of the effects of light on polymeric materials. The photochemistry of polymers is complex. However, it is well established that PFR occurs in some aromatic polyesters [211–222] and polyamides [223,224]. This process can take place either in the main chain of the polyester or in the pendant groups.

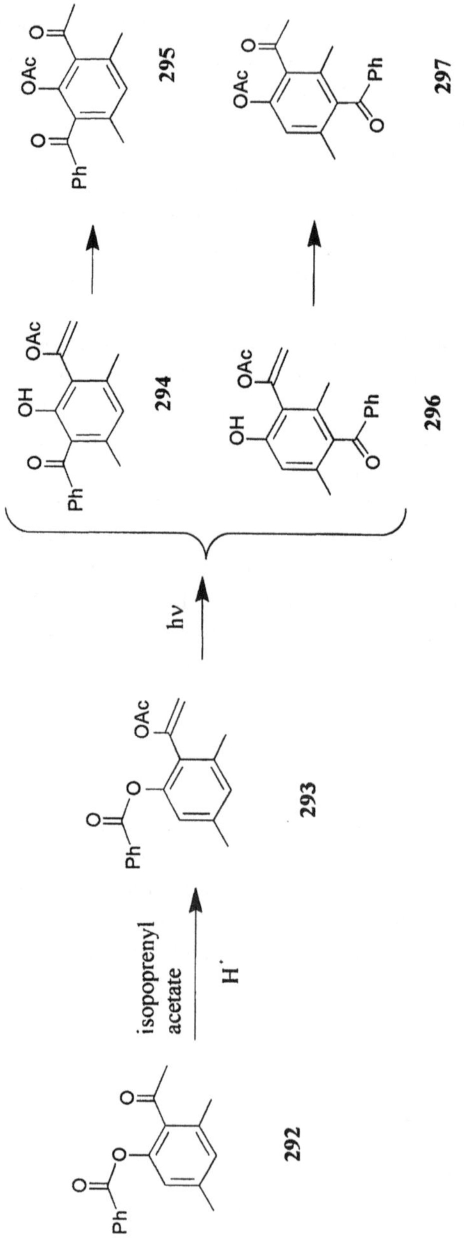

Scheme 73

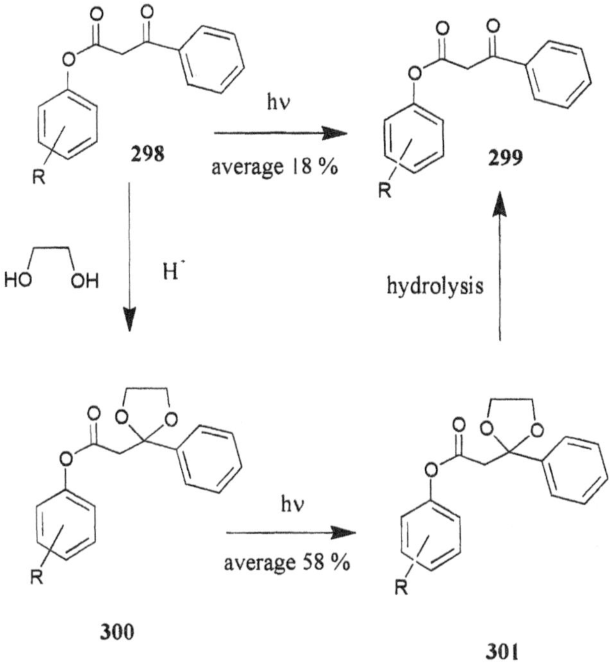

Scheme 74

An example of *main-chain rearrangement* (i.e., in the backbone of the polymer) is depicted in Scheme 75 [225].

Polycarbonates can be included in this category. In fact, the *bisphenol A-based polycarbonates* are one of the most deeply studied classes of polymers in relation to its photochemical stability. To date, it is well established that the photochemical processes occurring in these polymers are wavelength dependent [226–230]. When photolyzed with λ > 300 nm, several radical and oxidative reactions take place, whereas with λ < 300 nm, the PFR as shown in Scheme 76 becomes important. Analogous dual photochemistry has been shown recently for trimethylclohexane–polycarbonate [231].

Monitoring of the PFR can be made spectroscopically because the photoproducts have well-defined absorbtion bands in the UV-visible and infrared (IR) ranges [232]. Fluorescence spectroscopy allows the early detection of phenyl salicylate-type products in the photolysis of bisphenol A-based polycarbonates due to the characteristic emission of this chromophore around 470 nm [233].

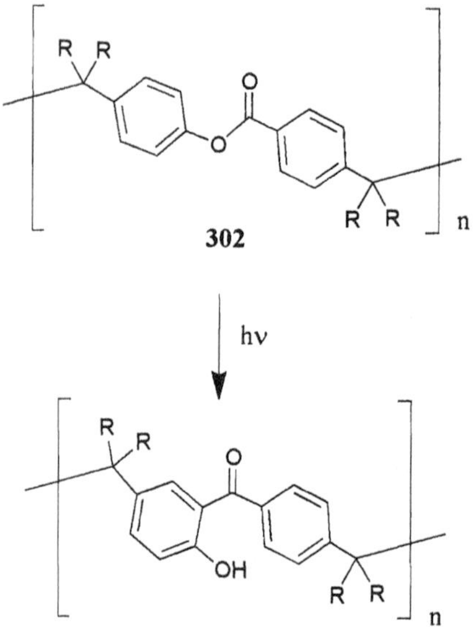

302

hv

303

Scheme 75

Figure 6 shows the kinetics of appearance of the ortho-rearranged product emission. It is noteworthy that after only few minutes, it is possible to detect photoproducts by means of this technique, whereas by UV-vis or IR. it usually takes longer to appreciate spectral changes.

Polyesters of cinnamic acid derivatives used as photorresists, whose main photochemical feature is $[2\pi + 2\pi]$ dimerization, are reported to experiment PFR along with trans–cis isomerization as secondary reactions [234–240].

Polyurethanes can also undergo PFR under irradiation, as has been demonstrated by means of UV-vis and FTIR spectrocopy [241,242].

As mentioned earlier, amides are less reactive than esters, and diamides are even less reactive. An example of this trend is the low reactivity of the dibenzoyl amides of 1,3- and 1,4-diaminobenzene, which are model compounds for the polymers Nomex [poly(1,3-phenylene isophthalamide)] and Kevlar B [poly (1,4-phenylene terephthalamide)] [243].

The other possibility is, for polymers with comblike structure, rearrangement in the *pendant groups*. Attending to the position of the aryl and acyl moieties relative to the backbone. it is possible to distinguish between two types of rearrangement. In poly(phenyl acrylate) [244] (**307**) and poly(phenyl methacrylate)

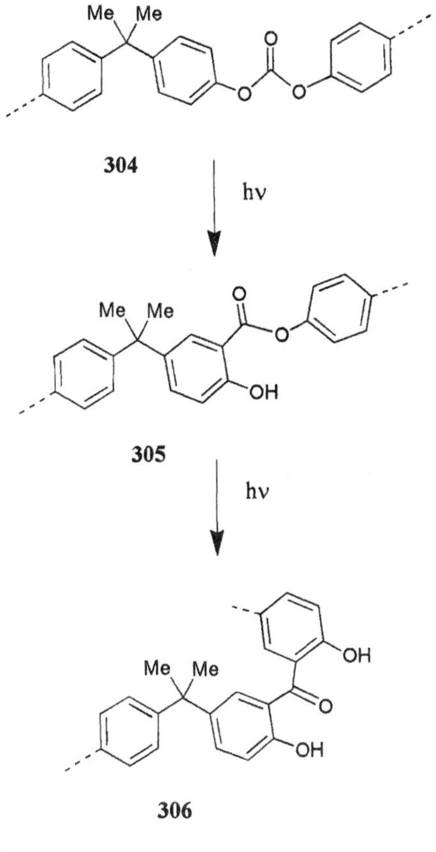

Scheme 76

[245] (**308**), the acyl substructure of the ester belongs to the main chain (Scheme 77). Analogous results are obtained with poly(1-naphthyl acrylate) [246], poly(2-naphthyl acrylate) [246], and poly(methacrylamides) [247]. The aryloxy group can also be attached to the main chain, as in the case of poly(*para*-acetoxystyrene) [248] and poly(*para*-acetamidostyrene) [248] (Scheme 78).

Not always is the PFR a side undesired reaction in polymers. In some cases, it has been used for practical purposes like *selective image development* [248]. Another important use of the PFR is the *photostabilization* of polymers [249]. Aromatic esters and polyesters are sometimes mixed with other polymers in order to protect them from light. This delays aging of the polymer because the light is

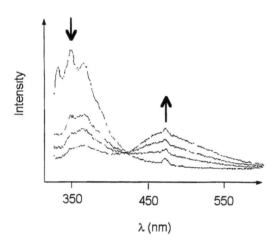

Figure 6 Emission spectra of polycarbonate film (excitation at 310 nm) after photolysis for 0, 15, 30, and 60 min. (Adapted from Ref. 233.)

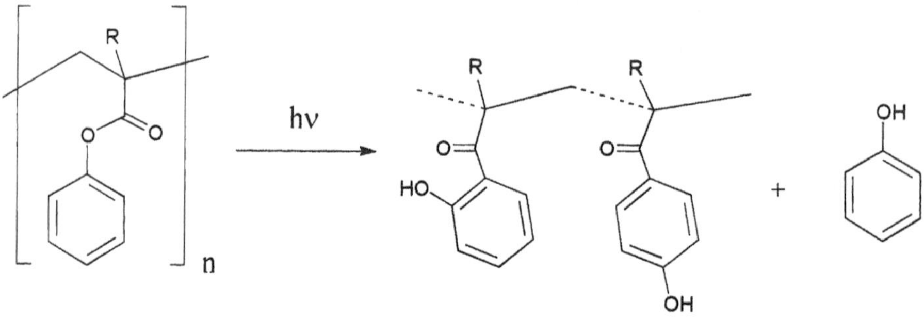

307 R=H
308 R=Me   **309**   **13**

Scheme 77

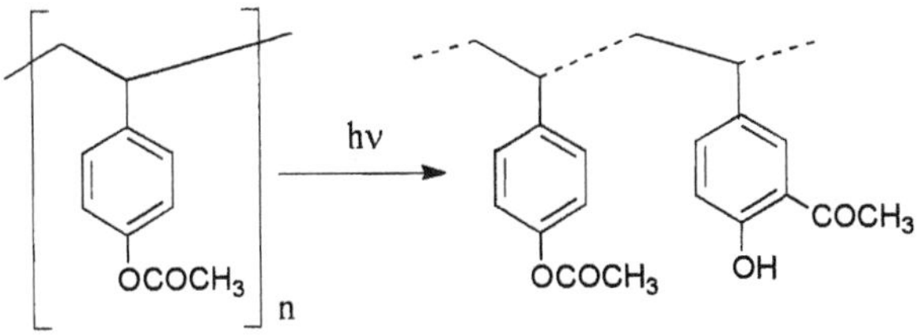

310   311

Scheme 78

absorbed by the polyester, which undergoes PFR. The resulting *ortho*-hydrox-yphenones are also stabilizers because they are good internal filters. An example of this application is the addition of poly(3-methoxyphenyl 4-vinylbenzoate) as the UV stabilizer for the copolymer of 4-vinylbenzaldehyde and malonic acid [250]. Polyethylene terephthalate can also be photostabilized via the PFR [251].

VI. THE PFR IN ORGANIZED MEDIA

It is well known that the environment may have a strong influence on the course of photochemical reactions. This influence ranges from small effects in nonpolar, low-viscosity solvents to strong interactions in the cages of inorganic microporous materials like zeolites. A number of articles and reviews deal with the host–guest interactions in the context of supramolecular photochemistry [252–255]. Because the photo-Fries rearrangement involves in-cage recombination versus diffusion of radicals, it has been chosen as a probe to study the effect of nonhomogeneous environments on the reactivity and selectivity of such kinds of reaction. Several types of host have been studied, namely micelles, cyclodextrins, zeolites, and polymeric matrixes. In general, enhancement of selectivity is achieved in those organized media but with little differences, which will be shown in the following subsections.

A. Cyclodextrins and Micelles

Cyclodextrins are cyclic oligosaccharides made up of several D-glucose units. Depending on the number of units, these doughnut-shaped molecules are able to host different sized molecules or the same molecule with different association constants. α-Cyclodextrin contains 6 units of D-glucoses, β-cyclodextrin contains seven units, and γ-cyclodextrin contains 8 units [256]. Although early studies on complexation and photochemistry of phenyl acetate [257] and phenyl benzoate [258] stated that cyclodextrins bring about selectivity toward the formation of the para isomer, more recent studies have found that the ortho product is preferred. The reason of this discrepancy could be that different amounts of cyclodextrin were added. Ramamurthy and co-workers studied the photo-Fries rearrangement of phenyl esters within β-cyclodextrin, both in aqueous solutions and in the solid complex [259]. Irradiation of the solid complexes of β-cyclodextrin (1:1 stoichometry) resulted in remarkable ortho selectivity (sometimes 99%). In solution, the dissociation constant for the complex β-cyclodextrin:phenyl acetate was determined to be 9.33×10^{-3} M/L, which explains the high selectivity in aqueous solutions with excess of β-cyclodextrin. Moreover, irradiation of meta-substituted esters and amides resulted in preferential rearrangement to one of the free ortho positions (para to the substituent).

Quantum yields determinations lead to analogous conclusions, although differences in the ortho-to-para ratio are found probably because of experimental reasons. However, the ortho selectivity seems to be well established. Phenyl acetate irradiated in water gives the following quantum yields of product formation: 0.16 (ortho), 0.067 (para), and 0.048 (phenol); in the presence of an excess of β-cyclodextrin, they change to 0.23 (ortho), 0.053 (para), and 0.27 (phenol) [260]. As can be seen, the ortho product is favored in the hydrophobic microenvironment of the cyclodextrin. Phenol quantum yield is enhanced with respect to the irradiation without cyclodextrin, which has been interpreted in terms of H abstraction from the inner walls of the host oligosaccharide.

Other studies on the PFR in the presence of cyclodextrins include substrates with different acyl moieties (phenyl propionate and phenyl valerate) [261] 1-naphthyl acetate [262,263], 1-naphthyl benzoate [263], sulfonate esters, [264,265], benzenesulfonylanilides [266], acetanilide [259,267], and benzanilide [259,268].

Micellar media created by sodium dodecyl sulfate or cetyltrimethylammonium bromide have also a positive effect on the ortho selectivity [262,269–271]. In a recent article, the PFR of 1-naphthyl acetate in aqueous solutions of an antenna polyelectrolyte like poly(sodium styrenesulfonate-co-2-vinylfluorene) has been reported, which, in addition to the micellar effect, also provides singlet sensitization by means of fluorene chromophores. In this particular case, the occurrence of in-cage versus out-of-cage processes can be clearly correlated with the different packing of the microdomains, which depends on the molar fraction of monomers in the antenna copolymer [272].

B. Zeolites

Zeolites are microporous inorganic materials that possess channels and cavities able to accommodate small organic molecules. Those channels and cavities could be considered as microscopic reactors because they are able to host single molecules [273,274]. Due to the special conditions of restricted mobility, the distribution of photoproducts is expected to be influenced by the constraint conditions. As with cyclodextrins, the reports dealing with this subject allow one to establish some general trends in the photochemistry of esters and related systems within zeolites. Early experiments were carried out by de Mayo and co-workers two decades ago with esters [275] and anilides [276] in the presence of silica gel. The conversions in SiO_2-slurry and in dry-coated silica gel are higher than those in solution; for example, 2,6-diisopropylphenyl benzoate rearranges photochemically to the para product in pentane solution in 9% yield, in SiO_2-slurry in 40% yield, and in dry SiO_2 in 70% yield [275].

Recently, the PFR of phenyl acetate and phenyl benzoate has been studied within zeolites of the faujasite and pentasil families, which are structurally very different. [277,278] It was found that for phenyl acetatet in faujasites X and Y, the predominating product is the ortho isomer, whereas in the pentasils ZSM-5 and ZSM-11, the preferred product is the para isomer. These differences have been explained taking into account the different sizes and shapes of the cavities and channels of the two families. The faujasites are made up of channels of ~ 8 Å diameter, leading to supercages of ~ 13 Å diameter, where the reaction is expected to occur, whereas the pentasils are made up of narrow channels of ~ 6 Å diameter. In the X and Y faujasites, the ortho position would be favored due to the restricted mobility, whereas in the channels of ZSM-5 and ZSM-11, there would be a shape limitation because the ortho isomer does not fit but the para isomer does (Table 15). With phenyl benzoate inside X or Y zeolites, the same

Table 15 Product Distribution in the Irradiation of Phenyl Acetate and Phenyl Benzoate in Several Zeolites and in Hexane Solutions

Conditions	Phenyl acetate			Phenyl benzoate		
	Phenol	Ortho	Para	Phenol	Ortho	Para
Hexane	20	53	27	12	49	37
LiY, hexane	7	87	6	1	92	7
NaY, hexane	5	91	4	1	93	6
KY, hexane	17	83	—	2	98	—
LiX, hexane	15	77	8	1	96	3
NaX, hexane	21	79	—	5	93	2
KX, hexane	27	73	—	7	93	—
Na ZSM-5, hexane	27	35	38	10	34	27
Na ZSM-5, isooctane	32	5	63	17	38	24
Na ZSM-11, isooctane	18	8	74	8	48	29

Source: Ref. 277.

selectivity was found, but no selectivity was found with ZSM-5 and ZSM-11 simply because phenyl benzoate is larger than the diameter of the channel and does not enter into the zeolite [277].

The selectivity has been found to depend on the size of the cation of the faujasite. Thus, with large cations like K^+ no para rearrangement is observed, whereas with small cations like Na^+ or Li^+, a bit of it is observed [277].

Similar results have been found with 2,5-disubstituted phenyl acetates [270], naphthyl esters [279], and amides [280].

With phenyl phenylacetates, there is also PFR, but the decarbonylation reaction is an important process [281,282]; product selectivity depends not only on the environment but also, in a very important manner, on the substituents of the ester molecule.

The great selectivity toward the ortho product in NaY zeolite is one of the most remarkable findings. The same selectivity has been observed in 1-naphthyl phenylacylates not only in Na^+- exchanged Y zeolite but also using Li^+, K^+, and Rb^+ cations. However with Cs^+ and Tl^+, the photorearrangement was not observed, as a consequence of the heavy-atom effect induced by these large cations [283]. More intriguing is the fact that decarbonylation of such 1-naphthyl phenylacylates does not take place in the supercages of the Y zeolites, which are large enough to allow a rapid process like the loss of CO. However, spatial considerations do not seem to play a special role in this case. Ramamurthy and Weiss have pointed out that such behavior could be explained considering binding of phenylacyl radicals to the exchanging cations, which would make them reluctant to loose CO.

The selectivity of the Lewis-acid-catalyzed Fries reaction is also strongly influenced by the inclusion into zeolites [284].

C. Other Ordered Media

The photo-Fries rearrangement has been conducted in many other environments different from micelles, cyclodextrins, and zeolites. Thus, PFR of diphenylcarbonate has been reported to be insensitive whether it occurs in homogeneous solution or embedded within an *acrylic polymer*, or whether the polymeric matrix is below or above the glass transition temperature [284] Phenyl phenylacetates included in *Nafion membranes* rearrange exclusively to the ortho position, showing a remarkable regioselectivity [77,285]. 2-Naphthyl myristate irradiated in several isomeric *alkyl alkanoates* exhibits higher selectivity in ordered phases than in disordered ones [77]. The phase order determines not only the regioselectivity but also the efficiency of competing reactions. The relative efficiency of trans–cis isomerization, PFR and $[2\pi + 2\pi]$ dimerization in phenyl cinnamates depends on the order of the phase in which the reaction takes place [286].

Usually, the PFR serves as a probe reaction for the study of the morphology of the matrix. This has been demonstrated by Weiss and co-workers in a series of recent articles on the photoreactivity of esters included in *several polyethylene films*. Low-density polyethylene (LDPE) films hosting 2-naphthyl esters bring about different selectivity in the PFR as compared with the reaction in solution. In addition, the selectivity is different if the polymer is stretched [286,287]. Table 16 indicates the different product distributions upon irradiation of 2-naphthyl esters, depending on the nature of the solvent or matrix. The most striking fact is that irradiation of 2-naphthyl myristate leads to the coupling at the position 1 in *tert*-butanol with 86% yield, whereas this product is absent in the irradiations in polyethylene films. Moreover, the product of coupling at position 3 is absent

Table 16 Product Distribution in the Irradiation of 2-Naphthyl Acylates in Solution and in LDPE Films

Substrate	Medium	Conversion (%)	Naphthol	C-1 coupling	C-3 coupling	C-6 coupling
R = Me	*tert*-BuOH	16	32	31	11	17
	LDPE (u)	12	21	31	20	28
	LDPE (s)	8	32	—	31	37
R = —(CH$_2$)$_{12}$—CH$_3$	*tert*-BuOH	12	—	86	—	14
	LDPE (u)	< 7	—	—	75	25
	LDPE (s)	< 7	—	—	92	8

Note. u = unstretched: s = stretched.
Source: Ref. 286.

in alcoholic medium but represents 75% and 92% in unstretched and stretched LDPE films, respectively. Also, the product resulting from coupling at the C-8 position was not detected. The effect of stretching can be clearly observed in the PFR of 2-naphthyl acetate in LDPE film. This is a nice example of how a *macroscopic* force (stretching) can influence *microscopically* the pathway followed in a photochemical reaction.

From the above data, it appears that the average sites experienced by reaction intermediates in the polymer are somewhat cylindrical, because the preferred photoproducts are those derived from coupling at C-3 and C-6 (i.e., those with a more extended shape) [286].

An analogous method has been followed to study the behavior of 1-naphthyl acylates [288], phenyl phenylacylates [289,290], and 1-naphthyl phenylacylates [291,292] in polyethylene films. In the case of phenylacylates, a competitive reaction like decarbonylation can only occur in the anisotropic viscous space provided by the polyolefin. This is due to the fact that in the homogeneous environment, loss of CO is too slow to compete with recombination; however, in the polyethylene environment, recombination is considerably slower. This allows one to use the decarbonylation reaction as a radical clock to measure absolute rates of radical-pair coupling inside the cavities provided by the polyolefin by knowing the concentrations of final products and the rate of decarbonylation, which is assumed to be independent of the surrounding medium. In this way, recombinations at position 2 for 1-naphthyl phenylacylates occur at an average rate in the order of $\sim 10^8-10^9$ sec^{-1}, whereas the rates of couplings at position 4 are one order of magnitude lower ($\sim 10^7-10^8$ sec^{-1}) [291].

Fluorescence decay measurements have also allowed one to propose a qualitative model of film occupancy in which both amorphous sites and crystalline–amorphous interfaces would be occupied by ester molecules, with predominance of the former ones [291].

Polyethylene films influence the product distribution in a different way, depending on the nature of the aryloxy radical. Thus, the ortho/para ratio in the PFR of phenyl esters is ~ 2, both in hexane and films. However, this ratio is very different for 1-naphthyl esters: ~ 2 in hexane and > 6 in the films. This has been ascribed to the more available rotational movement for phenoxyl than for naphthoxyl radicals. As a matter of fact, van der Waals volumes are quite different, being 81 Å^3 for the former and 124 Å^3 for the latter [292].

Time-resolved fluorescence has been used for the study of the PFR on *poly(methyl methacrylate)*. It has been found that the photoproducts of 2-naphthyl acetate in some zones of the material act as long-range quenchers for the unreacted esters in other zones of the polymer. A nonrandom chromophore distribution is generated in this way [293].

Finally, the PFR of 1-naphthyl acetate has been used to demonstrate the occurrence of solvent–solute clustering in supercritical carbon dioxide. This is a nice example of how the paradigmatic nature of the PFR allows one to use it as a chemical probe for the investigation of physical interactions [294].

VII. THE PFR OF BIOLOGICALLY ACTIVE COMPOUNDS

Among the parameters to be considered for the design of *pesticides*, photostability is important because these compounds are exposed to solar light or short-wavelength artifical light [295]. The herbicide *Propanil* [**312**, *N*-(3,4-dichlorophenyl) propanamide, Scheme 79] undergoes photochemical-Fries rearrangement along with other reactions [296].

Propoxur (**313**) is another herbicide which can be eliminated from the environment by means of photochemical treatments. In fact, it has been shown that direct irradiation of aerated aqueous solutions of propoxur leads to formation of PFR products, disappearing almost completely the starting material. By laser flash photolysis, it has been demonstrated that the key intermediate in this reaction is the 2-isopropoxyphenoxy radical [297].

Irradiation of isoprocarb and promecarb resulted in PFR to *ortho-* and *para-* hydroxybenzamides. The photodegradation of analogous carbamate pesticides (bendiocarb, *bendiocarb, ethiofencarb, furathiocarb, fenoxycarb,* and *pirimicarb*) has also been examined, but in these cases, the most general result was formation of the corresponding phenols [298].

Many *pharmacological substances* possess aromatic ester or amide groups capable of undergoing PFR if excited with the appropriate light. *Flutamide* (**314**) is a nonsteroidal antiandrogen drug which is commonly used in advanced prostate cancer. Being an anilide, it could, in principle, undergo PFR. However, in aqueous solution (phosphate buffer), the only reaction detected is photoconversion of the nitro into an hydroxyl group [299]. Using β-cyclodextrins as a biological mimicking environment, two additional reactions have been found to occur: (1) photoreduction of the nitro group and (2) cleavage of the amide bond to give the corre-

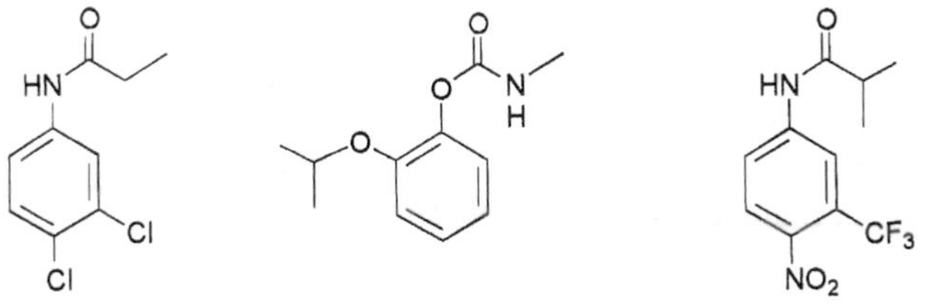

312   313   314

Scheme 79

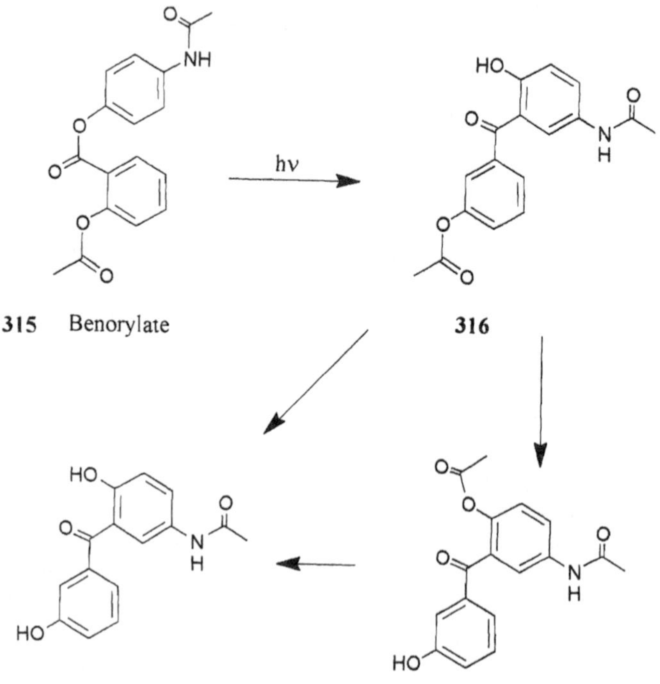

315   Benorylate

316

317

318

Scheme 80

sponding aniline but no rearrangement product. The involvement of radical intermediates in these reactions could explain the reported phototoxicity displayed by flutamide.

Benorylate (315) [4'-(acetamido)phenyl-2-acetoxybenzoate] is another example. It is the ester between two well-known antiinflamatory drugs, aspirin and paracetamol, and is employed in rheumatoid arthritis therapy. In view of the chemical structure with three photolabile groups (two esters and one amide), its possible phototoxicity has been investigated. From the preparative irradiations, it has been concluded that the PFR takes place with breaking of the central C—O bond to yield 5-acetamido-2'-acetoxy-2-hydroxybenzophenone (316). This product undergoes transacetylation to 5'-acetamido-2'-acetoxy-2-hydroxy-benzophenone (318) (Scheme 80) [300].

VIII. CONCLUDING REMARKS

The knowledge accumulated in the past four decades on the photo-Fries rearrangement has been reviewed in this work. It is one of the *classical photochemical reactions*, which bears a close resemblance to one of the *classical acid-catalyzed reactions*. The PFR is a suitable probe reaction for investigating the microenvironments provided by a wide variety of materials and media. In addition to its fundamental importance in photochemistry, the PFR is interesting because of its application in different fields, including organic synthesis and polymer chemistry.

REFERENCES

1. Anderson, J. C.; Reese C. B. *Proc. Chem. Soc. London* **1960**, 217.
2. Miranda, M. A. In *Handbook of Organic Photochemistry and Photobiology*, Horspool, W. M.; Song, P. S., eds.; CRC Press: Boca Raton, FL, **1995**; p. 570.
3. Miranda, M. A.; García, H. In *Rearrangements, Supplement B: The Chemistry of Acid Derivatives*, Patai, S., ed.; Wiley: New York, **1992**; Vol. 2., p. 1271.
4. Stenberg, V. I. *Org. Photochem.* **1967**, *1*, 127.
5. Bellus, D. *Adv. Photochem.* **1971**, *8*, 109.
6. Bellus, D.; Hrdlovic, P. *Chem. Rev.* **1967**, *67*, 599.
7. Martin, R. *Org. Prep. Proc. Int.* **1992**, *24*, 369.
8. Blatt, A. H. *Chem. Rev.* **1940**, *27*, 413.
9. Blatt, A. H. *Org. React.* **1942**, *1*, 342.
10. Kobsa, H. *J. Org. Chem.* **1962**, *27*, 2293.
11. Anderson, J. C.; Reese C. B. *J. Chem. Soc.* **1963**, 1781.
12. Sandner, M. R.; Trecker, D. J. *J. Am. Chem. Soc.* **1967**, *89*, 5725.
13. Sandner, M. R.; Hedaya, E.; Trecker, D. J. *J. Am. Chem. Soc.* **1968**, *90*, 7249.
14. Stone, T. J.; Waters, W. A. *J. Chem. Soc.* **1964**, 213.
15. Bellus, D.; Hrdlovic, P.; Slama, P. *Collect. Czech., Chem. Commun.* **1968**, *33*, 2646.
16. Bellus, D.; Hrdlovic, P.; Slama, P. *Collect. Czech., Chem. Commun.* **1968**, *33*, 3752.
17. Kalmus, C. E.; Hercules, D. M. *Tetrahedron Lett.* **1972**, *16*, 1575.
18. Kalmus, C. E.; Hercules, D. M. *J. Am. Chem. Soc.* **1974**, *96*, 449.
19. Meyer, J. W.; Hammond, G. S. *J. Am. Chem. Soc.* **1970**, *92*, 2187.
20. Meyer, J. W.; Hammond, G. S. *J. Am. Chem. Soc.* **1972**, *94*, 2219.
21. Haga, N; Takayanagi, H.; *J. Org. Chem.* **1996**, *61*, 735.
22. Pohlers, G.; Grime, S.; Deeskamp, H. *J. Photochem. Photobiol. A: Chem.* **1994**, *79*, 153.
23. Gorodetsky, M; Mazur, Y., *J. Am. Chem. Soc.* **1964**, *86*, 5213.
24. Shizuka, H.; Morita, T.; Mori, Y.; Tanaka, I. *Bull. Chem. Soc. Jpn.* **1969**, *42*, 1831.
25. Finnegan, R. A.; Mattice, J. J. *Tetrahedron* **1965**, *21*, 1015.
26. Lally J. M.; Spillane W. J., *J. Chem. Soc., Chem. Commun.*, **1987**, 8.
27. Lally J. M.; Spillane, W. J. *J. Chem. Soc., Perkin Trans. 2* **1991**, 803.
28. Gritsan, N. P.; Tsentalovich, Y. P.; Yurkovskaya, A. V.; Sagdeev, R. *J. Phys. Chem.* **1996**, *100*, 4448.
29. Coppinger, G. M., Bell, E. R. *J. Phys. Chem.* **1966**, *70*, 3479.
30. Shizuka H., Tanaka I. *Bull. Chem. Soc. Jpn.* **1969**, *42*, 909.
31. Plank, D. A., *Tetrahedron Lett.* **1968**, 5423.
32. Hageman, H. J. *Tetrahedron* **1969**, *25*, 6015.
33. Finnegan, R. A.; Knutson, D. *Tetrahedron Lett.* **1968**, 3429.
34. Finnegan, R. A.; Knutson, D. *J. Am. Chem. Soc.* **1967**, *89*, 1970.

35. Finnegan, R. A.; Knutson, D. *J. Am. Chem. Soc.* **1968**, *90*, 1670.
36. Rao, D. V.; Lamberti, V. *J. Org. Chem.* **1967**, *32*, 2896.
37. Bradshaw, J. S.; Loveridge, E. L.; White, L. *J. Org. Chem.* **1968**, *33*, 4127.
38. Humphrey, J. S.; Roller, R. S. *Mol. Photochem.* **1971**, *3*, 35.
39. Molokov, I. F.; Tsentalovich, Yu. P.; Yurkovskaya, A. V.; Sagdeev, R. Z. *J. Photochem. Photobiol. A: Chem.* **1997**, *110*, 159.
40. Arai, T.; Tobita, S.; Shizuka, H. *Chem. Phys. Lett.* **1994**, *223*, 521.
41. Arai, T.; Tobita, S.; Shizuka, H. *J. Am. Chem. Soc.* **1995**, *117*, 3968.
42. Woodward, R. B.; Hoffmann, R. *J. Am. Chem. Soc.* **1965**, *87*, 2511.
43. Kimura, Y.; Kakiuchi, N.; Shizuka, H. *J. Chem. Soc., Faraday Trans.* **1998**, *94*, 3077.
44. Jiménez, M. C.; Miranda, M. A.; Scaiano, J. C.; Tormos, R. *J. Chem. Soc. Chem. Commun.* **1997**, 1487.
45. Quinkert, G.; Kleiner, E.; Freitag, B.; Glenneberg, J.; Bilhardt, U.; Cech, F.; Schmieder, K. R.; Schudok, C.; Steinmetzer, H.; Bats, J. W.; Zimme.mann, G.; Dürner, G.; Rehm, D. *Helv. Chim. Acta.* **1986**, *69*, 469.
46. Adam, W.; de Sanabia, J. A.. H. Fischer, *J. Org. Chem.* **1973**, *38*, 2971.
47. Kaptein, R. *J. Chem. Soc., Chem. Commun.* **1971**, 732.
48. Schutte, L.; Havinga, E. *Tetrahedron*, **1967**, *23*, 2281.
49. Shine, H. J.; Subotkowski, W. *J. Org. Chem.* **1987**, *52*, 3815.
50. Nakagaki, R.; Hiramatsu, M.; Watanabe, T.; Tanimoto, Y.; Nagakura, S. *J. Phys. Chem.* **1985**, *89*, 3222.
51. Turro, N. J.; Kraeutler, B. In *Isotopes in Organic Chemistry*: Buncel, E.; Lee, C. C., eds.; Elsevier: Amsterdam, **1984**; Vol. 6, p. 107.
52. Beck, S. M.; Brus, L. E. *J. Am. Chem. Soc.* **1982**, *104*, 1805.
53. Rosenthal, I.; Mossoba, M. M.; Riesz, P. *Can. J. Chem.* **1982**, *60*, 1486.
54. McRae, J. A.; Symons, M. C. R. *J. Chem. Soc. B* **1968**, 428.
55. Sun, H.; Frei, H. *J. Phys. Chem. B*. **1997**, *101*, 205.
56. Vasenkov, S.; Frei, H. *J. Am. Chem. Soc.* **1998**, *120*, 4031.
57. Brown, C. E.; Neville, A. G.; Rayner, D. M.; Ingold, K. U.; Lusztyk, J. *Aust. J. Chem.* **1995**, *48*, 365.
58. Jacox, M. *Chem. Phys* **1982**, *69*, 407.
59. Gutsche, C. D.; Oude Alink, B. A. M. *J. Am. Chem. Soc.* **1968**, *90*, 5855.
60. Penn, J. H.; Liu, F. *J. Org. Chem.* **1994**, *59*, 2608.
61. Ghibaudi, E.; Colussi, A. *J. Chem. Phys. Lett.* **1983**, *94*, 121.
62. Churio, M. S.; Ghibaudi, E.; Colussi, A. J. *J. Photochem. Photobiol. A: Chem.* **1988**, *44*, 133.
63. Jiménez, M. C.; Leal, P.; Miranda, M. A.; Tormos, R. *J. Chem. Soc., Chem. Commun.* **1995**, 2009.
64. Galindo, F.; Miranda, M. A.; Tormos, R. *J. Photochem. Photobiol. A: Chem.* **1998**, *117*, 17.
65. Grimme, S. *Chem. Phys.* **1992**, *163*, 313.
66. Alvaro, M.; García, H.; Miranda, M. A.; Primo, J. *Recl. Trav. Chim. Pays-Bas* **1986**, *105*, 233.
67. Kende, A. S.; Belletire, J. L. *Tetrahedron Lett.* **1972**, 2145.

68. Obara, H; Takahashi, H. *Bull. Chem. Soc. Jpn.* **1967**, *40*, 1012.
69. Obara, H; Takahashi, H.; Hirano, H. *Bull. Chem. Soc. Jpn.* **1969**, *42*, 560.
70. García, H; Iborra, S.; Primo, J.; Miranda, M. A. *J. Org. Chem.* **1986**, *51*, 4432.
71. Trifonov, L. S.; Orahovats, A. S.; Prewo, R.; Bieri, J. H.; Heimgartner *J. Chem. Soc., Chem. Commun.* **1986**, 708.
72. Bellus, D.; Schaffner, K.; Hoigne, J. *Helv. Chim. Acta* **1968**, *51*, 1980.
73. Ohto, Y.; Shizuka, H.; Sekiguchi, S.; Matsui, K. *Bull. Chem. Soc. Jpn.* **1974**, *47*, 1209.
74. Wang, Z.; Holden, D. A.; McCourt, F. R. W. *Macromolecules* **1990**, *23*, 3773.
75. Holden, D. A.; Jordan, K.; Safarzadeh-Amiri, A. *Macromolecules* **1986**, *19*, 895.
76. Cui, C.; Wang, X.; Weiss, R. G. *J. Org. Chem.* **1996**, *61*, 1962.
77. Baldvins, J. E.; Cui, C., Weiss, R. G. *Photochem. Photobiol.* **1996**, *63*, 726.
78. Pathak, V. P.; Sainai, T. R.; Khanna, R. N. *Monatsh. Chem.* **1983**, *114*, 1269.
79. Thal, C.; Papacosta, D.; Beugelmans, R. *Bull. Soc. Chem. Fr.* **1972**, 1106.
80. Le Goff, M.-T.; Beugelmans, R.; *J. Am. Chem. Soc.* **1973**, *95*, 8472.
81. Le Goff, M.-T.; Beugelmans, R. *Bull. Soc. Chim. Fr.* **1974**, 69.
82. Le Goff, M.-T.; Beugelmans, R. *Bull. Soc. Chim. Fr.* **1972**, 1115.
83. Schwartz, A.; Pal, Z.; Szabo, L. *J. Heterocycl. Chem.* **1987**, *24*, 651.
84. Chan, A. C.; Hilliard, P. R., Jr. *Tetrahedron Lett.* **1989**, *30*, 6483.
85. Kulshrestha, S. K.; Dureja, P.; Mukerjee, S. K. *Ind. J. Chem.* **1984**, *23B*, 1064.
86. Miranda, M. A.; Primo, J.; Tormos, R. *Tetrahedron* **1989**, *45*, 7593.
87. Shizuka, H.; Tanaka, I. *Bull. Chem. Soc. Jpn.* **1968**, *41*, 2343.
88. Shizuka, H. *Bull. Chem. Soc. Jpn.* **1969**, *42*, 52.
89. Shizuka, H. *Bull. Chem. Soc. Jpn.* **1969**, *42*, 57.
90. Elad, V.; Rao, D. V.; Stenberg, V. I. *J. Org. Chem.* **1965**, *30*, 3252.
91. Bradshaw, J. S.; Knudsen, R. D.; Loveridge, E. L. *J. Org. Chem.* **1970**, *35*, 1219.
92. Katsuhara, Y.; Maruyama, H.; Shigemitsu, Y.; Odaira, Y. *Tetrahedron Lett.* **1973**, 1323.
93. Kan, R. O., Furey, R. L. *Tetrahedron Lett.* **1966**, 2573.
94. Heerklotz, J. A.; Fu, C.; Linden, A.; Hesse, M. *Helv. Chim. Acta* **2000**, *83*, 1809.
95. Fischer, M. *Tetrahedron Lett.* **1968**, 4295.
96. Fischer, M.; Mattheus, A. *Chem. Ber.* **1969**, *102*, 342.
97. Ishida, S.; Hashida, Y.; Shizuka, H.; Matsui, K. *Bull. Chem. Soc. Jpn.* **1979**, *52*, 1135.
98. Bouchet, P.; Jonceray, G.; Jacquier, R.; Elguero, J. *J. Heterocycl. Chem.* **1978**, *15*, 625.
99. Badr, M. Z. A.; Aly, M. M.; Abdel-Latif, F. F. *J. Org. Chem.* **1979**, *44*, 3244.
100. Staudenmayer, R.; Roberts, T. D. *Tetrahedron Lett.* **1974**, 1141.
101. Gunn, B. C.; Stevens, M. F. G. *J. Chem. Soc., Chem. Commun.* **1972**, 835.
102. Yang, N. C.; Morduchowitz, A. *J. Org. Chem.* **1964**, *29*, 1654.
103. Lenz, G. R. *Synthesis* **1978**, 489.
104. Ogata, Y.; Takagi, K.; Ishino, I. *J. Org. Chem.* **1971**, *36*, 3975.
105. Cleveland, P. G.; Chapman, O. L. *J. Chem. Soc., Chem. Commun.* **1967**, 1064.
106. Chapman, O. L.; Adams, W. R. *J. Am. Chem. Soc.* **1968**, *90*, 2333.
107. Ogata, M.; Matsumoto, H. *Chem. Pharm. Bull.* **1972**, *20*, 2264.

108. Kanaoka, Y.; Nakao, S.; Hatanaka, Y. *Heterocycles* **1976**, *5*, 261.
109. Kanaoka, Y.; Itoh, K.; Hatanaka, Y. Flippen, J. L.; Karla, I. L.; Witkop, B. *J. Org. Chem.* **1975**, *40*, 3001.
110. Tominga, T.; Odaira, Y.; Tsutsumi, S. *Bull. Chem. Soc. Jpn.* **1966**, *39*, 1824.
111. Ninomiya, I; Kiguchi, T.; Yamauchi, S., Naito, T. *J. Chem. Soc., Perkin Trans. I* **1980**, 197.
112. Edward, J. T.; Mo L. Y. S. *J. Heterocycl. Chem.* **1973**, *6*, 1047.
113. Itoh, K.; Kanaoka, Y. *Chem. Pharm. Bull.* **1974**, *22*, 1431.
114. Somei, M.; Natsume, M. *Tetrahedron Lett.* **1973**, 2451.
115. Carruthers, W.; Evans, N. *J. Chem. Soc., Perkin Trans. I* **1974**, 1523.
116. Ghosh, S.; Das, T. K.; Datta, D. B.; Mehta, S. *Tetrahedron Lett.* **1987**, *28*, 4611.
117. Erra-Balsells, R.; Frasca, A. R., *Tetrahedron Lett.* **1984**, *25*, 5363.
118. Bonesi, S. M.; Erra-Balsells, R. *J. Photochem. Photobiol. A: Chem.* **1991**, *56*, 55.
119. Shizuka, H.; Kato, M.; Ochiai, T.; Matsui, K.; Morita, T. *Bull. Chem. Soc. Jpn.* **1970**, *43*, 67.
120. Bonesi, S. M.; Erra-Balsells, R. *J. Photochem. Photobiol.* **1997**, *110*, 271.
121. Zander, M. *Chem. Ber.* **1981**, *114*, 2665.
122. Davis, A. Golden, J. H. *J. Chem. Soc. B* **1968**, 425.
123. Horspool, W. M.; Pauson, P. L. *J. Chem. Soc.* **1965**, 5162.
124. Davis, A.; Golden, J. H.; Mc Rae, J. A.; Symons M. C. R. *J. Chem. Soc., Chem. Commun.* **1967**, 398.
125. Mc Rae, J. A.; Symons, M. C. R. *J. Chem. Soc. B* **1968**, 428.
125. Pac, C.; Tsutsumi, S. *Bull. Chem. Soc. Jpn.* **1964**, *37*, 1392.
126. Caress, E. A.; Rosenberg, I. E. *J. Org. Chem.* **1972**, *37*, 3160.
127. Caress, E. A.; Rosenberg, I. E. *J. Org. Chem.* **1971**, *36*, 769.
128. Trecker, D. J.; Foote, C. S.; Osborn, C. L. *J. Chem. Soc., Chem. Commun.* **1968**, 1034.
129. Bellus, D.; Schaffner, K. *Helv. Chim. Acta* **1968**, *51*, 221.
130. Schwetlick, K.; Noack, R.; Schmieder, G. *Z. Chem.* **1972**, *12*, 107.
131. Schwetlick, K.; Noack, R.; Schmieder, G. *Z. Chem.* **1972**, *12*, 109.
132. Schwetlick, K.; Noack, R.; Schmieder, G. *Z. Chem.* **1972**, *12*, 140.
133. Schwetlick, K.; Noack, R.; Schmieder, G. *Z. Chem.* **1972**, *12*, 143.
134. Schwetlick, K.; Noack, R.; Schmieder, G. *Z. Chem.* **1972**, *12*, 199.
135. Stumpe, J.; Mehlhorn, A.; Schwetlick, *J. Photochem.* **1978**, 1.
136. Schultze, H. *Z. Naturforsch.* **1973**, B28, 339.
137. Hageman, H. J. *Recl. Trav. Chim. Pays-Bas*, **1972**, *91*, 362.
138. Mehlhorn, A.; Schwenzer, B., Schwetlick, K. *Tetrahedron* **1977**, *35*, 63.
139. Masilamani, D.; Hutchins, R. O.; Ohr, J. *J. Org. Chem.* **1976**, *41*, 3687.
140. Herweh, J. E.; Hoyle, C. E. *J. Org. Chem.* **1980**, *45*, 2195.
141. Osawa, Z.; Chen, E.; Ogiwara, Y. *J. Polym. Sci., Polym. Lett. Ed.* **1975**, *13*, 535.
142. Schwetlick, K.; Stumpe, J.; Noack, R. *Tetrahedron* **1979**, *35*, 63.
143. Noack, R.; Schwetlick, K. *Tetrahedron* **1974**, *30*, 3799.
144. Somei, M.; Natsume, M. *Tetrahedron Lett.* **1973**, 2451.
145. Lahti, P. M.; Modarelli, D. A.; Rossitto, F. C.; Inceli, A. L.; Ichimura, A. S.; Ivatury, S. *J. Org. Chem.* **1996**, *61*, 1730.

146. Inoue, T.; Shigemitsu, Y.; Odaira, Y. *J. Chem. Soc., Chem. Commun.* **1972**, 668.

147. Stratenus, J. L.; Havinga, E. *Recl. Trav. Chim. Pays-Bas* **1966**, *85*, 434.

148. Ogata, Y.; Takagi, K.; Yamada, S. *J. Chem. Soc., Perkin Trans.* 2 **1977**, 1629.

149. Andraos, Y.; Barclay, G. G.; Medeiros, D. R.; Baldovi, M. V.; Scaiano, J. C.; Sinta, R. *Chem. Mater.* **1998**, *10*, 1694.

150. Snell, B. K. *J. Chem. Soc. C* **1968**, 2367.

151. Nasielski-Hinkens, Y.; Maeck, J.; Tenvoorde, M. *Tetrahedron* **1972**, *28*, 5025.

152. Nozaki, H. Okada, T.; Noyori, R.; Kawanishi, M. *Tetrahedron* **1966**, *22*, 2177.

153. Chakrabarti, A.; Biswas, G. K.; Chakraborty, D. P. *Tetrahedron* **1989**, *45*, 5059.

154. Kricka, L. J.; Lambert, M. C.; Ledwith, A. *J. Chem. Soc. Chem. Commun.* **1973**, 244.

155. Loveridge, E. L.; Beck, B. R.; Bradshaw, J. S. *J. Org. Chem.* **1971**, *36*, 221.

156. Rungwerth, D.; Schwetlick, K. *Z. Chem.* **1974**, *14*, 17.

157. J. Martens, Praefcke, K. Simon, H. *Z. Naturforsch.* **1976**, *31B*, 1717.

158. Martens, J.; Praefcke, K.; *H. Chem. Zig.* **1978**, *102*, 108.

159. Höhne, G, Lohner, W., Praefcke, K. *Tetrahedron Lett.* **1978**, *7*, 613.

160. Lohner, W.; Martens, J.; Praefcke, K.; Simon, H. *J. Organomet. Chem.* **1978**, *154*, 263.

161. Finnegan, R. A.; Hagen, A. W. *Tetrahedron Lett.* **1963**, 365.

162. Pappas, S. P.; Alexander, J. E.; Long, G. L.; Zehr, R. D. *J. Org. Chem.* **1972**, *37*, 1258.

163. Veierov, D.; Bercovici, T.; Fischer, E.; Mazur, Y.; Yogev, A. *Helv. Chim. Acta* **1975**, *58*, 133.

164. Begley, M. J.; Mellor, M.; Pattenden, G. *J. Chem. Soc. Perkin Trans. 1* **1983**, 1905.

165. Seto, H.; Kosemura, H.; Fujimoto, Y. *J. Chem. Soc., Chem. Commun.* **1992**, 908.

166. Hoffmann, R. W.; Eicken, K. R. *Tetrahedron Lett.* **1968**, 1759.

167. Iida, H.; Aoyagi, S.; Kibayashi, C. *Heterocycles* **1976**, *4*, 697.

168. van Es, J. J. G. S.; Koek, J. H.; Erkelens, C.; Lugtenburg, J. *Recl. Trav. Chim. Pays-Bas* **1986**, *105*, 360.

169. Bochu, C.; Couture, A.; Grandclaudon, P.; Lablanche-Combier, A.; *J. Chem. Soc., Chem. Commun.* **1986**, 839.

170. Couture, A.; Denian, E.; Grandclaudon, P.; Lebrun, S. *Tetrahedron Lett.* **1996**, *37*, 7749.

171. Veierov, D.; Mazur, Y.; Fischer, E. *J. Chem. Soc. Perkin Trans. 2* **1980**, 1659.

172. García, H.; Martinez-Utrilla, R.; Miranda, M. A. *Tetrahedron Lett.* **1980**, *21*, 3925.

173. Alvaro, M.; Baldovi, V.; García, H.; Miranda, M. A.; Primo, J. *Tetrahedron Lett.* **1987**, *28*, 3613.

174. Diaz-Mondejar, M. R.; Miranda, M. A. *Heterocycles* **1984**, *22*, 1125.

175. Dhar, D. N. *The Chemistry of Chalcones and Related Compounds*; Wiley; New York, **1981**.

176. Wagner, H.; Farkas, L. In *The Flavonoids*; Harborne, J. B., ed.; Academic Press; New York, **1975**; p. 138.

177. Bowers, W. S.; Ohta, T.; Cleere, J. S.; Marsella, P. A. *Science* **1976**, *193*, 542.

178. Pratt, G. E.; Bowers, W. S. *Nature* **1977**, *265*, 548.

179. Miranda, M. A.; Primo, J.; Tormos, R. *Heterocycles* **1988**, *27*, 673.

180. Miranda, M. A.; Primo, J.; Tormos, R. *Heterocycles* **1991**, *32*, 1159.
181. Miranda, M. A.; Primo, J.; Tormos, R. *Tetrahedron* **1987**, *43*, 2323.
182. Baldwin, J. E. *J. Chem. Soc., Chem. Commun.* **1976**, 734.
183. García, H.; Iborra, S.; Primo, J.; Miranda, M. A. *J. Org. Chem.* **1986**, *51*, 4432.
184. Alvaro, M.; García, H.; Iborra, S.; Miranda, M. A.; Primo, J. *Tetrahedron* **1987**, *43*, 143.
185. Fariña, F. Martinez-Utrilla, R.; Paredes, M. C. *Tetrahedron* **1982**, *38*, 1531.
186. Diaz-Mondejar, M. R.; Miranda, M. A. *Tetrahedron* **1982**, *38*, 1523.
187. Belled, C.; Miranda, M. A.; Simon-Fuentes, A. *An. Quim. C* **1989**, *85*, 39.
188. Belled, C.; Miranda, M. A.; Simon-Fuentes, A. *An. Quim.* **1990**, *86*, 431.
189. Martinez-Utrilla, R.; Miranda, M. A. *Tetrahedron Lett.* **1980**, *21*, 2281.
190. Fillol, L.; Martinez-Utrilla, R.; Miranda, M. A.; Morera, I. M. *Monatsh. Chem.* **1989**, *120*, 863.
191. Martinez-Utrilla, R.; Miranda, M. A. *Tetrahedron Lett.* **1981**, *37*, 2111.
192. García, H.; Martínez-Utrilla, R.; Miranda, M. A. *Tetrahedron Lett.* **1981**, *22*, 1749.
193. García, H.; Martínez-Utrilla, R., Miranda, M. A.; Roquet-Jalmar, M. F. *J. Chem. Res. (S)* **1982**, 350.
194. Katagiri, N.; Nakano, J.; Kato, T. *J. Chem. Soc., Perkin Trans. 1* **1981**, 2710.
195. Lewis, J. R.; Paul, J. G. *J. Chem. Soc., Perkin Trans. 1* **1981**, 770.
196. Crouse, D. J.; Hurlbut, S. L.; Wheeler, D. M. S. *J. Org. Chem.* **1981**, *46*, 374.
197. Kende, A. S.; Belletire, T. J.; Hume, E.; Airey, J. *J. Am. Chem. Soc.* **1975**, *97*, 4425.
198. Marriot, K.-S. C.; Anderson, M.; Jackson Y. A. *Heterocycles* **2001**, *55*, 91.
199. Ishii, H.; Sekiguchi, F.; Ishikawa, T. *Tetrahedron* **1981**, *37*, 285.
200. Suau, R.; Valpuesta, M.; Torres, G. *Tetrahedron Lett.* **1995**, *36*, 1315.
201. Ban, Y.; Yoshida, K.; Goto, J.; Oishi, T. *J. Am. Chem. Soc.* **1981**, *103*, 6990.
202. Magnus, P.; Lescop, C. *Tetrahedron Lett.* **2001**, *42*, 7193.
203. Okada, K.; Suzuki, R.; Yokota, T. *Biosci. Biotechnol. Biochem.* **1999**, *63*, 257.
204. García, H.; Primo, J.; Miranda, M. A. *Synthesis* **1985**, 901.
205. García, H.; Martinez-Utrilla, R.; Miranda, M. A. *Tetrahedron Lett.* **1985**, *41*, 3131.
206. García, H.; Miranda, M. A.; Roquet-Jalmar, M. F.; Martínez-Utrilla, R. *Liebigs Ann. Chem.* **1982**, 2238.
207. García, H.; Miranda, M. A.; Primo, J. *J. Chem. Rese. (S)* **1986**, 100.
208. García, H.; Iborra, S.; Miranda, M. A.; Primo, J. *Heterocycles* **1986**, *24*, 2511.
209. García, H.; Iborra, S.; Miranda, M. A.; Primo, J. *Heterocycles* **1985**, *23*, 1983.
210. Primo, J.; Tormos, R.; Miranda, M. A. *Heterocycles* **1982**, *19*, 1819.
211. Bellus, D.; Hrdlovic, P.; Manasek, Z. *Polym. Lett.* **1966**, *4*, 1.
212. Humphrey, J. S.; Jr., Schultz, A. R.; Jacquiss, D. B. G. *Macromolecules* **1973**, *6*, 305.
213. Gupta, A.; Liang, R.; Moacanin, J.; Goldbeck, K.; Kliger, D. *Macromolecules* **1980**, *13*, 262.
214. Torikai, A.; Murata, T.; Fueki, K. *Polym. Photochem.* **1984**, *4*, 255.
215. Gupta, M. C.; Tahilyani, G. V. *Colloid Polym. Sci.* **1988**, *266*, 620.
216. Webb, J. D.; Czanderna, A. W. *Macromolecules* **1986**, *19*, 2810.
217. Gupta, A.; Rembaum, A.; Moacanin, J. *Macromolecules* **1978**, *11*, 1285.

218. Factor, A.; Lynch, J. C.; Greenberg, F. H. *J. Polym. Sci. A: Polym. Chem.* **1987**, *25*, 3413.
219. Lemaire, J.; Gardette, A.; Rivaton, A.; Roger, A. *Polym. Degrad. Stab.* **1986**, *15*, 1.
220. Rivaton, A.; Sallet, D.; Lemaire, J. *Polym. Degrad. Stab.* **1986**, *14*, 23.
221. Mullen, P. A.; Searle, N. Z. *J. Appl. Polym. Sci.* **1970**, *14*, 765.
222. Lo, J.; Lee, S. N.; Pearce, E. M. *J. Appl. Polym. Sci.* **1984**, *29*, 35.
223. Beachell, H.; Chang, I. *J. Polym. Sci., Polym. Chem.* **1972**, *10*, 503.
224. Schultze, H. *Makromol. Chem.* **1973**, *172*, 57.
225. Cohen, S. M.; Young, R. H.; Markhart, A. H. *J. Polym. Sci. A* **1971**, *9*, 3263.
226. Rivaton, A.; Sallet, D.; Lemaire, J. *Polym. Degrad. Stab.* **1986**, *14*, 1.
227. Rivaton, A.; Sallet, D.; Lemaire, J. *Polym. Photochem.* **1983**, *3*, 463.
228. Factor, A.; Chu, M. L. *Polym. Degrad. Stab.* **1980**, *2*, 203.
229. Rivaton, A.; Gardette. J.-L. *Angew. Makromol. Chem.* **1998**, *262*, 173.
230. Factor, A.; Ligon, W. V.; May, R. J. *Macromolecules* **1987**, *20*, 2461.
231. Rivaton, A.; Mailhot, B.; Soulestin, J.; Varghese, H.; Gardette, J. L.; *Polym. Degrad. Stab.* **2002**, *75*, 17.
232. Clark, D. T.; Munro, H. S. *Polym. Degrad. Stab.* **1982**, *4*, 441.
233. Hoyle, C. E.; Shah, H.; Nelson, G. L. *J. Polym. Sci. A: Polym. Chem.* **1992**, *30*, 1525.
234. Sapich, B.; Stumpe, J.; Krawinkel, T.; Kricheldorf, H. R. *Macromolecules* **1998**, *31*, 1016.
235. Stumpe, J.; Merterdorf, C. Ringsdorf, H. *Liq. Cryst.* **1991**, *9*, 337.
236. Stumpe, J.; Selbmann, C.; Kreysig, D. *J. Photochem. Photobiol. A: Chem.* **1991**, *58*, 15.
237. Subramanian, P.; Creed, D., Griffin, A. C.; Hoyle, C. E., Venkataram, K. *J. Photochem. Photobiol. A: Chem.* **1991**, *61*, 317.
238. Creed, D., Griffin, A. C.; Hoyle, C. E., Venkataram, K. *J. Am. Chem. Soc.* **1990**, *112*, 4049.
239. Haddleton, D. M.; Creed, D.; Griffin, A. C.; Hoyle, C. E., Venkataram, K. *Macromol. Chem., Rapid Commun.* **1989**, *10*, 391.
240. Creed, D., Griffin, A. C.; Gross, J. R. D.; Hoyle, C. E., Venkataram, K. *Mol. Cryst. Liq. Cryst.* **1988**, *155*, 57.
241. Wilhelm, C.; Rivaton, A.; Gardette, J.-L. *Polymer* **1998**, *39*, 1223.
242. Liaw, D. J.; Lin, S. P.; Liaw, B. Y. *J. Polym. Sci. A: Polym. Chem.* **1999**, *37*, 1331.
243. Carlsson, D. J.; Gan, L. H.; Wiles, D. M. *Can. J. Chem.* **1975**, *53*, 2337.
244. Li, S. K. L.; Guillet, J. E. *Macromolecules*, **1977**, *10*, 840.
245. Bellus, D.; Slama, P.; Hrdlovic, Manasek. Z. Durisinova, L. *J. Polym. Sci. C* **1969**, *22*, 629.
246. Merle-Aubry, L., Holden, D. A.; Merle, J.; Guillet, J. E. *Macromolecules* **1980**, *13*, 1138.
247. Stumpe, J; Zaplo, O.; Kreysig, D.; Niemann, M.; Ritter, H. *Makromol. Chem.* **1992**, *193*, 1567.
248. Tessier, T. G.; Frechet, J. M. J.; Willson, C. G.; Ito, H. *ACS Symp. Ser.* **1984**, *266*, 269.

249. Allen, N. S. *Polym. Photochem.* **1983**, *3*, 167.
250. Pinazzi, C. P.; Fernandez, A. *ACS Symp. Ser.* **1976**, 25.
251. Lin, M. S.; Lee, C. T. *Abstr. Papers of ACS*, **2001**, *221*, 298.
252. Ramamurthy, V., ed. *Photochemistry of Organized and Constrained Media*, VCH: New York, 1991.
253. Ramamurthy, V.; Weiss, R. G.; Hammond, G. S. In *Advances in Photochemistry*; Volman, D. H.; Neckers, D., Hammond, G. S., eds.; Wiley–Interscience: New York, **1993**; Vol. 18, p. 67.
254. Ramamurthy, V.; Weiss, R. G.; Hammond, G. S. *Acc. Chem. Res.* **1993**, *26*, 530.
255. Balzani, V.; Scandola, F. *Supramolecular Photochemistry*; Ellis Horwood: Chichester, **1991**.
256. Bender, M. L.; Komiyama, M. *Cyclodextrin Chemistry*; Springer-Verlag: Berlin, **1978**.
257. Ohara, M.; Watanabe, K. *Angew. Chem. Int. Ed. Engl.* **1975**, *14*, 820.
258. Chenevert, R.; Voyer, N. *Tetrahedron Lett.* **1984**, *25*, 5007.
259. Syamala, M. S.; Rao, B. N.; Ramamurthy, V. *Tetrahedron* **1988**, *44*, 7234.
260. Veglia, A. V.; Sanchez, A. M.; de Rossi, R. H. *J. Org. Chem.* **1990**, *55*, 4083.
261. Veglia, A. V.; de Rossi, R. H. *J. Org. Chem.* **1993**, *58*, 4941.
262. Xie, R.-Q.; Liu, Y.-C.; Lei, X.-G. *Res. Chem. Intermed.* **1992**, *18*, 61.
263. Banu, H. S.; Pitchumani, K.; Srinivasan, C. *Tetrahedron* **1999**, *55*, 9601.
264. Pitchumani, K.; Velusamy, P.; Manickam, M. C.; Durai, M. C.; Srinivasan, C. *Proc. Indian Acad. Sci., Chem. Sci.* **1994**, *106*, 49.
265. Pitchumani, K.; Manickam, M. C.; Durai, M. C.; Srinivasan, C., *Indian J. Chem. B*, **1993**, *32B*, 1074.
266. Pitchumani, K.; Durai, M. C.; Srinivasan, *Tetrahedron Lett.* **1991**, *32*, 2975.
267. Nassetta, M.; de Rossi, R. H.; Cosa, J. J. *Can. J. Chem.* **1988**, *66*, 2794.
268. Chenevert, R.; Plante, V. *Can. J. Chem.* **1983**, *61*, 1092.
269. Singh, A. K.; Raghuraman, T. S. *Tetrahedron Lett.* **1985**, *26*, 4125.
270. Suau, R.; Torres, G.; Valpuesta, M. *Tetrahedron Lett.* **1995**, *36*, 1311.
271. Singh, A. K.; Raghuraman, T. S. *Synth. Commun.* **1986**, *16*, 485.
272. Nowakowska, M.; Storsberg, J.; Zapotoczny, S.; Guillet, J. E. *New J. Chem.* **1999**, *23*, 617.
273. van Bekkum, H.; Flanigen, E. M.; Jansen, J. C. *Introduction to Zeolite Science and Practice*; Elsevier: Amsterdam, **1991**.
274. Meier, W. M.; Olsen, D. H. *Atlas of Zeolites Structure Types*; Butterworths, London, **1988**.
275. Avnir, D.; de Mayo, P.; Ono, I. *J. Chem. Soc., Chem. Commun.* **1978**, 1109.
276. Abdel-Malik, M. M.; de Mayo, P. *Can. J. Chem.* **1984**, *62*, 1275.
277. Pitchumani, K.; Warrier, M.; Ramamurthy, V. *J. Am. Chem. Soc.* **1996**, *118*, 9428.
278. Pitchumani, K.; Warrier, M.; Ramamurthy, V. *Res. Chem. Intermed* **1999**, *25*, 623.
279. Pitchumani, K.; Warrier, M.; Cui, C.; Weiss, R. G.; Ramamurthy, V. *Tetrahedron Lett.* **1996**, *37*, 6251.
280. Balkus, J. K.; Khanmamedova, A. K.; Woo, R. *J. Mol. Catal. A: Chem.* **1998**, *134*, 137.
281. Tung, C.-H.; Ying, Y.-M.; *J. Chem. Soc. Perkin. Trans. 2* **1997**, 1319.

282. Tung, C.-H.; Wu, L.-Z.; Zhang, L.-P.; Li, H.-Ru.; Yi, X.-Y.; Song, K.; Xu, M.; Yuan, Z.-Y.; Guan, J.-Q.; Wang, H.-W.; Ying, Y.-Ming; Xu, X.-H. *Pure Appl. Chem.* **2000**, *72*, 2289.

283. Gu, W.; Warrier, M.; Ramamurthy, V.; Weiss, R. G. *J. Am. Chem. Soc.* **1999**, *121*, 9467.

284. Cundy, C. S.; Higgins, R.; Kibby, S. A. M.; Lowe, B. M.; Paton, R. M. *Tetrahedron Lett.* **1989**, *30*, 2281.

285. Tung, C.-H., Xu, X.-H. *Tetrahedron Lett.* **1999**, *40*, 127.

286. Weiss, R. G.; Cui, C. *J. Am. Chem. Soc.* **1993**, *115*, 9820.

287. Cui, C.; Naciri, J.; He, Z.; Jenkins, R. M.; Lu, L.; Ramesh, V.; Hammond, G. S., Weiss, R. G. *Quim. Nova* **1993**, *16*, 578.

288. Gu, W.; Hill, A. J.; Wang, X.; Cui, C.; Weiss, R. G. *Macromolecules* **2000**, *33*, 7801.

289. Gu, W.; Weiss, R. G. *J. Org. Chem.* **2001**, *66*, 1775.

290. Gu, W.; Abdallah, D. J.; Weiss, R. G. *J. Photochem. Photobiol, A: Chem.* **2001**, *139*, 79.

291. Gu, W.; Weiss, R. G. *Tetrahedron*, **2000**, *56*, 6913.

292. Gu, W.; Weiss, R. G. *J. Photochem. Photobiol. C: Photochem. Rev.* **2001**, *2*, 117.

293. Wang, Z.; Holden, D. A.; McCourt, F. R. W. *Macromolecues* **1990**, *23*, 3773.

294. Andrew, D.; Des Islet, B. T.; Margaritis, A.; Weedon, A. C. *J. Am. Chem. Soc.* **1995**, *117*, 6132.

295. Meallier, P. *Handbook of Environmental Chemistry*; Boule, P., ed.; Springer-Verlaq: Berlin, **1999**; Vol. 2.

296. Sturini, M.; Fasani, E.; Prandi, C.; Albini, A. *Chemosphere* **1997**, *35*, 931.

297. Sanjuan, A.; Aguirre, G.; Alvaro, M.; García, H.; Scaiano, J. C. *Appl. Catal. B: Environ.* **2000**, *25*, 257.

298. Climent, M. J.; Miranda, M. A. *J. Chromatogr. A*, **1996**, *738*, 225.

299. Sortino, S.; Giuffrida, S.; De Guidi, G; Chillemi, R.; Petralia, S.; Marconi, G.; Condorelli, G.; Sciuto, S. *Photochem. Photobiol.* **2001**, *73*. 6.

300. Castell, J. V.; Gómez-L., M. J.; Mirabet, V.; Miranda, M. A. Morera, I. M. *J. Pharm. Sci.* **1987**, *76*, 374.

3

The Characterization and Reactivity of Photochemically Generated Phenylene Bis(diradical) Species as Revealed by Matrix Isolation Spectroscopy and Computational Chemistry

Athanassios Nicolaides
University of Cyprus, Nicosia, Cyprus

Hideo Tomioka
Mie University, Mie, Japan

I. INTRODUCTION

Reactive intermediates are species of fleeting existence which appear during the course of chemical reactions. Because they determine the outcome of a reaction, understanding their properties is inextricably linked to our knowledge of chemical transformations. Their ephemeral existence has been a major obstacle to their study and the driving force for the development of specialized techniques.

Time-resolved laser experiments make possible the detection of species with short lifetimes [1–6]. Usually, such experiments are carried out under "normal" conditions and a quick enough snapshot can capture the intermediate in action. A different philosophy is to prolong the lifetime of the reactive species by generating it in an inert environment at low temperature. Matrix isolation techniques coupled with some conventional spectroscopic method are used to characterize reactive species and to explore their reactivity [7–14].

Although the experimental methods have advanced impressively in "handling" highly reactive species, the data that have been collected would be difficult to process without the help of computational chemistry. Nowadays, advances in both hardware and software have popularized the application of computational methods to molecular systems [15,16]. Computational data are not as exact as the experimental ones, but they are accurate enough to expedite, confirm, and generally aid in the interpretation of the available experimental information.

The combination of specialized experimental techniques and computational tools has been applied amply in the study of (mono)carbenes and (mono)nitrenes. Thus, many aspects of their rich chemistry are now fairly well understood [17–25]. On the other hand, the properties of polycarbenes and polynitrenes are not necessarily the simple sum of their subunits. Thus, arylcarbenes and arylnitrenes have been used as building blocks for the synthesis of organic polycarbenes and polynitrenes with impressively high-spin ground states [26], which could potentially give rise to organic magnetic materials. Although the realization of this goal seems to be quite far, study along these lines has revealed the importance of understanding spin control in such systems, a topic closely related to the electronic structure of molecules.

Phenylene-linked bis(diradicals), where the diradical is a carbene or a nitrene, are relatively simple and experimentally accessible and it is possible to vary the carbene site by chemical substitution. Thus, such compounds are reasonable models for the spin–spin interactions across conjugated systems and, in particular, across aromatic rings. This chapter is devoted in the recent efforts to generate and study such systems in order to progress our understanding of more complex cases.

II. GENERAL ASPECTS

A. Theoretical Framework

One important feature of carbenes is the presence of two nonbonding electrons and two available orbitals, which are nominally located on the carbon atom. In bent carbenes, the two orbitals have different energies and are often denoted as σ and π, with σ being the in-plane orbital and π the out-of-plane orbital. Within this simple picture, four electronic states can be envisioned, which are depicted in Fig. 1a. Singlets S_1 and S_3 have the two electrons in the same orbital (σ and π, respectively) and are often characterized as closed-shell singlets. On the other hand, singlet S_2 has the same open-shell electronic occupation ($\sigma\pi$) as the triplet (T) state.

According to Fig. 1a, state S_1 is likely to be the lowest-energy singlet, because both electrons lie in the lower-energy σ orbital. However, the relative

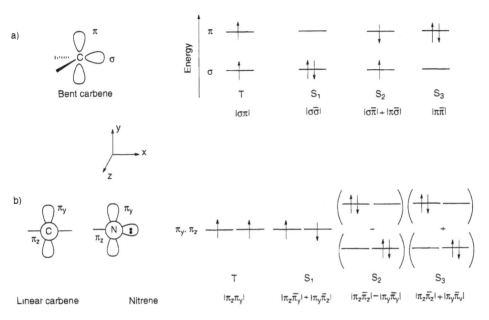

Figure 1 Electronic states of diradical species.

order of S_1 and T, and, therefore, the ground-state multiplicity cannot be predicted with confidence from such a simple diagram. In terms of orbital energies, T is expected to be higher in energy than S_1, but in the former state, better electron correlation may reduce its electron repulsion sufficiently to make it the ground state. This is the case with the parent compound, methylene (**1**), which has a triplet ground state and a T–S gap of 9.0 kcal/mol [27–29].

In a linear carbene with cylindrical symmetry (C_{∞} or $D_{\infty h}$), instead of one σ and one π, there are two π orbitals. These are shown as π_y and π_z in Fig. 1b. The π_z orbital is perpendicular to the plane of the paper and only its front lobe is shown. By symmetry, π_y and π_z are degenerate in energy, and, thus, Hund's rule predicts a triplet ground state. In this particular case, the S_1 and S_2 states are degenerate in energy, whereas S_3 lies higher.

A similar type of analysis is applicable to nitrenes. If the nitrene has cylindrical symmetry ($C_{\infty v}$), then, assuming sp hybridization, the nitrogen lone pair and its bonding electron pair can be placed each in one of the two hybrids. This leaves two degenerate π orbitals for the nonbonding electrons and, therefore, a high-spin ground state is expected. In agreement with this description, the parent

Structures 1–7

nitrene **2** has a triplet ground state and a ΔE_{ST} of 36.0 kcal/mol [30] (Structures 1–7).

Phenylcarbene (**3**) [31–36] and phenylnitrene (**4**) [37–43], although more complex, can still be understood qualitatively by using the above simple picture. Both species are well studied and their similarities and differences [31,44,45], as well as some substituent effects [36,46,47] on their electronic structures, have been examined. Both **3** and **4** have triplet ground states, but although the $T–S$ splitting of the former is approximately 4 kcal/mol, that of the latter is 18 kcal/mol. Thus, in both cases, the phenyl group stabilizes the singlet with respect to the triplet (compared to the parent compounds **1** and **2**), but the stabilization is much more significant in the case of phenylnitrene. Another difference is that the lowest singlet ($^1A'$) of **3** is closed shell (S_1 of Fig. 1a), but of **4** ($^1A_2$ in C_{2v} symmetry or $^1A''$ in C_s), it is open shell (S_1 of Fig. 1b) [40]. Perhaps it should be added that in phenylnitrene, the lack of cylindrical symmetry lifts the degeneracy between states S_1 and S_2 (Fig. 1b), and the former becomes more stable [40].

Although phenylcarbene has a triplet ground state, replacement of the carbenic hydrogen by a halogen like F, Cl, or Br (**5**, **6**, and **7**, respectively) switches the ground-state multiplicity to a singlet. This switching of electronic configurations can be attributed to the higher electronegativity of the halogen, which inductively stabilizes the σ-orbital of the carbene, and/or to the π-donating ability of the lone pair of the halogen, which destabilizes the π-orbital of the carbene. Both effects act in the same direction, increasing the $\sigma–\pi$ separation and, thus, stabilizing the singlet with respect to the triplet state. Although the $T–S$ splittings in the halogen-substituted cases are not as well established as that for the parent **3**, computed estimates of -19, -7, and -7 kcal/mol for **5**, **6**, and **7**, respectively, have been reported (Table 1) [48].

The number of important electronic states increases significantly when two diradicals are present in a molecule, as is the case with dicarbenes, dinitrenes, and carbenonitrenes. Treating these molecules as four-electrons-in-four-orbitals cases, there are 70 different possible electronic configurations which give rise to 36 electronic states [50]. However, not all of these states need be considered. Half of these are of the "donor-acceptor" or "ionic" type, corresponding to

Table 1 Singlet–triplet Splittings (ΔE_{ST}, kcal/mol) of
Diradicals (**4**, **5**, **6**, and **7**) at Various Levels of Theory[a]

	B3LYP	MCSCF	CASPT2	E_{QCI}[b]
4		17.8	19.4	
5[c]	−16.3	−18.3	−19.2	−17.4
6[c]	−7.6	−6.8	−6.7	−8.6
7[c]	−7.8	−6.5	−7.4	−7.3

[a] With the 6-31G(d) basis set.
[b] $E_{QCI} = E_{QCISD(T)/6-31G(d)} + E_{RMP2/6-311-G(3d1,2p)} - E_{RMP2/6-31G(d)}$. For details, see Ref. 49.
[c] Corrected values based on the assumption that $\Delta E_{ST}(3) = 4$ kcal/mol. For details, see Ref. 48.

formal electron transfer from one site to the other. These states are expected be of higher energy than the "neutral" or "covalent" ones. From the "neutral" states which are left (1 quintet, 7 triplets and 10 singlets), 7 states (2 triplets and 5 singlets) contain configurations that are locally doubly excited [two electrons in the π-orbital in the case of bent carbenes (S_3 of Fig. 1a)] and should also be of high-energy content [25]. Still, 11 states are left (1 quintet, 5 triplets, and 5 singlets), which may need to be considered. This variety in potentially accessible states, of different multiplicities and/or electronic configurations, is what makes the study of these systems difficult and attractive at the same time. An obvious challenge is the chemical design of systems with specific ground-state electronic configurations. To approach this in a rational way, a conceptual framework, which can predict (at least qualitatively) the electronic configuration of a target system based on its building blocks, is needed.

The rather wide range of $T–S$ splittings (ranging from +19 to −19 kcal/mol) on going from phenylnitrene (**4**) to phenylcarbene (**3**) and its halogen analogs (**5–7**) (Table 1) suggests that these can be used as basic units for constructing a variety of phenylene-linked bis(diradicals). From the systematic study of such model systems, one hopes to understand something about spin–spin interactions in general, and the factors that control the electronic configuration of the ground state in particular. In Section III, several of these model systems will be detailed. The diradical sites (or centers) are sometimes referred to as "subunits" and the phenylene group as "linker." As will be seen, both the nature of the diradical site (nitrene, carbene, or halocarbene) and the topology (connectivity) of the linker (ortho, meta, or para) are important in determining the properties of the system.

B. Computational Procedures

Quantum chemical calculations have become a valuable tool in the research of reactive intermediates. Unfortunately, there is no unique computational method that can be uniformly applied in all cases and give an accurate answer at a practical cost. A variety of computational methods are available, each with its own weaknesses and advantages. The species that are of interest in this chapter and which often have unpaired electrons, pose specific problems in calculating their properties, and some care in choosing appropriate methods is necessary for obtaining meaningful results. In this respect, an excellent guide for calculations on open-shell molecules has been recently published [51].

In the study of biscarbenes and related species, density functional theory (DFT) [52,53] and so-called multiconfigurational (MC) [54] methods have been widely used.

Among DFT methods, B3LYP [55,56] in conjunction with the 6-31G(d) basis set [57] is very popular. One of the criticisms against B3LYP, and DFT methods in general, is that there is no way for improving systematically any single computational result. As a consequence, there is a relatively high uncertainty as to the quality and the reliability of a given result. On the other hand, such calculations are seldom carried out in an absolute sense without some point of reference.

For the systems under consideration, DFT has been widely applied in calculating vibrational frequencies. The computed frequencies are then compared (directly or after scaling by some appropriate factor) [58] with experimental infrared (IR) spectra of the observed species for identification purposes. The empirical finding, that B3LYP and related methods reproduce the observed IR patterns reasonably well, is the main justification for their use and what has made them so popular [59]. In addition, DFT methods can be applied to "large" molecules (50–100 heavy atoms), at a computational cost that is slightly higher than that of Hartree–Fock (HF) level of theory [16]. Unrestricted HF (UHF) wave functions in open-shell species often suffer from artifacts related to spin contamination, whereas unrestricted DFT methods seem to be less prone to this kind of problem. UB3LYP calculations have performed well also in the assignment of hyperfine couplings in electron spin resonance (ESR) spectra [51].

One of the drawbacks of DFT is its inability to treat multiconfigurational problems properly. In these cases, some type of configuration interaction (CI) method is needed [60]. Ideally, one would like to carry out full CI calculations, but this is generally not possible for practical reasons. Thus, some kind of approximation is needed and, in this respect, the so-called complete active self-consistent field (CASSCF) procedure is often used.

The CASSCF procedure (a special case of MC methods) is a full CI calculation among all possible configuration state functions (CSFs), which arise from the distribution of a certain number of electrons (active electrons) in a certain number of occupied and virtual molecular orbitals (active orbitals). Obviously, the number of CSFs rises rapidly based on the size of active space and this soon becomes a serious limitation of the method. Because of this limitation, the choice of the active space [i.e., which electrons and which molecular orbitals (MOs) are to be included in the CI part of the CASSCF calculation] is difficult and can be problematic. Furthermore, this type of calculation requires considerable effort in preparing the necessary input and the use of some extra program (to visualize MOs) is extremely helpful. Despite these difficulties, CASSCF is the only reasonable way for treating multiconfigurational problems and modeling excited states in a uniform and consistent way.

The CASSCF wave functions are eigenfunctions of the total spin operator S^2 and, so, do not suffer from spin-contamination problems. On the other hand, CASSCF recovers only a small part of the correlation energy. This is a problem when dealing with excited states, as their relative energies often depend on the so-called dynamic correlation. This shortcoming has been improved by the application of many-body perturbation theory (MBPT) to such wave functions. The first such implementation resulted in a procedure called CASPT2 [61], which has turned out to be very successful in the balanced treatment of excited states. Due to a lack of analytical gradients, CASPT2 is not generally recommended for geometry optimizations, but is usually used for single-point energy calculations.

As already mentioned, choosing the active space for CASSCF calculations is not always a trivial matter. In the systems under consideration, there is a plane of symmetry (that of the phenylene linker), which helps in classifying the MOs as σ and π (or A' and A'', using group-theory notation). Experience shows that a reasonably balanced active space is made of the π system of the linker and one σ-orbital and one π-orbital per reactive site (carbene or nitrene) (Fig. 2).

Although CASSCF is formally better suited for treating multiconfigurational cases, there is a class of singlet biraricals where DFT can still give good results. In general, singlet biradicals require two Slater determinants for their

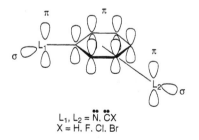

Figure 2   Active space for phenylene-linked carbenes and nitrenes.

proper description [62]. Within the UHF formalism, as applied to DFT calculations (UDFT), one can still obtain "singlet" wave functions by carrying out symmetry-broken calculations. Essentially, this amounts to using only one of the two Slater determinants for the calculation and the resulting wave function is highly spin contaminated with a $\langle S^2 \rangle$ value close to unity instead of zero. This suggests that the wave function is approximately a 50:50 mixture of singlet and triplet states and devoid of any physical meaning. However, when the singlet–triplet splitting is very small (as is the case with the quinonoid radicals to be described later), this spin-contaminated singlet seems to be a reasonable approximation to the true singlet. Thus, properties like the geometry and IR spectrum of the singlet state can be reasonably approximated by computing the corresponding ones for the triplet.

Interestingly, despite their formal shortcomings, UDFT calculations seem to give the correct order of states for weakly interacting biradicals. Thus, in the quinonoid radicals to be described in Sections III.A.2 and III.C.2, UDFT methods correctly predict that the singlet is energetically below the triplet state. The electronic component of the $T–S$ gap is the energy difference between singlet and triplet states. However, as mentioned earlier, in this particular case the UDFT "singlet" is more like an "equimolar" mixture of singlet and triplet wave functions. In this respect, the "raw" $T–S$ gap computed with UDFT methods is likely to be underestimated by roughly a factor of 2. It is not uncommon to apply this empirical correction when reporting results at this level of theory [63]. However, UDFT calculations on singlet biradicals are likely to give reasonable results only when the $T–S$ gap is small and the odd electrons are localized at distant orbitals.

Overall, DFT methods are user-friendly and helpful in identifying reactive species based on their IR features. For relative energies of the various electronic states, CASSCF and, in particular, CASPT2 calculations are recommended. Especially the latter method is expected to be semiquantitatively correct. It should be noted, though, that in several cases, DFT relative energies come surprisingly close to the CASPT2 results.

C. Experimental Considerations

With the exception of thermodynamically stabilized [64] or sterically protected [65] carbenes, these species and their hetero-analogs, nitrenes, are very reactive and therefore special conditions are required for their direct observation. Fast spectroscopic techniques capable of characterizing species with lifetimes of a few picoseconds have been used [1–3]. More recently, time-resolved IR (TRIR) experiments have been used to characterize species with lifetimes of microseconds and even nanoseconds [4–6].

An alternative approach in the study of reactive intermediates is to isolate the species in an inert matrix at low temperature [10–12]. This seems to be the most convenient method for the direct observation of biscarbenes and related species, which we are going to deal with in this chapter. Matrix isolation coupled with some conventional spectroscopic method (usually ESR, UV, and IR) [7–9,25] is used to detect and characterize the species of interest. There are several advantages in studying the chemical behavior of reactive species under matrix-isolation conditions [13,14]. Thus, the inert material of the matrix essentially shuts down any intermolecular reactions, giving the opportunity to examine the intrinsic reactivity of the reactive intermediate. Under such conditions, the photochemistry of reactive intermediates can be studied rather easily, whereas in solution, two-laser two-color methods are required to achieve something similar. By annealing the matrix, it is possible to see if low-barrier isomerizations are available to the investigated species. Doping the matrix with some reagent and/ or using less inert material for the matrix (e.g., oxygen, methane) gives insight into several intermolecular chemical pathways that may be available. Finally, molecules in an inert matrix can be essentially considered as isolated, making the comparison of experimental and computational data more straightforward.

Synthetically, carbenes and nitrenes are accessible by the photochemical decomposition of appropriate nitrogenous precursors. Commonly diazo [66] and diazirine [67] groups are utilized for the generation of carbene centers and azido [68] groups for nitrenes. In the matrix-isolation experiments, the precursor is deposited in high dilution in an inert matrix (typical examples are Ar and N_2) and irradiated. The photoproduct can then be characterized by UV/vis [8,9] IR [11,25], and ESR [7–9] spectroscopy. Although UV/vis spectroscopy can be used in monitoring the reaction and giving some general information about the species involved, most of the useful structural information usually comes from the IR and ESR data.

With ESR spectroscopy, open-shell species can be observed and characterized as long as their total spin differs from zero. With variable-temperature ESR spectroscopy, it is possible to deduce whether the observed multiplicity is a thermally populated excited state or is the ground state [69]. From such experiments, the $T–S$ splittings of a variety of biscarbene and bisnitrenes have been determined. ESR spectroscopy is very sensitive to paramagnetic species, and because it does not "see" any singlet impurities or by-products, it is relatively easy to pick out the desired signals. At the same time, analysis of ESR spectra is not trivial and special simulations are required for their interpretation.

Infrared spectroscopical data encode a lot of structural information and can be analyzed with the help of computational methods (*vide supra*) aiding in the identification of the observed species. Sometimes, two different electronic states may lie very close in energy and have similar geometries. In such cases (e.g., the quinonoid radicals to be described in Section II.B.), the predicted differences in the IR spectra are too small to allow an unambiguous assignment of the ground-state multiplicity. In this respect, ESR spectroscopy provides valuable complementary information. On the other hand, there are examples where different electronic states have sufficiently different IR spectra, making assignment possible.

Often, the observed species may undergo rearrangement either thermally or photochemically. The end products of such isomerizations tend to have singlet ground states, rendering IR spectroscopy the most powerful tool for elucidating these processes. If the end product is the result of a thermal reaction, then due to the low temperature, not many pathways are available. This facilitates identification and also allows for the possibility of carrying out kinetic measurements.

Kinetics in polycrystals differ from those in solution phase, because in the former, the thermal reactions usually follow a nonexponential rate law, something that is attributed to a multiple-site problem. In contrast to a first-order reaction in solution, the rate constant of a nonexponential process in the solid state is time dependent; molecules located in the reactive site will have decayed during the warmup procedure and/or the initial stage of the reaction at the given temperature. These considerations need to be taken into account when the decay of the intensity of the IR signals in a matrix at low temperature are used for kinetic measurements [70].

The products of photochemical rearrangements are occasionally quite different from what one may intuitively expect and this creates difficulty in their identification. In such cases, computational chemistry is perhaps our only resort. Several possible structures can be screened computationally with rather little cost in terms of time, effort, and money. Unfortunately, computational chemistry cannot predict a priori the structure of the unknown. However, if a good match between theoretically derived and experimental IR spectrum is found, then this constitutes a strong case and often is taken as proof of identification.

III. PHENYLENE-LINKED SYSTEMS

In the following subsections, T–S splittings and ΔE_{ST}'s imply energy differences in the sense: $E_S - E_T$. Thus, positive values indicate that the triplet is lower in energy than the singlet. The same convention is used for other splittings (e.g., $Q - T$ and ΔE_{TQ} stand for $E_T - E_Q$, etc.). Unless otherwise noted, DFT results imply the B3LYP/6-31G(d) method. Finally, the nitrogenous precursors are numbered and followed by one or more code letters, depending on what groups are present in the molecule: N for azido (as, for example, 18-N of Scheme 3), D for diazo (and R for diazirine). A bisazido compound is noted by N_2 (as, for example, 13-N_2 of Scheme 2), a diazo-azido by DN (17-DN of Scheme 3), and so on.

A. 1,4-Phenylene Linker

1. General Remarks

As mentioned in Section II.A, several possible electronic states can arise from the presence of two diradical sites (carbenes and/or nitrenes) in a molecule. When the two diradical centers are coupled via 1,4-phenylene (*p*-phenylene), then some of the resulting states are depicted in Fig. 3. Within this simple approximation, a minimal basis of two orbitals per radical center and the π system of the benzene ring are considered. The orbitals on the substituents are classified as σ and π, with the former lying in the plane of the benzene ring and the latter perpendicular to it. In terms of electrons, each diradical center (be it a carbene or a nitrene) contributes two electrons and the phenylene unit contributes six electrons. In CASSCF parlance, one would say that the active space is made of 10 electrons in 10 orbitals (8 π and 2 σ MOs).

There are several ways of classifying the resulting states:

1. Overall multiplicity (singlet, triplet, or quintet).
2. "Local" multiplicity of each diradical center (singlet or triplet as approximated by the VB structures of Fig. 3).
3. Number of σ and π electrons contributed by each diradical center. For example, $\sigma^2/\sigma\pi$ means that one center has a σ^2 closed-shell singlet "local" configuration and the other has $\sigma\pi$ occupancy. The latter may correspond to triplet or to open-shell singlet "local" configuration.
4. A' or A'' depending on the number of π-electrons (even or odd, respectively).

The quintet state (Q) arises when each diradical center is "locally" in a triplet state with high-spin coupling between the centers (Fig. 3). On the other hand, there are several ways of forming a singlet state. The two $^1A'$ states (S_1 and S_4) shown in Fig. 3 differ in that, in S_1, there are two σ active electrons, but four in S_4. In S_1, it is expected that the two π electrons of L_1 and L_2 along with the six π-electrons of the benzene ring will form a (para) quinonoid system with localized double bonds. In contrast, in S_4, the benzene π system should be much less perturbed and the whole system should resemble a para-substituted benzene ring with the expected bond delocalization inside the ring. The two A'' singlets (S_2 and S_3) of Fig. 3 have one diradical center in a σ^2 "local" configuration.

The electronic occupancies of the triplet states shown in Fig. 3 are similar to those of the singlets. Thus, T_1 represents a quinonoid system for the same reason as S_1 does. The two states S_1 and T_1 differ in the spin coupling of the two depicted σ electrons. Triplet states T_2 and T_3 have also one σ^2 center like the A'' singlets.

From the simplistic pictures of Fig. 3, maximum bonding occurs for the quinonoid biradicals S_1 and T_1, although this is achieved with some loss of aromaticity. Both states are expected to be very close in energy, and deciding which one is lower is not easy. With symmetrical substituents ($L_1 = L_2$), the two odd electrons of S_1 and T_1 lie in orbitals that are essentially degenerate. Thus, based on Hund's rule, T_1 is expected to be below S_1. However, in this case, the MOs

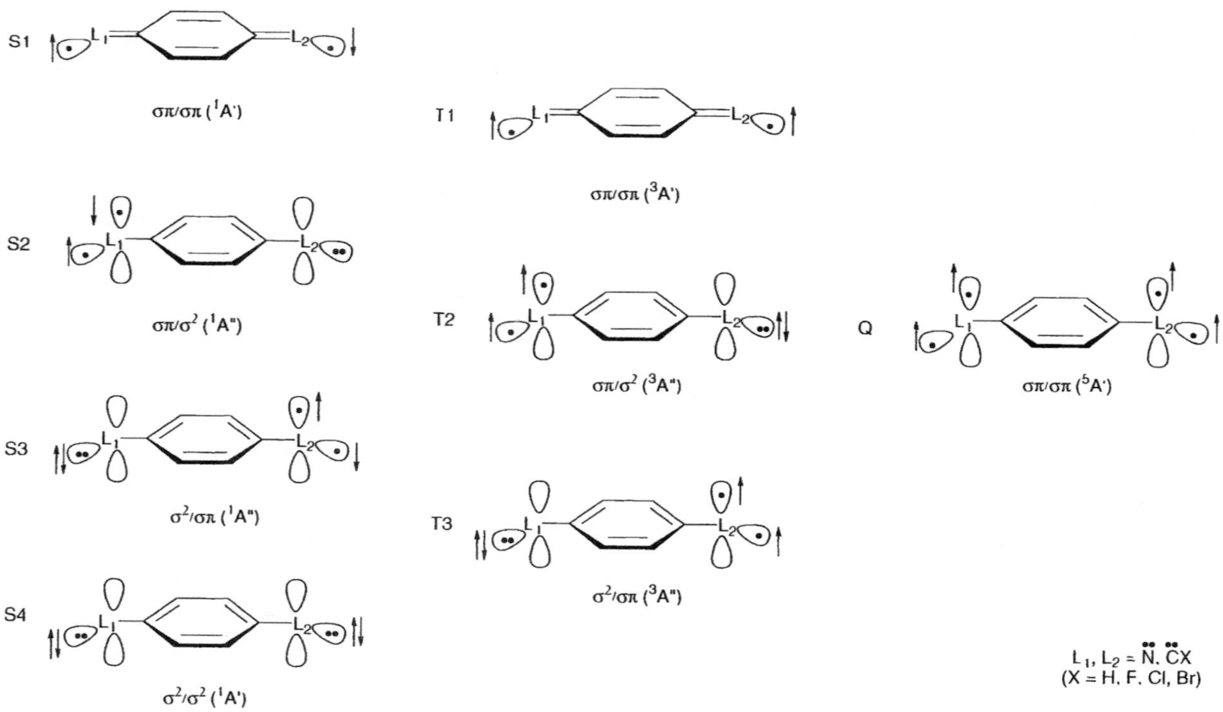

Figure 3 Valence-bond depictions of electronic states arising from the coupling of two reactive centers (carbene, nitrene) via the 1.4-phenylene linker.

are (spatially) disjoint and electronic polarization will favor the singlet more than the triplet, perhaps making $S1$ lower in energy than $T1$. Spin-polarization arguments based on valence-bond considerations clearly predict that $S1$ is below $T1$, in agreement with computational results. Thus, these molecules can be thought of as formal violations of Hund's rule [71]. When L_1 and L_2 are different, Hund's rule is not applicable any more and the low-spin state should be preferred.

In the case of carbenes, the σ^2 local configuration may be intrinsically preferred to the $\sigma\pi$ one, if there is a π-donor (such as F, Cl or Br) directly attached to it. In this respect, the relative energies of the A'' (singlet or triplet) and the $S4$ states (Fig. 3) are likely to be significantly affected by chemical substitution at the carbene center. If the perturbation is sufficiently large, it is not unreasonable to expect that these states may compete effectively in terms of energy with the quinonoid biradicals described earlier. Thus, by tuning the "local" electronic configurations of the subunits, it may be possible to manipulate the configuration of the ground state.

2. 1,4-Phenylene-Linked Systems

The formation of the parent system p-phenylenebismethylene (**8**; Scheme 1) was first attempted in the gas phase from the pyrolysis of ^{13}C-labeled 1,4-bis(5-tetrazo-lyl)benzene. Under such conditions, it was not possible to detect the intermediate directly and specify it in detail, but its formation was deduced from the product analysis [72]. In 1998, though, irradiation of the bisdiazo precursor **8**-D$_2$ made possible the characterization of **8** by IR and UV/vis spectroscopy [73]. The identification was based on trapping experiments with HCl (to form **9**) and oxygen (Scheme 1) and by simulating the IR spectrum of **8** [UB3LYP/6-31G(d,p)] [73].

The ground state of **8** is a singlet of A' symmetry with localized double bonds ($S1$ in Fig. 3). Thus, this species is best thought of as a quinonoidal biradical. This is an example where application of DFT theory needs to be carried out within the UHF formalism (UDFT) and the resulting singlet state is highly spin contaminated by the corresponding triplet wave function (see Sect. II.B). The lowest-lying excited state is computed to be the $^3A'$ one (about 2 kcal/mol higher in energy) [63,73,74], as might have been expected. The quintet state lies about 27 kcal/mol above the $^1A'$ (Table 2). The good agreement between approaches so different as DFT and CASSCF inspires certain confidence in these computational results. On the other hand, it is difficult to pinpoint the relative energy of the $^3A''$ state, because the answers of the two approaches differ by about 8 kcal/mol. It is likely that the truth is somewhere in between because B3LYP is likely to underestimate the $^1A' - {}^3A''$ energy difference and CASPT2 to overestimate it. However, more sophisticated calculations will be required before this issue is resolved. Finally the σ^2/σ^2 singlet is placed by DFT calculations about 31 kcal/mol above the ground state, a value that is likely to be an underestimate [63].

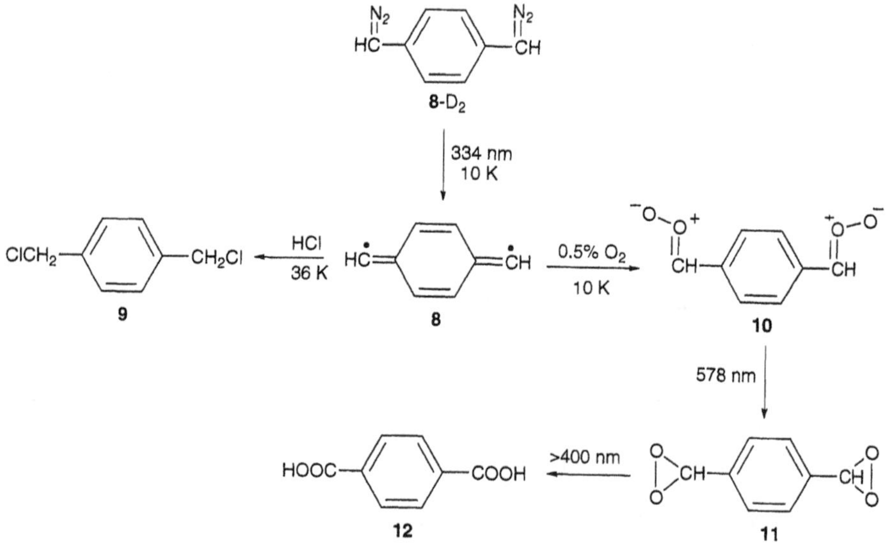

Scheme 1 Photochemical formation of **8** and its chemical reactivity toward HCl and O$_2$.

Attempts to detect a thermally populated triplet state ($^3A'$) of **8** by ESR spectroscopy were unsuccessful. This was attributed to the high reactivity of the diradical, which presumably easily abstracts hydrogen atoms in hydrocarbon matrices (to form p-quinodimethane), even at very low temperatures. In this context, the triplet state of p-phenylenebis(phenylmethylene) has been observed. Apparently, substitution of the carbenic hydrogens of **8** by phenyl groups confers sufficient stability (thermodynamic and perhaps kinetic) to the biradical, which allows its observation. According to variable-temperature ESR spectroscopy, the triplet state of p-phenylenebis(phenylmethylene) is thermally populated and the singlet state lies 0.5–1 kcal/mol lower in energy [76–79].

In contrast to phenylcarbene, which ring-expands photochemically, **8** was found to be "remarkably inert" under similar conditions [73]. This supports indirectly the quinonoid structure of **8**, as opposed to a biscarbene structure (such as represented by *S4* of Fig. 3) with weakly or noninteracting divalent carbon centers.

The reaction of **8** with oxygen (Scheme 1) leads to the bis-carbonyl-*O*-oxide **10** and can be seen either as a reaction of two carbene centers or as a reaction of two radical centers, giving a bis-peroxide. The IR and UV data suggest that there is little interaction between the two carbonyl centers via the linker,

Table 2 Relative Energies[a] (kcal/mol) of Various States of **8**,[b] **13**,[c] **16**, **17**, **18**, **19**, **22**,[b] and **27**.[b]

Species	B3LYP	MCSCF	CASPT2
8 ($^1A'$, σ^2/σ^2)	31.6		
8 ($^3A''$)	20.2	28.3	27.6
8 ($^5A'$)	25.9	27.0	26.6
8 ($^3A'$)	2.3	1.6	2.2
8 ($^1A'$, $\sigma\pi/\sigma\pi$)	0	0	0
13 ($^5A'$)	33.8	35.7	32.3
13 ($^3A'$)	2.2	1.3	1.9
13 ($^1A'$, $\sigma\pi/\sigma\pi$)	0	0	0
16 ($^3A''$)	26.8	34.6	34.0
16 ($^5A'$)	29.5	30.6	29.3
16 ($^3A'$)	1.4	1.5	2.0
16 ($^1A'$, $\sigma\pi/\sigma\pi$)	0	0	0
17 ($^3A''$)	6.7	12.1	11.2
17 ($^5A'$)	29.6	30.6	29.0
17 ($^3A'$)	1.3	1.5	1.9
17 ($^1A'$, $\sigma\pi/\sigma\pi$)	0	0	0
18 ($^3A''$)	14.2	22.9	21.1
18 ($^5A'$)	28.3	29.8	28.1
18 ($^3A'$)	1.4	1.5	2.0
18 ($^1A'$, $\sigma\pi/\sigma\pi$)	0	0	0
19 ($^3A''$)	14.0	23.3	21.1
19 ($^5A'$)	28.3	29.9	28.2
19 ($^3A'$)	1.3	1.5	2.1
19 ($^1A'$, $\sigma\pi/\sigma\pi$)	0	0	0
22 ($^1A'$, σ^2/σ^2)	10.0	18.2	20.3
22 ($^3A''$)	7.8	17.4	14.2
22 ($^5A'$)	24.5	25.6	22.8
22 ($^3A'$)	2.3	1.7	2.2
22 ($^1A'$, $\sigma\pi/\sigma\pi$)	0	0	0
27 ($^1A'$, σ^2/σ^2)	0	0	1.7
27 ($^3A''$)	6.8	13.3	5.3
27 ($^5A'$)	33.8	27.3	24.8
27 ($^3A'$)	10.1	2.1	1.8
27 ($^1A'$, $\sigma\pi/\sigma\pi$)	7.9	0.7	0

[a] With the 6-31G(d) basis set, unless otherwise noted. Results from Ref. 48, unless otherwise noted.

[b] With the 6-31+G(d) basis set. Results from Ref. 63.

[c] With the 6-31G(d) basis set; Ref. 84. For the results with the 6-31+G(d) basis set, see Ref. 63.

which implies that the biradical character of such species should be limited [73,80]. The oxide 10 isomerizes photochemically to bisdioxirane 11, which, upon further photolysis, is converted to terephthalic acid (12), in accordance with the behavior of other carbonyl oxides. [81].

p-Phenylenebisnitrene (13; Scheme 2) [63,69,75,76,82–84] is the nitrene analog of 8, but has been known for a longer time, presumably because synthesis of azides and generation of nitrenes from them is more facile. The isoelectronic nature of 8 and 13 is likely to be responsible for many of the similarities between the two species. Thus, the lowest singlet and triplet A' states of 13 are expected to have quinonoid geometries and the quintet benzenoid, in agreement with computational predictions.

Calculations find that 13, like 8, has a singlet biradical ground state ($^1A'$, $\sigma\pi/\sigma\pi$) and a low-lying triplet ($^3A'$, $\sigma\pi/\sigma\pi$) about 2 kcal/mol higher in energy (Table 2). The quintet state ($^5A'$, $\sigma\pi/\sigma\pi$) is above the ground state by about 32 kcal/mol, somewhat higher than in 8 [63,84]. The three states ($^1A'$, $^3A'$, $^5A'$) have the same ($\sigma\pi/\sigma\pi$) electronic occupancy in both 8 and 13. Thus, the similarity in relative energies suggests that these are determined primarily by the different spin coupling and, to a lesser extent, by the presence of the nitrogen heteroatom.

In reasonable agreement with the calculations are the ESR data, which give a T–S splitting of -0.6 to -0.8 kcal/mol for 13 [69,83]. It has been pointed out that these experimental T–S gaps should be compared with computationally derived free energies (and not enthalpies), in which case the disagreement between theory and experiment is reduced to within 1 kcal/mol [63].

Irradiation of the precursor 13-N$_2$ in an Ar matrix quickly gives rise to a compound with strong IR bands at 1758 and 1772 cm^{-1} [75, 84]. However, careful monitoring of the irradiation by IR spectroscopy revealed that this was not the primary photoproduct. As shown in Fig. 4, during irradiation, the concentration of the starting material (monitored by its absorption at 2084 cm^{-1}) decreases and gives rise to a new component A. However, before consumption of 13-N$_2$ is completed, the concentration of A (monitored by its absorption at 1063 cm^{-1}) decreases and signals due to components B and C (monitored by their absorptions at 1758 and 3280 cm^{-1}, respectively) appear.

With DFT calculations, identification of A as the bisnitrene 13 was rather easy because 13 is the expected product and an excellent match between theoretical predictions and experimental IR data was found. Identification of the secondary photoproduct (B) was more challenging, because a pathway leading to it is not immediately obvious. The presence of a weak absorption of B at 2210 cm^{-1} hinted at the possible presence of a cyano group, which was helpful in considering possible structures. However, it was the ease of carrying out the calculations that made possible the efficient screening of several candidate structures for B. Component B was identified as the substituted cyclopropene 14 (Scheme 2), and,

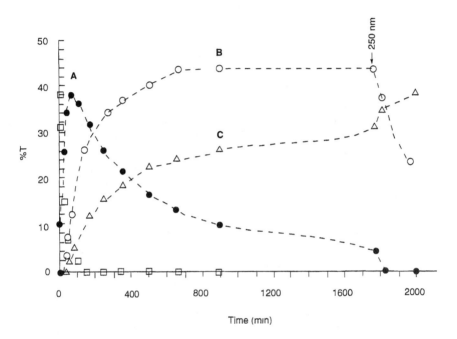

Figure 4 Chronological profile of the photolysis of **13**-N₂.

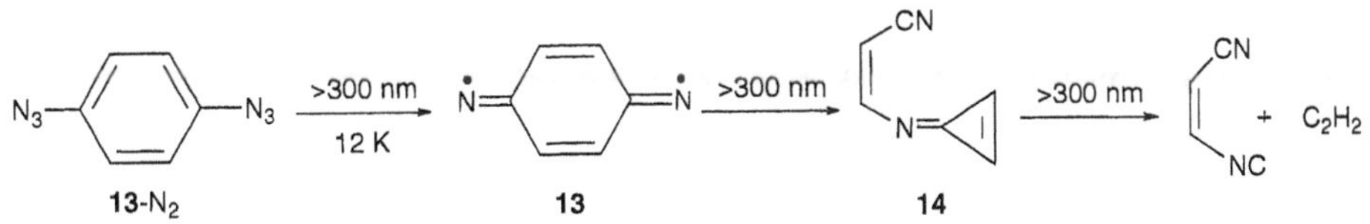

Scheme 2 Photochemical formation of **13** and photoproducts derived from it.

although there is some ambiguity as to its exact conformation, the assignment is believed to be correct.

Upon further irradiation with a shorter wavelength ($\lambda > 250$ nm), the concentration of component **C** rapidly increases at the expense of **14**, whereas further thawing of the matrix to 35 K did not result in any significant changes in the IR spectrum. It was established that component **C** is a mixture of acetylene and 1-cyano-2-isocyanoethylene (Scheme 2) [84].

According to the DFT calculations, the IR intensities for the C≡N stretch in **13** are two to three orders of magnitude weaker than in the secondary photoproduct **14**, and this can explain the dominance of the IR signals of **14** at the early stages of the irradiation [75,81].

Thus, unlike **8**, bisnitrene **13** is very unstable under its conditions of formation and isomerizes readily to **14**, which can decompose even further. Although the mechanism of the **13**-to-**14** transformation remains unknown, it was put forward that it is a photochemical process.

An interesting application of laser technology was the laser photolysis of **13**-N$_2$ in hexane. This lead to the formation of dark purple precipitates in $> 90\%$ yield [75]. The formed polymers, which were found to be hopelessly insoluble in most usual solvents, were characterized spectroscopically (NMR, UV, Raman) and assigned as polykis azobenzenes (**15**) (Structure 15), with unspecified terminating groups. Presumably, it is the intense irradiation of the excimer laser that is responsible for the formation of such oligomers and polymers by producing locally high concentration of the reactive monomer **13**.

The "mixed" *p*-phenylenecarbenonitrene (**16**, Scheme 3) [49,85] has also been reported. In this case, due to the difference of electronegativities between nitrogen and carbon, there is no reason to expect degeneracy of the frontier molecular orbitals and, thus, Hund's rule is not applicable. In line with biscarbene **8** and bisnitrene **13**, the ground state of **16** is also a singlet ($^1A'$-**16**) and there is a low-lying triplet ($^3A'$-**16**) about 2 kcal/mol higher in energy (Table 2). The Q–S spitting in **16** is -29 kcal/mol, a value intermediate of the Q–S gaps in **8** and **13**, but essentially of the same order. There is a disagreement of 7 kcal/mol between DFT and CASPT2 as far as the relative energy of the $\sigma^2/\sigma\pi$ triplet ($^3A''$)

15

Structure 15

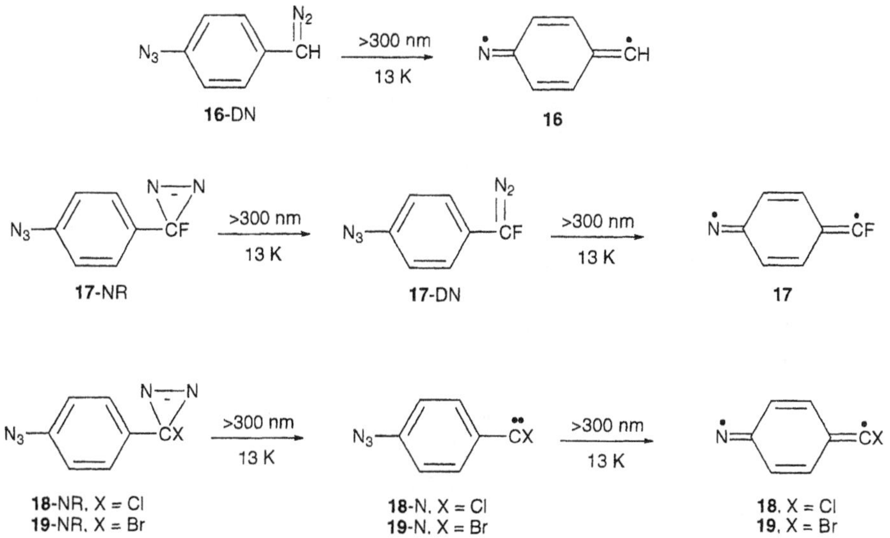

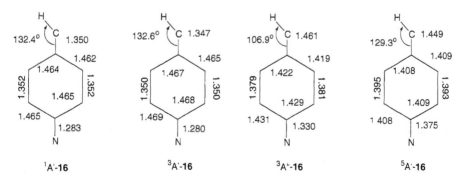

Scheme 3   Photochemical formation of **16** and its halogen analogues **17, 18,** and **19.**

is concerned. As discussed earlier for **8**, the correct answer for the relative energy of $^3A''$ must be somewhere in between the DFT and CASPT2 results.

As expected from the valence-bond descriptions of Fig. 3, the computed geometries of the lowest A' singlet and triplet states of **16** have quinonoid geometries displaying bond localization (Fig. 5). On the other hand, the quintet $^5A'$-**16**

Figure 5   Selected geometrical parameters of the $^1A'$, $^3A'$, $^3A''$, and $^5A'$ electronic states of **16** calculated at the CASSCF(10,10)/6–31G(d) level of theory.

has a benzenoid geometry. Perhaps the most distinct geometrical feature of $^3A''$-16 is the bond angle at the carbon divalent center which is significantly smaller compared to the other three states. This smaller bond angle is compatible with the valence-bond description of $^3A''$-16 as the coupling of a singlet carbene with a triplet nitrene center ($\sigma^2/\sigma\pi$, Fig. 3).

The IR spectra of several electronic states of 16 were computed (Fig. 6). It is interesting to note that the major IR bands of the $^3A''$ state of 16 are predicted to be about three times stronger compared to those of the other states. The predicted IR intensities of the quintet state are quite similar to those of the $^1A'$ and $^3A'$ states, but the pattern is sufficiently different. On the other hand, the latter two states cannot be discriminated easily by IR because their spectra are very similar, something that is intimately related to the similar geometries of the two states.

Irradiation of 16-DN afforded 16 (Scheme 3) which was characterized by UV/vis and IR spectroscopy [48]. The UV spectrum of 16 shows a long-wavelength absorption in the area of 375–440 nm, which is similar to that of the parent 8 (365–460 nm). The experimental IR spectrum is compatible with the formation of 16 in its $^1A'$ or $^3A'$ state.

Halogen substitution at the carbene center of 16 is likely to affect significantly the relative energy of the $^3A''$ state by changing the "local" T–S splitting of the carbene subunit. To test this, the azidodiazirine compounds 17-NR, 18-NR, and 19-NR were synthesized [48]. These precursors were irradiated (Scheme 3) under conditions very similar to those of 16-DN and the products of the photolysis were characterized by IR.

During the irradiation of 17-NR, a new absorption characteristic of a diazo group was observed and it was postulated that 17-NR isomerizes at least partially to 17-DN. On the other hand, during the irradiation of the chloro and bromo analogs (18-NR and 19-NR, respectively) the new azido absorptions at the early stages of the irradiation were attributed to the intermediate formation of azidocarbenes 18-N and 19-N, respectively. In all three cases, the final photoproducts were identified as carbenonitrenes (17, 18, and 19, Scheme 3) [48].

According to the calculations, the various electronic states of 17, 18, and 19 share the same geometrical characteristics as the parent 16 in terms of bond alternation and width of the bond angle at the divalent carbon (Fig. 5). The ground states of 17, 18, and 19 are calculated to be singlet biradicals. The $^3A'$ and $^5A'$ are about 2 and 29 kcal/mol higher in energy, irrespectively of the halogen (Table 2). Thus, these three electronic states are hardly affected by the halogen perturbation and this is shown graphically in Fig. 7.

On the other hand, the relative energy of the $^3A''$ is greatly reduced upon halogen substitution, especially in the case of fluorine. Based on valence-bond approximations (Fig. 3), it was argued that the $^3A''$–$^1A'$ splitting in compounds 16–19 should be closely related to the T–S splitting of the correspondingly substi-

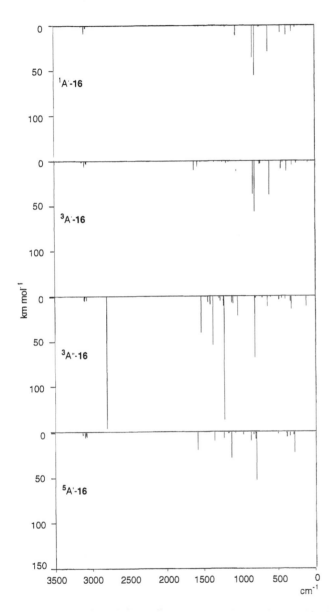

Figure 6 Computed IR spectra in the region 0–3500 cm^{-1} of the $^1A'$, $^3A'$, $^3A''$, and $^5A'$ electronic states of **16**.

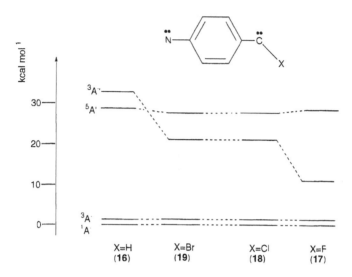

Figure 7 CASPT2/6–31G(d) relative energies (kcal/mol) of $^1A'$, $^3A'$, $^3A''$, and $^5A'$ electronic states of **16–19**, with respect to $^1A'$.

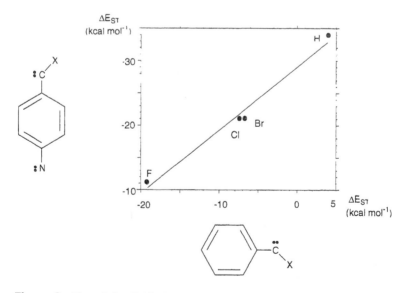

Figure 8 Plot of the $T(A'')$–S splitting (ΔE_{ST}, kcal/mol) of N—C_6H_4—CX (**16–19**) against the S–T splitting (ΔE_{ST}, kcal/mol) of the corresponding phenylcarbene (C_6H_5 —CX) [X = H (**3**), F (**5**), Cl (**6**), Br (**7**)].

tuted phenylcarbene (**3**, **5**, **6**, and **7**, respectively). This relationship was found to be linear as the plot of Fig. 8 shows.

From this plot, it was deduced that if the carbene subunit has $\Delta E_{ST} < -30$ kcal/mol, then the corresponding p-phenylenecarbenitrene would have a $^3A''$ ground state. Because the ΔE_{ST} of phenylfluorocarbene (about -19 kcal/mol, Table 1) is not small enough to satisfy this requirement, (4-nitrenophenyl)fluoromethylene (**17**) end up having a siglet ground state. Evidence that this relationship may be semiquantitatively correct was given by calculations on (4-nitrenophenyl)-fluorosilylene (**20**) (Structures 20 and 21), which is computed to have a triplet ($^3A''$) ground state. ΔE_{ST} for phenyl fluorosilylene (**21**) is about -53 kcal/mol, satisfying the above requirement.

The quintet states of **16–19** lie about 30 kcal/mol higher in energy than the corresponding $^1A'$ states. This energy difference can be attributed to the extra π bond present in $^1A'$ states. Thus, the ground-state electronic configuration in these systems is determined mainly by the interplay of two factors. One factor requires that both reactive centers offer one π-electron, implying that, "locally," they have triplet ($\sigma\pi$) electronic configurations, so that bonding is maximized. The other factor requires that the reactive center retain its inherent preferable electronic configuration, which can be either a singlet (σ^2) or a triplet ($\sigma\pi$), depending on the case. When the two factors act in the same direction, then $^1A'$ ground states are expected. However, if the two factors oppose each other, then the preference for a local closed-shell configuration (σ^2) should be at least as large as the magnitude of the extra bond that is sacrificed (i.e., ~ 30 kcal/mol), in order for $^3A''$ to compete energetically with $^1A'$.

From the above results, it seems that halogen substitution at one carbene center does not suffice to change the ground-state electronic configuration. Obviously, there is a better chance to achieve this if two carbene centers are substituted at the same time. In this respect, the reports on **22** and **27** (halogenated derivatives of **8**) are significant.

Irradiation of bischlorobisdiazirine **22**-R_2 in an inert matrix like N_2 [86] or Ar [87] causes stepwise elimination of two nitrogen molecules and leads to the formation of bischlorocarbene **22** (Scheme 4). Identification of **22** was achieved indirectly by trapping it with HCl (to form **23**) and O_2 (giving **25** and **26**) and

20   **21**

Structures 20 and 21

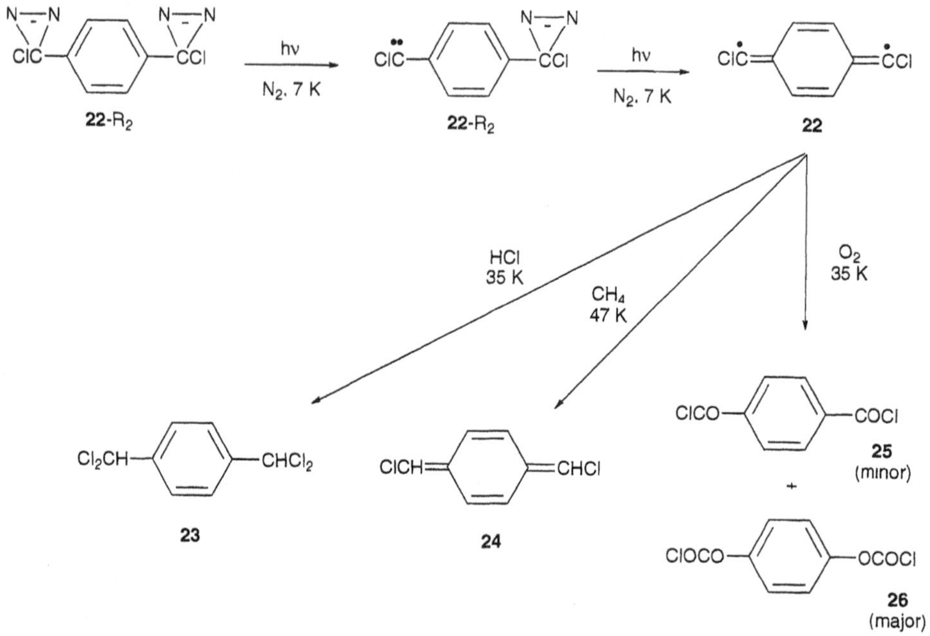

Scheme 4 Photochemical generation of **22** and its reaction products with HCl, methane, and oxygen.

comparing the IR spectra of the products with those of authentic samples. Further-more, **22** was found to react with methane, producing methyl radicals and **24**. The question of whether **22** is best thought of as a biscarbene (σ^2/σ^2) or as a biradical ($\sigma\pi/\sigma\pi$) was addressed. However, at that time, DFT was not used routinely for identification of such species and the arguments were focused on the spectroscopical characteristics and chemical reactivity of **22**.

Thus, the UV spectrum of **22**, was interpreted as incompatible with a σ^2/σ^2 configuration. At the same time, the IR data were interpreted in favor of the $\sigma\pi/\sigma\pi$ configuration. In addition, the reactivity pattern of **22** differs significantly from that of **6**, which has a σ^2 ground-state configuration. For example, **22** does not appear to ring expand upon irradiation (as **6** is known to do) and reacts with hydrocarbons (like methane) via hydrogen-atom abstraction (rather than by C—H insertion, as is the case with **6**) [86,87]. Thus, the conclusion that **22** is a $\sigma\pi/\sigma\pi$ biradical was reached.

Early semiempirical calculations [87] supported the above interpretation of the experimental data for the biradical character of **22**. Later, more sophisticated *ab initio* calculations arrived at the same conclusion (Table 2) [63,69]. The triplet biradical ($^3A'$) is, as expected, about 2 kcal/mol above $^1A'$. The Q–S_1 splitting (around 23 kcal/mol) is similar to that of **8**, although somewhat lower. In contrast, the $^3A''$ ($\sigma^2/\sigma\pi$) and the (σ^2/σ^2)–$^1A'$ states are stabilized by about 12 and 22 kcal/ mol (B3LYP values, Table 2) as compared to **8**. Based on the valence-bond approximate representations of Fig. 3, these two states [$^3A''$ and (σ^2/σ^2)–$^1A'$] have at least one of the divalent centers in a σ^2 local electronic configuration and they are most influenced by the chlorine substitution. However, in **22**, the gain in covalency overrides the preference of the chlorocarbene subunits for a closed-shell (σ^2) configuration and the (σ^2/σ^2)–$^1A'$ remains an excited state.

This result can be rationalized, if, as mentioned earlier, the gain in covalency is taken to be roughly 30 kcal/mol. By approximating the intrinsic preference of both chlorocarbene subunits in **22** for a σ^2 local configuration, as twice the ΔE_{ST} of phenyl chlorocarbene (about -7 kcal/mol, Table 1), the energetic price to be paid for the promotion of each divalent carbon center to a $\sigma\pi/\sigma\pi$ configuration is about 14 kcal/mol, significantly less than the benefit due to the extra π bond that is formed in the ($\sigma\pi/\sigma\pi$)–$^1A'$ state.

Preparation of **27** (Scheme 5) [88] was carried out by irradiating bisdiazirine **27-R$_2$** in a nitrogen matrix. A stepwise loss of the nitrogen molecules was observed and the intermediate carbenodiazirine (**27-R**) was characterized by IR and UV/vis spectroscopy. This species exhibited the characteristic IR and UV absorptions expected of the diazirine group as well as the UV/vis bands (strong absorption at 300 nm and broad band at 450–750 nm with $\lambda_{max} = 580$ nm) expected of arylhalocarbenes. The final product of the photolysis was identified as biscarbene **27** based on its trapping with HCl (to form **28**) and the good agreement between its experimental IR spectrum and calculated one.

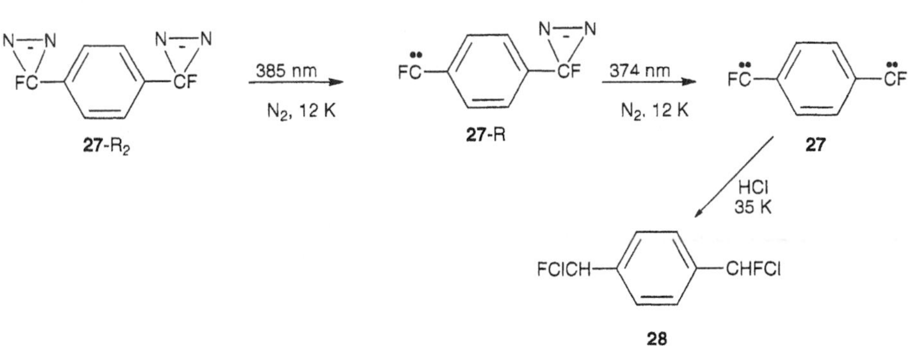

Scheme 5 Photochemical generation of **27** and its reaction product with HCl.

The interpretation of the IR data suggests that **27** is a para-disubstituted aromatic and not a quinonoid compound [88]. Accordingly, the UV/vis spectrum of **27** is essentially identical to that of its monodiazirinemonocarbene immediate precursor and to that of phenylfluorocarbene (**5**). Biscarbene **27** does not react with oxygen under conditions where **8** and **22** do. Also, no hydrogen-atom abstraction reactions in 3-methylpentane and CH_4 matrices by **27** were observed, in contrast to the high reactivity of **8** and **22**. All of these data were interpreted in favor of **25** being a singlet benzenoid biscarbene of the σ^2/σ^2 type and not a singlet quinonoid diradical.

Ab initio calculations [63,74] support the presence of the $\sigma^2/\sigma^2-{}^1A'$ **27** in the above experiments. However, as shown in Table 2, the two singlet A' states (σ^2/σ^2 and $\sigma\pi/\sigma\pi$) are very close in energy and which one is the ground state depends on the method and level of calculation used. This small energy difference decreases even further by improving the basis set. Thus, it has been suggested that both electronic states were actually observed experimentally and that some weak IR bands originally assigned to combination bands of the σ^2/σ^2 biscarbene reveal the accompanying presence of the $(\sigma\pi/\sigma\pi)-{}^1A'$ biradical [63].

Phenylfluorocarbene has a strong preference for a closed-shell singlet ground state (by about 19 kcal/mol, Table 1). If twice this amount is taken to approximate the preference of both carbene subunits in **27** for σ^2 configurations, then it is seen that in this case, the gain in covalency (approximated as about 30 kcal/mol) does not suffice to overcome the energy required for the internal promotion. Thus, one would have predicted that the σ^2/σ^2 of **27** is lower in energy than the $(\sigma\pi/\sigma\pi)-{}^1A'$ state, in qualitative agreement with the interpretation of the more rigorous computational results [63].

3. Summary

In this series of *p*-phenylene-linked bis(diradicals), the most systematically studied systems are the carbenonitrenes **16–19**. From these data, along with what is known for the parent compounds **8** and **13** and two of the halogenated derivatives of **8** (**22** and **27**), the following general features of *p*-phenylene-linked bis(diradicals) can be derived.

The quinonoidal $\sigma\pi/\sigma\pi$ singlet (*S1* of Fig. 3) is likely to be the ground state in such species because of the additional π bond present, which appears to have a bond strength approximately equal to the Q–S1 splitting (roughly 30 kcal/mol). This is qualitatively the gain in covalency that needs to be overridden for a switch in ground-state electronic configuration to take place.

In the case of carbenonitrenes, the nitrene subunit has an intrinsic preference for a triplet configuration, and, thus, for a $^3A''$ ground state, a carbene with a very strong preference for a closed-shell singlet ground state is required. Such a case has not been realized yet. On the other hand, if both subunits are carbenes, then, there is a better chance to achieve this switching, as demonstrated by bis(fluoro-carbene) **27**.

Finally the valence-bond depictions of Fig. 3, although not perfect, seem to be useful intuition guides in the study of these systems and reasonable approximations to estimating the relative energies of the different electronic states.

B. 1,3-Phenylene Linker

1. General Remarks

In Fig. 9, valence-bond approximations of various electronic states that arise from the interaction of two diradical centers via the m-phenylene linker are displayed. The conventions used in classifying the electronic states are the same as in Fig. 3 for the para topology (see Section III.A.1). The most important difference between the meta and para connectivity is that the former does not allow for conjugative coupling between the two radical centers. Thus, in the meta topology, quinonoid structures like $S1$ and $T1$ of Figure 3 are not possible and there is no driving force for the relative stabilization of the corresponding electronic configurations. Indeed, at first glance, the valence-bond approximations of Fig. 9 seem to predict that if two triplets are coupled via this linker, then $S1$ and $Q1$ should have approximately the same energy.

According to simple Hückel MO theory for $L_1 = L_2 = CH$ (and ignoring differences in α and β values between di- and tri-substituted atoms), the systems of Fig. 9 have eight π MOs: three bonding, two nonbonding (NBMO), and three antibonding (Fig. 10). Thus, for eight π electrons, the energetically lowest configuration would be a triplet [89]. Taking into account the two σ NBMOs, a quintet ground state is expected (assuming that both diradical centers have $\sigma\pi$ local electronic configurations) [90–92]. Such predictions have been verified experimentally, also leading to the generation of polyradicals (like **29**) [93] and polycarbenes (like **30**) [94] with impressively high ground-state multiplicities (Structures 29 and 30).

Among the other states depicted in Fig. 9, singlet $S1$ has the same "local" electronic configuration as $Q1$, but the two states differ in the overall coupling of the two triplets. Thus, the energy difference between these states can be taken as a measure for the preference of the linker for high-spin states. As with the para topology, the A'' states ($S2$, $S3$, $T3$, and $T4$ of Fig. 9) have an odd number of π electrons, and in these states, one of the radical centers is formally a closed-shell (σ^2) singlet. When this diradical center is a carbene, these states and the σ^2/σ^2 A' singlet ($S4$ of Fig. 9) are expected to be the ones most influenced by chemical perturbation, as discussed in Section III.A.1.

2. 1,3-Phenylene-Linked Systems

Biscarbenes **30** ($n = 0$) and **31** (Structures 31–33) were the first hydrocarbons for which quintet states were demonstrated [79,95]. Because the species were

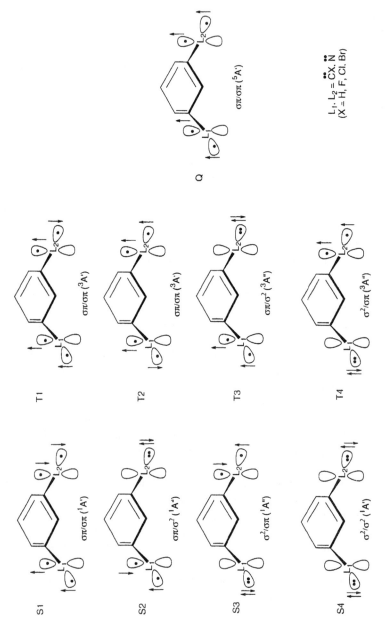

Figure 9 Valence-bond depictions of electronic states arising from the coupling of two reactive centers (carbene, nitrene) via the 1,3-phenylene linker.

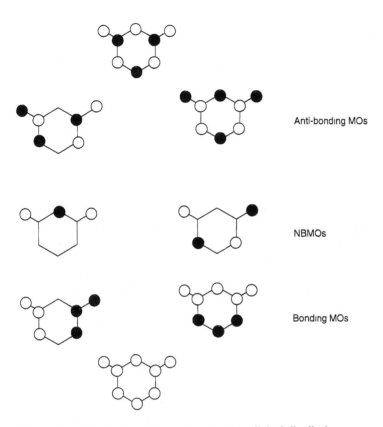

Anti-bonding MOs

NBMOs

Bonding MOs

Figure 10 Qualitative π MOs of *m*-phenylene-linked diradicals.

Structures 29 and 30

Structures 31–33

generated at 4 K, it was proposed that the observed quintet state is either the ground state or it lies energetically very close to it. Fitting the ESR data to an appropriate Hamiltonian gave $|D/hc| = 0.0701$ and 0.0844 cm^{-1} and $|E/hc| = 0.020$ and 0.0233 cm^{-1} for **30** ($n = 0$) and **31**, respectively [95]. These results were also verified by calculations based on a simple MO approximation [78,96]. The bisnitrene analog **32** was reported along with biscarbene **31**[95]. It also has a quintet ground state, with $|D/hc| = 0.156$ cm^{-1} and $|E/hc| = 0.029$ cm^{-1}. At a later time, the "mixed" carbenonitrene **33** was assigned a quintet ground state with $|D/hc| = 0.124$ cm^{-1} and $|E/hc| = 0.002$ cm^{-1}[97]. An interesting aspect of **33** is that the two NBMOs are no longer degenerate (as in the "symmetric" cases of **31** and **32**), because of the difference in electronegativity between carbon and nitrogen. So, there is no a priori reason to expect a high-spin ground state.

The finding that, despite the chemical perturbation, Hund's first rule is inviolate reinforced the idea that the ground-state multiplicity of alternant non-Kekulé molecules is mainly determined by good connectivity [98,99] or by topological symmetry [100] and that the perturbation by heteroatoms is not important. Thus, m-phenylene has been thought of as a robust ferromagnetic linker between radicals and carbene-related species and this has been a major driving force for the design of high-spin organic molecules. Although this is true in general, it has been shown recently, that geometry distortion [101–103] or appropriate chemical substitution [104] of m-phenylene-linked diradicals can reduce the ability of this linker for high-spin coupling leading to low-spin ground states.

In m-phenylene-linked dicarbenes, geometrical distortion may not be effective, but chemical substitution can be successfully applied to generate singlet ground states. Irradiation of bisdiazirine **34**-R$_2$ in the low-temperature (14 K) N$_2$ matrix gave mainly the monocarbene **34**-R (Scheme 6) [49]. Further irradiation with slightly shorter wavelength gave its diazo isomer (**34**-D) and the desired biscarbene **34**. The former could be transformed to the latter by selective irradiation at 480 nm. Compound **34** was trapped with HCl to give **35**, which was identified by comparing its IR spectrum with that of an authentic sample.

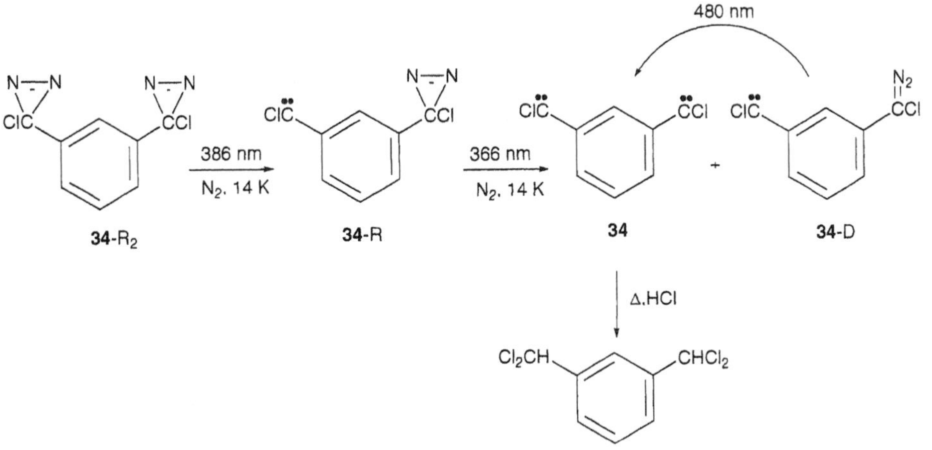

Scheme 6 Photochemical formation of **34** and its reaction product with HCl.

Biscarbene **34** was characterized by IR and UV/vis spectroscopy [49]. The analysis of the experimental data showed that these are compatible with the presence of two phenylchlorocarbene (**6**) subunits in **34**. This interpretation was further supported by the reactivity behavior of **34**, which, like **6**, is unreactive toward oxygen under conditions where triplet carbenes react fast. In contrast to its para isomer (**22**), **34** appears to undergo photochemical ring expansion analogous to that of **6**[105]. In addition, the computed [RHF/6–31G(d)] IR spectrum of **34**, which is in good agreement with the observed one, is based on the wave function for the singlet (σ^2/σ^2) biscarbene (*S4* of Fig. 9).

The multiplicity of the ground state of **34** has been investigated computationally beyond the SCF level, but originally with ambiguous conclusions [74]. DFT methods [BLYP and LSDA with the 6–31G(d) basis set] find the σ^2/σ^2 singlet to be the ground state. On the other hand, at the CASSCF level of theory, a quintet or singlet (σ^2/σ^2) ground state was predicted depending on the size of the active space [74]. Subsequent calculations at definite levels of theory [106] agree with the DFT results and predict a singlet ground state.

As mentioned in the beginning of this section, *m*-phenylenecarbenonitrene (**33**) was found by ESR spectroscopy to have a quintet ground state. At that experiment, the carbenonitrene was produced by irradiation of **33**-DN in a 2-methyltetrahydrofuran (MTHF) matrix at 18 K [97]. Interestingly enough, when the same precursor was irradiated under similar conditions (Ar matrix, 13 K;

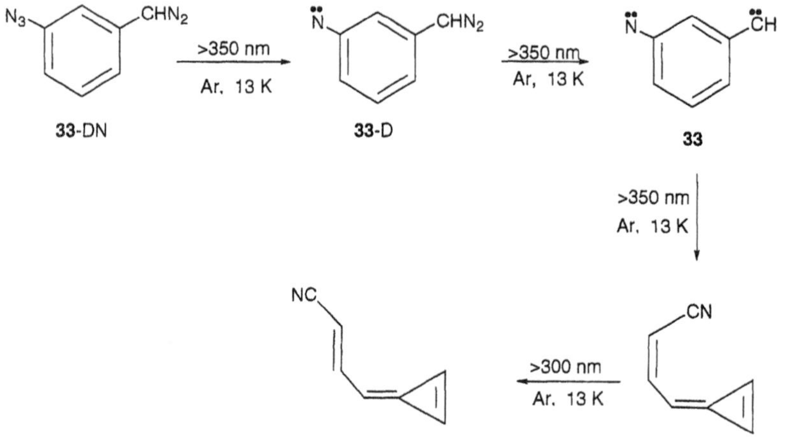

Scheme 7   Photochemical formation of **33** and its rearrangement to **36**.

Scheme 7) and the reaction was monitored by IR spectroscopy, no significant amount of **33** could be detected [106]. Instead, at the initial stage of the irradiation, an intermediate diazo compound was observed, which was assigned the structure **33**-D Further irradiation gave rise to a compound, which was identified with the help of calculations as substituted cyclopropene Z-**36** (Scheme 7). Although it is possible that **36** is formed directly from **33**-D, it is more likely that **33** is an intermediate of the reaction, as the ESR data imply. ESR spectroscopy is generally more sensitive than IR, and the failure of the latter to detect **33** is likely due to its inherently weak-intensity vibrational absorptions (as indicated by calculations) and/or its high photoreactivity.

The formation of the cyclopropene derivative Z-**36** is very similar to the isomerization of p-phenylenebisnitrene (**13**) to **14** (Scheme 2) and both processes are likely to be photochemical, rather than thermal. Further irradiation of Z-**36** causes its isomerization to another compound, which according to DFT simulations, is compatible with the structure of its E-isomer (E-**36**, Scheme 7).

Irradiation of the nitrogenous precursors **37**-NR, **38**-NR, and **39**-NR was carried out essentially under the same experimental conditions as for **33**-DN [106]. At the initial stage of the irradiation, in the fluorine case (**37**-NR, Scheme 8a), partial isomerization of the diazirine precursor to its diazo isomer (**37**-DN) was inferred from the IR spectral changes. In the cases of the chloro and bromo derivatives (**38**-NR and **39**-NR, respectively), the formation of azidocarbenes (**38**-N and **39**-N, respectively, Scheme 8b) is compatible with the new azido IR peaks

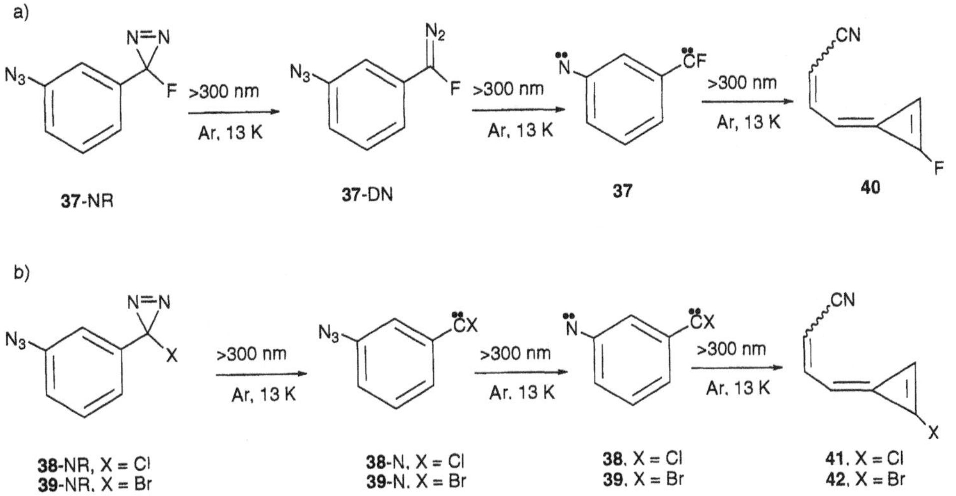

Scheme 8 Photochemical formation of carbenonitrenes **37**, **38**, and **39** and photoproducts derived from them.

that were observed. In all three cases, the carbenonitrenes **37–39** were detectable by IR spectroscopy, unlike their parent **33** (Scheme 7). Interestingly, the computed spectra of the $^3A''$ states of **37–39** were in much better agreement with the experimental data than those of the quintet ($^5A'$) states, implying that the ground-state multiplicites of **37–39** differ from that of the parent **33**. Further evidence for this was provided by irradiating precursors **37-NR**, **38-NR**, and **39-NR** in the MTHF matrix at 77 K and detecting signals due to triplet nitrene species.

According to the calculations, the IR intensities of the $^3A''$ states of **33** and **37–39** are two to three times stronger than those of the $^5A'$ ones [106]. This can explain, at least in part, the finding that the latter were detectable by IR spectroscopy, but not the former.

The halocarbenonitrenes **37–39** are photolabile and give rise to substituted cyclopropenes (**40–42**, respectively; Scheme 8) in a reaction that seems to be the same as for the parent **33** (Scheme 7). The photoreactivity of the *meta*-carbenonitrenes **33** and **37–39** contrasts the relative stability of their para isomers (**16–19**). In the former compounds, the reactive centers are not strongly coupled and they retain much of their carbenic and nitrenic identities, exhibiting photolability similar to that of phenylecarbene and phenylnitrene. In contrast, in the para isomers, the strong conjugative coupling between the radical centers, which transforms them to vinyl and iminyl moieties (as suggested by the structures of

Scheme 3) seems to infer some extra stability similar to that of *p*-phenylenebis-methylene (**8**).

Five different electronic states of **33** and **37–39** were considered computationally: one quintet (*A'*), two triplets (*A'* and *A"*), and two singlets (*A'* and *A"*). The $^5A'$ and $^3A"$ have little multiconfigurational character, but the other states need MCSCF wave functions for their proper description.

Figure 11 shows some of the geometrical parameters (computed at the MCSCF level of theory) for the five electronic states of **33**. The local geometries of the carbene and nitrene subunits (bond angle at the carbene center, C—H bond length of the carbene and the bond lengths between the diradical centers and the benzene ring) of the quintet state ($^5A'$) are very similar to that of triplet phenylcarbene and phenylnitrene computed at the same level of theory (Fig. 12). The situation is less clear for the other two *A'* states ($^1A'$ and $^3A'$), but the geometry of

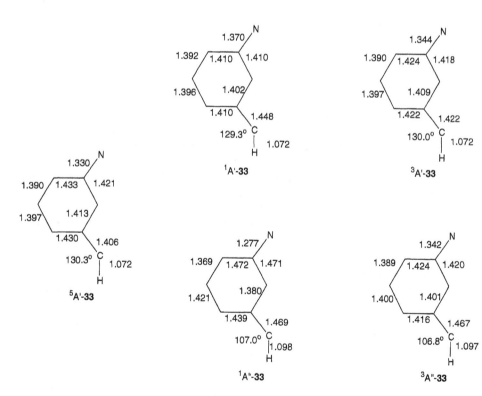

Figure 11 Selected bond lengths (Å) and bond angles (deg) of various electronic states of **33**.

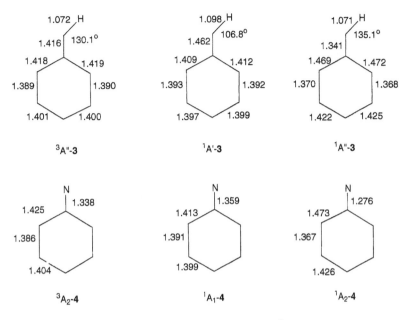

Figure 12 Selected MCSCF/6-31G(d) bond lengths (Å) and bond-angles (deg) of various electronic states of **3** and **4**.

the carbene subunit is compatible with a local triplet configuration. The computed geometries of the two A'' states ($^1A''$ and $^3A''$) are compatible with the presence of a closed-shell singlet (σ^2) carbene subunit combined with an open-shell singlet (in $^1A''$-**33**) or triplet (in $^3A''$-**33**) nitrene subunit. Similar correlations between computed geometrical features and electronic states were found for the halogen analogs **37–39** as well.

The relative energies of the five different electronic states of **33** and **37–39** at various levels of theory are shown in Table 3. Halogen substitution has the most dramatic effect on the A'' states in which the divalent carbon center has approximately a σ^2 configuration. Indeed in the case of the fluorine analog **37**, a triplet ($^3A''$) ground state is predicted in contrast to the parent compound, but in agreement with the interpretation of the IR data. For the chloro and bromo analogs (**38** and **39**, respectively), DFT and MCSCF give mixed results. At higher levels of theory, [QCISD(T) combined with a reasonable approximation for the basis set effect], all three halogen-substituted carbenonitrenes are predicted to have triplet ground states with the $Q-T(^3A'')$ splittings decreasing (in absolute magnitude) on going down the periodic table. In addition, a linear relationship (Fig. 13) was found between the $Q-T(^3A'')$ gap in **33, 37, 38**, and **39** and the $T-S$

Table 3 Relative Energies (kcal/mol) of Various States of **33** and **37–39** with Respect to the $^5A'$ State (Hartree)

	B3LYP[a,b]	MCSCF[b]	CASPT2[b]	E_{QCI}[c]
33 ($^1A''$)		33.4	38.0	
33 ($^1A'$)		10.1	10.3	
33 ($^3A''$)	12.3	15.0	18.0	11.9
33 ($^3A'$)		5.8	5.3	
33 ($^5A'$)	−324.20098	−322.37862	−323.30523	−323.58826
37 ($^1A''$)		10.2	14.3	
37 ($^1A'$)		10.0	9.8	
37 ($^3A''$)	−8.5	−7.9	−5.3	−11.8
37 ($^3A'$)		5.9	5.2	
37 ($^5A'$)	−423.43976	−421.22559	−422.32602	−422.72349
38 ($^1A''$)		21.9	25.3	
38 ($^1A'$)		9.6	9.5	
38 ($^3A''$)	−0.4	3.9	5.8	−2.9
38 ($^3A'$)		5.7	5.1	
38 ($^5A'$)	−783.81262	−781.28135	−782.35371	−782.72518
39 ($^1A''$)		22.3	25.5	
39 ($^1A'$)		9.7	9.5	
39 ($^3A''$)	0.4	4.2	5.9	−1.5
39 ($^3A'$)		5.7	5.0	
39 ($^5A'$)	−2895.32284	−2891.68888	−2892.76209	−2895.56012

[a] Includes zero-point energy corrections.
[b] With the 6-31G(d) basis set.
[c] $E_{QCI} = E_{QCISD(T)/6-31G(d)} + E_{RMP2/6-311-G(3df,2p)} - E_{RMP2/6-31G(d)}$. For details, see Ref. 106.

gap of the corresponding phenylcarbene (**3, 5, 6,** and **7**, respectively). From this plot, a switching in ground-state multiplicity (from quintet to triplet) can take place if the intrinsic preference of the carbene subunit for a σ^2 configuration is at least 6 kcal/mol greater than for a triplet $\sigma\pi$ configuration.

Interestingly, the energy difference between the two A'' states (in **33** and **37–39**) is 18–20 kcal/mol (Table 3) *independently* of the substituent at the carbene center (H, F, Cl, or Br). This is not surprising within the valence-bond approximations of Fig. 9. Thus, as shown in Fig. 14, the difference between the electronic configurations of the singlet and triplet A'' states can be approximated by a spin-flip at the nitrene center, which, in terms of energy, should be roughly equal to the singlet–triplet gap in phenylnitrene (**4**).

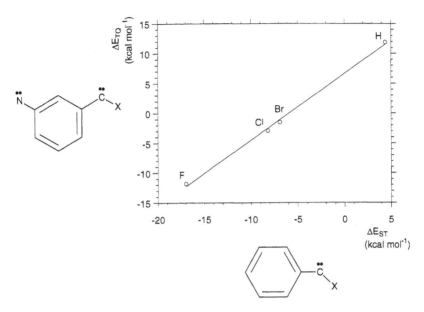

Figure 13   Plot of the Q–$T(A'')$ splitting (ΔE_{TQ}, kcal/mol) of 1,3-N—C_6H_4—CX [X = H (**33**), F (**37**), Cl (**38**), Br (**39**)] against the T–S splitting (ΔE_{ST}, kcal/mol) of the corresponding phenylcarbene (C_6H_5—CX) [X = H (**3**), F(**5**), Cl (**6**), Br (**7**)].

The singlet A' state of these systems can be thought of as the low-spin coupling of a triplet carbene with a triplet nitrene subnit ($S1$ of Fig. 9). These states lie approximately 10 kcal/mol higher in energy than the $^5A'$ states irrespectively of the substituent X (X = H, F, Cl, Br) at the carbene site (Fig. 14, Table 3). It seems reasonable to identify this Q–$S(A')$ splitting with the amount of energy required to overcome the preference of the m-phenylene linker for ferromagnetic coupling by imposing a low-spin coupling in its π space. On the other hand, according to the linear relationship of Fig. 13, 6 kcal/mol are required in order to overcome the linker's preference for ferromagnetic coupling. The two quantities are not expected to be the same, because in the first case, the multiplicity is lowered due to a spin-flip in the π space, but in the second case, the spin-flip is associated with a change in the number of π-electrons. Nevertheless, it seems safe to assume that the m-phenylene linker tends to promote high-spin coupling by 6–10 kcal/mol.

The $^3A'$ states can be approximated as the combination of a triplet carbene with a nitrene in which its σ- and π-electrons have opposite spins (Fig. 14). If the latter configuration is taken to correspond to an open-shell singlet nitrene (as

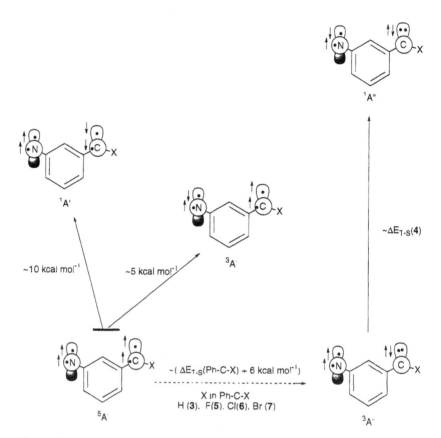

Figure 14   Relative energies of various electronic states with respect to $^5A'$ for **33** and **37–39**.

was done with the $^1A''$ state), then the $^3A'$ state should be higher in energy than the corresponding $^5A'$ state by about 20 kcal/mol (i.e., the energy difference between $^1A''$ and $^3A''$ states, as discussed earlier). However, the computed energy difference between $^3A'$ and $^5A'$ states is only about 5 kcal/mol and is essentially independent of the substituent at the carbene center. To resolve the issue, it was proposed that the two electrons attributed to the nitrene center are better correlated in the $^3A'$ states of the *m*-phenylene-carbenonitrenes of Fig. 14 than in open-shell singlet phenylnitrene ($^1A_2$-**4**), due to the tendency of the linker to promote high-spin coupling in its π space.

The nitrene center of $^1A_2$-**4** is described qualitatively by two determinants: $|\ldots \pi_N(\alpha)\sigma_N(\beta)| - |\ldots \sigma_N(\alpha)\pi_N(\beta)|$; however, in the $^3A'$ states of **33** and

37–39, only the first determinant is important because the m-phenylene linker promotes ferromagnetic coupling between the nitrene and carbene subunits. In other words, in the $^3A'$ states of the m-phenylenecarbenonitrenes, if the carbene is locally in a triplet configuration ($|\ldots\pi_C(\alpha)\sigma_C(\alpha)|$), then a reasonably good qualitative description of these states is $|\ldots\pi_C(\alpha)\sigma_C(\alpha)\pi_N(\alpha)\sigma_N(\beta)|$, whereas the contribution of ($|\ldots\pi_C(\alpha)\sigma_C(\alpha)\sigma_N(\alpha)\pi_N(\beta)|$ is not as important. This is also supported from the leading configurations of the $^3A'$ MCSCF wave functions, which show that two π-electrons are triplet coupled [106].

The wave function of two electrons described by a determinant of the type $|\pi(\alpha)\sigma(\beta)|$ corresponds to neither a pure singlet nor a pure triplet wave function, but to a "mixed" wave function made of 50% singlet and 50% triplet character. By extending this analogy to the $^3A'$ states of **33** and **37–39**, the "local" nitrene subunit is expected to lie energetically about "halfway" between triplet and (open-shell) singlet phenylnitrene. This means that these states should be roughly 9 kcal/mol (i.e., half of the T–S gap in phenylnitrene) higher in energy than the corresponding $^5A'$ states. Consistent with this simplistic approach is the computationally derived $^5A'$–$^3A'$ gap of 5 kcal/mol (Table 1).

3. Summary

m-Phenylene, unlike p-phenylene, does not allow for a direct conjugative interaction between the (di)radical centers. One consequence is the tendency of this linker to promote high-spin ground states, as exemplified by the parent systems **31**, **32**, and **33**, all of which have quintet ground states. The intrinsic preference of this linker for ferromagnetic coupling is estimated, to be about 6–10 kcal/mol. Not surprisingly then, it is rather easy to overcome this preference by chemical perturbation, and several systems with low-spin ground states have been identified.

The two chlorocarbene subunits of **34**, with a combined intrinsic preference for σ^2 configurations of about 14 kcal/mol, are capable of overriding the intrinsic preference of the m-phenylene linker for high-spin configurations, resulting in a singlet ground state. Even one chlorocarbene subunit, as exemplified by carbenonitrene **38**, stabilizes the local σ^2 configuration effectively. These examples contrast sharply with the corresponding p-phenylene-linked isomers.

The rather weak coupling between m-phenylene-linked diradical sites gives rise to several low-lying excited states. The relative energies of these states are not known experimentally, making it difficult to calibrate the computational results. In most cases, CASPT2 relative energies are compatible with what can be inferred indirectly from the experimental IR data. In a few cases, comparison of the CASPT2 Q–$T(A'')$ splittings with those obtained at higher levels of theory suggests that CASPT2 overestimates the relative stability of the quintet by 6–8 kcal/mol. This gives roughly the magnitude and direction of the error expected from CASPT2 calculations in similar cases. Interestingly, the relative energies

of several of the electronic states can be rationalized by valence-bond approximations, such as shown in Figs. 9 and 14.

C. 1,2-Phenylene Linker

1. General Remarks

The electronic properties of the ortho and para topologies share a lot in common
due to their similar mesomeric effects. Thus, electronic configurations, analogous
to the ones shown in Fig. 3 for p-phenylene-linked bis(diradicals), are expected
to be operative for the systems discussed in this section. An important difference,
though, between the ortho topology, on one hand, and the para and meta topologies, on the other, is the proximity of the two reactive centers in the former,
which allows for intramolecular reactions not possible in the latter. This extra
complication makes the study of o-phenylene-linked bis(diradicals) more challenging, and, perhaps, is responsible that such systems have not been as widely
studied.

2. 1,2-Phenylene-Linked Systems

1,2-Phenylenebisnitrene (**43**) has been generated at low temperatures [107,108]
from the corresponding bisazide (**43**-N$_2$) and characterized by IR spectroscopy
(Scheme 9). No intermediates, resulting from elimination of one nitrogen molecule, were observed under these conditions. Thus, no IR bands attributable to
nitrene **45** (Structures 45 and 46), which was suggested as an intermediate in the
formation of dicyanobutadiene **44** (mucononitrile) from the thermolysis of bisazide **43**-N$_2$ [109] or the oxidation of **46** [110], were observed.

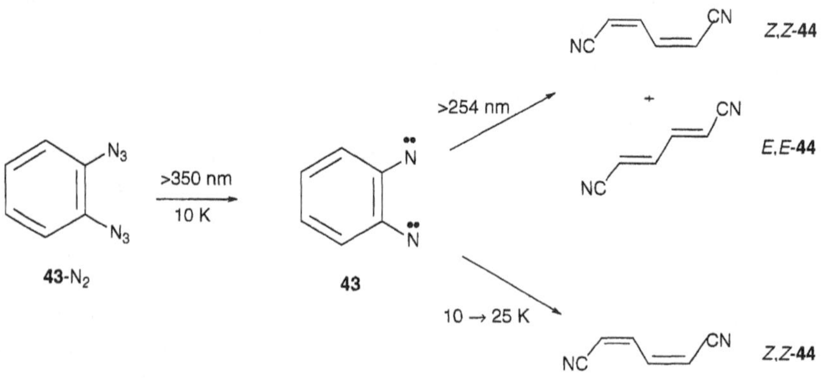

Scheme 9   Generation of **43** and its rearrangement to **44**.

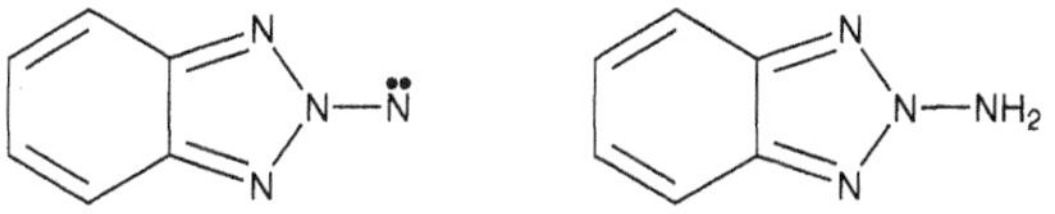

45   46

Structures 45 and 46

Bisnitrene **43** easily undergoes thermal isomerization to **44** when the matrix is warmed in the temperature range 10–25 K (Scheme 9). The same isomerization can be achieved photochemically upon irradiation with shorter-wavelength light, which gives rise to *Z,Z* and *E,E* isomers of **44**. From the kinetics of the thermal isomerization at low temperature, the activation energy was estimated at 2.8 kcal/mol [108].

In a similar fashion, *o*-phenylenecarbenonitrene (**47**) was generated by photolysis of the azido–diazo precursor **47**-DN (Scheme 10) and identified by IR spectroscopy with the help of calculations. Further irradiation caused a ring opening of carbenonitrene **47** to its isomer **48**, which was formed as a mixture of *Z,Z* and *E,E* isomers. The same isomerization can be achieved thermally, and from the kinetic analysis in the temperature range 40–50 K, the activation energy was found to be 5.1 kcal/mol [108].

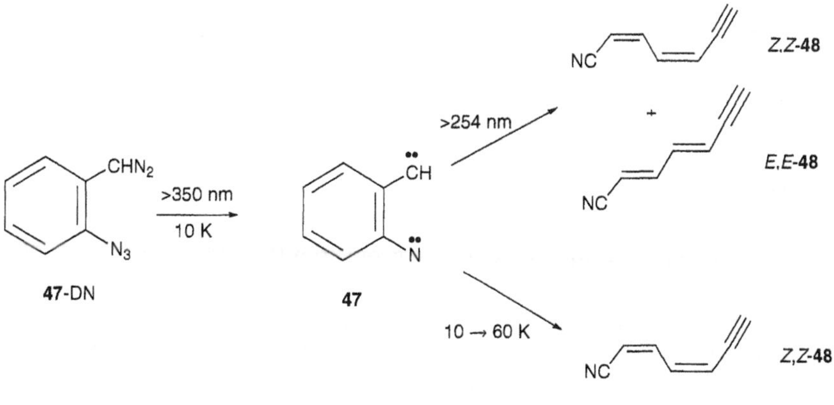

Scheme 10   Generation of **47** and its rearrangement to **48**.

49   50   51

52   53

Structures 49–53

Pyrolysis of sulfoximide **49** (Structures 49–53) at 500°C also gives cyanoacetylene **48** (Z,Z isomer) along with biphenylene [111]. The intermediate for this process was assumed to be indazole nitrene **50**, which was postulated to ring expand to benzotriazine **51**. Significant fragmentation of the latter under the vigorous conditions employed would give benzyne, the dimerization of which can explain the presence of biphenylene in these experiments [111,112]. However, under matrix-isolation conditions, indazole nitrene **50** was not detected by IR [108]. Also, irradiation of benzotriazine **51** at low temperature, instead of benzyne, afforded benzazete **52** [108].

A convenient precursor for the matrix isolation study of *o*-phenylenebiscarbene (**53**) is not available yet, so this species is not well known. Its presence has been postulated in the gas-phase pyrolysis of the corresponding precursor for *p*-phenylenebiscarbene **8** (Sect. III.A.2). Thus formation of benzocyclobutadiene dimers in the pyrolysate is interpreted as indicating that the para isomer **8** undergoes carbene- to-carbene rearrangement (known to occur under these conditions) to generate **53**, which eventually undergoes ring closure to form the cyclobutadiene [72].

Computational studies on the three model compounds **43**, **47** and **53** have been reported.[108,113]. The computed geometries (Fig. 15) of the lowest singlet ($^1A'$) and triplet ($^3A'$) states have a quinonoid character, as expected due to conjugative coupling of the reactive centers. On the other hand, the quintet states ($^5A'$) have benzenoid geometries. In **47** and **53**, the bond angle at the divalent carbon center is similar to that expected of a vinyl radical or triplet carbene. The relative orientation of the CH with respect to the neighboring radical center gives rise to

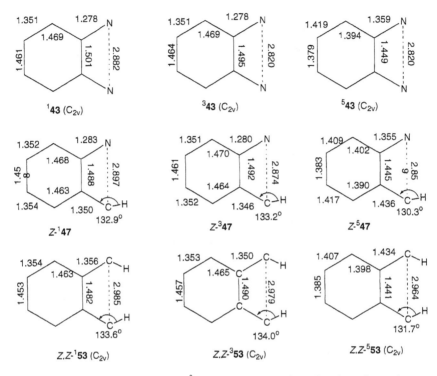

Figure 15 Selected bond lengths (Å) and bond-angles (deg) of various electronic states of **43, 47,** and **53.**

different conformations (*E*- and *Z*-conformers in the case of **47**, and *Z,Z*-, *Z,E*-, and *E,E*- in the case of **53**). These conformations have very similar geometries and energies (Table 4) and the barrier for their interconversion is estimated around 4 kcal/mol, essentially the same as that for inversion in vinyl radical.

Like their para isomers, **43,47,** and **53** have open-shell singlet ($^1A'$) ground states (Table 4). The triplet ($^3A'$) and quintet ($^5A'$) states are predicted to be about 2 and 20–25 kcal/mol higher in energy, respectively. Thus, the S–Q splittings are approximately 5 kcal/mol smaller than the corresponding ones of the para isomers (**13, 16** and **8,** respectively), but qualitatively very similar, as expected due to the similar electronic properties of the *o*- and *p*-phenylene linkers.

The IR spectra of **43, 47,** and **53** cannot be used alone for predicting the ground-state multiplicity, because of the similarity between the $^1A'$ and $^3A'$ states. However, the computational results in conjunction with the low temperature used in the experiments and the lack of ESR signals when **47** was generated were interpreted in favor of singlet biradical ground states [108].

Table 4 Relative Energies (kcal/mol) of Various States of **43**, **47**, and **53** with Respect to the Singlet State (Hartree)[d]

	B3LYP[b]	MCSCF	CASPT2
^{5}**43**	25.0	29.5	25.5
^{3}**43**	2.7	1.5	2.7
^{1}**43**	−340.31041	−338.36713	−339.31035
E-^{5}**47**	23.2	25.8	23.0
Z-^{5}**47**	22.2	25.3	22.2
E-^{3}**47**	3.1	2.2	2.7
Z-^{3}**47**	2.2	1.5	1.9
E-^{1}**47**	1.2	0.2	0.9
Z-^{1}**47**	−324.22239	−322.31416	−323.23648
E,E-^{5}**53**	19.8	21.2	19.5
Z,Z-^{5}**53**	19.2	21.5	19.3
E,Z-^{5}**53**	18.9	20.8	18.7
E,E-^{3}**53**	3.7	2.8	2.7
Z,Z-^{3}**53**	3.4	2.8	2.5
E,Z-^{3}**53**	3.0	2.4	2.1
Z,Z-^{1}**53**	0.9	0.7	0.6
E,Z-^{1}**53**	0.6	−306.25722	−307.15799
E,E-^{1}**53**	−308.12897	c	c

[a] With the 6-31G(d) basis set. Results from Ref. 16.
[b] Includes zero-point energy corrections.
[c] No minimum located at the MCSCF/6-31G(d) level of theory.

Intramolecular reaction pathways available for the o-phenylene system are especially intriguing because they also provide insight into the nature of these exotic species. Both bisnitrene ^{1}**43** and carbenonitrene ^{1}**47** undergo, rather easily, ring opening (giving **44** and **48**, respectively), with the latter process requiring a higher activation energy than the former. Calculations reproduce these results in a semiquantitative way (Table 5). Furthermore, calculations predict that ring opening of the biscarbene **53** will be less facile. This trend in reactivity reflects to some extent the greater tendency of p-phenylenebisnitrene (**13**) for rearrangement as compared to p-phenylenebisarbene (**8**).

The reactivity trend in terms of the ring-opening reaction for the series **43**, **47**, and **53** can be analyzed in terms of the geometries of imino and vinyl radicals. Most of the spin density in imino radicals is concentrated in the p-orbital on nitrogen, but vinyl radicals have a nonlinear, sp^2-like structure, where the spin density is in the sp^2-orbital of the carbon. In the o-quinone diradicals **43,47**, and **53**, the p-orbitals of the nitrogens (accomodating the "odd" electrons) are better

Table 5   Barriers for Ring Opening ($\Delta H_{ro}{}^{\ddagger}$, kcal/mol) and Ring Closure ($\Delta H_{rc}{}^{\ddagger}$, kcal/mol) Reactions and Enthalpies of the Ring Opening (ΔH_{ro}, kcal/mol) and Ring Closure (ΔH_{rc}, kcal/mol) Reactions for ^{1}**43**, ^{1}**47**, and ^{1}**53**.[a]

	$\Delta H_{ro}{}^{\ddagger}$	ΔH_{ro}	$\Delta H_{rc}{}^{\ddagger}$	ΔH_{rc}
^{1}**43**	0.6 (2.8)	− 55.2	18.9	−7.3
^{1}**47**	4.3 (5.1)	− 44.4	6.8	−49.9
^{1}**53**	5.2	− 36.1	− 0	− 77.8

[a] B3LYP/6-31G(d). Experimental activation energies in parenthesis. Results from Ref. 106.

aligned with the orbital of the breaking C—C bond than the sp^2-orbital of the vinyl radical, which is oriented somewhat away from the C—C bond. Because of this better overlap, the length of the C—C σ bond in question increases (Fig. 15), and to the extent that these effects are present in the corresponding transition state structures, the ring-opening isomerization is more facile on going from the biscarbene **53** to the bisnitrene **43**. The calculated trend is also in agreement with Hammond's postulate in the sense that the more exothermic reaction (Table 5) is predicted to be faster.

An alternative mode of reaction available to the o-quinone diradicals **43**, **47**, and **53** is ring closure to benzocyclobutadienes **54**, **52** and **55**, respectively (Scheme 11). The computed values for the barriers for this reaction follow the opposite trend and so do the associated enthalpies. Thus, biscarbene **53** is predicted to be the most reactive towards ring closure. Furthermore, **53** is predicted to prefer this mode of reaction, in contrast to **43** and **47**, for which calculations find that ring opening is more facile. The available experimental results are compatible with the computational data, as far as **43** and **47** are concerned. On the

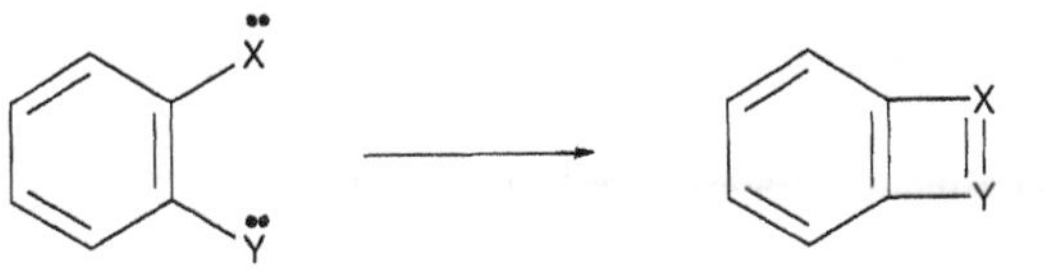

43. X = Y = N
47. X = CH. Y = N
53. X = Y = CH

54. X = Y = N
52. X = CH. Y = N
55. X = Y = CH

Scheme 11   Ring-closure reaction of o-phenylene linked bis(diradicals) **43**, **47**, and **53**.

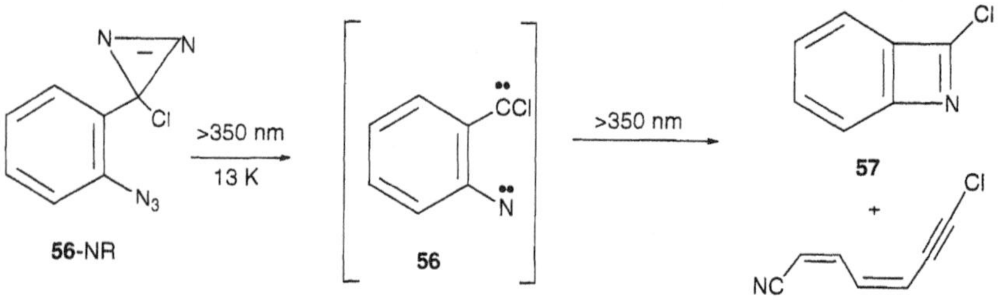

Scheme 12   Photochemical decomposition of **56**-NR.

other hand, there is only scant experimental evidence that biscarbene **53** may prefer to ring close than to ring open and this evidence comes from high-temperature gas-phase studies [72].

Some more evidence, that the prediction about **53** may be fulfilled, comes from attempts to generate its chloro derivative **56** (Scheme 12). Irradiation of chlorodiazirine **56**-NR at low temperature resulted in the formation of a mixture of products, which were identified, with the help of calculations, as benzazete **57** and nitrile Z, Z-**58**. Although the target carbenonitrene (**56**) was not detected by IR under the experimental conditions, speculation of its intermediacy is reasonable. The formation of both ring-closure (**57**) and ring-opening (**58**) products implies that these two paths are competing, presumably because the energies of the corresponding transition states are similar [114]. Indeed, DFT calculations find that the barriers for the ring-opening and ring-closure reactions are 6.4 and 6.9 kcal/mol, respectively, of a chlorine atom in **56** can be considered as a perturbation or fine-tuning of the potential energy surface of **47**. Although, more experimental data are clearly desirable, from the current ones it can be seen that chemical perturbation (as by the Cl atom in the case of **56**) can modify the potential energy surface of **47**, fine-tuning the outcome of the reaction.

3. Summary

Electronically, the *o*-phenylene-linked carbenes and nitrenes are very similar to their para isomers. However, the proximity of the reactive centers in the former case allows for intramolecular reactions to take place revealing, thus, some aspects of their intrinsic chemical reactivity. *o*-Phenylene-linked carbenes and nitrenes are less known compared to their meta and para isomers, due to difficulties of preparing appropriate precursor molecules. From the available data, the following general remarks can be made.

Generation of *o*-phenylene-linked bis(diradicals) under matrix-isolation conditions gives information about possible intramolecular reaction pathways, which are available. From the two dominant paths that have been observed, one leads to ring opening of the o-phenylene linker, giving 1,4-disubstituted butadienes as products. The other pathway (ring-closure) leads to benzocyclobutadiene type products. Which path is to be preferred depends on the nature of the diradical center. Thus, ring opening is the only pathway observed for bisnitrene **43**, presumably due to the formation of C—N triple bonds. The ring-closure path becomes increasingly more favored by substituting a nitrene center with a carbene, and this is predicted to be the preferred path for **53**; although, solid experimental evidence for this is lacking. Preliminary results suggest that chemical perturbation at the carbene center can also affect the relative importance of the two pathways leading to mixtures of ring-opened and ring-closed products.

From the limited data on these systems, it is seen that DFT methods are useful not only in predicting reasonably well the IR spectra of such compounds but also in simulating the barriers of their intramolecular rearrangements.

Overall, it is desirable that more of the o-phenylene-linked systems are generated so that the above conclusions can be tested more rigorously.

IV. CONCLUDING REMARKS

Matrix isolation is a powerful tool for studying highly reactive species. The matrix provides an inert environment, which increases the lifetime of the "reactive" species to the extent that it may be thought of as a "stable" species. This allows for its spectroscopical identification and/or characterization using normal-type spectrometers. It is also possible to learn something about the intrinsic reactivity of the species, because, under such conditions, intermolecular reactions can be excluded.

For species of higher than singlet multiplicity, matrix isolation coupled with ESR spectroscopy is often used for detection and characterization purposes. This is how the first examples of "exotic" organic compounds with quintet ground states were obtained. Since then, this kind of spectroscopy has been closely linked to the developments in the area of carbenes and nitrenes.

Matrix isolation coupled with IR spectroscopy offers structural information on the observed species. This method has become increasingly popular in the recent years due to developments in computational chemistry. In the years before, because of intractable difficulties in unambiguous interpretations of IR spectra, the most common way of using IR spectroscopy for identification of reactive species was the generation of the desired species from more than one different precursor molecules. This time-consuming and painstaking procedure was a serious limitation of the method. However, nowadays, IR spectra can be predicted computationally with enough accuracy that structural and vibrational assignments can be done with high confidence. The relative ease with which such calculations can be carried out, undoubtedly, has been a major factor in popularizing matrix isolation IR spectroscopy.

High-energy-content molecules, like carbenes and nitrenes, can exhibit photochemical and sometimes thermal reactivity, even when isolated in matrices at cryogenic temperatures. Because the products of such reactions are often singlet species, IR spectroscopy is currently the most useful method for gaining insight into this type of chemistry. In this connection, calculations are also invaluable. Because the structure of the products is not always intuitively obvious, computational chemistry provides a reasonably fast and rather effortless way to screen over a variety of potential candidates. In addition, computed transition states and energy barriers can supplement possible experimental data of thermal reactions, providing a more detailed picture of the intrinsic chemical behavior of such reactive intermediates.

In the study of phenylene-linked carbenes and nitrenes, the combination of experimental and computational techniques outlined here is required to achieve a comfortable level of confidence when interpreting experimental data or predicting physicochemical properties. Most of the parent carbenes and nitrenes, which arise from the coupling of two diradical sites via a phenylene linker (o-, m-, or p-phenylene), are now reasonable well known. Much of the data can be rationalized in terms of simple valence-bond representations. According to this picture, the overall system may be viewed as a composite of the two diradical subunits and the phenylene linker. Depending on the topology of the linker, the coupling between the reactive centers can be stronger or weaker.

Chemical modification of the substrate can modify the overall electronic configuration of the system, leading to compounds with different physical and chemical properties from the parent systems. The available data show that halogen substitution at the carbene center can be successful in certain cases in altering the electronic nature of the ground state. Such electronic configuration switching has been achieved for p- and m-phenylene-linked systems, but has not been reported yet for o-phenylene-linked ones. The induced changes in the ground-state electronic configuration can be qualitatively understood if the intrinsic preference of the radical subunits for closed-shell or open-shell configuration is taken into account. The changes have also been quantified, but further research in this direction is needed in order to verify whether these quantitative relations are restricted to halogen substitution or have wider implications.

Several of the phenylene-linked carbenes and nitrenes exhibit photochemical or thermal reactivity in the matrix. The photolabile p- and m-phenylene-linked species give products of rather unexpected structures via mechanisms that are not understood yet. The o-phenylene-linked species isomerize rather easily either via ring opening of the phenylene linker or by an apparent direct reaction of the two proximal diradical centers to give ring-closure products. The available data imply that it may be possible to control the relative importance of the isomerization paths by chemical manipulation of the diradical site.

ACKNOWLEDGMENTS

The research presented in this chapter has been supported by the Ministry of Education, Science, Culture, Sports, Science and Technology, Japan and Nishida Foundations. Several of the calculations were carried out at the Computer Center of IMS, Okazaki.

REFERENCES

1. Eisental, K. B. In *Ultrashort Light Pulse*; Shapiro, S., ed; Spinger-Verlag: Berlin, 1977; Chap. 5.
2. Sitzman, E. V.; Eisental, K. B. *Application of Picosecond Spectroscopy to Chemistry*; Reidel 1984.
3. Platz, M. S.; Maloney, V. M. In *Kinetics and Spectroscopy of Carbene and Biradicals*, Platz, M. S., ed.; Plenum: New York, 1990.
4. Sun, X. Z.; Virrels, I. G.; George, M. W.; Tomioka, H. *Chem. Lett.* **1996**, 1089.
5. Wang, Y.; Yuzawa, T.; Hamaguchi, H.; Toscano, J. P. *J. Am. Chem. Soc.* **1999**, *121*, 2875.
6. Wang, Y.; Toscano, J. P.; *J. Am. Chem. Soc.* **2000**, *122*, 4512.
7. Trozzolo, A. M. *Acc. Chem. Res.* **1968**, *1*, 329.
8. Trozzolo, A. M.; Wasserman, E. In *Carbenes*; Moss, R. A.; Jones, M., Jr., eds; Wiley: New York, 1975; Vol. II, p. 185.
9. Sander, W.; Bucher, G.; Wierlacher, S. *Chem. Rev.* **1993**, *93*, 1583.
10. Mayer, B. *Low Temperature Spectroscopy*; Elsevier: New York, 1971.
11. Hallam, H. E. *Vibrational Spectroscopy of Trapped Species*; Wiley: London, 1973.
12. Craddock, S.; Hinthclife, A. J. *Matrix Isolation*; Cambridge University Press: Cambridge, 1975.
13. Dunkin, I. R. *Chem. Soc. Rev.* **1980**, *9*, 1.
14. Sheridan, R. S. In *Organic Photochemistry*; Padwa, A., ed.; Marcel Dekker: New York, 1987; Vol. 8. pp. 159–248.
15. Hehre, W. J.; Radom, L.; Pople, J. A.; Schleyer, P. v. R. *Ab Initio Molecular Orbital Theory*; Wiley: New York, 1986.
16. Jensen, F. *Introduction to Computational Chemistry*; Wiley: Chichester, 1999.
17. Kirmse, W. *Carbene Chemistry*, 2nd ed.; Academic Press: New York, 1971
18. Moss, R. A., Jones, M., Jr., eds.; *Carbenes*; Wiley: New York, 1973, 1975; Vols. I and II.
19. Jones, W. M. In *Rearrangements in Ground and Excited States*; de Mayo, P., ed.; Academic Press: New York, 1980; Vol. 1, Chap. 3.
20. Wentrup, C. *Reactive Molecules*; Wiley–Interscience: New York, 1984; Chap. 4.
21. Wentrup, C. In *Azides and Nitrenes*; Scriven, E. F. V., ed.; Academic Press: New York, 1984; pp. 395–432.

22. Gaspar, P. P.; Hsu, J.-P.; Chari, S.; Jones, M. *Tetrahedron* **1985**, *41*, 1479.
23. Regitz, M., ed.; *Carbene (oide), Carbine*; Houben-Weyl, Thieme: Stuttgart, 1989; Vol. E19b.
24. Brinker, U., ed.; *Advances in Carbene Chemistry*; JAI Press: Greewich, CT, 1989; Vols. 1 and 2.
25. Zuev, P. S.; Sheridan, R. S. *Tetrahedron* **1995**, *51*, 11,337.
26. Iwamura, H. *Adv. Phys. Org. Chem.* **1990**, *26*, 179.
27. Schaefer, H. F. *Science* **1986**, *231*, 1100.
28. Jensen, P.; Bunker, R. P. *J. Chem. Phys.* **1988**, *89*, 1327.
29. Shavitt, I. *Tetrahedron* **1984**, *41*, 1531.
30. Engelking, P. C.; Lineberger, W. C. *J. Chem. Phys.* **1976**, *65*, 4323.
31. Platz, M. S. *Acc. Chem. Res.* **1995** *28*, 487.
32. Matzinger, S.; Bally, T.; Patterson, E. V.; McMahon, R. J. *J. Am. Chem. Soc.* **1996**, *118*, 1535.
33. Wong, M. W.; Wentrup, C. *J. Org. Chem.* **1996**, *61*, 7022.
34. Schreiner, P. R.; Karney, W. L.; Schleyer, P. v. R.; Borden, W. T.; Hamilton, T. P.; Schaefer, H. F. III. *J. Org. Chem.* **1996**, *61*, 7030.
35. Poutsma, J. C.; Nash, J. J.; Paulino, J. A.; Squires, R. R. *J. Am. Chem. Soc.* **1997**, *119*, 4686.
36. Geise, C. M.; Hadad, C. M. *J. Org. Chem.* **2000**, *65*, 8348.
37. Hayes, J. C.; Sheridan, R. S. *J. Am. Chem. Soc.* **1990**, *112*, 5879.
38. Travers, M. J.; Cowles, D. C.; Clifford, E. P.; Ellison, G. B. *J. Am. Chem. Soc.* **1992**, *114*, 8699.
39. Kim, S.-J.; Hamilton, T. P.; Schaefer, H. F. *J. Am. Chem. Soc.* **1992**, *114*, 5349.
40. Hrovat, D. A.; Waali, E. E.; Borden, W. T. *J. Am. Chem. Soc.* **1992**, *114*, 8698.
41. McDonald, R. N.; Davidson, S. J. *J. Am. Chem. Soc.* **1993**, *115*, 10,857.
42. Castell, O.; García, V. M.; Bo, C.; Caballol, R. *J. Comput. Chem.* **1996**, *17*, 42.
43. Borden, W. T.; Gritsan, N. P.; Hadad, C. M.; Karney, W. L.; Kemnitz, C. R.; Platz, M. S. *Acc. Chem. Res.* **2000**, *23*, 765.
44. Karney, W. L.; Borden, W. T. *J. Am. Chem. Soc.* **1997**, *119*, 1378.
45. Gritsan, N. P.; Likhotvorik, I.; Tsao, M.-L.; Flebi, N.; Platz, M. S.; Karney, W. L.; Kemnitz, C. R.; Borden, W. T. *J. Am. Chem. Soc.* **2001**, *123*, 1425.
46. Karney, W. L.; Borden, W. T. *J. Am. Chem. Soc.* **1997**, *119*, 3347.
47. Kemnitz, C. R.; Karney, W. L.; Borden, W. T. *J. Am. Chem. Soc.* **1998**, *120*, 3499.
48. Nicolaides, A.; Enyo, T.; Miura, D.; Tomioka, H. *J. Am. Chem. Soc.* **2001**, *123*, 2628.
49. Zuev, P. S.; Sheridan, R. S. *J. Org. Chem.* **1994**, *59*, 2267.
50. Dougherty, D. *Acc. Chem. Res.* **1991**, *24*, 88.
51. Bally, T.; Borden, W. T. In *Reviews in Computational Chemistry*; Lipowitz, K. B.; Boyd, D. B., eds.; Wiley: New York, 1999; Vol. 13, p. 1.
52. Kohn, W.; Becke, A. D.; Parr, R. G. *J. Phys. Chem.* **1996**, *100*, 12,974.
53. Baerends, E. J.; Grisenko, O. V. *J. Phys. Chem. A* **1997**, *101*, 5383.
54. Roos, B. O. In *Lecture Notes in Quantum Chemistry*; Roos, B. O., ed.; Springer-Verlag: Berlin, 1992; Vol. 58, p. 177.
55. Becke, A. D. *J. Chem. Phys.* **1992**, *97*, 9173.

56. Lee, C.; Yang, W.; Parr, R. G. *Phys. Rev. B.* **1988**, *37*, 785.
57. Hariharan, P. C.; Pople, J. A. *Chem. Phys. Lett.* **1972**, *16*, 217.
58. Scott, A. P.; Radom, L. *J. Phys. Chem.* **1996**, *100*, 16,502.
59. Johnson, B. G.; Gill, P. M. W.; Pople, J. A. *J. Chem. Phys.* **1993**, *98*, 5612.
60. Shavitt, I. In *Methods of Electronic Structure Theory*; Schaefer, H. F., ed.; Plenum Press: New York, 1977; p. 189.
61. Andersson, K.; Roos, B. O. In *Modern Electronic Structure Theory, Part I*; Advanced Series in Physical Chemistry, Yarkony, D. R., ed.; World Scientific: singapore, 1995; Vol. 2, p. 55.
62. Borden, W. T. In *Diradicals*; Borden, W. T., ed.; Wiley: New York, 1982; pp. 1–72.
63. Flock, M.; Pierloot, K.; Nguyen, M. T.; Vanquickenborne *J. Phys. Chem. A* **2000**, *104*, 4022.
64. Bourissou, D.; Guerret, O.; Gabbaï, F. P.; Bertrand, G. *Chem. Rev.* **2000**, *100*, 39.
65. Tomioka, H. *Acc. Chem. Res.* **1997**, *30*, 315.
66. Regitz, M.; Maas, G. *Diazo Compounds. Properties and Syntheses*, Academic Press: Orland O, FL, 1986.
67. Liu, M. T. H. *Chemistry of Diazirines*; LRC Press: Boca Raton, FL, 1987.
68. Smith, P. A. S. *Org. React.* **1962**, *3*, 337.
69. Minato, M.; Lahti, P. M. *J. Am. Chem. Soc.* **1997**, *119*, 2187.
70. Wierlacher, S.; Sander, W.; Liu, M. T. H. *J. Am. Chem. Soc.* **1993**, *115*, 8943.
71. Hrovat, D.; Borden, W. T. *J. Mol. Struct. (THEOCHEM)* **1997**, *398–399*, 211.
72. Baum, M. W.; Font, J. L.; Meislich, M. E.; Wentrup, C.; Jones, M., Jr. *J. Am. Chem. Soc.* **1987**, *109*, 2534.
73. Subhan, W.; Rempala, P.; Sheridan, R. S. *J. Am. Chem. Soc.* **1998**, *120*, 11528.
74. Trindle, C.; Datta, S. N.; Mallik, B. *J. Am. Chem. Soc.* **1997**, *119*, 12947.
75. Ohana, T.; Ouchi, A.; Moriyama, H.; Yabe, A. *J. Photochem. Photobiol. A: Chem.* **1993**, *72*, 83.
76. Trozzolo, A. M.; Murray, R. W.; Smolinsky, G.; Yager, W. A.; Wasserman, E. *J. Am. Chem. Soc.* **1963**, *85*, 2526.
77. Sixl, H.; Mathes, R.; Schaupp, A.; Ulrich, K.; Huber, R. *Chem. Phys.* **1986**, *107*, 105.
78. Higuchi, J. *J. Chem. Phys.* **1963**, *38*, 1237.
79. Itoh, K. *Chem. Phys. Lett.* **1967**, *89*, 235.
80. Warner, P. M. *J. Org. Chem.* **1996**, *61*, 7192.
81. Sander, W. *Angew. Chem., Int. Ed. Engl.* **1990**, *29*, 344.
82. Singh, B.; Brinen, J. S. *J. Am. Chem. Soc.* **1971**, *93*, 540.
83. Nimura, S.; Kikuchi, O.; Ohana, T.; Yabe, A.; Kaise, M. *Chem. Lett.* **1996**, 125.
84. Nicolaides, A.; Tomioka, H.; Murata, S. *J. Am. Chem. Soc.* **1998**, *120*, 11530.
85. Koseki, S.; Tomioka, H.; Toyota, A. *J. Phys. Chem. A* **1994**, *98*, 13203.
86. Zuev, P.; Sheridan, R. S. *J. Am. Chem. Soc.* **1993**, *115*, 3788.
87. Tomioka, H.; Komatsu, K.; Nakayama, T.; Shimizu, M. *Chem. Lett.* **1993**, 1291.
88. Zuev, P.; Sheridan, R. S. *J. Am. Chem. Soc.* **1994**, *116*, 9381.
89. Longuet-Higgins, H. C. *J. Chem. Phys.* **1950**, *18*, 265.
90. Borden, W. T. *Mol. Cryst. Liq. Cryst* **1993**, *232*, 219.

91. Alexander, S. A.; Klein, D. J. *J. Am. Chem. Soc.* **1988**, *110*, 3401.
92. Ovchinnikov, A. A. *Theoret. Chim. Acta* **1978**, *47*, 297.
93. Rajca, A. *Polym. News* **2001**, *26*, 43.
94. Fujita, I.; Teki, Y.; Takui, T.; Kinoshita, T.; Itoh, K.; Miko, F.; Sawaki, Y.; Iwamura, H. *J. Am. Chem. Soc.* **1990**, *112*, 4074.
95. Wasserman, E.; Murray, R. W.; Yager, W. A.; Trozzolo, A. M.; Smolinsky, G. *J. Am. Chem. Soc.* **1967**, *89*, 5076.
96. Higuchi, J. *J. Chem. Phys.* **1963**, *39*, 1847.
97. Tukada, H.; Mutai, K.; Iwamura, H. *J. Chem. Soc., Chem. Commun.* **1987**, 1159.
98. Rule, M.; Matlin, A. R.; Dougherty, D. A.; Hilinski, E.; Berson, J. A. *J. Am. Chem. Soc.* **1979**, *101*, 5098.
99. Seeger, D. E.; Lahti, P. M.; Rossi, A. R.; Berson, J. A. *J. Am. Chem. Soc.* **1986**, *108*, 1251.
100. Itoh, K. *Pure Appl. Chem.* **1978**, *50*, 1251.
101. Fang, S.; Ming-Shi, L.; Hrovat, D. A.; Borden, W. T. *J. Am. Chem. Soc.* **1995**, *117*, 6727.
102. Kanno, F.; Inoue, K.; Koga, N.; Iwamura, H. *J. Am. Chem. Soc.* **1993**, *115*, 847.
103. Dvolaitzky, M.; Chiarelli, R.; Rassat, A. *Angew. Chem., Int. Ed. Engl.* **1992**, *3*, 180.
104. West, A. P., Jr.; Silverman, S. K.; Dougherty, D. A. *J. Am. Chem. Soc.* **1996**, *118*, 1452.
105. Sander, W. *Spectrochim. Acta* **1987**, *43A*, 637.
106. Enyo, T.; Nicolaides, A.; Tomioka, H. *J. Org. Chem.* **2002**, *67*, 5578.
107. Yabe, A. *Bull. Chem. Soc. Jpn.* **1979**, *51*, 789–795.
108. Nicolaides, A.; Nakayama, T.; Yamazaki, K.; Tomioka, H; Koseki, S.; Stracener, L. L.; McMahon, R. J. *J. Am. Chem. Soc.* **1999**, *121*, 10563.
109. Hall, H. J.; Patterson, E. *J. Am. Chem. Soc.* **1967**, *89*, 5856.
110. Campbell, C. D.; Rees, C. W. *J. Chem. Soc. C* **1969**, 742.
111. Adger, B. M.; Keating, M.; Rees, C. W.; Storr, R. C. *J. Chem. Soc., Perkins Trans. I* **1975**, 41.
112. Adger, B. M.; Rees, C. W.; Storr, R. C. *J. Chem. Soc., Perkins Trans. I* **1975**, 45.
113. Koseki, S.; Tomioka, H.; Yamazaki, K.; Toyota, A. *J. Phys. Chem. A* **1997**, *101*, 3377.
114. Tomioka, H., unpublished results.

4

Photoinduced-Electron-Transfer Initiated Reactions in Organic Chemistry

Philip Schmoldt, Heiko Rinderhagen, and Jochen Mattay
Universität Bielefeld, Bielefeld, Germany

I. INTRODUCTION

Electronic excitation of molecules lead to a drastic change of their reactivities. One effect of the excitation is the powerful change of the redox properties, a phenomenon which may lead to photoinduced electron transfer (PET) [1–4] The electron-donating as well as the electron-accepting behavior of the excited species are approximately enhanced by excitation energy. This can be explained by means of a simple orbital scheme. By excitation of either the electron donor (D) or the acceptor (A) of a given pair of molecules, the former thermodynamically unfavorable electron transfer process becomes exergonic (k_{et}) (Scheme 1).

In contrast to thermal electron-transfer processes, the back-electron transfer (BET) (k_{bet}) in the PET is generally exergonic as well. The apparent contradiction can be resolved by the cyclic process: *excitation–electron transfer–back-electron transfer* in which the excitation energy is consumed. The back-electron transfer is not the formal reverse reaction of the photoinduced-electron-transfer step and so not necessarily endergonic. This has different influences on PET reactions. On the one hand, BET is the reason for energy consumption and low quantum yields. On the other hand, it can cause more complex reaction mechanisms if the

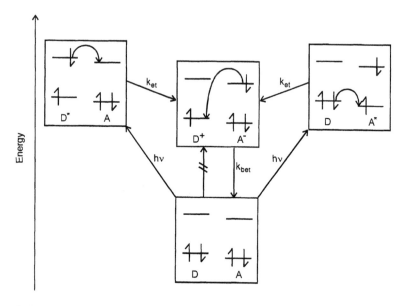

Scheme 1 Energetics of photoinduced electron transfer between a donor molecule and an acceptor molecule.

back electron transfer takes place after the primary reactions of the oxidized or reduced substrates.

Normally, the reaction partners in PET reactions are neutral molecules. That is why a donor radical cation—acceptor radical anion pair is obtained by the PET step. These highly reactive intermediates can be used for triggering interesting reactions. Since the PET is not restricted to neutral molecules PET reactions of donor anions and neutral acceptors or neutral donors and acceptor cations resulting in radical—radical anion (cation) pairs are known as well. These reactions are also called "charge shift reactions" due to the fact that the overall number of charged species is kept constant throughout the PET step. Finally, a PET process of a donor anion and a acceptor cation is possible as well (Scheme 2).

The free-enthalpy change ΔG of a PET process can be described by the simplified Rehm–Weller equation [5]:

$$\Delta G = F\,[E_{1/2}^{ox}\,(D)_{\mathrm{SolvA}} - E_{1/2}^{red}\,(A)_{\mathrm{SolvA}}] - \Delta E_{\mathrm{excit}} + \Delta E_{\mathrm{coul}} \qquad [1]$$

where

$E_{1/2}^{ox}\,(D)_{\mathrm{SolvA}}$ = oxidation potential of the donor (measured in the solvent A)

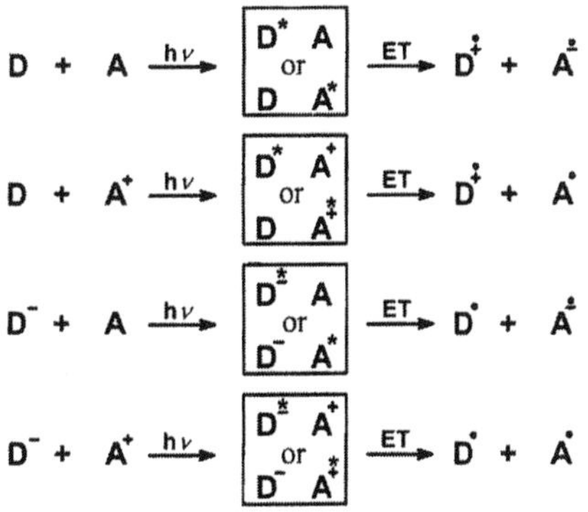

$$D + A \xrightarrow{h\nu} \boxed{\begin{array}{cc} D^* & A \\ & or \\ D & A^* \end{array}} \xrightarrow{ET} D^{\cdot +} + A^{\cdot -}$$

$$D + A^+ \xrightarrow{h\nu} \boxed{\begin{array}{cc} D^* & A^+ \\ & or \\ D & A^{+*} \end{array}} \xrightarrow{ET} D^{\cdot +} + A^{\cdot}$$

$$D^- + A \xrightarrow{h\nu} \boxed{\begin{array}{cc} D^{-*} & A \\ & or \\ D^- & A^* \end{array}} \xrightarrow{ET} D^{\cdot} + A^{\cdot -}$$

$$D^- + A^+ \xrightarrow{h\nu} \boxed{\begin{array}{cc} D^{-*} & A^+ \\ & or \\ D^- & A^{+*} \end{array}} \xrightarrow{ET} D^{\cdot} + A^{\cdot}$$

Scheme 2 Photoinduced – electron – transfer, charge shift, and charge recombination processes.

$E_{1/2}^{red} (A)_{SolvA}$ = reduction potential of the acceptor (measured in the solvent A)

ΔE_{excit} = excitation energy

ΔE_{coul} = the Coulomb term

In the Coulomb term, the electrostatic and solvation effects depending on the polarity of the media are summarized. For solvent-separated ion pairs (solvation energy calculated by the Born equation), it is given by

$$\Delta E_{coul} = \frac{e^2 N_a}{4\pi\epsilon_0 a} \left(\frac{1}{\epsilon} - \frac{2}{\epsilon_{SolvA}} \right) \qquad [2]$$

where a is the radical ion pair distance and ϵ is the dielectric constant of the reaction PET solvent. It shows that the thermodynamic driving force of a PET reaction increases with solvent polarity. In charge shift reactions, the Coulomb term is nearly negligible, because of the absence of strong ionic electrostatic interactions and the partial compensation of solvation energy of the donor and the acceptor redox pairs.

The primary products of a PET reaction are generally radicals or radical ions. These reactive species can react in different ways to produce new molecules. The most common reactions of radicals and radical ions are the mutual reverse reactions *addition* and *fragmentation*. In contrast to radicals, there are two different ways for radical ions to fragment. One way is the fragmentation into a radical ion and a neutral molecule. The second way leads to two different reactive intermediates: a radical and an ion (Scheme 3).

Other reaction pathways in PET-initiated reactions are secondary electron-transfer reactions and additions of radicals and radical ions (Scheme 4).

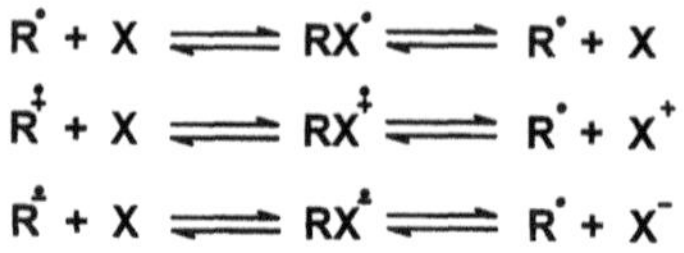

$$R^{\bullet} + X \rightleftharpoons RX^{\bullet} \rightleftharpoons R^{\bullet} + X$$

$$R^{\ddagger} + X \rightleftharpoons RX^{\ddagger} \rightleftharpoons R^{\bullet} + X^{+}$$

$$R^{\bullet}_{\bullet} + X \rightleftharpoons RX^{\bullet}_{\bullet} \rightleftharpoons R^{\bullet} + X^{-}$$

Scheme 3 Reactions of radicals and radical ions.

$$R^{+} \xleftarrow{-e^{-}} R^{\bullet} \xrightarrow{+e^{-}} R^{-}$$

$$R^{2+} \xleftarrow{-e^{-}} R^{\ddagger} \xrightarrow{+e^{-}} R$$

$$R \xleftarrow{-e^{-}} R^{\bullet}_{\bullet} \xrightarrow{+e^{-}} R^{2-}$$

$$R1^{\bullet} + R2^{\bullet} \longrightarrow R1R2$$

$$R1^{\ddagger} + R2^{\bullet} \longrightarrow R1R2^{+}$$

$$R1^{\bullet}_{\bullet} + R2^{\bullet} \longrightarrow R1R2^{-}$$

$$R1^{\ddagger} + R2^{\bullet}_{\bullet} \longrightarrow R1R2$$

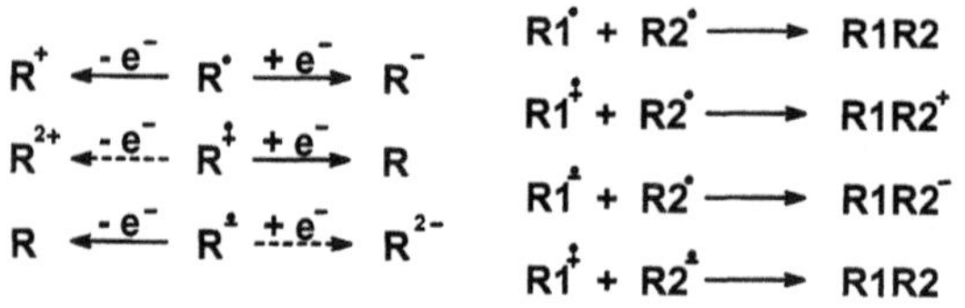

Scheme 4 Electron-transfer and addition reactions of radicals and radical ions.

To simplify the complex reaction pathways of PET reactions and to make them more transferable to multiple substrates it is sometimes advisable to carry them out in a sensitized way. The sensitizer has three characteristics: the substrate is excited for the primary PET process, its resulting radical ion or radical is so inert that it does not react with the substrate, and, in most cases, the sensitizer is regenerated by back-electron transfer. A simplified mechanism of a sensitized PET reaction is shown in Scheme 5.

Except for the back-electron-transfer step, these reactions are similar to electron-transfer reactions with oxidizing or reducing agents and are often used

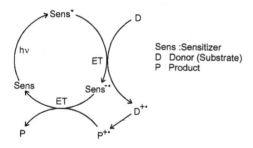

Sens :Sensitizer
D   Donor (Substrate)
P   Product

Scheme 5 Simple sensitized PET process.

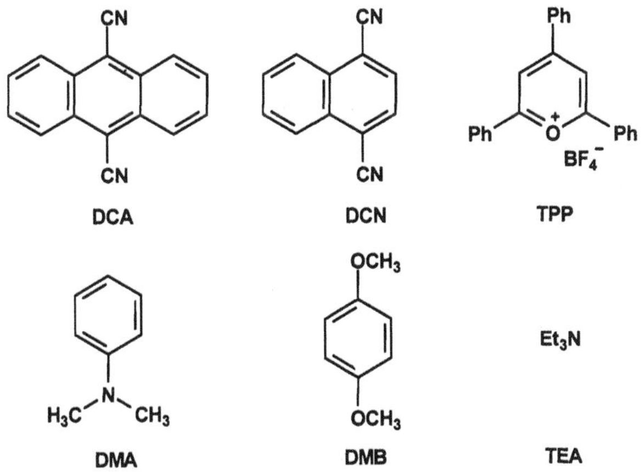

Scheme 6 PET sensitizers.

as an alternative for these reactions. For oxidative purposes, electron-deficient aromatics (e.g., dicyanoanthracene, dicyanonaphthaline, or triphenylpyrillium salts) are often used. For reductive PET reactions, electron-donor-substituted aromatics or amines are applied. The latter are not sensitizers in a strict way but sacrificial cosubtrates, because they are consumed during the reaction and, in most cases, they are not the excited reaction partners in the PET (Scheme 6).

Sensitized PET reactions are often very slow and have low quantum yields due to dominating back-electron transfer. In these cases, the addition of cosubtrates (e.g., biphenyl or phenanthrene to DCA- or DCN-sensitized reactions) is useful. The use of such an additive is called cosensitization. In these reactions, the substrate is not oxidized (or reduced) by the excited sensitizer but by the radical ion of the cosensitizer (ET_b). This is a thermal electron-transfer step without the problems of back-electron transfer. The key step is the primary PET process (ET_a) in which the cosensitizer radical ion is formed. The main characteristic of cosensitization systems is the high quantum yield of the free-radical ion (e.g., $\Phi_{Fri} = 0.83$ for DCA/biphenyl) [6]. For this reason, the overall quantum yield is high and the reaction is fast (Scheme 7).

An additional effect of the cosensitization may lead to different products or product ratios, which is caused by the efficient sensitizer radical ion—substrate ion separation. This separation inhibits the early back-electron transfer to the substrate radical ion or early intermediates and favors products of complex reaction pathways (late ET_t)

In the following section, we give examples of preparative and mechanistic investigations of PET reactions covering the last 6 years of research. As mentioned

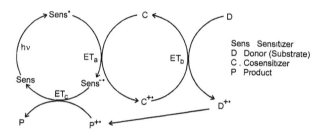

Scheme 7 Simple cosensitized PET process.

earlier, the primary products of a PET reaction (radicals and radical ions) can undergo various basic reactions. We will start with simple examples where one basic reaction dominates and will proceed to more complex reactions where many steps are involved.

II. FRAGMENTATION

Though the fragmentation is one of the basic reactions of radical ions, this destructive reaction pathway seems to be synthetically useless at first sight. Nevertheless, the electron-transfer-induced bond cleavage can be specific and, for this reason, synthetically useful (e.g., for ring enlargement reactions).

An useful alternative to the already known retropinacol reactions is presented by Liu and co-workers [7]. This works demonstrates that pinacols bearing (dimethylamino)phenyl substituents can be subjected to fast oxidative fragmentation via photoinduced electron transfer with chloroform as the electron acceptor in yields up to 80%. The extremely fast dechlorination of the chloroform radical anion inhibits back-electron transfer and thus leads to effective fragmentation of the pinacol radical cation (Scheme 8).

An illustrative example for photochemical cleavage of strained ring compounds is provided by Cossy and co-workers. The photoinduced single-electron

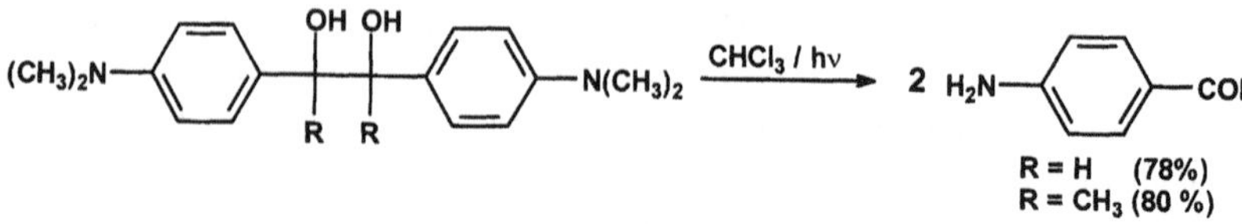

Scheme 8 Photoinduced retropinacol reaction.

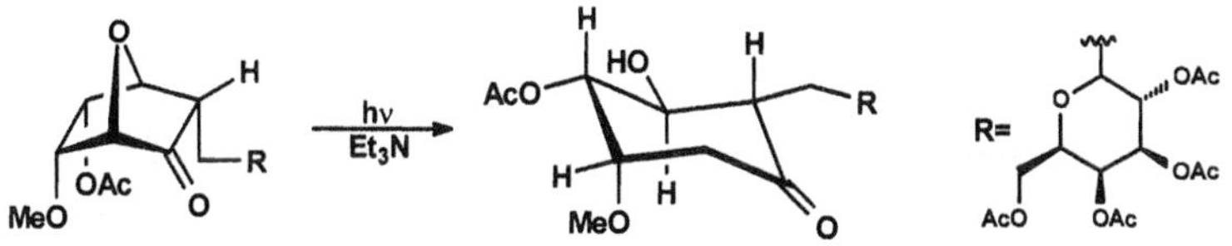

Scheme 9   PET-induced cleavage of 7-oxabicyclo[2.2.1]heptan-2-ones.

transfer from triethylamine to 7-oxabicyclo[2.2.1]heptan-2-ones resulted in the formation of 3-hydroxycyclohexanones in moderate to good yields, thus providing a suitable alternative to procedures employing samarium diiodide. Applying this method for the preparation of more complex structures was achieved with the synthesis of α-C-galactopyranosides of carbapentopyranoses derivatives, a new class of disaccharide mimics (Scheme 9) [8].

The regioselectivity in ketyl-radical-promoted ring cleavage of configurationally restricted tricyclo[3.3.0.0$^{2.8}$]octanone derivatives was investigated by Maiti and Lahiri, revealing new and additional information for this class of compounds in contrast to simple and unrestrained cyclopropyl ketones [9].

An interesting synthetic example involving the reductive transformation of α,β-epoxy ketones via PET employing 1,3-dimethyl-2-phenylbenzimidazoline (DMPBI) as the electron donor was presented by Hasegawa et al. [10]. Thereby, the cleavage of the epoxides produced high yields and complete conversion to the corresponding β-hydroxy ketones, making this PET reaction procedure a powerful substitute for conventional ones (Scheme 10).

The same group investigated the photoreaction of halomethyl-substituted benzocyclic ketones with amines yielding ring expansion products. The crucial reaction intermediate formed after cleavage of the carbon–halogen bond through intramolecular electron transfer is a cyclopropyloxy radical which then opens to an ring-enlarged ethoxy carbonyl-substituted carbon radical. Moderate yields of 40–50% are obtainable, making this photochemical reaction a useful alternative

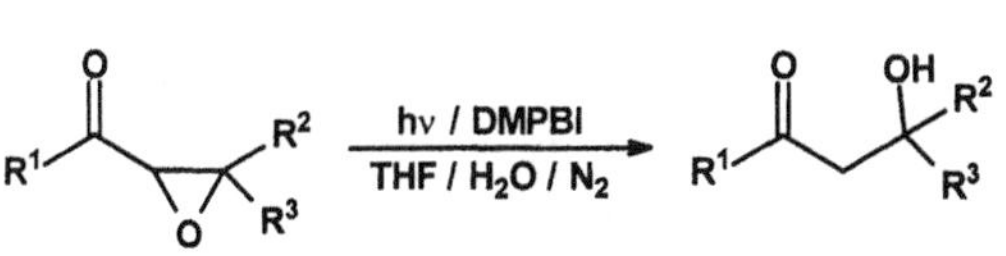

Scheme 10   Reductive cleavage of α,β-epoxy ketones.

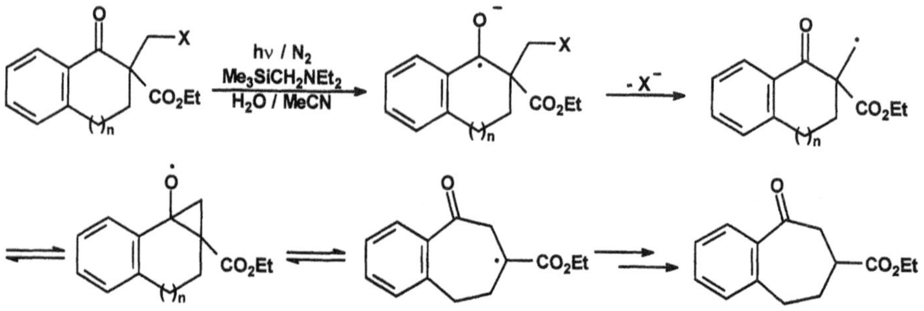

Scheme 11   Photochemical ring expansion reaction.

to the already know ring enlargement reaction promoted by tributyltin hydride (Scheme 11) [11].

In comparison to the above examples, PET reactions are not only useful for fragmentation of cyclopropyl systems but are also successfully employed in the photoinduced reduction of *gem*-dichlorocyclopropane as reported by Ogawa and co-workers [12]. Upon irradiation with visible light ($\lambda > 300$ nm), *gem*-dichlorocyclopropanes undergo reductive dechlorination with samarium diiodide and thiophenol to provide the corresponding cyclopropanes in yields in the range 60–70%. Although thermal samarium diiodide-promoted dechlorination is already well known, excess amounts of HMPA are normally required. This major disadvantage, the toxicity of HMPA, is avoided by using this new photochemical procedure.

A novel ring enlargement strategy employing a [2+2] cycloaddition followed by oxidative radical decarboxylation was reported by Piva et al. The cyclobutane framework was formed by irradiation of an oxoacid and a cycloalkene, yielding mainly the cis-anti-cis isomer. The ring enlarged product was obtained by subsequent irradiation of the adduct in toluene with small amounts of acridine, continuous bubbling of oxygen, and treatment with dimethyl sulfide in 30–50% yield (Scheme 12) [13].

III.   DIMERIZATION

Radicals produced in the course of a PET reaction can also yield dimerization products. This kind of reaction is interesting not only from the mechanistic point of view but also from the products derived from such a reaction.

Mechanistic insight into the PET reaction of aryl-substituted tropylium cations to aryl-substituted bitropyls via reduction as well as the oxidative counterpart

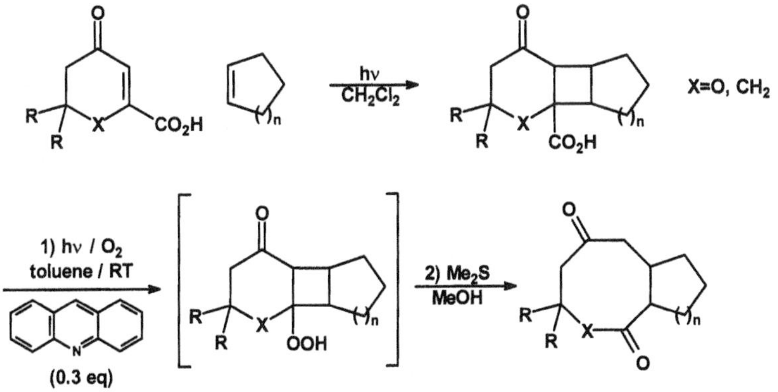

Scheme 12   Ring enlargement reaction by cyclobutane ring cleavage.

starting from aryl-substituted cycloheptatrienes was reported by Abraham et al. using laser flash and electron spin resonance (ESR) spectroscopy [14–16].

Nakamura and co-workers provided detailed mechanistic information for the photoinduced electron transfer from tri-1-naphthyl phosphate and related compounds to 9,10-dicyanoanthracene yielding binaphthyls The intramolecular nature of the reaction could be established by using laser flash photolysis experiments as well as fluorescence measurements [17].

Moriwaki et al. studied the photochemical behavior of various substituted homonaphthoquinones under PET conditions using amine and arene compounds as electron donors. Thereby, the main reaction path found was the opening of the cyclopropyl system, following dimerization [18].

Xu and co-workers studied the photoinduced electron-transfer reaction of flavone with triethylamine, yielding 2,2′-biflavone and another bilflavonid derivative in a combined yield of 80–90% [19]. These products derive from the radical addition of an 1,4-ketyl or an 1,2-ketyl anion to flavone, respectively. This investigation is not only worth mentioning because of the biological and physiological aktivities of flavone but also because of flavone serving as a suitable compound for studying the photochemistry of α,β-unsaturated carbonyls (Scheme 13) [19].

IV. NUCLEOPHILIC CAPTURE

Radical cations as well as cations are good electrophiles dependent on steric and electronic effects and, therefore, potent targets for nucleophilic capture.

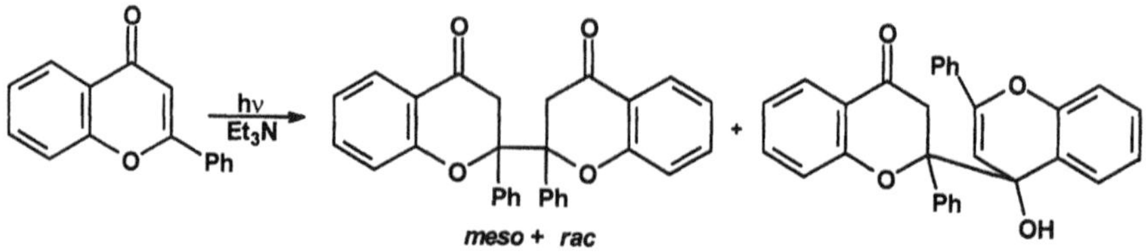

Scheme 13   Reductive PET flavone dimerization.

The stereoselective reactivity of radical cations of 7-benzhydrylidenenor-bornene and its derivatives toward the addition of water and methanol was investigated by Hirano et al., revealing that among the π–π interactions in these radical cations, the interaction between the benzhydrylidene group and the *endo* olefin induces the efficient nucleophilic capture of the radical cation at the position 7 with the anti selectivity to the endo olefin. Generation of the radical cations was achieved by using the sensitizer system phenanthrene/1,4-dicyanobenzene at an irradiation wavelength above 334 nm. Yields, conversion rates, and selectivities were highly dependent on the derivatization and thus emphasize favorable attack of nucleophiles anti to the endo olefin of the norbornene skeleton (Scheme 14) [20].

A more theoretical work is presented by Kojima and co-workers [21], who investigated the regioselective 1,4-addition of ammonia to 1-arylalka-1,3-dienes and 1-aryl-4-phenylbuta-1,3-dienes by photoinduced electron transfer. This photoamination proceeds by nucleophilic addition of NH_3 to the radical cations of the diene generated by photoinduced electron transfer to 1,4-dicyanobenzene (DCB). The resulting regiochemistry is related to the positive-charge distribution in the radical cations of the starting material calculated by the PM3-UHF/RHF method, the stability of the aminated radicals formed by NH_3 addition, and the stability of the aminated anions formed by reduction of the aminated radicals by

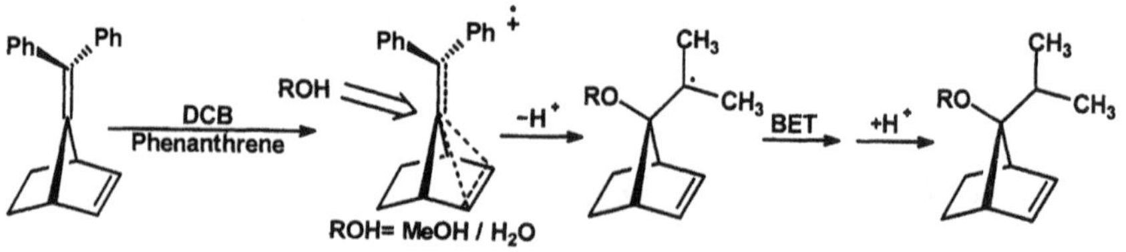

Scheme 14   Nucleophilic capture of 7-benzhydrylidenenorbornene cations.

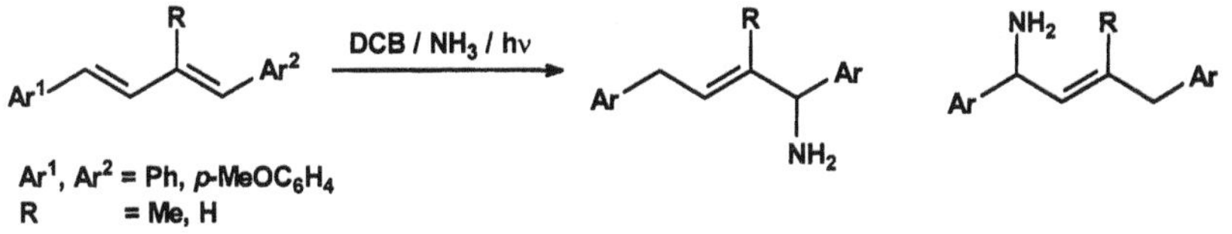

Scheme 15   Photoamination of 1-aryl-4-phenylbuta-1.3-dienes.

DCB radical anions. Stabilities of the intermediates were estimated by calculated heats of formation. Employing these, the theoretical results were in good agreement with the experiment (Scheme 15) [21].

V.   PHOTOINDUCED RADICAL REACTIONS

Photoinduced single-electron transfer followed by fragmentation of the radical cation is an efficient method for generating carbon-centered radicals under exceptionally mild conditions. The fate of the thus formed radicals depends primarily on their interaction with the acceptor radical anions. Typically observed reactions are either back-electron transfer or radical coupling, but from the synthetic point of view, another most intriguing possibility is the trapping of the radical with suitable substrates such as olefins (Scheme 16).

A.   Intermolecular Radical Additions to Unsaturated Compounds

Albini and co-workers were able to trap radicals by alkenes giving rise to two processes, namely the radical olefin addition–aromatic substitution and the addi-

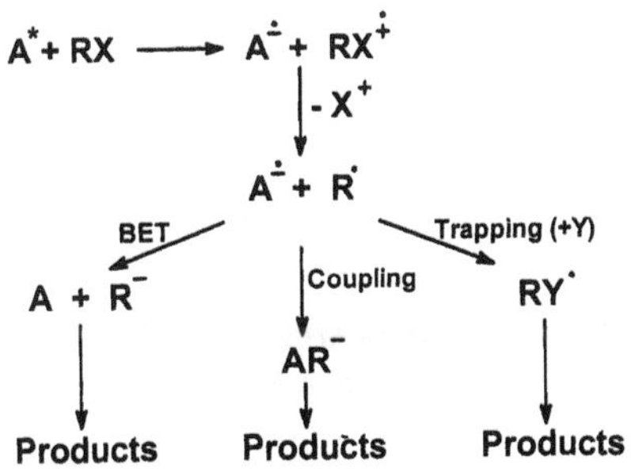

Scheme 16   Formation of radicals by oxidative electron transfer and the following reactions.

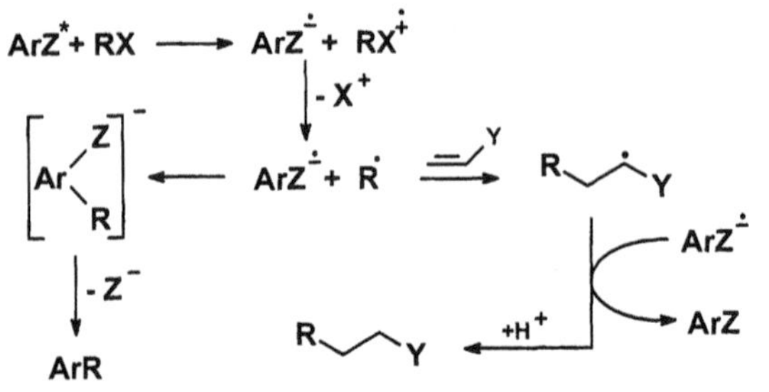

Scheme 17 Radical addition to alkenes versus radical olefin addition–aromatic substitution reaction.

tion to the alkenes used. Depending on the radical precursors and experimental conditions, these reactions predominate [22]. Further studies explored the scope of this reaction sequence by using different radical precursors like *tert*-butyl- and *iso*-propyltrimethylstannanes, various unsaturated substrates including alkynes as radical traps, photosensitizers, and the benefits of cosensitization (Scheme 17) [23].

Recent studies conducted by the same group revealed that the radical cation of toluene generated by photoinduced electron transfer can be deprotonated in a protic cosolvent and thus efficient trapping by electrophilic alkenes is feasible, yielding benzylation products. Secondary hydrogen abstraction by the benzyl radical from methanol generates hydroxymethyl radicals, which can also be used for preparative hydroxymethylation of alkenes (Scheme 18) [24].

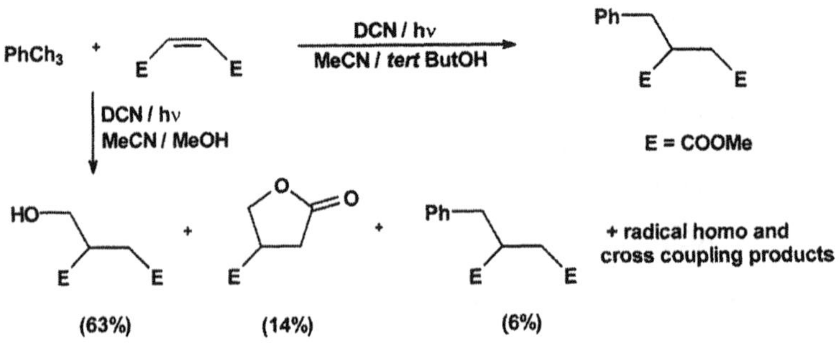

Scheme 18 DCA-sensitized reactions of toluene.

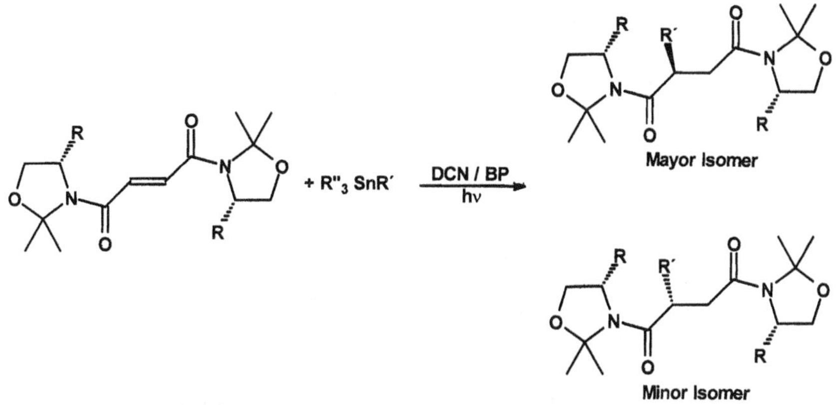

Scheme 19 Diastereoselective radical addition.

In addition, 2-substituted chiral succinic acid derivatives can be obtained via diastereoselective radical addition of 2-dioxolanyl radicals to fumaric acid diamides with a *de* up to 98% (Scheme 19) [25].

Highly efficient and stereoselective addition of tertiary amines to electron-deficient alkenes is used by Pete et al. for the synthesis of necine bases [26,27]. The photoinduced electron transfer of tertiary amines like *N*-methylpyrrolidine to aromatic ketone sensitizers yield regiospecifically only one of the possible radical species which then adds diastereospecifically to (5*R*)-5-menthyloxy-2-(5*H*)-furanone as an electron-poor alkene. For the synthesis of pyrrazolidine alkaloids in approximately 30% overall yield, the group uses a second PET step for the oxidative demethylation of the pyrrolidine. The resulting secondary amine react spontaneously to the lactam by intramolecular aminolysis of the lactone (Scheme 20) [26,27].

The addition of PET generated α-aminoalkyl radicals followed by an oxidative dealkylation of the amine can be carried out in an one pot reaction as well. Although only low yields in the range of 20–30% could be obtained by the anthraquinone—sensitized photoreaction of *N*-allylamines with α,β-unsaturated esters leading to lactams in an one pot reaction as reported by Das et al., this method generates α-aminoalkyl radicals without prior derivatization of the amines needed. One major advantage of using tertiary amines is that thermal Michael addition does not interfere with the photocatalyzed reaction (Scheme 21) [28].

Another illustrative example where captodative olefins could be functionalized which is otherwise difficult to achieve or even impossible is the photochemical alkylation of ketene dithioacetal *S,S*-dioxides. Because substituted α-thiosul-

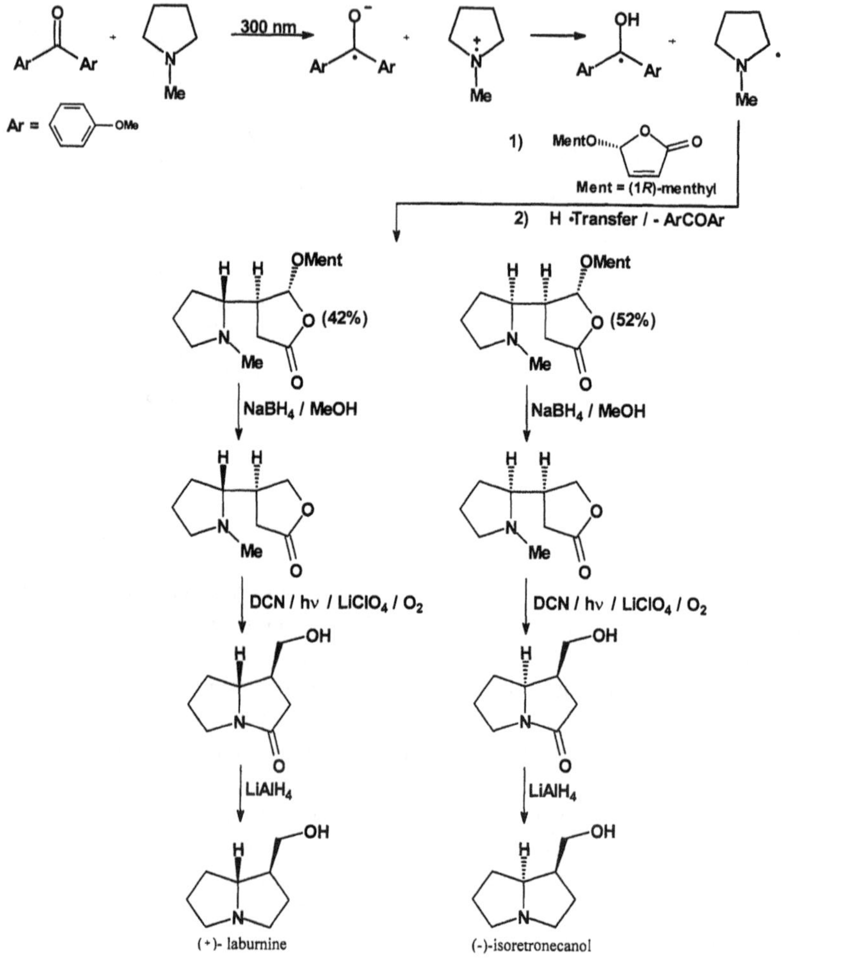

Scheme 20 PET-supported synthesis of necine bases.

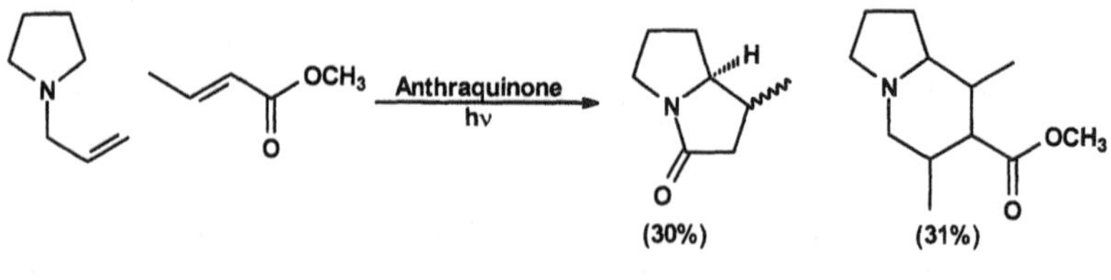

Scheme 21 Addition of PET-generated α-aminoalkyl radicals to α,β-unsaturated esters.

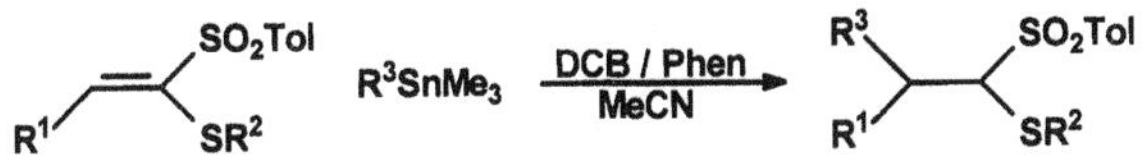

Scheme 22   Photoalkylation of ketene dithioacetal *S,S*-dioxides.

fones are convenient synthetic intermediates, their photochemical preparation is desirable and thus affording an alternative to direct alkylation of α-thiosulfones via enolates where the corresponding alkyl halides are not easily available (Scheme 22) [29].

Investigations by Griesbeck et al. concerning the regioselective and diastereoselective formation of 1,2-azidohydroperoxides by photo-oxygenation of alkenes in the presence of azide anions not only follow a basically unexplored route toward 1,2-N,O-functionalization of alkenes but also provide new mechanistic details helpful especially for biochemical investigations becuase azide salts are routinely applied to suppress singlet oxygen reactions (Scheme 23) [30].

α-Stannyl ethers and α-stannyl sulfides are also useful precursors for the carbon–carbon bond formation. These compounds yield the respective cations upon irradiation which then cleave into α-alkyloxo and thio radical intermediates with the elimination of the stannyl group and were efficiently added to 2-cyclohexen-1-one and other α,β-unsaturated ketones in up to 80% yield. These studies also revealed that the corresponding silyl derivatives were hardly reactive under the reaction conditions employed [31,32]. Nevertheless Mizuno et al. succeeded in the regioselective photoallylation and benzylation of 2,3-dicyanopyrazines by use of allylic and benzylic silanes (Scheme 24) [33].

The photoinduced thioselenation of allenes represent an additional example of a highly efficient and selective introduction of two different chalcogen func-

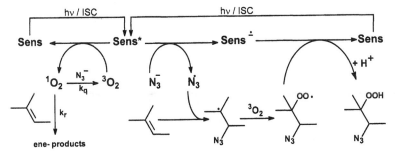

Scheme 23   Mechanism of 1,2-azidohydroperoxide formation.

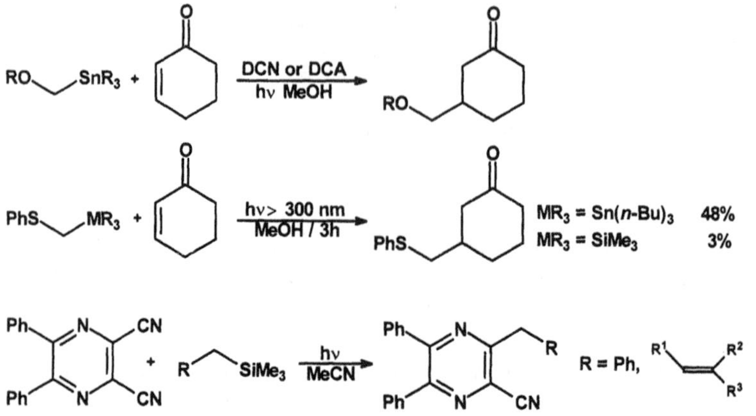

Scheme 24 Radical formation and addition by photoinduced cleavage of stannyl and silyl derivatives.

tions into allenes by using the binary system $(PhS)_2/(PhSe)_2$. Irradiation with light of 400 nm employing the system $(PhS)_2/(PhTe)_2$ leads to dithiolation of allenes, which could not be achieved by the radical reaction of $(PhS)_2$ alone (Scheme 25) [34].

In this context, an elegant alkylative carbonylation of alkenes accompanied by phenylselenyl transfer is described by Ryu and Sonoda. This three-component coupling involves the addition of a methoxycarbonylmethyl radical to an alkene, the trapping of the produced alkyl free radical by CO, and termination of the reaction by phenylselenenyl group transfer from the starting material. Yields are

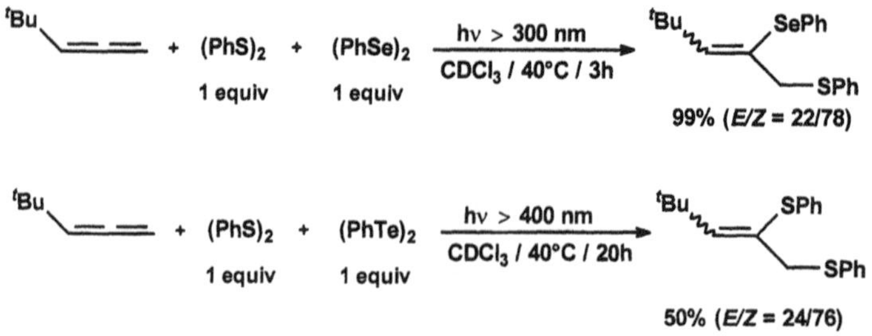

Scheme 25 Photochemical transformations of allenes.

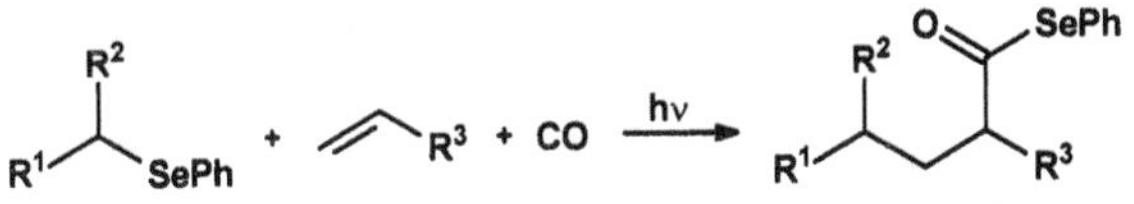

Scheme 26 Alkylative carbonylation of alkenes.

in the range of 20–70%, depending on the alkenes and selenides employed (Scheme 26) [35].

B. Cyclization Reactions

The synthesis of complex ring-systems is still a challenge in organic chemistry. With the exception of cycloadditions, the ring-building key step is always an intramolecular bond formation. In PET-initiated reactions, this step can be an addition (or recombination) reaction of a radical or radical ion or a nucleophilic capture of a cation or radical cation.

1. Cyclization Initiated by Radical Cations

Concerning their structure and reactions, organic radical cations have been the focus of much interest. Among bimolecular reactions, the addition to alkenes and their nucleophilic capture by alcohols, which lead to C—C and C—O bond formation, respectively have been investigated in detail. Unimolecular reactions like geometric isomerization and several other rearrangements have also attracted attention.

Mattay et al. examined the regioselective and stereoselective cyclization of unsaturated silyl enol ethers by photoinduced electron transfer using DCA and DCN as sensitizers. Thereby the regiochemistry (6-endo versus 5-exo) of the cyclization could be controlled because in the absence of a nucleophile, like an alcohol, the cyclization of the siloxy radical cation is dominant, whereas the presence of a nucleophile favors the reaction pathway via the corresponding α-keto radical. The resulting stereoselective cis ring juncture is due to a favored reactive chair like conformer with the substituents pseudoaxial arranged (Scheme 27) [36,37].

An interesting 6-*endo* cyclization of α-silylamino-enones and ynones was described by Mariano et al. 9,10-Dicyanoanthracene-sensitized photoreaction of α-silylamines generates carbon-centered α-amino radicals. Via intramolecular 6-endo addition to an olefin substituted by an electron-withdrawing group, a piperidine ring containing products can be constructed in moderate yields with a high degree of stereochemical and regiochemical control (Scheme 28) [38].

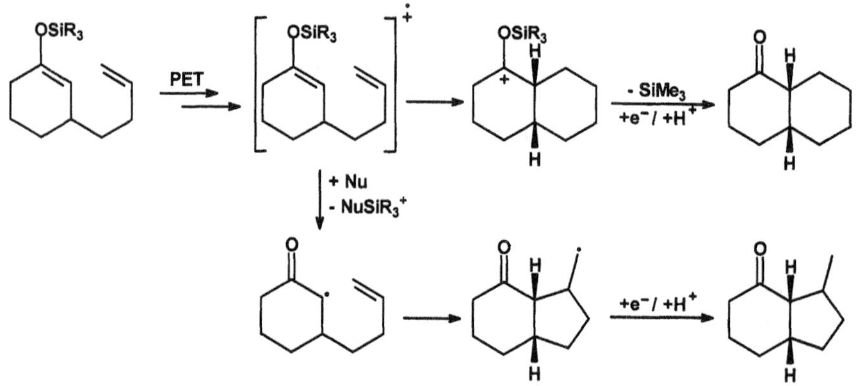

Scheme 27 Regioselective and stereoselective cyclization of unsaturated silyl enol ethers.

In addition to the former example, Pandey et al. achieved efficient α-arylation of ketones by the reaction of silyl enol ethers with arene radical cations generated by photoinduced electron transfer from 1,4-dicyanonaphthalene. Using this strategy various five-, six-, seven-, and eight-membered benzannulated compounds are accessible in yields in the range 60–70% [39].

Another example of stereoselective radical cation addition was presented by Hirano and co-workers. The reaction of 1,1-diphenyl-1,n-alkadienes employing 1,4-dicyanobenzene as a sensitizer yielded intramolecular tandem cyclization products in up to 60% yield (Scheme 29) [40].

Zimmerman and Hoffacker also observed a regioselective reaction subjecting various aryl-substituted 1,4-pentadienes to photoinduced electron transfer using DCN and DCA. The radical cations produced underwent a regioselective cyclization wherein one electron-deficient aryl group of one diarylvinyl moiety bonds to the β-carbon of the second diarylvinyl group (Scheme 30) [41].

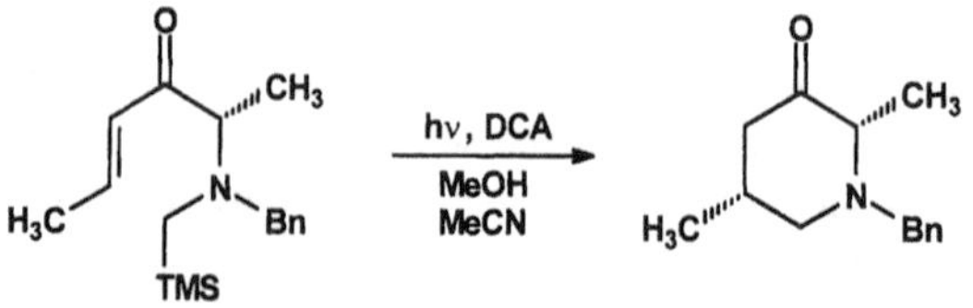

Scheme 28 6-Endo cyclization of α-silylamino-enones.

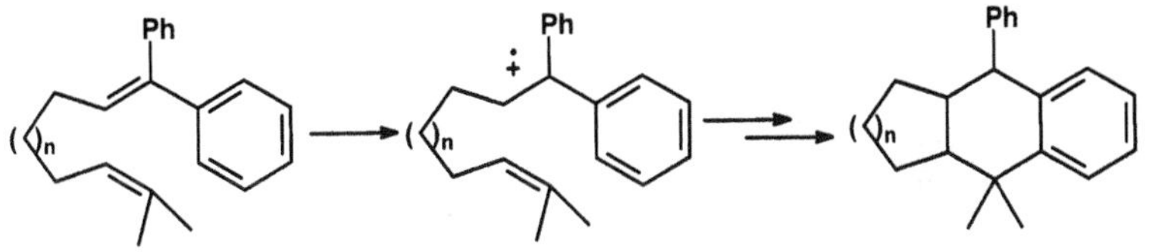

Scheme 29 Intramolecular tandem cyclizations of 1,1-diphenyl-1, *n*-alkadienes.

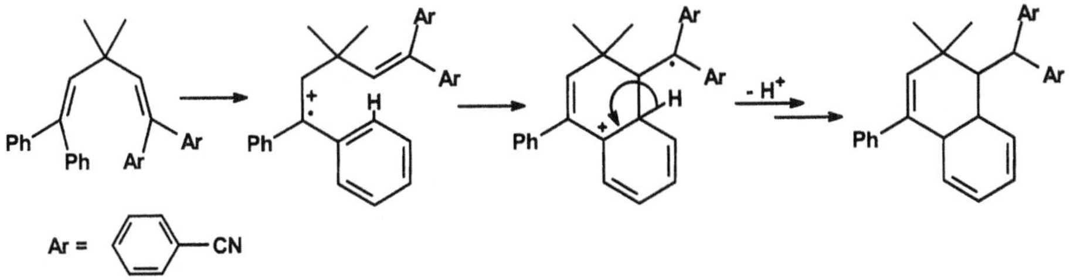

Ar = —CN

Scheme 30 Regioselective PET reactions of arylsubstituted 1,4-pentadienes.

A new cyclization strategy via PET-generated electrophilic selenium species starting from diphenyldiselenide was introduced by Pandey et al. They succeeded in the cyclization of enyne compounds in good yields in the range 50–60% as well as in synthesizing α,α'-*trans* dialkyl cyclic ethers via diastereoselective oxyselenation of 1,*n*-diolefins (Scheme 31) [42,43].

In this context, the investigations by Yasuda et al. concerning the cyclization of 1,2-distyrylbenzene should be mentioned. Photoamination of the former unsat-

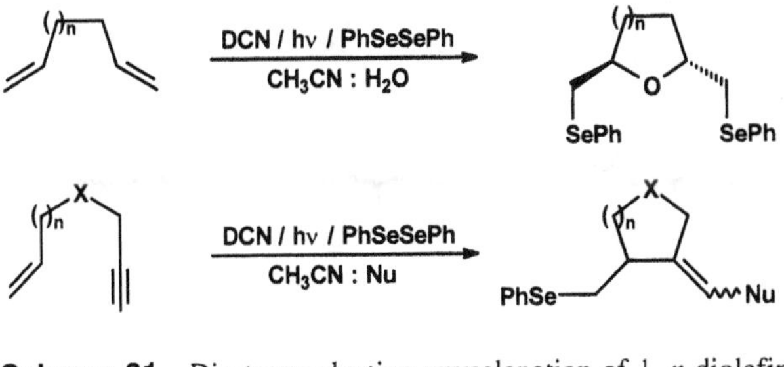

Scheme 31 Diastereoselective oxyselenation of 1, *n*-diolefins.

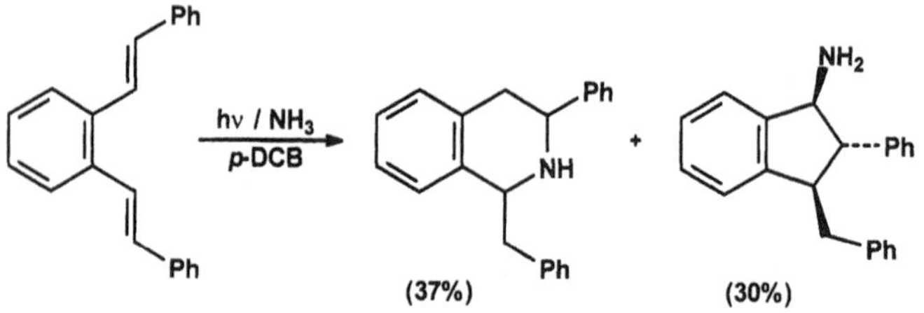

Scheme 32 Photoamination cyclization reaction of 1,2-distyrylbenzene.

urated compound with NH_3 in the presence of p-dicyanobenzene led to a tetrahy-droisoquinoline and an indane derivative, respectively, in a combined yield of ~ 60% (Scheme 32) [44].

Interesting results from a mechanistic point of view are reported by Roth and co-workers for the electron-transfer photochemistry of geraniol and farnesol employing DCA and DCN as sensitizers [45]. Product analysis revealed that the course of the reaction is governed by the free energy of electron transfer. As a consequence, a marginal driving force leads to contact radical ion pairs (CRIPs). The close proximity to the sensitizer causes the predominant formation of the cis-fused cyclopentadiyl radical cation and suppresses further cyclization by fast back-electron transfer. Solvent-separated radical ion pairs (SSRIPs) are derived from exergonic electron transfer. Under these reaction conditions, the radical cation tends to form the more stable trans-fused cyclopentadiyl radical cation, and due to slow back-electron transfer, a second cyclization by intramolecular nucleophilic capture of the cation takes place (Scheme 33) [45].

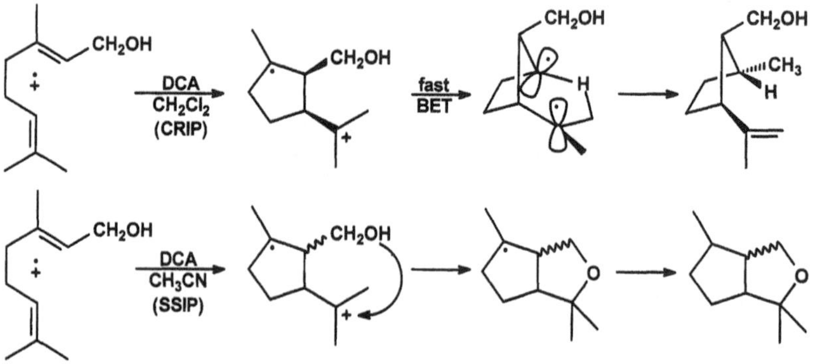

Scheme 33 Mechanistic studies on the PET-sensitized cyclization of geraniol.

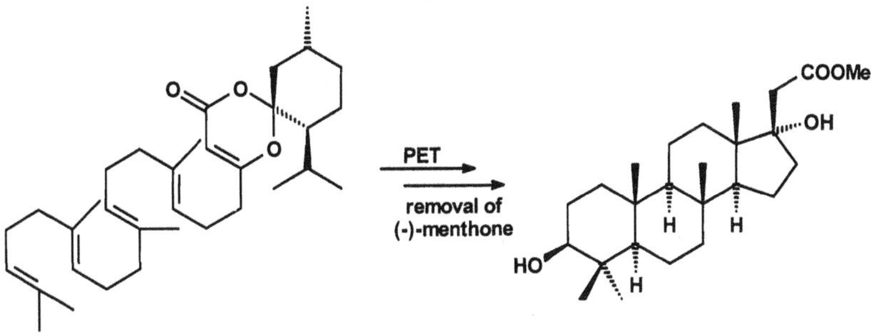

Scheme 34   Enatioselective formation of terpenoids.

Investigations in the same direction using suitably functionlized polyal-kenes readily available from geraniol and farnesol for the radical cationic cycliza-tion were conducted by Demuth et al.. The transformation is, without exception, highly stereoselective and chemoselective and the substitution pattern of the poly-alkene allows, optionally, the construction of five- and six-membered rings [46,47]. The same group provided an exemplary strategy to access enantiomer-ically pure tricyclic and tetracyclic terpenoids by the use of (−)-menthone as chiral auxiliary. Thereby, the efficient asymmetric induction is provided by a chiral spirocyclic dioxinone (Scheme 34) [48].

Transformations of molecules containing strained moieties like cyclopropyl and cyclobutyl rings have been the focus of interest not only from a synthetic point of view but also because of mechanistic considerations [49–52].

For example, during the last few years, Roth and co-workers have studied the electron-transfer chemistry of vinylcyclopropane [53,54]. Using 1,4-dicyano-benzene as a sensitizer, the corresponding radical cation is formed, which then can be trapped by nucleophiles. A detailed product analysis elucidated not only the regioselectivity of this cyclization but also the charge distribution of the intermediately formed radical cations. *Ab initio* calculations were conducted as well and were in good agreement with the experimental results [53]. Another molecule, a bridged norcaradiene, serving as a good probe for radical cation nucleophilic capture was investigated by the same group using similar techniques [54].

As a consequence of the above observations, the intramolecular capture of the vinylcyclopropane radical cation was attempted next. Thereby, chrysanthemol and homochrysanthemol both yielded the expected cyclization products (Scheme 35) [55].

The synthetic potential of reactions involving the cyclopropyl moiety is further shown by Cha and co-workers. One electron oxidation of readily available

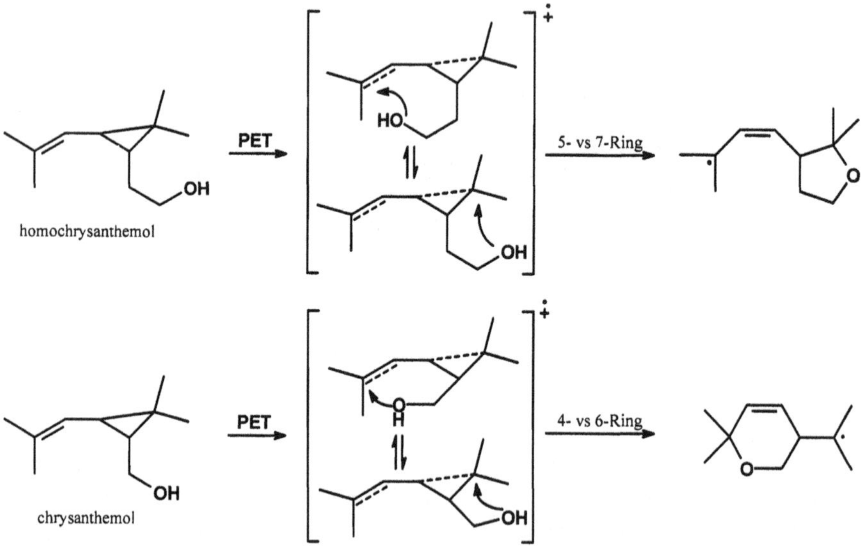

Scheme 35 Cyclization by intramolecular capture of vinylcyclopropane radical cations.

olefin-tethered cyclopropylamines to excited 1,4-dicyanobenzene cleaves the cyclopropyl ring and the two following 5-exo radical cyclizations produce the formal [3+2] annulation products in moderate yields (Scheme 36) [56].

2. Cyclizations Initiated by Radical Anions

Pandey and co-workers developed two photosystems useful for initiating one-electron reductive chemistry and applied them to activate α,β-unsaturated ketones. The resulting carbon-centered radicals cyclize stereoselectively with proximate olefins. Their concept involved a secondary and dark electron transfer from

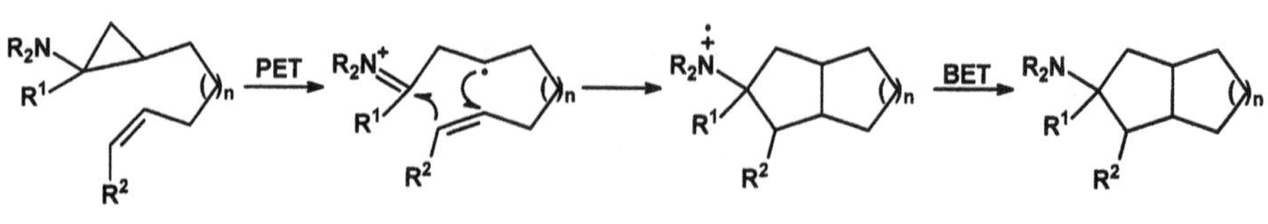

Scheme 36 Radical cascade reactions of substituted cyclopropylamines.

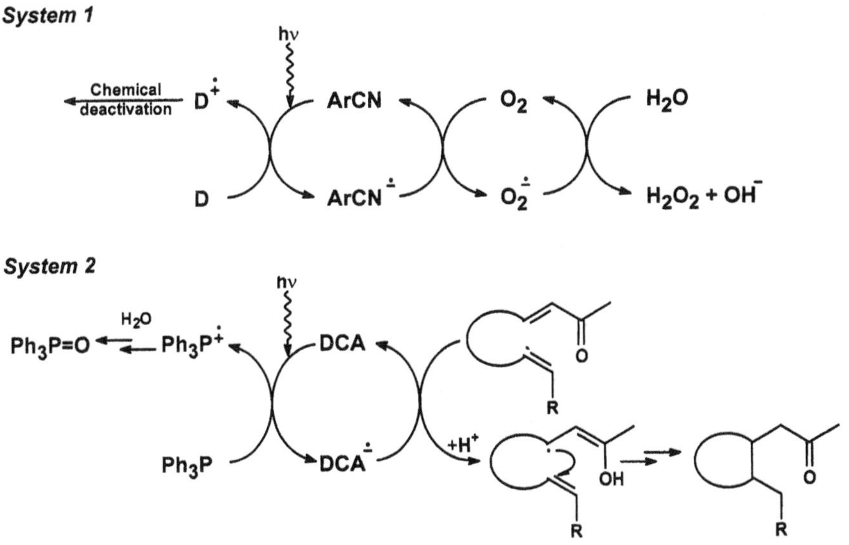

System 1

System 2

Scheme 37 PET-reductive photosystems.

a 9,10-dicyanoanthracene radical anion generated through a primary electron-transfer process employing different sacrificial electron donors (Scheme 37).

By using either one of these photosystems, one-electron β-activation of α,β-unsaturated carbonyl compounds produced carbon-centered radical precursors which cyclize efficiently and stereoselectively to tethered activated olefins or carbonyl groups. The 1,2-anti-stereochemistry observed contrasts with the general trend of syn-stereochemistry expected in 5-hexenyl radical cyclizations. Application of this methodology was successfully demonstrated by the stereoselective synthesis of optically pure C-furanoside, starting from L-tartaric acid (Scheme 38) [57,58].

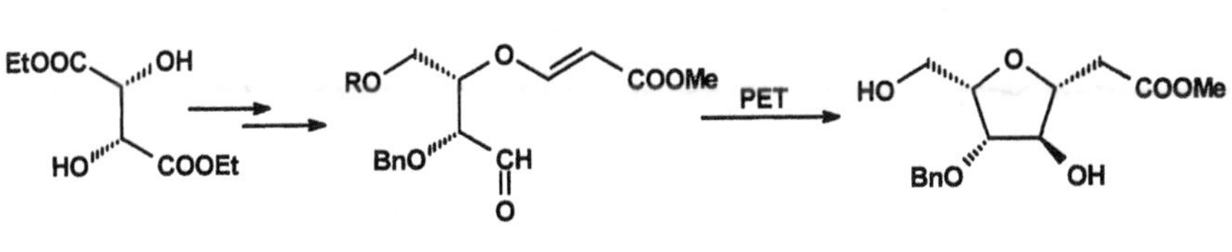

Scheme 38 PET-supported synthesis of optically pure C-furanosides.

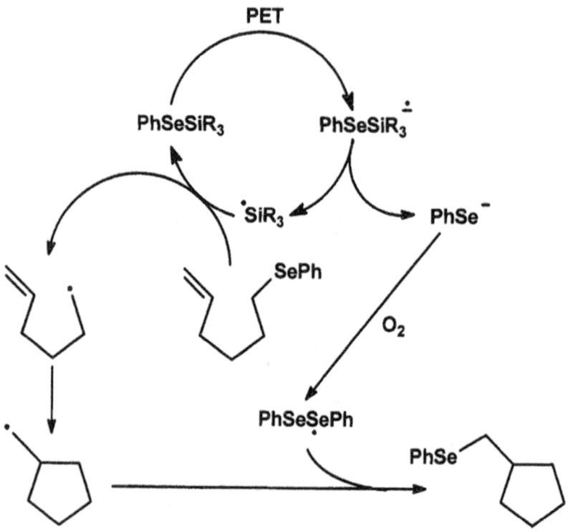

Scheme 39 Radical reactions of selenosilanes.

Investigations conducted by the same group using laser flash photolysis techniques elucidated details of the PET-reductive activation of selenosilanes and the application of this chemistry to a bimolecular group-transfer radical reaction and intermolecular radical chain-transfer addition [59]. Based on this new concept, a catalytic procedure utilizing $PhSeSiR_3$ for radical reactions such as cyclization, intermolecular addition and tandem anellation was designed (Scheme 39) [60].

A photoreductive strategy for a short total synthesis of the sesquiterpene ($\pm$)isoafricanol was developed by Cossy et al. Thereby, the hydroazulenol skeleton was generated by photoreductive cyclization of an unsaturated ketone employing triethylamine as the electron donor (Scheme 40) [61].

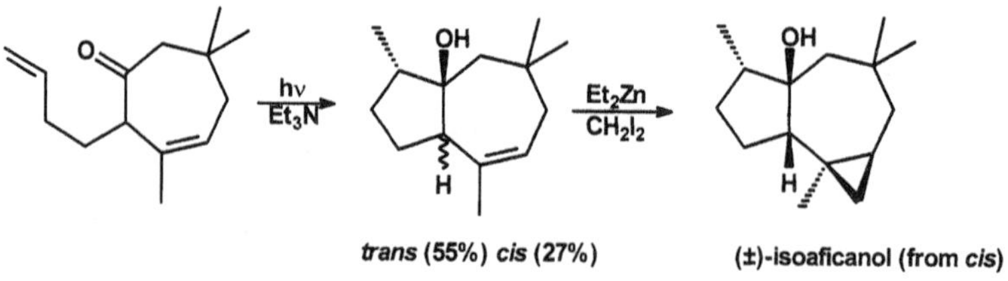

trans (55%) *cis* (27%)   (±)-isoaficanol (from *cis*)

Scheme 40 ($\pm$)Isoafricanol synthesis.

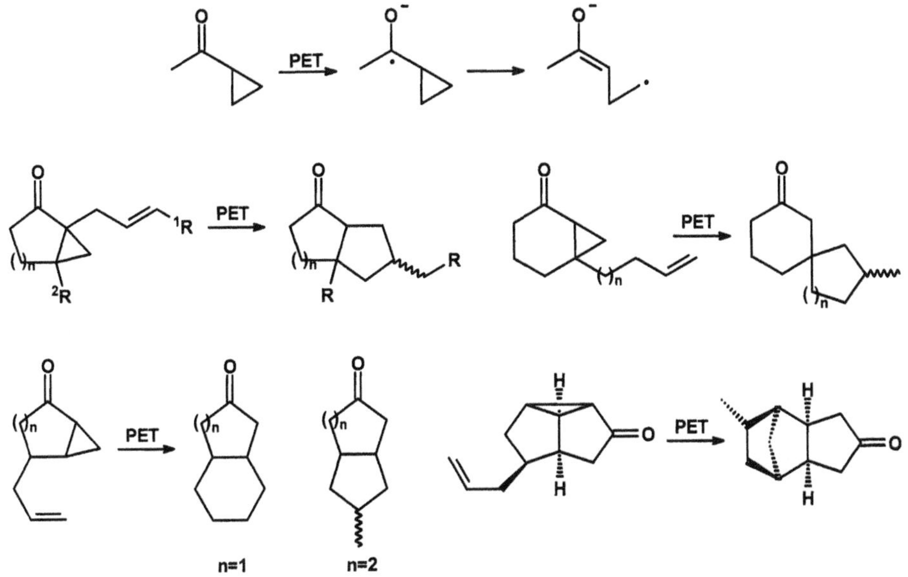

Scheme 41 PET-reductive fragmentation cyclization reactions.

Triethylamine as the electron donor was also used by Mattay and co-workers in tandem fragmentation cyclization reactions of α-cyclopropylketones. The initial electron transfer on the ketone moiety is followed by the fast cyclopropyl-carbinyl–homoallyl rearrangement, yielding a distonic radical anion. With an appropriate unsaturated side chain within the molecule both annealated and spirocyclic ring systems are accessible in moderate yields (Scheme 41) [62].

Vivona and co-workers studied the transformation of 5-aryl-1,2,4-oxadiazoles to quinazolin-4-one derivatives [63]. Irradiation of the 1,2,4-oxadiazole derivatives in the presence of either different donor sensitizers like diphenylacetylene and 9,10-diphenylanthracene or ground-state donors like triethylamine produces the key intermediate, an oxadiazole radical anion, from which breaking of the N—O bond takes place. Depending on the reaction conditions, quinazolin-4-ones are formed by heterocyclization together with varying amounts of ring-opened products resulting from a formal reduction at the ring O—N bond in moderate to good yields (Scheme 42) [63].

VI. DONOR-ACCEPTOR REACTIONS

In many PET reactions, the electron-transfer step is used for initiating reactions of the donor or the acceptor or both of them but on different pathways. In these

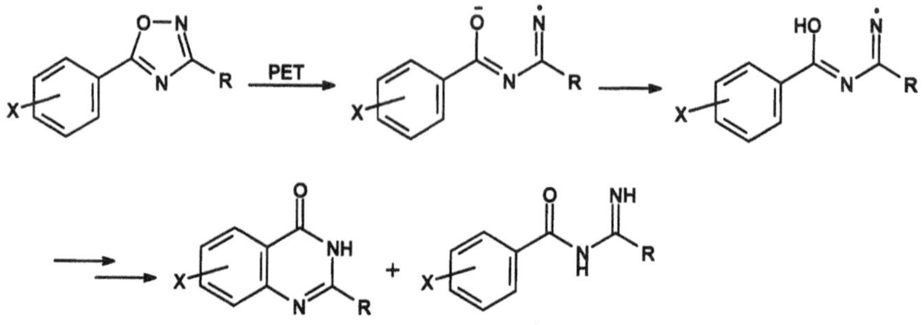

Scheme 42 Fragmentation and heterocyclization of oxadiazole radical anions.

cases, the undesirable direct reactions are suppressed (e.g., by polar solvents for fast ion pair separation). On the other hand, the mutual reactions of the donor and acceptor radical ions or their subsequent intermediates can intentionally be used for synthesizing interesting substrates.

An elegant procedure for the synthesis of γ-hydroxy ester derivatives employing cyclopropanone acetals was presented by Abe and Oku [64]. The oxidative ring opening of unsymmetrically substituted cyclopropanone acetals via a PET process with carbonyl compounds generates a β-carbonyl and ketyl radical pair. The following C—C bond formation reaction occurs at the sterically crowded β-position of the propanoates and yields the corresponding products in up to 80%. The addition of Mg(ClO₄)₂ to the reactants has a profound effect on product yield and suppression of side reactions (Scheme 43) [64].

Recently, Sugimoto et al. presented a strategy toward the synthesis of tricyclic lactones containing a 2-azabicyclo[3.3.1]nonane skeleton via tandem intra-

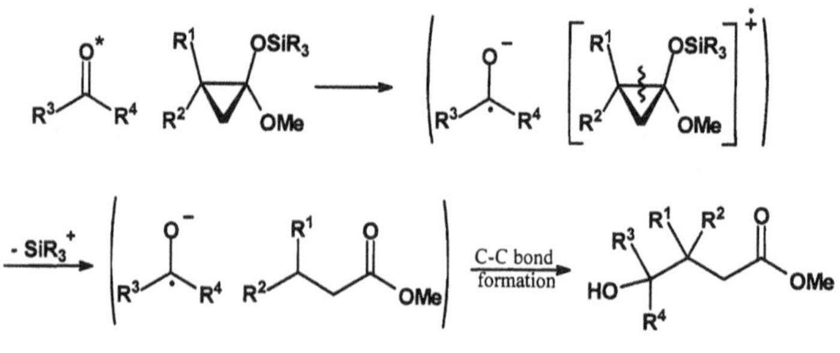

Scheme 43 PET synthesis of γ-hydroxy ester.

molecular photocyclization of 2-(1-naphthyl)ethyl-ω-anilinoalkanoate. Depending on the length of the polymethylene chain between the anilino and the naphthyl groups, yields were in the range 10–40%. Mechanistically, the reaction probably proceeds via a primary single-electron transfer from the anilino group to the naphthyl group, followed by proton transfer. The subsequent coupling leads to an intermediate that easily undergoes secondary electron transfer from the anilino group to the dihydronaphthyl moiety (Scheme 44) [65].

The photochemistry of phthalimide systems was thoroughly investigated by many groups over the last two decades. This chromophore shows a broad spectrum of reactivity leading mainly to cycloaddition and photoreduction products by either intermolecular or intramolecular processes. In the presence of electron donors, the electronically excited phthalimide could also undergo electron transfer and act as an electron acceptor.

Griesbeck et al. successfully transformed ω-phthalimidoalkanoates via PET with concomitant decarboxylation and C,C combination leading to medium- and large-ring compounds with yields in the range 60–80%. Thereby, the solvent system acetone/water and K_2CO_3 employed for the deprotonation of the carboxylic acids were crucial (Scheme 45) [66].

An intermolecular addition on phthalimides employing heteroatom-substituted carboxylates was reported by the same group. In that study, α-thioalkyl- and α-oxoalkyl-substituted carboxylates were decarboxylated in the presence of

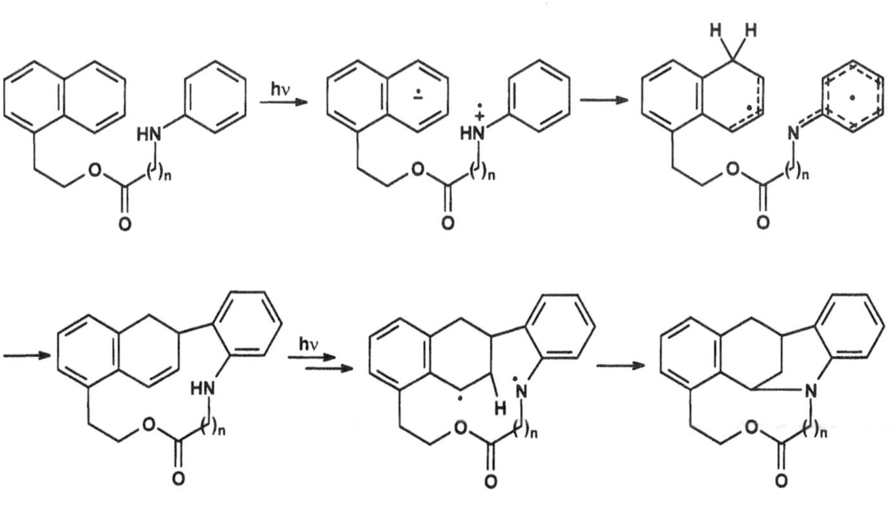

Scheme 44 Tandem intramolecular photocyclization of 2-(1-naphthyl)ethyl ω-anilinoalkanoate.

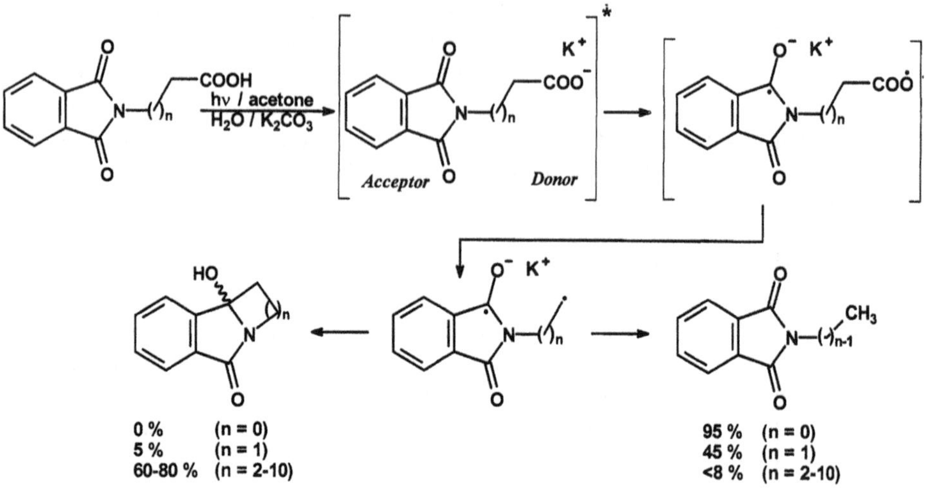

Scheme 45 Photochemistry of ω-phthalimidoalkanoates.

electronically excited phthalimides and led to radical coupling products in moderate yields (Scheme 46) [67].

Other electron-donating groups as well as spacers were also analysed. Alternatively, the (trimethylsilyl)methoxy substituent has been investigated by Yoon and co-workers as a versatile electron-donor group which, after oxidation via

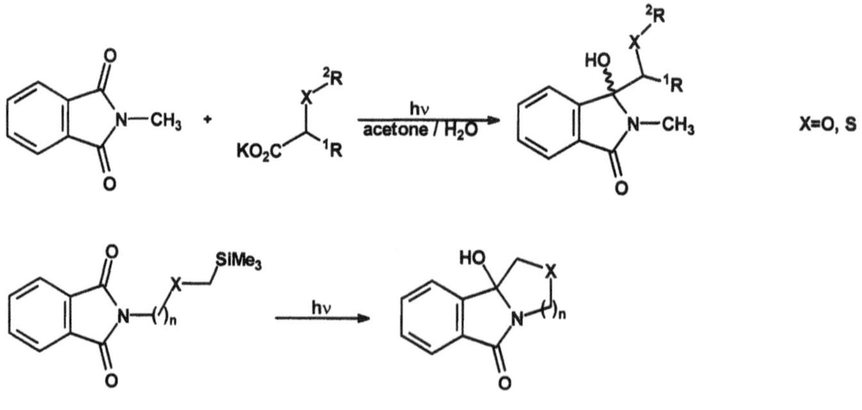

Scheme 46 Intermolecular addition to phthalimides.

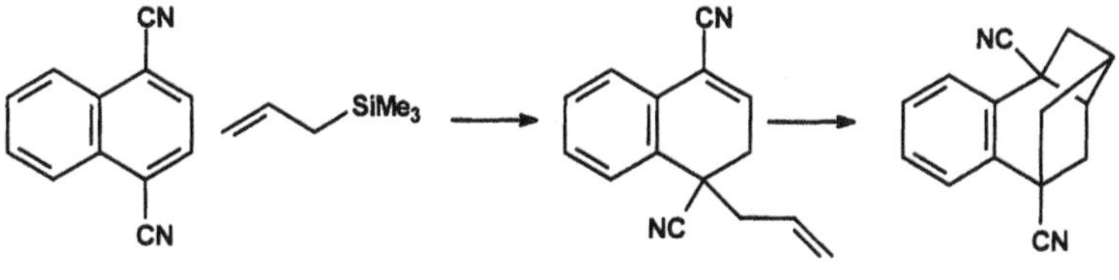

Scheme 47   Reductive photoallylation of mono-cyanonaphthalines and dicyanonaphthalines.

PET and desylilation, was converted into an oxy-stabilized C radical similar to that of Griesbeck et al. [68].

Mizuno et al. successfully synthesized benzotricyclo[4.2.1.0$^{3.8}$]nonenes by reductive photoallylation of monocyanonaphthalines and dicyanonaphthalines with allyltrimethylsilane, followed by an intramolecular [2 + 2] photocycloaddition (Scheme 47).

Although yields for the reactions with monocyanonaphthalines could exceed 70%, 2,3-disubtituted-1,4-dicyanonaphthalines failed to cyclize after the initial allylation reaction [69].

The photochemical nucleophile–olefin combination aromatic substitution (photo-NOCAS) reaction received considerable attention from many groups not only because of its synthetic value because the yields of nucleophile–olefin–arene (1:1:1) adducts can be high but also because of interesting mechanistic details (Scheme 48).

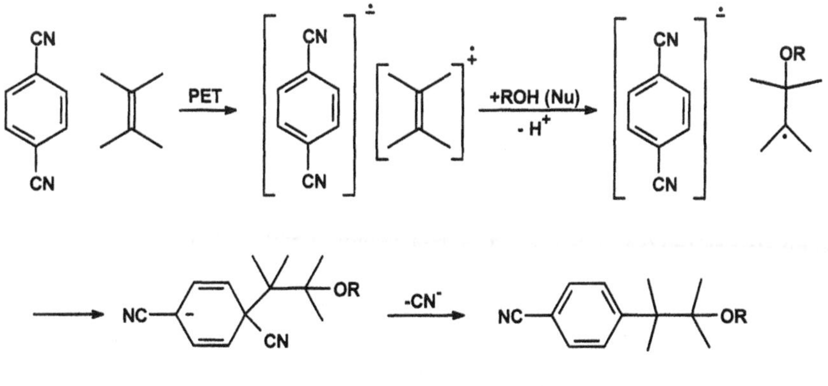

Scheme 48   Basic mechanism of photo-NOCAS reactions.

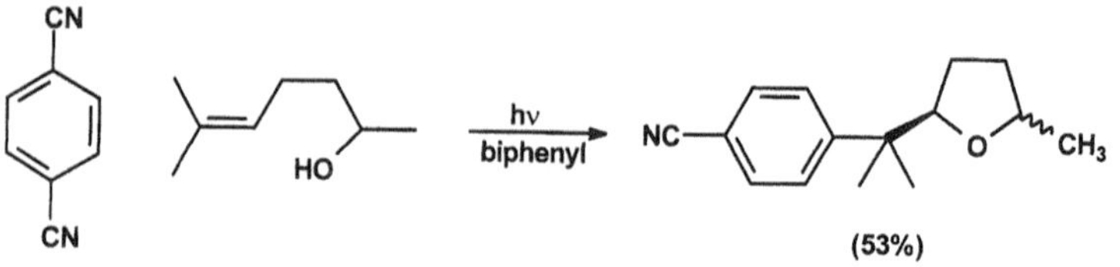

Scheme 49 Intramolecular cyclization by photo-NOCAS initiated reaction.

Among these are the factors controlling the regiochemistry of the reaction, the employed diene compounds, and the possible intramolecular cyclization of various alk-4-enol and ene–diene radical cations (Scheme 49) [70–72]. Also, by changing the nucleophile from the generally used methanol to, for instance, acetonitrile can yield photo-NOCAS products in good yields [73,74].

Recently, Kochi et al. described a novel photochemical synthesis for α-nitration of ketones via enol silyl ethers. Despite the already well-known "classical" methods, this one uses the photochemical excitation of the intermolecular electron-donor–acceptor complexes between enol silyl ethers and tetranitromethane. In addition to high yields of nitration products, the authors also provided new insights into the mechanism on this nitration reaction via time-resolved spectroscopy, thus providing, for instance, an explanation of the disparate behavior of α- and β-tetralone enol silyl ethers [75]. In contrast to the more reactive cross-conjugated α-isomer, the radical cation of β-tetralone enol silyl ether is stabilized owing to extensive π-delocalization (Scheme 50).

VII. CYCLOADDITION REACTIONS

There has been considerable interest in various photocycloaddition reactions over the last years which not only broadened the number of useful photochemical applications but also revealed further mechanistic insight into these reactions [76,77]. Among these reactions, reports focusing on either the [2 + 2] or the [4 + 2] cycloaddition, are numerous. Also the efforts toward the enantiodifferentiating photosensitization in photocyclization reactions have to be mentioned [78].

Blechert and co-workers successfully employed the [4 + 2] cycloaddition for the transformation of indole derivatives [79–81]. For instance, using 2-vinyl-indole derivatives as heterodienes, β-acceptor-substituted cyclic and acyclic ena-mines (dienophile), and triarylpyrylium tetrafluoroborate as the photosensitizer, the corresponding Diels–Alder adducts were formed in moderate to good yields with complete regiochemical and stereochemical control [79]. Alternatively, good results could be obtained in the reaction of indoles and exocyclic 1,3-dienes, thus providing an easy excess to multifunctionalized carbazoles [80]. Quantum

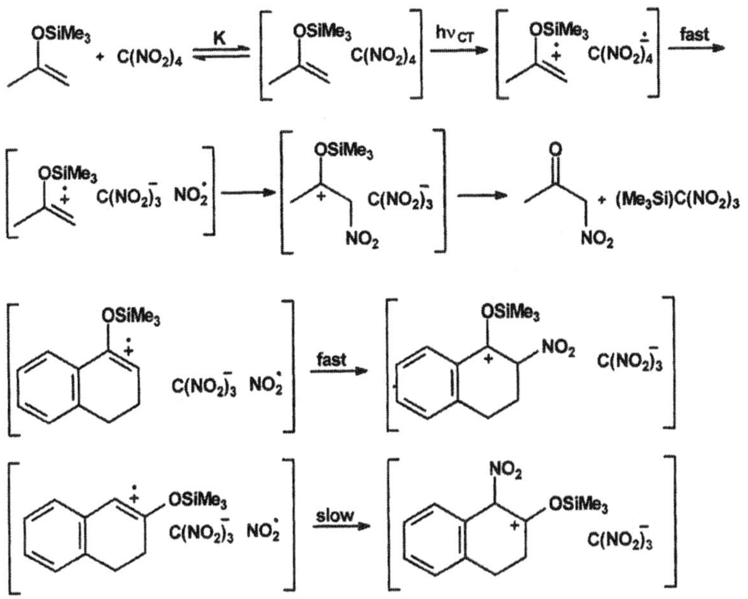

Scheme 50   Photochemical α-nitration of ketones via enol silyl ethers.

chemical calculations supply evidence for a nonsynchronous and nonconcerted reaction and lead to a reaction model consistent with the observed products (Scheme 51) [81].

The same group studied the radical cation cycloaddition of 2-vinylbenzofurans with various alkene and diene compounds initiated by photoinduced electron transfer. Depending on the unsaturated compound used, yields up to 60% were feasible. In contrast to 2-vinylindoles, 2-vinylbenzofurans prefer to react as dienophiles, very similar to styrenes [82].

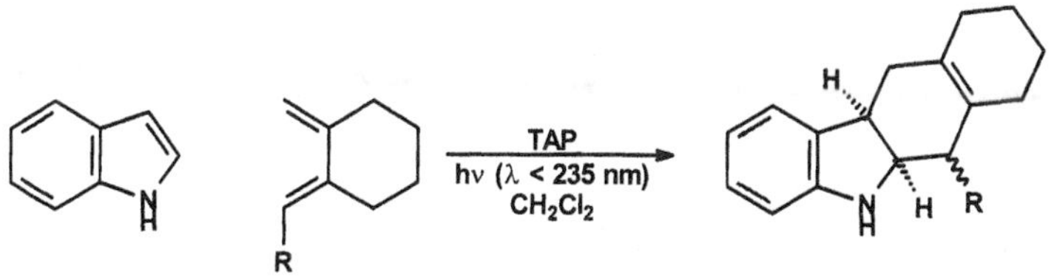

Scheme 51   [4 + 2] Cycloadditions of indole derivatives.

Kochi and co-workers studied photoinduced Diels–Alder cycloadditions via direct photoexcitation of anthracene as a diene with maleic anhydride and various maleimides as dienophiles. Here, fluorescence-quenching experiments, time-resolved absorption measurements, and the effect of solvent polarity provide striking evidence for an ion–radical pair to be the decisive intermediate [83].

The [3 + 2] cycloaddition of aziridines and dipolarophiles, like dimethyl acetylenedicarboxylate or dimethyl fumarate and maleate, was investigated by Gaebert and Mattay. Via C—C and C—N bond cleavage five-membered heterocycles are formed in moderate yields. The different product ratios dependent on the reaction conditions (PET/direct excitation/thermal reaction) gave insights to the reaction details and are summarized in the proposed mechanism (Scheme 52) [84].

A novel 1,8-photoaddition of dimethyl 1,4-naphthalenedicarboxylate and 1,4-dicyanonaphthalene to alkenes, a formal [3 + 2] cycloaddition, was reported by Kubo and co-workers. Although yields of the desired products were low, $Cu(OAc)_2$ showed a profound impact on the reaction by increasing the yields considerably [85].

The photoinduced reaction of chloranil with various 1,1-diarylethenes is another example of an intramoleclar [2 + 2] cycloaddition as reported by Xu and co-workers [86]. Although not interesting from the preparative point of view, the diverse reaction outcomes caused by parallel reaction pathways with and without single-electron transfer and various secondary reactions of the primary products show that the photochemistry involving haloquinones is far from being explored. Another interesting example in this context is the reaction of dichlorobenzoquinone with various diarylacetylenes in the solid phase via photoinduced electron transfer as reported by Kochi and co-workers [87]. Here, time-resolved spectroscopy revealed the radical ion pair of the two reactants to be the first reactive intermediate that then underwent coupling.

Detailed mechanistic information concerning an intramolecular arylalkene cycloaddition yielding cyclobutane derivatives via a radical cation process gener-

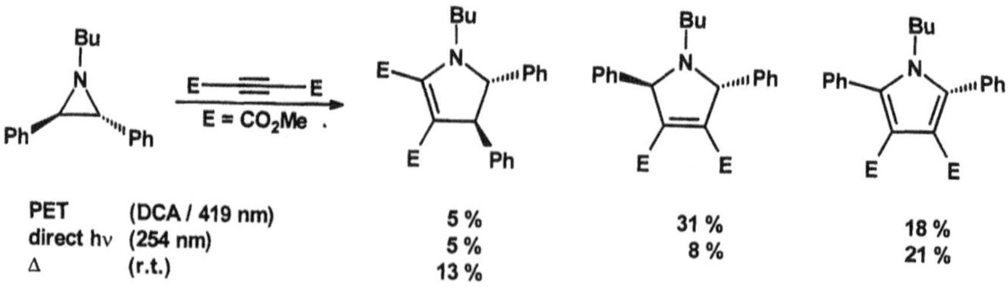

	PET	(DCA / 419 nm)	5 %	31 %	18 %
direct hv	(254 nm)		5 %	8 %	21 %
Δ	(r.t.)		13 %		

Scheme 52   [3 + 2] Cycloaddition of aziridines and dipolarophiles.

ated by either photoionization or photoinduced electron transfer was reported by Bauld and co-workers [88]. The results of the laser flash photolysis experiments on the picosecond and nanosecond time scale illustrate that limitation in the development and calibration of intramolecular radical cation probes based on aryalkene cycloaddition may be encountered.

In the context of photoinduced [2 + 2] cycloadditions, the reaction of an olefin with a carbonyl center, known as the Paterno–Büchi reaction, has to be mentioned. The resulting oxetanes are thereby produced with high regioselectivity and stereoselectivity.

For instance, Kochi and co-workers [89,90] reported the photochemical coupling of various stilbenes and chloranil by specific charge-transfer activation of the precursor donor–acceptor complex (EDA) to form *trans*-oxetanes selectively. The primary reaction intermediate is the singlet radical ion pair as revealed by time-resolved spectroscopy and thus establishing the electron-transfer pathway for this typical Paterno–Büchi reaction. This radical ion pair either collapses to a 1,4-biradical species or yields the original EDA complex after back-electron transfer. Because the alternative cycloaddition via specific activation of the carbonyl compound yields the same oxetane regioisomers in identical molar ratios, it can be concluded that a common electron-transfer mechanism is applicable (Scheme 53) [89,90].

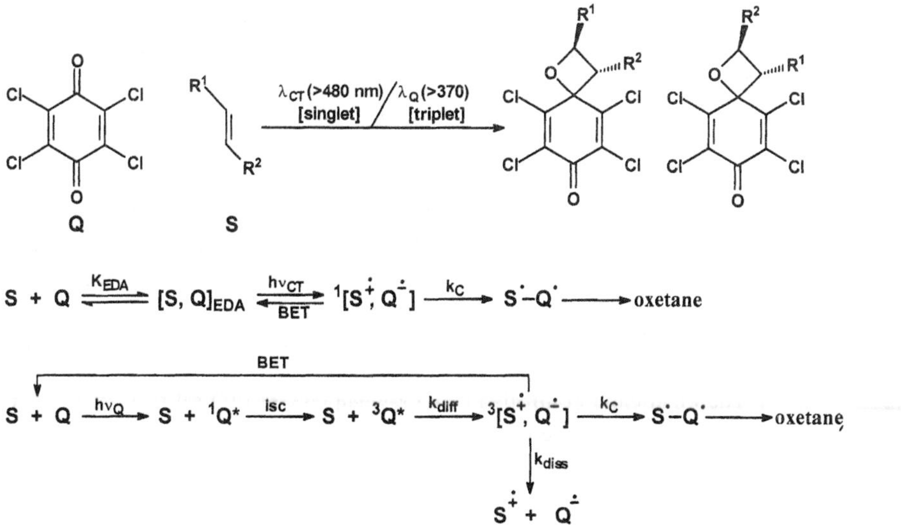

$$S + Q \underset{K_{EDA}}{\rightleftharpoons} [S, Q]_{EDA} \underset{BET}{\overset{h\nu_{CT}}{\rightleftharpoons}} {}^1[S^{+\cdot}, Q^{-\cdot}] \xrightarrow{k_C} S^{\cdot}-Q^{\cdot} \longrightarrow oxetane$$

$$S + Q \xrightarrow{h\nu_Q} S + {}^1Q^* \xrightarrow{isc} S + {}^3Q^* \xrightarrow{k_{diff}} {}^3[S^{+\cdot}, Q^{-\cdot}] \xrightarrow{k_C} S^{\cdot}-Q^{\cdot} \longrightarrow oxetane$$

$$\downarrow k_{diss}$$

$$S^{+\cdot} + Q^{-\cdot}$$

Scheme 53 Mechanistic details of the Paterno–Büchi reaction.

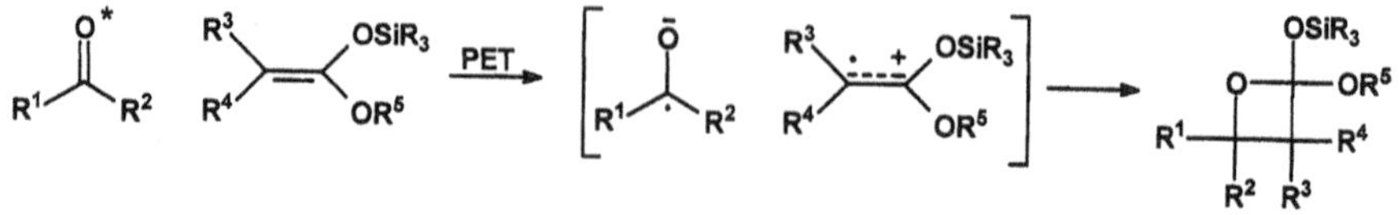

Scheme 54   Photoreaction of aromatic carbonyl compounds with β,β-dimethyl ketene silyl acetals.

Another good and synthetically useful example is provided by Abe and co-workers, who succeeded in the regioselective formation of 2-alkoxyoxetanes via the photoreaction of aromatic carbonyl compounds with β,β-dimethyl ketene silyl acetals [91]. This specific formation of 2-alkoxyoxetanes in the range from 70% to 90% presents an addition to the manifold reports, yielding 3-alkoxyoxetanes as products [92]. Thereby, solvent polarity and bulkiness of the silylgroups are critical parameters for the regioselectivity in the oxetane formation and are investigated in detail (Scheme 54).

VIII. MISCELLANEOUS

From the great variety of photoinduced reactions that are either interesting from the mechanistical or preparative point of view and do not fit into the above scheme, some striking examples will be presented in this section. Although gaining more and more attention, the subject of optical on/off switching via intramolecular photoinduced charge separation, the possible applications and photobiological studies are not covered as well [93–99].

An example of a photolabile mask for alcohols and thiols employing o-benzoylbenzoate esters is reported by Porter and co-workers [100]. Primary and secondary alcohols as well as thiols can be easily masked via the formation of the corresponding 2-benzoylbenzoate esters and converted back into alcohols/thiols under PET-reductive conditions. Irradiation in isopropanol/benzene 1:1 leads to 3-phenylphthalide dimers as the coproduct (together with acetone), whereas more potent electron donors (e.g., amines) result in the monomers and the corresponding imine. Yields generally range from 60% to 90% (Scheme 55) [100].

The kinetically controlled Cope rearrangement of 2,5-bis(4-methoxyphenyl)hexa-1,5-dienes induced by photosensitized electron transfer to DCA was examined by Miyashi and co-workers [101–103]. Remarkable in this context was the temperature-dependent change of the photostationary ratio of this rearrangement, yielding the thermodynamically less stable compound at −80°C in 96%. A radical cation-cyclization diradical cleavage mechanism (RCCY–DRCL) is

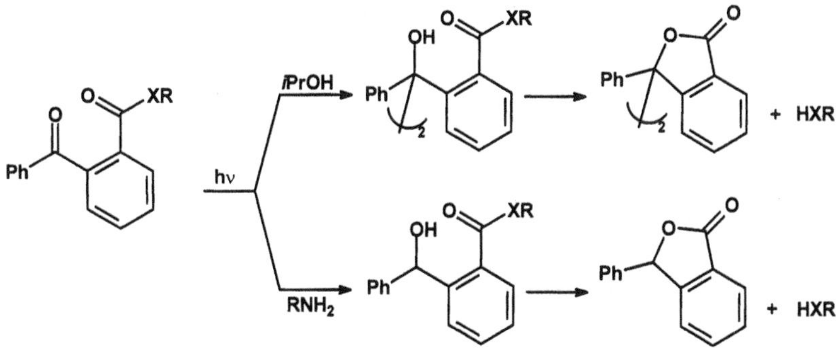

Scheme 55 Photolabile protecting groups for alcohols and thiols.

proposed and supports the observations made [101]. By studying the degenerate Cope rearrangement of 2,5-diaryl-3,3,4,4-tetradeuterio-1,5-hexadienes, further mechanistic details were gathered supporting the above mechanism (Scheme 56) [102,103].

The photochemistry of 1,4-unsaturated systems has been studied intensively over the last three decades, as it can be exemplified by reactions like the di-π-methane rearrangement or the oxa-di-π-methane rearrangement. In this context, Armesto and co-workers reported a novel 1-aza-di-π-methane rearrangement of 1-substituted-1-aza-1,4-dienes promoted by DCA sensitization. Because of the

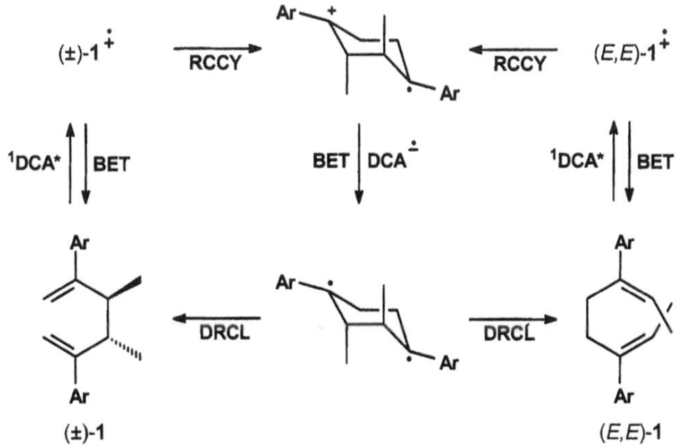

Scheme 56 Mechanism of the kinetically controlled Cope rearrangement.

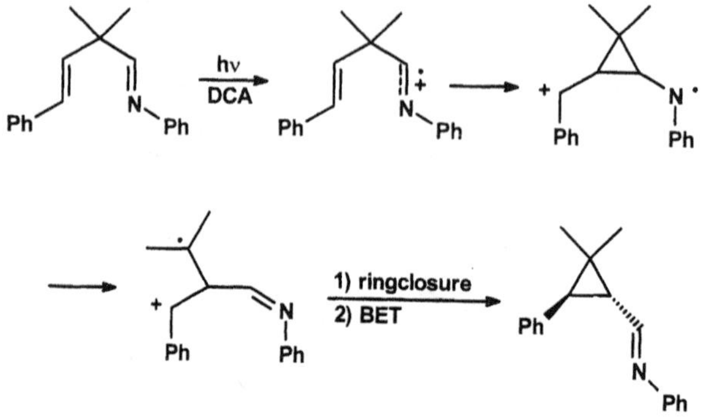

Scheme 57 Aza-di-π-methane rearrangement.

fact that di-π-methane rearrangements have been considered to result from the excited states of molecules, the proposed mechanism via a substrate radical cation adds a new dimension to these kind of reactions (Scheme 57) [104].

Optically active benzene(poly)carboxamides and benzene(poly)carboxylates were used by Inoue and co-workers as sensitizers for the geometrical photoisomerization of (Z)-cyclooctene and (Z,Z)-cyclooctadienes in various solvents at different temperatures. Under energy-transfer conditions, enantiomeric excesses up to 64% ee in unpolar solvents like pentane were reported. The use of polar solvents diminished the product ee's due to the intervention of a free or solvent-separated radical ion pair generated through the electron transfer from the substrate to the excited chiral sensitizer (Scheme 58) [105–109].

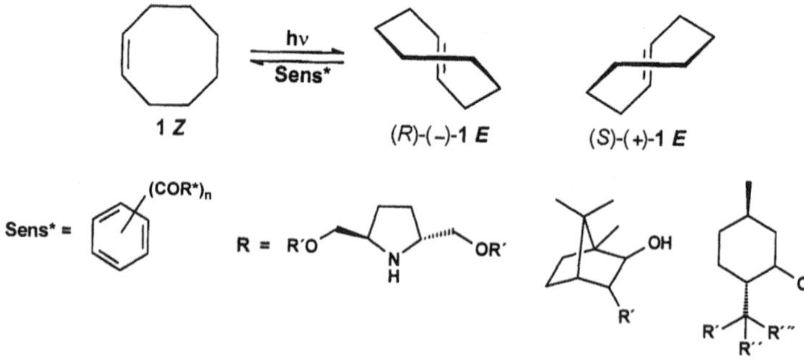

Scheme 58 Enantioselective geometrical photoisomerization of (Z)-cyclooctene.

A new PET-based chemosensor for uronic and sialic acids utilizing the cooperative action of boronic acid and metal chelate was reported by Shinkai and co-workers. This group synthesized a novel fluorescent chemosensor molecule bearing both an *o*-aminomethylphenylboronic acid group for diol binding to a saccharide and a 1,10-phenanthroline-Zn(II)chelate moiety for the carboxylate binding, which enables this sensor to discriminate between neutral monosaccharides and acidic compounds [110].

ACKNOWLEDGMENTS

Financial support of our own research was provided by the Federal Department of Science, Research and Technology (BMBF), the Deutsche Forschungsgemeinschaft (DFG), the Volkswagen Stiftung, and the AG-Solar project of NRW. H. R. thanks the Studienstiftung des deutschen Volkes for a predoctoral scholarship. Special thanks go to Heinrich Luftmann (University of Muenster) and to Christian Wolff (University of Kiel) for their important contributions to mass spectrometry and NMR spectrometry, respectively.

REFERENCES

1. Balzani, V. (ed.) *Electron Transfer in Chemistry*; Wiley–VCH: Weinheim, **2001**, Vols. 1–5.
2. Kavarnos, G. J. *Fundamentals of Photoinduced Electron Transfer*; Wiley–VCH: Weinheim, **1993**.
3. Mattay, J. (ed.) *Topics Curr. Chem.* **1990–1993**, *156, 158, 159, 163, 168*.
4. Fox, M. A.; Chanon, M. *Photoinduced Electron Transfer*; Elsevier; Amsterdam, **1988**, Parts A–D.
5. Rehm, D.; Weller, A. *Ber. Bunsenges. Phys. Chem.* **1969**, *73*, 834–839.
6. Gould, I. R.; Ege, D.; Moser, J. E.; Farid, S. *J. Am. Chem. Soc.* **1990**, *112*, 4290–4301.
7. Zhang, W.; Yang, L.; Wu, L.-M.; Liu, Y.-C.; Liu, Z.-L. *J. Chem. Soc., Perkin Trans. 2* **1998**, 1189–1193.
8. Cossy, J.; Ranaivosata, J.-L.; Bellosta, V.; Ancerewicz, J.; Ferritto, R.; Vogel, P. *J. Org. Chem.* **1995**, *60*, 8351–8359.
9. Maiti, B. C.; Lahiri, S. *Tetrahedron* **1998**, *54*, 9111–9122.
10. Hasegawa, E.; Yoneoka, A.; Suzuki, K.; Kato, T.; Kitazume, T.; Yanagi, K. *Tetrahedron* **1999**, *55*, 12.957–12.968.
11. Hasagawa, E.; Tamura, Y.; Tosaka, E. *Chem. Commun.* **1997**, 1895–1896.
12. Ogawa, A.; Ohya, S.; Hirao. T. *Chem. Lett.* **1997**, 275–276.
13. Piva-Le Blanc, S.; Hénon, S.; Piva, O. *Tetrahedron Lett.* **1999**, *39*, 9683–9684.
14. Jakobi, D.; Abraham, W.; Pischel, U.; Grubert, L.; Stösser, R.; Schnabel, W. *J. Chem. Soc., Perkin Trans. 2* **1999**, 1695–1702.
15. Jakobi, D.; Abraham, W.; Pischel, U.; Stösser, R.; Schnabel, W. *J. Photochem. Photobiol. A: Chem.* **1999**, *128*, 75–83.

16. Jakobi, D.; Abraham, W.; Pischel, U.; Grubert, L.; Schnabel, W. *J. Chem. Soc., Perkin Trans. 2* **1999**, 1241–1248.
17. Nakamura, M.; Dohno, R.; Majima, T. *J. Org. Chem.* **1998**, *63*, 6258–6265.
18. Moriwaki, H.; Oshima, T.; Nagai, T. *J. Chem. Soc., Perkin Trans. 1* **1997**, 2517–2523.
19. Chen, C.-F.; Zhu, Y.; Liu, Y.-C.; Xu, J.-H. *Tetrahedron Lett.* **1995**, *36*, 2835–3838.
20. Ishii, H.; Shiina, S.; Hirano, T.; Niwa, H.; Ohashi, M. *Tetrahedron Lett.* **1999**, *40*, 523–526.
21. Kojima, R.; Yamashita, T.; Tanabe, K.; Shiragami, T.; Yasuda, M.; Shima, K. *J. Chem. Soc., Perkin Trans. 1* **1997**, 217–222.
22. Fagnoni, M.; Mella, M.; Albini, A. *Tetrahedron* **1995**, *51*, 859–864.
23. Fagnoni, M.; Mella, M.; Albini, A. *J. Org. Chem.* **1998**, *63*, 4026–4033.
24. Fagnoni, M.; Mella, M.; Albini, A. *Eur. J. Org. Chem.* **1999**, 2137–2142.
25. Campari, G.; Fagnoni, M.; Mella, M.; Albini, A. *Tetrahedron: Asymmetry* **1998**, *11*, 1891–1906.
26. Bertrand, S.; Hoffmann, N.; Pete, J.-P. *Eur. J. Org. Chem.* **2000**, 2227–2238.
27. Bertrand, S.; Glapski, C.; Hoffmann, N.; Pete, J.-P. *Tetrahedron Lett.* **1999**, *40*, 3169–3172.
28. Das, S.; Dileep Kumar, J. S.; Shivaramayya, K.; George, M. V. *Tetrahedron* **1996**, *52*, 3425–3434.
29. González-Cameno, A. M.; Mella, M.; Fagnoni, M.; Albini, A. *J. Org. Chem.* **1999**, *65*, 297–303.
30. Griesbeck, A. G.; Hundertmark, T.; Steinwäscher, J. *Tetrahedron Lett.* **1996**, *37*, 8367–8370.
31. Mikami, T.; Harada, M.; Narasaka, K. *Chem. Lett.* **1997**, 425–426.
32. Ikeno, T.; Harada, M.; Arai, N.; Narasaka, K. *Chem. Lett.* **1997**, 169–170.
33. Mizuno, K.; Konishi, G.-i; Nishiyama, T. Inoue, H. *Chem. Lett.* **1995**, 1077–1078.
34. Ogawa, A.; Obayashi, R.; Doi, M.; Sonoda, N.; Hirao, T. *J. Org. Chem.* **1998**, *63*, 4277–4281.
35. Ryu, I.; Muraoka, H.; Kambe, N.; Komatsu, M.; Sonoda, N. *J. Org. Chem.* **1996**, *61*, 6396–6403.
36. Hintz, S.; Fröhlich, R.; Mattay, J. *Tetrahedron Lett.* **1996**, *37*, 7349–7352.
37. Hintz, S.; Mattay, J.; van Eldi, R.; Fu, W.-F. *Eur. J. Org. Chem.* **1998**, 1583–1596.
38. Khim, S.-K.; Cederstrom, E.; Ferri, D. C.; Mariano, P. S. *Tetrahedron* **1996**, *52*, 3195–3222.
39. Pandey, G.; Karthikeyan, H.; Murugan, A. *J. Org. Chem.* **1998**, *63*, 2867–2872.
40. Ishii, H.; Yamaoka, R.; Imai, Y.; Hirano, T.; Maki, S.; Niwa, H.; Hashizume, D.; Iwasaki, F.; Ohashi, M. *Tetrahedron Lett.* **1998**, *39*, 9501–9504.
41. Zimmermann, H. E.; Hoffacker, K. D. *J. Org. Chem.* **1996**, *61*, 6526–6534.
42. Pandey, G.; Soma Sekhar, B. B. V. *Tetrahedron* **1995**, *51*, 1483–1494.
43. Pandey, G.; Sochanchingwung, R.; Tiwari, S. K. *Synlett* **1999**, 1257–1258.
44. Kojima, R.; Shiragami, T.; Shima, K.; Yasuda, M.; Majima, T. *Chem. Lett.* **1997**, 1241–1242.
45. Roth, H. D.; Wenig, H.; Zhou, D.; Herbertz, T. *Pure Appl. Chem.* **1997**, *69*, 809–814.

46. Warzecha, K.-D.; Xing, X.; Demuth, M. *Helv. Chim. Acta* **1995**, *78*, 2065–2076.
47. Warzecha, K.-D.; Xing, X.; Demuth, M. *Pure and Appl. Chem.* **1997**, *69*, 109–112.
48. Heinemann, C.; Xing, X.; Warzecha, K.-D.; Ritterskamp, P.; Görner, H.; Demuth, M. *Pure Appl. Chem.* **1998**, *70*, 2167–2176.
49. Ohba, Y.; Kubo, K.; Sakurai, T. *J. Photochem. Photobiol. A: Chem.* **1998**, *113*, 45–51.
50. Bally, T.; Bernhard, S.; Matzinger, S.; Truttmann, L.; Zhu, Z.; Roulin, J. L.; Marcinek, A.; Gebicki, J.; Williams, F.; Chen, G.-F.; Roth, H. D.; Herbertz, T. *Chem. Eur. J.* **2000**, *6*, 849–857.
51. Tamai, T.; Ichinose, N.; Tanaka, T.; Sasuga, T.; Haschida, I.; Mizuno, K. *J. Org. Chem.* **1998**, *63*, 3204–3212.
52. Bergmark, W.; Hector, S.; Jones, G., II; Oh, C.; Kumagai, T.; Hara, S.-i.; Segawa, T.; Tanaka, N.; Mukai, T. *J. Photochem. Photobiol. A: Chem.* **1997**, *109*, 119–124.
53. Herbertz, T.; Roth, H. D. *J. Am. Chem. Soc.* **1998**, *120*, 11.904–11.911.
54. Herbertz, T.; Blume, F.; Roth, H. D. *J. Am. Chem. Soc.* **1998**, *120*, 4591–4599.
55. Herbertz, T.; Roth, H. D. *J. Org. Chem.* **1999**, *64*, 3708–3713.
56. Ha, J. D.; Lee, J.; Blackstock, S. C.; Cha, J. K. *J. Org. Chem.* **1998**, *63*, 8510–8514.
57. Pandey, G.; Hajra, S.; Ghorai, M. K.; Kumar, K. R. *J. Am. Chem. Soc.* **1997**, *119*, 8777–8787.
58. Pandey, G.; Hajra, S.; Ghorai, M. K.; Kumar, K. R. *J. Org. Chem.* **1997**, *62*, 5966–5973.
59. Pandey, G.; Rao, K. S. S. S.; Palit, D. K.; Mittal, J. P. *J. Org. Chem.* **1998**, *63*, 3968–3978.
60. Pandey, G.; Rao, K. S. S. P.; Nageshwar Rao, K. V. *J. Org. Chem.* **2000**, *65*, 4309–4314.
61. Cossy, J.; BouzBouz, S.; Mouza, C. *Synlett* **1998**, 621–622.
62. Kirschberg, T.; Mattay, J. *J. Org. Chem.* **1996**, *61*, 8885–8896.
63. Buscemi, S.; Pace, A.; Vivona, N.; Caronna, T.; Galia, A. *J. Org. Chem.* **1999**, *64*, 7028–7033.
64. Abe, M.; Oku, A. *J. Org. Chem.* **1995**, *60*, 3065–3073.
65. Sugimoto, A.; Hayashi, C.; Omoto, Y.; Mizuno, K. *Tetrahedron Lett.* **1997**, *38*, 3239–3242.
66. Griesbeck, A. G.; Henz, A.; Kramer, W.; Lex, J.; Nerowski, F.; Oelgemöller, M.; Peters, K.; Peters, E.-M. *Helv. Chim. Acta* **1997**, *80*, 912–933.
67. Griesbeck, A. G.; Oelgemöller, M. *Synlett* **2000**, 71–72.
68. Yoon, U. C.; Kim J. W.; Ryu, J. Y.; Cho, S. J.; Oh, S. H.; Mariano, P. S. *J. Photochem. Photobiol. A: Chem.* **1997**, *106*, 145–154.
69. Nishiyama, T.; Mizuno, K.; Otsuji, Y.; Inoue, H. *Tetrahedron* **1995**, *51*, 6695–6706.
70. Arnold, D. R.; Chan, M. S. W.; McManus, K. A. *Can. J. Chem.* **1996**, *74*, 2143–2166.
71. McManus, K. A.; Arnold, D. R. *Can. J. Chem.* **1995**, *73*, 2158–2169.
72. Arnold, D. R.; Connor, D. A.; McManus, K. A.; Bakshi, P. K.; Cameron, T. S. *Can. J. Chem.* **1996**, *74*, 602–612.
73. de Ilijser, H. J. P.; Arnold, D. R. *J. Org. Chem.* **1997**, *62*, 8432–8438.
74. Torriani, R.; Mella, M.; Fasani, E.; Albini, A. *Tetrahedron* **1997**, *53*, 2573–2580.

75. Rathore, R.; Kochi, J. K. *J. Org. Chem.* **1996**, *61*, 627–639.
76. Reddy, G. D.; Wiest, O. *J. Org. Chem.* **1999**, *64*, 2860–2863.
77. Tung, C.-H.; Ying, Y.-M.; Yuan, Z.-Y. *J. Photochem. Photobiol. A: Chem.* **1998**, *119*, 93–99.
78. Asaoka, S.; Ooi, M.; Jiang, P.; Wada, T.; Inoue, Y. *J. Chem. Soc., Perkin Trans. 2* **2000**, 77–84.
79. Gürtler, C. F.; Steckhan, E.; Blechert, S. *J. Org. Chem.* **1996**, *61*, 4136–4143.
80. Peglow, T.; Blechert, S.; Steckhan, E. *Chem. Commun.* **1999**, 433–434.
81. Haberl, U.; Steckhan, E.; Blechert, S.; Wiest, O. *Chem. Eur. J.* **2000**, *6*, 849–857.
82. Botzem, J.; Haberl, U.; Steckhan, E.; Blechert, S. *Acta Scand. Chem.* **1998**, *52*, 175–193.
83. Sun, D.; Hubig, S. M.; Kochi, J. K. *J. Photochem. Photobiol. A: Chem.* **1999**, *122*, 87–94.
84. Gaebert, C.; Mattay, J. *Tetrahedron* **1997**, *53*, 14,297–14,316.
85. Kubo, Y.; Yoshioka, M.; Kiuchi, K.; Nakajima, S.; Inamura, I. *Tetrahedron Lett.* **1999**, *40*, 527–530.
86. Xue, J.; Xu, J.-W.; Yang, L.; Xu, J.-H. *J. Org. Chem.* **2000**, *65*, 30–40.
87. Bosch, E.; Hubig, S. M.; Lindemann, S. V.; Kochi, J. K. *J. Org. Chem.* **1998**, *63*, 592–601.
88. Schepp, N. P.; Shukla, D.; Sarker, H.; Bauld, N. L.; Jonston, L. J. *J. Am. Chem. Soc.* **1997**, *119*, 10,325–10,334.
89. Sun, D.; Hubig, S. M.; Kochi, J. K. *J. Org. Chem.* **1999**, *64*, 2250–2258.
90. Hubig, S. M.; Sun, D.; Kochi, J. K. *J. Chem. Soc., Perkin Trans. 2* **1999**, 781–788.
91. Abe, M.; Shirodai, Y.; Nojima, M. *J. Chem. Soc., Perkin Trans. 1* **1998**, 3253–3260.
92. Bach, T. *Liebigs Ann.* **1995**, 855–865.
93. de Silva, A. P.; Gunaratne, H. Q. N.; Gunnlaugsson, T.; McCoy, C. P.; Maxwell, P. R. S.; Rademacher, J. T.; Rice, T. E. *Pure Appl. Chem.* **1996**, *68*, 1443–1448.
94. Nagamura, T. *Pure Appl. Chem.* **1996**, *68*, 1449–1454.
95. de Silva, A. P.; Gunaratne, H. Q. N.; Gunnlaugsson, T.; Lynch, P. L. M. *New. J. Chem.* **1996**, *20*, 871–880.
96. Koshima, H.; Wang, Y.; Teruo, M.; Miyahara, I.; Mizutani, H.; Hirotsu, K.; Asahi, T.; Masuhara, H. *J. Chem. Soc., Perkin Trans. 2* **1997**, 2033–2038.
97. Endtner, J. M.; Effenberger, F. Hartschuh, A. Port, H. *J. Am. Chem. Soc.* **2000**, *122*, 3037–3046.
98. Adam, W.; Grimm, G. N.; Marquardt, S.; Möller, C. R. S. *J. Am. Chem. Soc.* **1999**, *121*, 1179–1185.
99. Maurer, T. D.; Kraft, B. J.; Lato, S. M.; Ellington, A. D.; Zaleski, J. M. *Chem. Commun.* **2000**, 69–70.
100. Jones, P. B.; Pollastri, M. P.; Porter, N. A. *J. Org. Chem.* **1996**, *61*, 9455–9461.
101. Ikeda, H.; Ishida, A.; Takasaki, T.; Tojo, S.; Takamuku, S.; Miyashi, T. *J. Chem. Soc., Perkin Trans. 2* **1997**, 849–850.
102. Ikeda, H.; Minegishi, T.; Abe, H.; Konno, A.; Goodman, J. L.; Miyashi, T. *J. Am. Chem. Soc.* **1998**, *120*, 87–95.
103. Ikeda, H.; Takasaki, T.; Takahashi, Y.; Konno, A.; Matsumoto, M.; Hoshi, Y.; Aoki, T.; Susuki, T.; Goodman, J. L.; Miyashi *J. Org. Chem.* **1999**, *64*, 1640–1649.

104. Ortiz, M. J.; Agarrabeitia, A. R.; Aparicio-Lara, S.; Armesto, D. *Tetrahedron Lett.* **1999**, *40*, 1759–1762.

105. Shi, M.; Inoue, Y. *J. Chem. Soc., Perkin Trans.* 2 **1998**, 1725–1729.

106. Inoue, Y.; Tsuneishi, H.; Hakasushi, T.; Tai, A. *J. Am. Chem. Soc.* **1997**, *119*, 472–478.

107. Inoue, Y.; Yamasaki, N.; Yokoyama, T.; Tai, A. *J. Org. Chem.* **1993**, *58*, 1011–1018.

108. Inoue, Y.; Yamasaki, N.; Yokoyama, T.; Tai, A. *J. Org. Chem.* **1992**, *57*, 1332–1345.

109. Inoue, Y.; Yokoyama, T.; Yamasaki, N.; Tai, A. *J. Am. Chem. Soc.* **1989**, *111*, 6480–6482.

110. Yamamoto, M.; Takeuchi, M. Shinkai, S. *Tetrahedron* **1998**, *54*, 3125–3140.

5

Design and Fine Control of Photoinduced Electron Transfer

Shunichi Fukuzumi and Hiroshi Imahori

Osaka University, Osaka, Japan

I. INTRODUCTION

Among many types of chemical reactions, an electron-transfer reaction is undoubtedly the most fundamental one because the electron is the minimal unit of the change in chemical reactions and any chemical bond is formed via electrons. This simplicity of the reaction has permitted the development of a relatively simple detailed quantitative theory, known as the Marcus theory of electron transfer [1–3]. With the aid of such an analytical theory, the field of electron-transfer reactions has undergone a remarkable expansion not only in chemistry but also in physics, biology, and advanced technology during the past half-century [4–7]. In particular, electron transfer has been recognized as the key step in a number of biological processes which are essential for life, such as photosynthesis and respiration [8–10]. Photoexcited states as well as the ground states are involved in photosynthesis, where multistep electron-transfer processes are remarkably well designed to optimize the efficiency of solar energy conversion [10].

The rates of electron-transfer reactions can be well predicted provided that the electron transfer is a type of adiabatic outer-sphere reaction and the free-energy change of electron transfer and the reorganization energy (λ) associated with the electron transfer are known [1–7]. This means that electron-transfer reactions can be designed quantitatively based on the redox potentials and the reorganization energies of molecules involved in the electron-transfer reactions.

In addition to simple electron transfers in which no chemical bond is either broken or formed, numerous organic reactions, previously formulated by "movements of electron pairs," are now understood as processes in which an initial electron transfer from a nucleophile (reductant) to an electrophile (oxidant) produces a radical ion pair, which leads to the final products via the follow-up steps involving cleavage and formation of chemical bonds [11–23]. The follow-up steps are usually sufficiently rapid to render the initial electron transfer the rate-determining step in an overall irreversible transformation [24]. In such a case, the overall reactivity is determined by the initial electron-transfer step, which can also be well designed based on the redox potentials and the reorganization energies of a nucleophile (reductant) and an electrophile (oxidant).

The fundamental properties in electron-transfer reactions such as the redox potentials and the reorganization energies of molecules involved in electron transfer can be modulated by interaction of the molecules with an environment including solvents [1–7]. The redox potentials and the reorganization energies can also be changed through specific interactions of an electron donor or an electron acceptor with a third component which is added to the electron-transfer system [24–26]. Noncovalent interactions such as hydrogen-bonding, π-stacking, and other electrostatic interactions have also played an important role in controlling electron transfer in biological systems [27]. Thus, there are a number of ways to control electron-transfer processes and redox reactions in which electron transfer is the rate-determining step.

In this review, we focus on ways to control photoinduced electron-transfer processes as well as photoinduced redox reactions in which the rates of the initial photoinduced electron transfer and the back-electron transfer govern the overall rate. First, photoinduced electron-transfer systems are presented to mimic the function of a photosynthetic reaction center, which may be the most finely designed electron-transfer system in nature. Then, we explore photoinduced redox systems in which the initial photoinduced electron transfer is the rate-determining step, leading to the selective formation of desired products by modulating the redox potentials and the reorganization energies of electron donors and/or electron acceptors. Finally, the photochemical redox reactions which would otherwise be unlikely to occur are made possible to proceed efficiently by the catalysis on the photoinduced electron-transfer steps. The mechanistic viability is described by showing a number of examples of photochemical reactions that involve catalyzed electron-transfer processes as the rate-determining steps.

II. CONTROL OF MULTISTEP PHOTOINDUCED ELECTRON-TRANSFER SYSTEMS

Natural photosynthesis applies electron transfer systems, where a relay of electron-transfer reactions evolves among chlorophyll and quinone moieties embed-

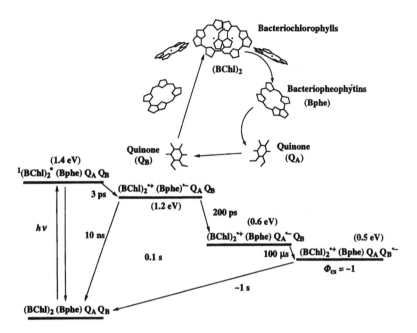

Scheme 1   Multistep photoinduced electron transfers in a natural photosynthetic system.

ded in a transmembrane protein matrix as shown in Scheme 1 [8–10]. Relatively little energy (0.2 eV) is consumed in the rapid initial electron-transfer step from bacteriochlorophyll dimer [(BChl)$_2$] to bacteriopheophytin (Bphe) on a time scale of 3 psec. In contrast, the back-electron transfer to the ground state has nearly 1.2 eV of driving force and, therefore, should be within an inverted region of the Marcus parabolic free-energy relationship where the rate of electron transfer decreases with increasing the driving force [1–3]. In such a case, a charge shift reaction from Bphe$^{\bullet-}$ to an electron-acceptor quinone (Q$_A$) occurs much faster on a time scale of 200 psec than the back-electron transfer, which occurs on a time scale of 10 nsec (Scheme 1). The further charge separation occurs, which achieves a nearly quantitative quantum yield of the final charge-separated state, which has an extremely long lifetime ($\sim$1 sec).

The mimicry of these complex and highly versatile processes has prompted the design of synthetic donor–acceptor-linked ensembles such as triads, tetrads, and pentads [28–46]. However, achieving the long lifetime and the high efficiency of charge separation like the natural photosynthesis has been mainly hampered by synthetic difficulties. A specific challenge involves attaining a finely tuned and directed redox gradient along donor–acceptor-linked arrays as shown in Scheme 1.

Successful mimicry of primary events in photosynthesis has recently been achieved using a well-designed ferrocene (Fc)–zincporphyrin (ZnP)–freebaseporphyrin (H_2P)–fullerene (C_{60}) tetrad (Fc–ZnP–H_2P–C_{60}) in which a ferrocene unit is tethered at the end of ZnP–H_2P–C_{60} [47]. The spherical shape of C_{60}, containing 60 delocalized π-electrons, renders these carbon allotropes ideal components for the construction of efficient electron-transfer systems with porphyrins [46–49]. The small reorganization energy, found for C_{60} in electron-transfer processes, results in an overall acceleration of the charge-separation (CS) step, whereas a deceleration was noted for the energy-wasting charge-recombination (CR) step [46–49]. The tetrad molecule, Fc–ZnP–H_2P–C_{60}, displays coupled photoinduced energy transfer and electron transfer, sequence, Fc–^{1}ZnP*–H_2P–C_{60} (2.04 eV) $\rightarrow$ Fc–ZnP–^1H_2P*–C_{60} (1.89 eV) $\rightarrow$ Fc–ZnP–$H_2P^{\bullet+}$–$C_{60}^{\bullet-}$ (1.63 eV) $\rightarrow$ Fc–ZnP$^{\bullet+}$–H_2P–$C_{60}^{\bullet-}$ (1.34 eV) $\rightarrow$ Fc$^+$–ZnP–H_2P–$C_{60}^{\bullet-}$ (1.11 eV) as shown in Scheme 2 [47]. Each energy has been determined by the one-electron redox potentials obtained from the differential pulse voltammetry measurements and the excited energies obtained from the photophysical data [47]. The final charge-separated state (Fc$^+$ and $C_{60}^{\bullet-}$ pair) is characterized by a large edge-to-edge distance (R_{ee}) value of 48.9 Å. Time-resolved techniques, including fluorescence lifetime and transient absorption measurements, have been employed to probe each electron-transfer (ET) step of Fc–ZnP–H_2P–C_{60} in Scheme 2 [47].

Considering the energy gap between the two singlet excited states (0.15 eV) with the lower state being that of H_2P, a rapid singlet–singlet energy transfer (EN) prevails with a time constant of $k_{EN} = 2.6 \times 10^{10}$ sec^{-1}. ^1H_2P* is starting point for the formation of Fc–ZnP–$H_2P^{\bullet+}$–$C_{60}^{\bullet-}$ (1.63 eV) via the first charge separation step (CS1) in Scheme 2. The formation kinetics of the π-radical cation of H_2P at 550 nm and the fullerene π-radical anion in the near-infrared region (i.e., 1000 nm) [50], are valuable aids in determining the absolute rate constants, providing similar values ($k_{ET(CS1)}$) of 7.1 $\times$ 10^9 and 6.7 $\times$ 10^9 sec^{-1}, respectively. The absorption of the H_2P π-radical cation, as it emerged from the CS process, is subject to a rapid intramolecular decay via the first charge shift reaction (CSH1) in Scheme 2. The product of this deactivation ($k_{ET(CSH1)} = 2.5 \times 10^9$ sec^{-1}) is a differential absorption spectrum, which displays the ZnP π-radical cation absorption with a maximum at 650 nm [51–53]. Taking the spectroscopic and kinetic evidence in concert, it is concluded that the resulting species evolves from a first CSH between the two porphyrin donors, yielding the Fc–ZnP$^{\bullet+}$–H_2P–$C_{60}^{\bullet-}$ radical pair (1.34 eV). It is imperative to note that this energy level (1.34 eV) falls substantially below that of any singlet or triplet excited state of ZnP, H_2P, and C_{60}, therefore ruling out the subsequent formation of an excited state as a product of charge recombination. Nanosecond transient

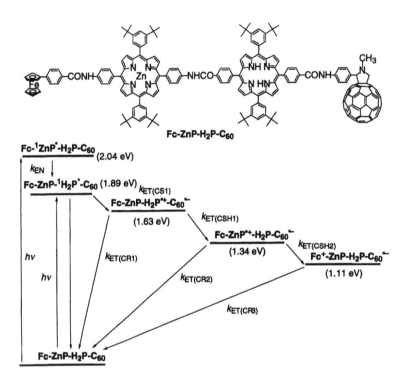

Scheme 2   Reaction scheme and energy diagram for Fc–ZnP–H$_2$P–C$_{60}$ in PhCN. (From Ref. 21.)

absorption studies following laser excitation of Fc–ZnP–H$_2$P–C$_{60}$ reveal unambiguously the C$_{60}$$^{\bullet-}$ fingerprint (~1000 nm) in the final CS state, Fc$^+$–ZnP–H$_2$P–C$_{60}$$^{\bullet-}$, as shown in Fig. 1 [47].

The total quantum yield [Φ_{CS}(total)] for CS is decreased to 0.17 in dimethylformamide (DMF) due to the competition of the CSH from Fc–ZnP–H$_2$P$^{\bullet+}$–C$_{60}$$^{\bullet-}$ (1.63 eV) to Fc–ZnP$^{\bullet+}$–H$_2$P–C$_{60}$$^{\bullet-}$ (1.34 eV) versus the decay of Fc–ZnP–H$_2$P$^{\bullet+}$–C$_{60}$$^{\bullet-}$ to the triplet states of the freebase porphyrin (1.40 eV) and the C$_{60}$ (1.50 eV) [47]. In contrast to the case of most donor–acceptor-linked systems, the decay dynamics of the charge-separated radical pair (Fc$^+$–ZnP–H$_2$P–C$_{60}$$^{\bullet-}$) does not obey first-order kinetics, but, instead, obeys second-order kinetics [47]. This indicates that the *intra*molecular electron transfer in Fc$^+$–ZnP–H$_2$P–C$_{60}$$^{\bullet-}$ is *too slow* to compete with the diffusion-limited *inter*molecular electron transfer in solution.

In order to segregate the *inter*molecular electron transfer from the *intra*molecular electron-transfer processes in Fc$^+$–ZnP–H$_2$P–C$_{60}$$^{\bullet-}$, electron spin reso-

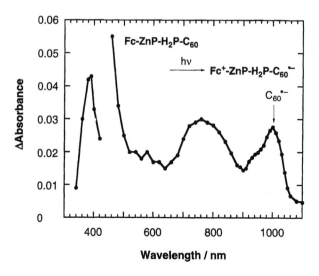

Figure 1 Differential absorption spectrum obtained upon nanosecond flash photolysis (532 nm) of 7.2×10^{-6} M solutions of Fc–ZnP–H$_2$P–C$_{60}$ triad in nitrogen saturated benzonitrile with a time delay of 50 nsec at 298 K. (From Ref. 47.)

nance (ESR) measurements were performed in a frozen matrix at variable temperatures using a low concentration in PhCN (1.0×10^{-5} M) under irradiation [47]. The ESR spectrum under irradiation at 203 K shows a characteristic broad signal attributable to the C$_{60}$·$^-$ moiety ($g = 2.0004$) [47]. The ESR signal grows in immediately upon "turning on" the irradiation of Fc–ZnP–H$_2$P–C$_{60}$. Upon "turning off" the irradiation source, the ESR signal exhibited a slow and nearly seconds lasting decay (Fig. 2a), which was best fitted by first-order kinetics yielding a rate constant of 3.0 sec^{-1} (Figure 2b) for the *intra*molecular CR process ($k_{ET(CR3)}$) in Scheme 2. The clean monoexponential dynamics corroborate that the origin of the Fc–ZnP–H$_2$P–C$_{60}$ ESR signal is indeed the *intra*molecular CR in the radical ion pair (Fc$^+$–ZnP–H$_2$P–C$_{60}$·$^-$). Furthermore, under similar experimental conditions, the ESR response of Fc–ZnP–H$_2$P–C$_{60}$ was much large than those of ZnP–C$_{60}$ and ZnP–H$_2$P–C$_{60}$ despite the lower quantum yield of CS for Fc–ZnP–H$_2$P–C$_{60}$ (0.24) in PhCN solution relative to those of ZnP–C$_{60}$ (0.85) [53] and ZnP–H$_2$P–C$_{60}$ (0.40) [53], as shown in Fig. 2a.

The temperature dependence of $k_{ET(CR3)}$ revealed only a moderate change (2.6–3.0 sec^{-1}) upon varying the temperature between 163 and 203 K [47]. The longest lifetime of the resulting charge-separated state (i.e., ferricenium ion C$_{60}$ radical anion pair) in frozen benzonitrile (PhCN) is determined as 0.38 sec [47], which is more than one order of magnitude longer than any other *intramolecular*

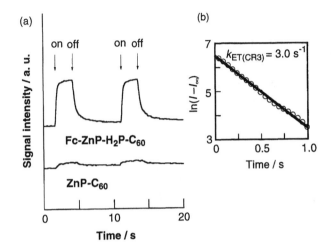

Figure 2 (a) ESR signal intensity response of Fc–ZnP–H$_2$P–C$_{60}$ (1.0 × 10^{-5} M, top) and ZnP–C$_{60}$ (1.0 × 10^{-5} M, bottom) in a frozen deaerated PhCN solution at the maximum of the ESR signal intensity due to the C$_{60}$$^{\cdot-}$ upon irradiation of ultraviolet-visible light from a high-pressure Hg lamp. (b) First-order plot for the decay of the ESR signal intensity (*I*) in Fc–ZnP–H$_2$P–C$_{60}$ ($k_{ET(CR3)}$ = 3.0 sec^{-1}). (From Ref. 47.)

charge-recombination processes of synthetic systems and is comparable to that observed for the bacterial photosynthetic reaction center in Scheme 1. The weak temperature dependence suggests that stepwise *intra*molecular CR processes of Fc$^+$–ZnP–H$_2$P–C$_{60}$$^{\cdot-}$ via a transient Fc–ZnP–H$_2$P–C$_{60}$$^{\cdot-}$ intermediate can be ruled out with a high degree of confidence [47].

The photoinduced electron-transfer dynamics has also been examined for a series of porphyrin–fullerene-linked molecules with the same spacer employed for Fc–ZnP–H$_2$P–C$_{60}$: ZnP–C$_{60}$ (edge-to-edge distance: R_{ee} = 11.9 Å), Fc–ZnP–C$_{60}$ (R_{ee} = 30.3 Å) and ZnP–H$_2$P–C$_{60}$ (R_{ee} = 30.3 Å), shown in Chart 1 [53]. The driving force dependence of the electron-transfer rate constants (k_{ET}) of these dyad, triads, and tetrad molecules is shown in Fig. 3, where log k_{ET} is plotted against the driving force ($-\Delta G^\circ_{ET}$) [47].

To quantify the driving-force dependence of k_{ET}, the Marcus equation [Eq. (1)] [2] is employed:

$$k_{ET} = \left(\frac{4\pi^3}{h^2 \lambda k_B T}\right)^{1/2} V^2 \exp\left(-\frac{(\Delta G^\circ_{ET} + \lambda)^2}{4 \lambda k_B T}\right) \qquad [1]$$

with λ being the reorganization energy, ΔG°_{ET} the free-energy change of ET, *V* the electronic coupling matrix element, and *T* the absolute temperature. The lines

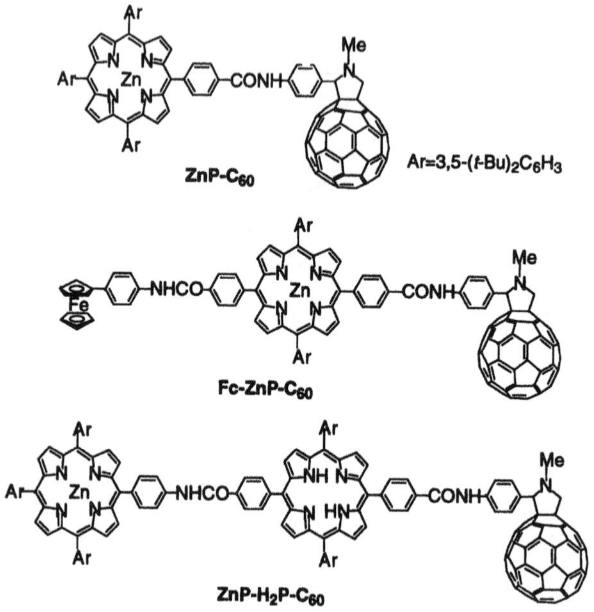

ZnP-C60

Ar=3,5-(*t*-Bu)₂C₆H₃

Fc-ZnP-C60

ZnP-H₂P-C60

Chart 1

in Fig. 3 represent the best fit to Eq. (1) (ZnP–C₆₀: $\lambda = 0.66$ eV, $V = 3.9$ cm^{-1}, [53]; Fc–ZnP–C₆₀, Fc–H₂P–C₆₀, and ZnP–H₂P–C₆₀: $\lambda = 1.09$ eV, $V = 0.019$ cm^{-1} [53]; Fc–ZnP–H₂P–C₆₀: $\lambda = 1.32$ eV, $V = 0.00017$ cm^{-1}) [47]. The λ value increases, whereas the V value decreases with increasing the edge-to-edge distance in the order of the dyad ($R_{ee} = 11.9$ Å), the triads ($R_{ee} = 30.3$ Å) and the tetrad ($R_{ee} = 48.9$ Å). The maximum k_{ET} value (k_{ETmax}) of each Marcus plot in Fig. 3 is correlated with the edge-to-edge distance (R_{ee}), separating the radical ions, according to Eq. (2) [47]:

$$\ln k_{ETmax} = \ln\left(\frac{2\pi^{3/2} V_0^2}{h(\lambda k_B)^{1/2}}\right) - \beta R_{ee} \qquad [2]$$

Hereby, V_0 refers to the maximal electronic coupling element and β is the decay coefficient factor (damping factor), which depends primarily on the nature of the bridging molecule. From the linear plot of $\ln k_{ETmax}$ versus R_{ee} the β value is obtained as 0.60 Å^{-1} [47]. This β value is located within the boundaries of nonadiabatic ET reactions for saturated hydrocarbon bridges (0.8–1.0 Å^{-1}) and unsaturated phenylene bridges (0.4 Å^{-1}) [1–4,54,55].

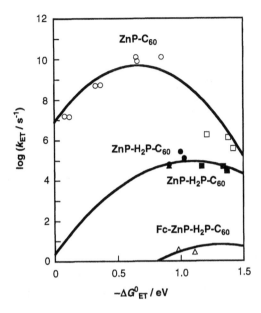

Figure 3 Driving force $(-\Delta G^{\circ}_{ET})$ dependence of intramolecular ET rate constants in ZnP–C$_{60}$ (CS: white circles; CR: white squares), Fc–ZnP–C$_{60}$ (black circles), Fc–H$_2$P–C$_{60}$ (black triangles), ZnP–H$_2$P–C$_{60}$ (black squares), and Fc–ZnP–H$_2$P–C$_{60}$ (white triangles). The lines represent the best fit to the Marcus relation. [Eq. (1)] (see text). (From Ref. 47.)

In polar solvents such as benzonitrile, the rates of photoinduced electron-transfer reactions of zincporphyrin–C$_{60}$-linked molecules can be well predicted by the Marcus theory of electron transfer as shown in Fig. 3. In nonpolar solvents such as benzene ($\varepsilon_t = 2.28$), however, the charge-separated state (ZnP$^{\bullet+}$–C$_{60}$$^{\bullet-}$) formed via photoinduced electron transfer from the excited singlet state of the porphyrin to the C$_{60}$ moiety (k_{CS1}) undergoes charge recombination to yield the C$_{60}$ singlet excited state (k_{CR2}), followed by intersystem crossing to the C$_{60}$ triplet excited state (k_{ISC}) as shown in Scheme 3, because the energy level of the charge-separated state is higher than that of the C$_{60}$ singlet excited state (1.75 eV) [56]. Formation of ZnP$^{\bullet+}$–C$_{60}$$^{\bullet-}$ in benzene is confirmed by picosecond laser photolysis when a new absorption band at 1020 nm due to the C$_{60}$ radical anion (C$_{60}$$^{\bullet-}$) appears clearly, accompanied by the rise in a broad absorption around 650 nm due to the zincporphyrin radical cation (ZnP$^{\bullet+}$) [56]. Because the rise rate (rate constant of 2.5×10^9 sec^{-1}) at 750 nm due to ^{1}C$_{60}$* in benzene does not match with the decay rate (rate constant of 5.0×10^9 sec^{-1}) at 610 nm due to ^{1}ZnP*,

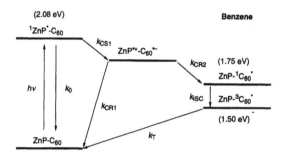

Scheme 3 Schematic energy levels and relaxation pathways for ^{1}ZnP* and ^{1}C$_{60}$* in ZnP-C$_{60}$ in benzene. (From Ref. 56.)

it is strongly suggested that the charge-separated state (ZnP$^{\bullet +}$–C$_{60}^{\bullet -}$) in benzene, produced via photoinduced ET, undergoes the CR to yield ^{1}C$_{60}$* [56]. The subsequent formation of the C$_{60}$ triplet excited state (^{3}C$_{60}$*) in benzene is confirmed by the transient spectrum at the nanosecond time scale, which exhibits a characteristic triplet–triplet absorption maximum at 700 nm [56]. More polar solvents such as anisole (ε_s = 4.33) render the energy level of the charge-separated state lower than the C$_{60}$ singlet excited state, resulting in the direct formation of the C$_{60}$ triplet excited state (1.50 eV) from the charge-separated state, formed by the photoinduced charge separation from the porphyrin to the C$_{60}$ singlet excited state as well as from the porphyrin excited singlet state to the C$_{60}$. In polar solvents such as PhCN (ε_s = 25.2), where the energy level of the charge-separated state (1.38 eV) is low compared to the C$_{60}$ triplet excited state, the charge-separated state, produced upon excitation of the both chromophores, decays directly to the ground state. Thus, the charge-separated state formed via photoinduced electron transfer from the excited singlet state of the porphyrin to the C$_{60}$ moiety decays to different energy states depending on the energy level of the charge-separated state relative to the singlet and triplet excited states of the C$_{60}$ moiety [56].

The fate of the charge-separated state of porphyrin-linked C$_{60}$ can also be altered by modifying conjugation of the porphyrin ligand, because the degree of macrocycle conjugation causes a considerable change in the excited states and the redox properties [57]. The number of reduced double bonds in the pyrrole rings is zero in the case of porphyrins, one in the case of chlorins, and two in the case of bacteriochlorins. The ionization potentials of the tetrapyrroles decrease with increasing peripheral saturation [58], consistent with the electrochemical data [59,60]. Because natural photosynthesis utilizes chlorophylls as antenna molecules, model compounds have been synthesized where a chlorophyll-like donor

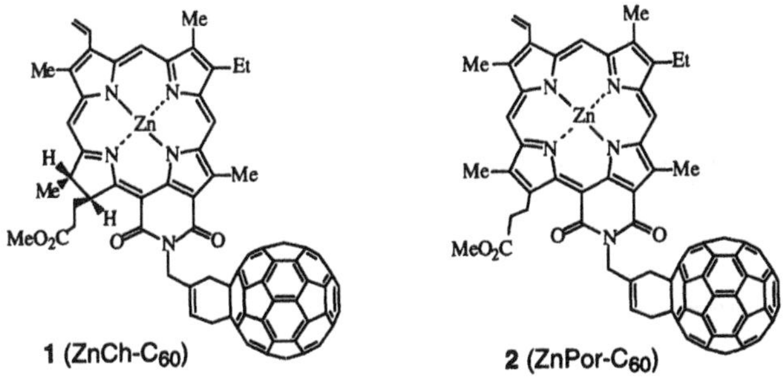

1 (ZnCh-C$_{60}$)   2 (ZnPor-C$_{60}$)

Chart 2

is linked with C$_{60}$ [61–65]. Electrochemical and photochemical properties of chlorin (H$_2$Ch)–C$_{60}$ and porphyrin (H$_2$Por)–C$_{60}$ dyads as well as the Zn analogs (ZnCh–C$_{60}$ and ZnPor–C$_{60}$), all of which have the same linkage (Chart 2) have been examined in order to quantitatively extract the results of utilizing a chlorophyll-like donor (a chlorin) instead of a fully conjugated porphyrin [65].

The E_{red}^0 values for reduction of linked C$_{60}$ which are attributed to the fullerene moiety are almost invariant irrespective of the type of linked macrocyclic ring, whereas the E_{ox}^0 values for oxidation of the macrocycle are shifted in a negative direction in the following order: H$_2$Por > H$_2$Ch > ZnPor > ZnCh [65]. Accordingly, the free-energy change of electron transfer from ZnCh to C$_{60}$ in ZnCh–C$_{60}$ (equivalent to the driving force of the back-electron transfer), obtained from the difference between E_{ox}^0 and E_{red}^0, is the smallest among the examined dyads. In this case, the radical ion pair state (ZnCh$^{\bullet+}$–C$_{60}^{\bullet-}$) is lower in energy (1.33 eV) than both the triplet excited state of C$_{60}$ (1.45 eV) [66] and ZnCh (1.36–1.45 eV) [67]. Thus, photoinduced electron transfer from the singlet excited state of the ZnCh (^{1}ZnCh*) to the C$_{60}$ part of the molecule occurs to produce the radical ion pair, ZnCh$^{\bullet+}$–C$_{60}^{\bullet-}$, in competition with the intersystem crossing to ^{3}ZnCh* and energy transfer to produce ZnCh–^{1}C$_{60}$* and ZnCh–^{3}C$_{60}$*, as shown in Scheme 4a [65]. Electron transfer from ^{3}ZnCh* to C$_{60}$ as well as electron transfer from ZnCh to ^{1}C$_{60}$* and ^{3}C$_{60}$* also results in the formation of ZnCh$^{\bullet+}$– C$_{60}^{\bullet-}$. Thus, in any case, the radical ion pair, ZnCh$^{\bullet+}$–C$_{60}^{\bullet-}$, is formed, which was detected as a transient absorption spectrum (λ_{max} = 1000 nm due to C$_{60}^{\bullet-}$ and 790 nm due to ZnCh$^{\bullet+}$) [65]. The radical ion pair ZnCh$^{\bullet+}$–C$_{60}^{\bullet-}$ decays via back-electron transfer (k_{BET}) to the ground state rather than to the triplet excited state (k_T) [65]. The BET rate was determined from the disappearance of the absorption band at 790 nm due to ZnCh$^{\bullet+}$ in ZnCh$^{\bullet+}$–C$_{60}^{\bullet-}$ as 9.1×10^3 sec^{-1}. This value is two to six orders of magnitude smaller than the value ever reported for any other porphyrin or chlorin donor–acceptor dyad system.

(a)

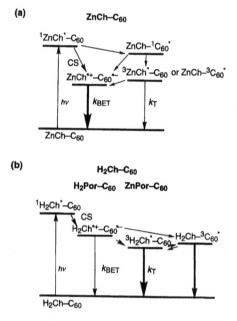

(b)

Scheme 4 Reaction schemes and energy diagrams for (a) $ZnCh-C_{60}$ and (b) H_2Ch-C_{60}, $ZnPor-C_{60}$ and H_2Por-C_{60} dyads in PhCN. (From Ref. 65).

In contrast to the case of $ZnCh-C_{60}$, no transient formation of $C_{60}^{\bullet-}$ was detected at 1000 nm for any other dyad in Scheme 4b [65]. In each case, only the triplet–triplet absorption due to the chlorin or porphyrin moiety was observed due to the higher energy of the radical ion pair as compared to the triplet excited state as is expected from the redox potentials. Thus, the energy level of the radical ion pair in reference to the triplet energy of a component is an important factor in determining the lifetime of the radical ion pair.

III. PHOTOCATALYTIC SYSTEMS OF UPHILL REDOX REACTIONS

The resulting high-energy CS state in fullerene-containing donor–acceptor-linked systems can be converted into chemical energy by combining an appropriate redox reaction. In the biological redox systems, the NADH/NAD$^+$ system (NAD$^+$ = nicotinamide adenine dinucleotide; NADH = reduced form of NAD$^+$) plays a key role in energy conversion and storage [68,69]. In the absence of oxygen, the photolytically generated $C_{60}^{\bullet-}$ moiety in $ZnP^{\bullet+}-C_{60}^{\bullet-}$ and $ZnP^{\bullet+}-H_2P-C_{60}^{\bullet-}$

radical ion pairs (Chart 1) undergoes one-electron oxidation by hexyl viologen (HV^{2+}), whereas the $ZnP^{\cdot+}$ moiety is reduced by NADH analogs, 1-benzyl-1,4-dihydronicotinamide (BNAH) and 10-methyl-9,10-dihydroacridine ($AcrH_2$) [70]. An NADH analog, 9-phenyl-10-methyl-9,10-dihydroacridine (AcrHPh) has previously been used as an electron source in the direct photochemical oxidation of AcrHPh by N,N'-dimethyl-4,4'-bipyridinium dication (methyl viologen, MV^{2+}) without a photocatalyst [71]. The amount of energy stored in this photochemical oxidation was estimated as 0.76 eV (73 kJ/mol) [71].

The photocatalytic system is shown in Scheme 5, where BNAH is oxidized by the $ZnP^{\cdot+}$ moiety in the radical ion pair $ZnP^{\cdot+}-C_{60}^{\cdot-}$ (k_1) produced upon photoirradiation of $ZnP-C_{60}$, whereas HV^{2+} is reduced to $HV^{\cdot+}$ by the $C_{60}^{\cdot-}$ moiety of $ZnP^{\cdot+}-C_{60}^{\cdot-}$ (k_2). These individual electron-transfer processes compete, however, with the BET in the radical ion pair (k_{BET}). This pathway was experimentally confirmed by photolysis of the $ZnP-C_{60}/BNAH/HV^{2+}$ and $ZnP-H_2P-C_{60}/BNAH/HV^{2+}$ systems with visible light (433 nm) in deoxygenated PhCN [70]. For instance, Fig. 4 depicts the steady-state photolysis in deoxygenated PhCN, in which the $HV^{\cdot+}$ absorption band ($\lambda_{max} = 402$ and 615 nm) increases progressively with irradiation time. By contrast, no reaction occurs in the dark or in the absence of the photocatalyst (i.e., $ZnP-C_{60}$ or $ZnP-H_2P-C_{60}$) under photoirradiation [70]. Once $HV^{\cdot+}$ is generated in the photochemical reaction, it was found to be stable in deoxygenated PhCN. The stoichiometry of the reaction is established as given by Eq. (3), where BNAH acts as a two-electron donor to reduce two equivalents of $HV^{\cdot+}$ [70]:

$$BNAH + 2HV^{2+} \xrightarrow[\substack{ZnP-C_{60} \\ ZnP-H_2P-C_{60}}]{hv} BNA^+ + 2HV^{\cdot+} + H^+ \qquad [3]$$

By applying the steady-state approximation to the concentrations of the

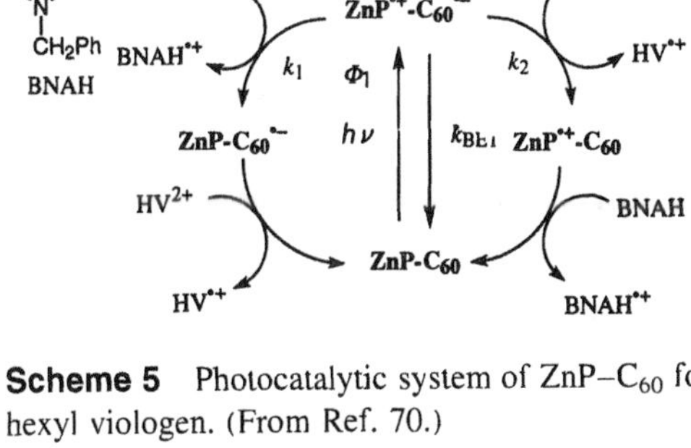

Scheme 5 Photocatalytic system of $ZnP-C_{60}$ for uphill photo-oxidation of BNAH by hexyl viologen. (From Ref. 70.)

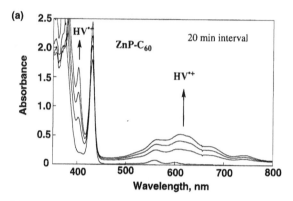

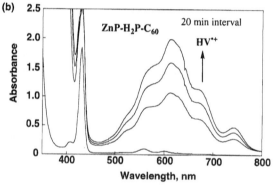

Figure 4 (a) Spectral change observed in the steady-state photolysis of a PhCN solution of BNAH (4.0×10^{-4} M). HV^{2+} (8.0×10^{-4} M) and ZnP-C_{60} (3.0×10^{-6} M) under irradiation of monochromatized light of $\lambda = 433$ nm. (b) Spectral change observed in the steady-state photolysis of a PhCN solution of BNAH (4.0×10^{-4} M), HV^{2+} (8.0×10^{-4} M) and ZnP-H_2P-C_{60} (3.0×10^{-6} M) under irradiation of monochromatized light of $\lambda = 433$ nm. (From Ref. 70.)

$ZnP^{\bullet+}$ and $C_{60}^{\bullet-}$ moieties in Scheme 5, the dependence of the quantum yield of formation of $HV^{\bullet+}$ (Φ_{obs}) on [BNAH] and [HV^{2+}] can be derived as given by Eq. (4):

$$\Phi_{obs} = \frac{\Phi_1(k_1[\text{BNAH}] + k_2[\text{HV}^{2+}])}{k_{BET} + k_1[\text{BNAH}] + k_2[\text{HV}^{2+}]} \quad [4]$$

where Φ_1 is the quantum yield of the radical ion pair formation. Under the experimental conditions (i.e., $k_1[\text{BNAH}]$ or $k_2[\text{HV}^{2+}] \gg k_{BET}$), the limiting quantum yield (Φ_∞) corresponds to Φ_1. The observed dependence of Φ_{obs} on [HV^{2+}] in the case of ZnP–C_{60} agrees with Eq. (4), from which the Φ_∞ value is determined as 0.99 [70]. This value agrees well with the Φ_1 value obtained independently from the direct kinetic analysis of the radical ion pair formation (0.99) [53]. In the case of ZnP–H_2P–C_{60}, the Φ_∞ value agrees also with the Φ_1 value (0.40) obtained from the transient absorption spectrum [53]. Such agreements confirm the validity of Scheme 5.

Upon replacing BNAH with $AcrH_2$, photoinitiated oxidation by HV^{2+} produces $AcrH^+$ (λ_{max} = 358 nm) and $HV^{\bullet+}$ (λ_{max} = 402 and 615 nm) in the presence of ZnP–H_2P–C_{60} as in the case of BNAH [70]. The amount of free energy (ΔG) stored in the photochemical oxidation of $AcrH_2$ by HV^{2+} is obtained as 1.28 eV (124 kJ/mol) based on the difference between the redox potentials of the $AcrH_2/AcrH^+$ (0.22 V versus SCE) [72] and $HV^{2+}/HV^{\bullet+}$ (-0.42 V versus SCE) redox couples [Eq. (5)] [70]. In the case of BNAH,

$$\Delta G = 2F[E^0(AcrH_2/AcrH^+) - E^0(HV^{2+}/HV^{\bullet+})] \tag{5}$$

[E^0(BNAH/BNA^+) = 0.02 V] [73], the corresponding value is obtained as 0.88 eV (85 kJ/mol). In the dark, the back reaction is extremely slow because of the kinetic barrier for a highly endergonic electron transfer from $HV^{\bullet+}$ (E_{ox} = -0.42 V versus SCE) to BNA^+ (E_{red} = -1.08 V versus SCE) [74], although the overall two-electron process is exergonic [75].

In the presence of oxygen, the lifetimes of both radical ion pairs (i.e., $ZnP^{\bullet+}$–$C_{60}^{\bullet-}$ and $ZnP^{\bullet+}$–H_2P–$C_{60}^{\bullet-}$) are decreased significantly due to oxygen-catalyzed back-electron transfer (BET) processes between $C_{60}^{\bullet-}$ and $ZnP^{\bullet+}$ [76]. The catalytic participation of O_2 in an intramolecular BET between $C_{60}^{\bullet-}$ and $ZnP^{\bullet+}$ in ZnP-linked C_{60} is depicted in Scheme 6 [76]. The intermolecular ET from $C_{60}^{\bullet-}$ to O_2 is facilitated by the partial coordination of O_2 to $ZnP^{\bullet+}$ in the transient state (denoted as $\neq$ in Scheme 6) [76]. Consequently, the one-electron reduction potential of the resulting $O_2^{\bullet-}$ is shifted toward positive values, namely in favor of the ET event. The strong coordination of $O_2^{\bullet-}$ to Zn(II) ion has been well established [77]. The complexation is then followed by a rapid intramolecular

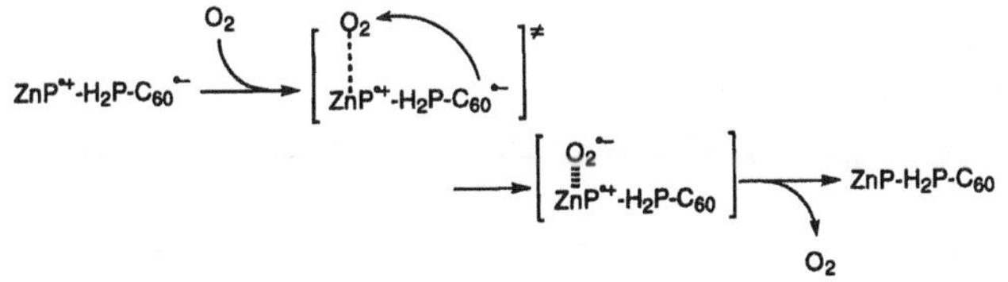

Scheme 6   Catalysis of O_2 in back-electron transfer in ZnP–linked C_{60}. (From Ref. 76.)

ET from $O_2^{\bullet-}$ to $ZnP^{\bullet+}$ in the $O_2^{\bullet-}$–$ZnP^{\bullet+}$ complex to regenerate O_2. Thus, an intermolecular ET from $C_{60}^{\bullet-}$ to O_2 occurs via the coordination of O_2 to $ZnP^{\bullet+}$ to yield $O_2^{\bullet-}$ bound to $ZnP^{\bullet+}$, followed by a rapid intramolecular ET from $O_2^{\bullet-}$ to $ZnP^{\bullet+}$ in the $O_2^{\bullet-}$–$ZnP^{\bullet+}$ complex to regenerate O_2 [76]. Such a binding of the radical anion to metal ion is known to accelerate the ET process [24,25,78].

Surprisingly, the Φ_{obs} value (0.14) of the triad (ZnP–H_2P–C_{60}) system for the photooxidation of $AcrH_2$ by HV^{2+} is little affected by molecular oxygen as compared to the value (0.15) found in the absence of O_2 [70]. In contrast to the triad, the Φ_{obs} value (0.02) of the dyad (ZnP–C_{60}) system becomes significantly smaller than the corresponding value (0.25) in the absence of O_2 [70]. It should be noted that the k_{BET} value of the triad increases from 4.8×10^4 to 3.2×10^5 sec^{-1} in the absence and presence of O_2, respectively [76]. This value is, however, still significantly smaller than the k_{BET} value of the dyad, even in the absence of O_2 (1.3×10^6 sec^{-1}) [76]. Furthermore, in the case of the triad (ZnP–H_2P–C_{60}), the rate of intermolecular electron transfer from $C_{60}^{\bullet-}$ to HV^{2+} (5.6×10^6 sec^{-1} at 8.0×10^{-3} M HV^{2+}) is much faster than the BET rate (3.2×10^5 sec^{-1}) in O_2-saturated PhCN [70]. This can be regarded as a potential rationale for the observation that the Φ_{obs} value of the triad (ZnP–H_2P–C_{60}) system is little affected by the presence of oxygen. Conversely, in the case of the dyad (ZnP–C_{60}), the BET rate, especially in an O_2-saturated PhCN solution (1.5×10^7 sec^{-1}) [76] is much faster than the electron-transfer rate from $C_{60}^{\bullet-}$ to HV^{2+}. Thus, the catalytic performance of the triad (ZnP–H_2P–C_{60}) system, which exhibits a much longer lifetime of the radical ion pair, is little affected by O_2, whereas the catalytic reactivity of the dyad (ZnP–C_{60}) is reduced significantly by the presence of O_2 [70].

IV. CONTROL OF PHOTOINDUCED ELECTRON-TRANSFER REACTIONS BY SOLVENT POLARITY

Electron-transfer reactions are normally performed in polar solvents such as acetonitrile (MeCN), in which the product ions of the electron transfer are stabilized by the strong solvation [6,7]. When a cationic electron acceptor (A^+) is employed in electron-transfer reactions with a neutral electron donor (D), the electron transfer from D to A^+ produces a radical cation ($D^{\bullet+}$) and a neutral radical ($A^{\bullet}$). In such a case, the solvation before and after the electron transfer may be largely canceled out when the free-energy change of electron transfer is expected to be rather independent of the solvent polarity. The solvent independent ΔG°_{ET} value is confirmed by determination of the E°_{ox} values of alkylbenzene derivatives (electron donors) and E°_{red} values of acridinium cations (electron acceptors) in solvents with different polarities [79]. The E°_{ox} values of alkylbenzene derivatives in a less polar solvent (CH_2Cl_2) are shifted to the positive direction by about 0.1 V due to the lower solvation of the radical cations as compared to that in MeCN [79]. Similar positive shifts are observed for the E_{red}^0 values of 10-methyl-9-substituted acridinium cations (RAcrH$^+$, R = methyl or decyl) in CH$_2$Cl$_2$ as compared to the E_{red}^0 values in MeCN, and the larger shifts are observed in CHCl$_3$ and benzene [79]. Thus, the ΔG_{ET}° values of photoinduced electron transfer from electron donors to the singlet excited state of RAcrH$^+$, obtained as the difference between E_{ox}^0 and E_{red}^0* [eq. (6)],

$$\Delta G_{et}^{\circ} = e(E_{ox}^0 - E_{red}^0*)$$ [6]

where e is the elementary charge, * denotes the excited state], become rather solvent independent because of the cancellation of the solvation. In such a case, photoinduced electron transfer from an electron donor (D) to the singlet excited state of RAcrH$^+$ (RAcrH$^+$*) occurs efficiently even in a nonpolar solvent such as benzene as well as in a polar solvent such as MeCN [79]. The mechanism of photoinduced electron transfer from an electron donor (D) to ^{1}RAcrH$^+$* is shown in Scheme 7, where k_{12} and k_{21} are diffusion and dissociation rate constants, respectively, in the encounter complex (D RAcrH$^+$*) formed prior to electron transfer, k_{ET} and k_{BET} are the rate constants of the forward-electron transfer from D to RAcrH$^+$* in the complex and the back-electron transfer to the ground state [80,81]. According to Scheme 7, the observed rate constant for intermolecular electron transfer (k_{et}) is given by Eq (7).

$$k_{et} = \frac{k_{ET} \, k_{12}}{k_{21} + k_{ET}}$$ [7]

The dependence of k_{ET} for electron transfer in the encounter complex in Scheme 7 on ΔG_{ET}° has been well established by Marcus as given by Eq. (1), where the pre-exponential factor should be replaced by $h/k_B T$ for adiabatic electron transfer [1–4].

The k_{et} values have been determined by fluorescence quenching of ^{1}RAcrH$^+$* by electron donors [79]. Figure 5 exhibits the dependence of the observed rate constant for intermolecular electron transfer (k_{et}) on ΔG_{et}° [Eq. (6)], which is assumed to be the same as ΔG_{ET}° (Scheme 7) in MeCN (part a) and in

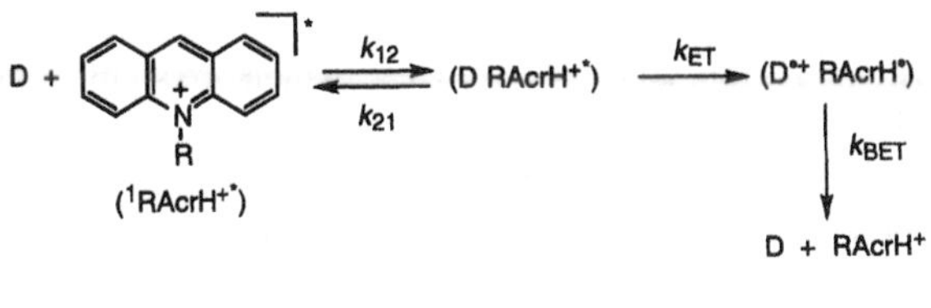

Scheme 7 Photoinduced eletron transfer an electron donor (D) to ^{1}RacrH.

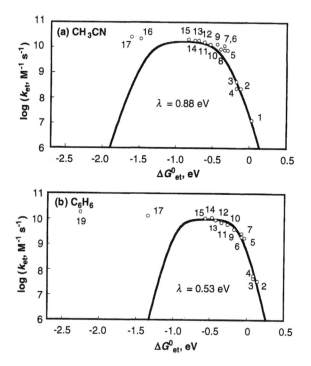

Figure 5 Plots of log k_{et} versus ΔG°_{et} for fluorescence quenching of RAcrH⁻ (2.0 × 10⁻⁴ M) by various electron donors (a) in MeCN and (b) in benzene at 298 K. Numbers refer to electron donors: benzene (1), toluene (2), ethylbenzene (3), cumene (4), m-xylene (5), o-xylene (6), 1, 3, 5-trimethylbenzene (7), p-xylene (8), 1,3-trimethylbenzene (9), 1, 2, 3, 4-tetramethylbenzene (11), 1, 2, 3, 5-tetramethylbenzene (12), 1,2,4,5-tetramethylbenzene (13), pentamethylbenzene (14), hexamethylbenzene (15), triphenylamine (16), N,N-dimethylaniline (17), ferrocene (18), and decamethylferrocene (19). (From Ref. 79.)

benzene (part b) [79]. The dependence of k_{et} on ΔG°_{et} for electron-transfer quenching of ¹RAcrH⁺* by electron donors in MeCN is calculated based on Eqs. (1) and (7) using the best-fit value of λ (0.88 eV) as shown by the solid line in Fig. 5a [79]. The k_{et} values agree well with the calculated values except for the k_{et} values in the highly exergonic region, $\Delta G^\circ_{et} \ll -1$ eV, which are significantly larger than the calculated values in Fig. 5a. The calculated dependence of k_{et} on ΔG°_{et} predicts a decrease in the k_{et} value from a diffusion-limited value with increasing the driving force of electron transfer $(-\Delta G^\circ_{et})$ when the k_{et} values become smaller than the diffusion-limited value in the Marcus-inverted region $(\Delta G^\circ_{et} < -\lambda)$, provided that the λ value is constant in a series of electron-transfer reactions [1–4]. The absence of a Marcus-inverted region has well been recognized in forward photoinduced electron-transfer reactions [82]. In the case of back-electron-transfer reactions, however, the observation of the Marcus-inverted region has been well established [83–87]. The absence of an inverted region in forward photoinduced electron-transfer reactions in the highly exergonic region ($\Delta G_{et}^{\circ} < -\lambda$) can be explained by an increase in the λ value from the value for a contact radical ion pair (CRIP) to the value for a solvent-separated radical ion pair (SSRIP), which has a larger distance between the radical ions in the highly exergonic region [87]. The existence of low-energy excited states of radical cations may also contribute to the nonexistence of an inverted region.

The dependence of k_{et} on ΔG_{et}° for electron-transfer quenching of RAcrH^{+}* by aromatic electron donors in benzene is also calculated based on Eqs. (1) and (7), as shown in Fig. 5b [79]. The best-fit λ value in benzene (0.53 eV) is significantly smaller than the λ value in MeCN (0.88 eV), as expected from the smaller change in solvation in benzene than in MeCN [79]. The decrease in the solvent reorganization energy of electron transfer with decreasing the solvent polarity has been demonstrated by determining the rate constants of electron-transfer self-exchange reactions between 9-phenyl-10-methylacridinium ion (MeAcrPh^{+}) and the corresponding one-electron reduced radical (MeAcrPh$^{•}$) in solvents with different polarity [79]. The MeAcrPh$^{•}$ radical was produced by the electron-transfer reduction of MeAcrPh^{+} by the tetramethylsemiquinone radical anion, as shown in Fig. 6a, where the hyperfine splitting constants and the maximum slope linewidths (ΔH_{msl}) were determined from a computer simulation of the ESR spectrum [79]. The ΔH_{msl} value thus determined increases linearly with an increase in the concentration of MeAcrPh^{+} in MeCN, as shown in Fig. 6b [79]. Such linewidth variations of the ESR spectra can be used to investigate the rate processes involving the radical species [88,89]. Then, the rate constants (k_{ex}) of the electron-transfer self-exchange reactions between MeAcrPh^{+} and MeAcrPh$^{•}$ were determined using Eq. (8),

$$k_{ex} = \frac{1.57 \times 10^{7}(\Delta H_{msl} - \Delta H_{msl}^{0})}{(1 - P_{t})\,[\text{MeAcrPh}^{+}]} \qquad [8]$$

where ΔH_{msl} and ΔH_{msl}° are the maximum slope linewidth of the ESR spectra in the presence and absence of MeAcrPh^{+}, respectively, and P_{t} is a statistical factor which can be taken as nearly zero. The reorganization energies (λ) of the electron-transfer reactions are obtained from the k_{ex} values using Eq. (9), where k_{diff} is the diffusion rate constant which corresponds to k_{12} in Scheme 1 ($k_{diff} = 2.0 \times 10^{10}$ M^{-1}sec^{-1} in MeCN,

$$[(k_{ex})^{-1} - (k_{diff})^{-1}] = Z^{-1}\exp\left(\frac{\lambda}{4k_{B}T}\right) \qquad [9]$$

2.0×10^{10} M^{-1} sec^{-1} in CH$_2$Cl$_2$, 1.2×10^{10} M^{-1} sec^{-1} in CHCl$_3$, and $1.1 \times$

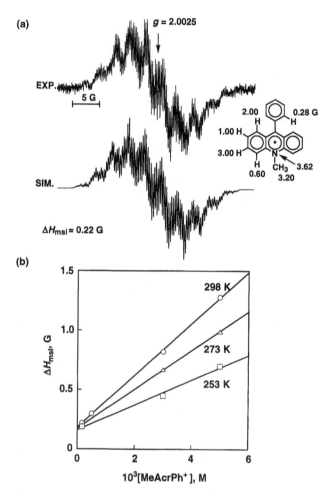

Figure 6 (a) ESR spectrum of MeAcrPh˙ in deaerated MeCN at 298 K and the computer simulation spectrum. (b) Plots of ΔH_{msl} versus [MeAcrPh⁻] for the ESR spectra of MeAcrPh˙ in deaerated MeCN at various temperatures. (From Ref. 79.)

10^{10} M⁻¹ sec⁻¹ in benzene, respectively) [80]. The λ values (eV) thus determined decrease with decreasing the solvent polarity: 0.34 (MeCN), 0.28 (CH₂Cl₂), 0.27 (CHCl₃), 0.21 (benzene) [79].

The k_{BET} value in Scheme 7 is expected to vary depending on the driving force ($-\Delta G°_{BET}$) of back-electron transfer as shown in Fig. 7, where the dependence of log k_{BET} versus $-\Delta G°_{BET}$ in MeCN and benzene is drawn schematically

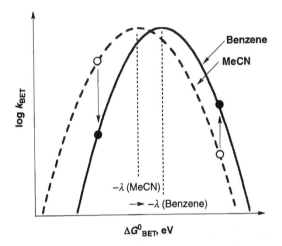

Figure 7   Dependence of log k_{BET} on $\Delta G°_{BET}$ in MeCN (solid line) and benzene (broken line). The lines are drawn schematically based on the Marcus equation [Eq. (1)]. The $\Delta G°_{BET}$ value at which k_{BET} is maximum corresponds to $-\lambda$ [Eq. (1)].

[1–4]. According to the Marcus equation (Eq. (1), where ET is replaced by BET), the k_{BET} value decreases with decreasing the λ value in the inverted region, whereas the k_{BET} value increases in the normal region (Fig. 7). Because the λ value decreases with decreasing the solvent polarity (*vide supra*), the k_{BET} value in MeCN becomes larger than the value in benzene when the back-electron transfer ($-\Delta G°_{BET}$) in MeCN is in the Marcus-inverted region. In contrast, the opposite trend may be obtained when the $-\Delta G°_{BET}$ value is in the normal region. In such a case, if the radical ion pair formed by photoinduced electron transfer affords products via bond-breaking and bond-forming processes which are insensitive to the solvent polarity, in competition with the back-electron transfer, the limiting quantum yield of product formation would vary depending on the solvent polarity. Such examples have indeed been reported as described below [79].

The photoalkylation of RAcrH$^+$ with 4-*tert*-butyl-1-benzyl-1,4-dihydronicotinamide (Bu'BNAH), which is known to act as an alkylating reagent [90], proceeds efficiently via photoinduced electron transfer from Bu'BNAH to ^{1}RAcrH$^+$*, as shown in Scheme 8 [79]. The photoinduced electron transfer from Bu'BNAH to ^{1}RAcrH$^+$* produces the radical cation/radical pair (Bu'BNAH$^{•+}$ RAcrH$^•$). The one-electron oxidation of Bu'BNAH is known to result in the selective C(4)—C bond cleavage of Bu'BNAH$^{•+}$ to produce Bu$^{r•}$ and BNA$^+$ [91,92]. Thus, cleavage of Bu'BNAH$^{•+}$ gives Bu$^{r•}$ that is coupled immediately with RAcrH$^•$ to yield the adduct (MeAcrHBu') in competition with the back-

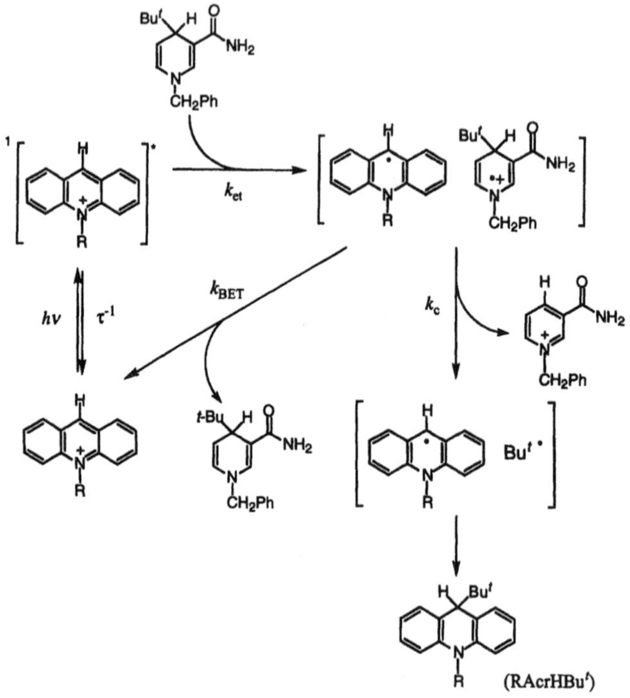

Scheme 8   Photoalkylation of RacrH $^+$ with Bu-BNAH via photoinduced electron transfer. (From Ref. 79.)

electron transfer from MeAcrH$^\bullet$ to Bu^{t}BNAH$^{\bullet+}$ (Scheme 8) [79]. By applying the steady-state approximation to the reactive species in Scheme 8, the dependence of Φ on [Bu^{t}BNAH] can be derived as given by Eq. (10),

$$\Phi = \left(\frac{k_c}{k_c + k_{BET}}\right) \frac{k_{et}\tau[\text{Bu}^t\text{BNAH}]}{(1 + k_{et}\,\tau[\text{Bu}^t\text{BNAH}])} \qquad [10]$$

which was confirmed experimentally [79]. Then, the limiting quantum yield Φ_∞ corresponds to $k_c/(k_c + k_{BET})$. In such a case, the Φ_∞ value is determined by the competition between the C(4)—C bond cleavage (k_c) and the back-electron transfer (k_{BET}). Because the reorganization energy (λ) of photoinduced electron-transfer reactions of BNAH and derivatives in MeCN is relatively small ($\lambda = 0.5$ eV) [93], the back-electron transfer from RAcrH$^\bullet$ to Bu^{t}BNAH$^{\bullet+}$ should be in the Marcus-inverted region ($-\Delta G^\circ_{BET} = 1.1$ eV $\gg 0.5$ eV). Thus, the solvent reorganization energy should decrease with decreasing the solvent polarity, lead-

Table 1 Electron-Transfer Rate Constants (k_{et}) Determined from the Fluorescence Quenching of RAcrH$^+$ by Bu^{t}BNAH and R$_4'$Sn, Determined from the Dependence of the Quantum Yields on [Bu^{t}BNAH] and [R$_4'$Sn] in the Photoalkylation of RAcrH$^+$ with Bu^{t}BNAH and R$_4'$Sn in MeCN, CHCl$_3$ and Benzene at 298 K, and the Driving Force of Back-Electron Transfer ($-\Delta G_{BET}^\circ$) from RAcrH$^\bullet$ to Bu^{t}BNAH$^{\bullet-}$ and R$_4'$Sn$^{\bullet+}$ in MeCN

Donor	Solvent	$k_{et}{}^a$ (M^{-1}/sec^{-1})	$k_{et}{}^b$ (M^{-1}/sec^{-1})	Φ_∞	$-\Delta G_{BET}^\circ$ (eV)
Bu^{t}BNAH	Benzene	2.3×10^{10}	2.1×10^{10}	0.81	
	CHCl$_3$	2.0×10^{10}	1.9×10^{10}	0.12	
	MeCN	2.1×10^{10}	2.3×10^{10}	0.04	1.10
Me$_4$Sn	Benzene	1.1×10^8	1.1×10^8	0.0088	
	MeCN	3.3×10^8	4.2×10^8	0.00042	2.11
Et$_4$Sn	Benzene	2.4×10^9	2.1×10^9	0.028	
	MeCN	2.0×10^{10}	2.4×10^9	0.007	1.67
Pr$_4'$Sn	Benzene	5.3×10^9	5.0×10^9	0.027	
	MeCN	7.7×10^9	8.0×10^9	0.028	1.43
Bu$_2'$Me$_2$Sn	Benzene	1.1×10^{10}	1.7×10^{10}	0.0055	
	MeCN	1.1×10^{10}	1.4×10^{10}	0.021	1.29

[a] Determined from the fluorescence quenching of RAcrH$^+$ by Bu^{t}BNAH and R$_4'$Sn.
[b] Determined from the dependence of the quantum yields on [Bu^{t}BNAH] and [R$_4$Sn].
Source: Ref. 79.

ing to the decrease in the back-electron transfer rate in the Marcus-inverted region, as shown in Fig. 7, and thereby an increase in the Φ_∞ value, as shown in Table 1 [79]. In contrast to the Φ_∞ values, the k_{et} values obtained from the fluorescence quenching and the dependence of Φ on [Bu^{t}BNAH] [Eq. (10)] in different solvents are diffusion-limited ones and independent of the solvent polarity (Table 1).

The photoalkylation of RAcrH$^+$ also proceeds in MeCN and benzene using R$_4'$Sn instead of Bu^{t}BNAH, yielding RAcrHR$'$ via photoinduced electron transfer (k_{et}) from R$_4'$Sn to ^{1}RAcrH^{+*}, as shown in Scheme 9 [79]. The Sn—C bond of R$_4'$Sn$^{\bullet+}$ in the radical cation–acridinyl radical pair (R$_4'$Sn$^{\bullet+}$ RAcrH$^\bullet$) produced by the photoinduced electron transfer is known to be cleaved to give the alkyl radical [94–97], which is coupled within the cage to yield the adduct selectively without dimerization of free RAcrH$^\bullet$ radicals escaped from the cage, in competition with the back-electron transfer to the ground state (k_{BET}).

The Φ_∞ value of the photoalkylation of RAcrH$^+$ with R$_4'$Sn varies drastically depending on the alkyl group (R$'$) of R$_4'$Sn and solvents from the smallest value (4.2×10^{-4}) for Me$_4$Sn in MeCN to the largest value (2.8×10^{-2}) for Pr$_4'$Sn in MeCN, as shown in Table 1. It is interesting to note that the Φ_∞ value of Me$_4$Sn in benzene is 21 times larger than the corresponding value in MeCN.

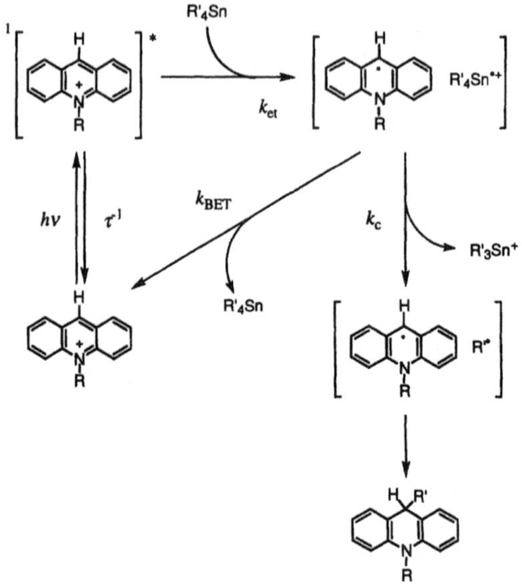

Scheme 9 Photoalkylation of RAcrH$^+$ with R$'_4$Sn via photoinduced electron transfer. (From Ref. 79.)

However, this is reversed for Bu$'_2$Me$_2$Sn$^\cdot$, the Φ_∞ value in a polar solvent (MeCN) is 38 times larger than the corresponding value in a nonpolar solvent (benzene). The ratio of the Φ_∞ value in benzene to that in MeCN changes systematically from Me$_4$Sn to Bu$'_2$Me$_2$Sn with decreasing the driving force ($-\Delta G^\circ_{BET}$) of back-electron transfer from RAcrH$^\cdot$ to R$'_4$Sn$^{\cdot+}$ (Table 1). According to Eq. (10), where [Bu$'$BNAH] should be replaced by [R$'_4$Sn], the Φ_∞ value is determined by the competition between the cleavage of C—Sn bond of R$'_4$Sn$^{\cdot+}$ in the radical cation/radical pair (k_c) and the back-electron transfer from RAcrH$^\cdot$ to R$'_4$Sn$^{\cdot+}$ (k_{BET}). The solvent effects on the cleavage rates of C—Sn bond of R$'_4$Sn$^{\cdot+}$ have been studied previously and it has been shown that the k_c values are rather independent of solvent polarity [97]. Thus, the Φ_∞ value is mainly determined by the back-electron transfer rate from RAcrH$^\cdot$ to R$'_4$Sn$^{\cdot+}$ (k_{BET}); the smaller the k_{BET} value, the larger the Φ_∞ value. The $-\Delta G^\circ_{BET}$ value is changed from the value in the Marcus-inverted region in the case of Me$_4$Sn to the normal region up to 1.2 eV, which is significantly smaller than the λ value (1.8 eV) in the case of Bu$'_2$Me$_2$Sn [79]. According to the Marcus equation [Eq. (1)], the k_{BET} value decreases with decreasing the λ value in the inverted region, whereas the k_{BET} value increases in the normal region (Fig. 7). This is the reason why the Φ_∞ value of Me$_4$Sn in benzene is much larger than the corresponding value in MeCN, whereas this is reversed for $Bu_2^tMe_2Sn$, the Φ_∞ value of which in MeCN is 38 times larger than the corresponding value in benzene (Table 1) [79]. Thus, whether the back-electron transfer is in the Marcus-inverted region or in the normal region determines whether the Φ_∞ value increases or decreases with decreasing the solvent polarity.

V. CONTROL OF PHOTOCATALYTIC REACTIVITY IN PHOTO-OXYGENATION REACTION

Selective oxygenation of ring-substituted toluenes to aromatic aldehydes has been one of the most important organic reactions in industrial chemistry because of useful applications of aromatic aldehydes as key chemical intermediates for the production of a variety of fine or specialty chemicals such as pharmaceutical drugs, dyestuffs, pesticides, and perfume compositions [98,99]. A number of methods using inorganic oxidants such as chromium(IV) [100,101], cobalt(III) [102], manganese(III) [103], cerium(IV) [104–106], benzeneseleninic anhydride [107], or peroxydisulfate/copper ion [108] have been reported for the oxygenation of ring-substituted toluenes to aromatic aldehydes. However, their synthetic utility has been limited because of low yield and poor selectivity. In addition, the use of stoichiometric amounts of inorganic oxidants results in the generation of copious amounts of inorganic waste, which causes serious pollution of the environment. Thus, the development of catalytic alternatives employing clean oxidants such as O_2 is highly desired. It has recently been reported that 10-methyl-9-substituted acridinium perchlorate [$AcrR^+$ ClO_4^- (R = H, Ph), green color] acts as an efficient photocatalyst for highly selective oxygenation of p-xylene to p-tolualdehyde under visible light irradiation via photoinduced electron transfer from p-xylene to the singlet excited state of $AcrR^+$ ($^1AcrR^+*$) [109]. In contrast to the oxidation by inorganic oxidants, photoinduced electron transfer is highly sensitive to the oxidation potentials of electron donors and, thus, no further oxidation of p-tolualdehyde occurs via photoinduced electron transfer from p-tolualdehyde to $^1AcrPh^+*$, leading to the formation of p-tolualdehyde as the sole oxygenated product of p-xylene.

The photo-oxygenation of ring-substituted toluenes to aromatic aldehydes proceeds via photoinduced electron transfer from toluenes to $^1AcrR^+*$ as shown in Scheme 10 for the case of p-xylene [81,109]. The photoinduced electron transfer from p-xylene to $^1AcrR^+*$ (k_{et}) is followed by the deprotonation of the p-xylene radical cation in competition with the back-electron transfer (k_{BET}) to the reactant pair to produce the p-xylenyl radical which couples with $AcrPh^•$ in the absence of oxygen to yield the adduct [$AcrR(CH_2C_6H_4CH_3\text{-}p)$] [109]. In the presence of oxygen, the p-xylenyl radical is readily trapped by oxygen to give the p-xylenylperoxyl radical that is reduced by back-electron transfer from $AcrR^•$ to yield p-xylenyl hydroperoxide, accompanied by the regeneration of $AcrR^+$

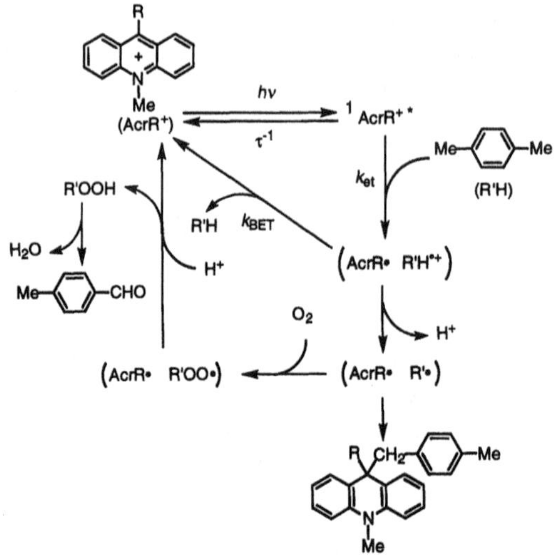

Scheme 10   Selective photo-oxygenation of p-xylene catalyzed by AcrR$^+$ via photoinduced electron transfer. (From Ref. 109.)

(Scheme 10). The hydroperoxide decomposes to yield p-tolualdehyde selectively [81,109].

The 100% selective photocatalytic oxygenation of p-xylene is made possible by the difference in the reactivity between p-xylene and the oxygenated product p-tolualdehyde, as indicated by the following fluorescence quenching experiments. The ^{1}AcrR$^+$* fluorescence was quenched efficiently by electron transfer from xylenes to ^{1}AcrR$^+$*, whereas no quenching was observed by p-tolualdehyde ($k_q \ll 1 \times 10^7$ M^{-1} sec^{-1}) [109]. The value of the fluorescence quenching rate constant (k_q) of ^{1}AcrH$^+$* decreases in order: p-xylene > o-xylene > m-xylene > o-tolualdehyde > toluene > m-tolualdehyde >> p-tolualdehyde (not observed) [109]. This order is consistent with the order of the mono-oxygenated and dioxygenated product yields [109]. In general, the faster the photoinduced electron transfer, the larger is the product yield. However, the k_q value for p-xylene determined in chloroform (4.2×10^9 M^{-1} sec^{-1}) is smaller than the value in MeCN (8.6×10^9 M^{-1} sec^{-1}), in contrast to the improved product yield in CHCl$_3$ (66%) as compared to the yield (37%) in a more polar solvent MeCN [109]. The improved product yield in CHCl$_3$ results from a decrease in the reorganization energy for the electron transfer with decreasing the solvent polarity (*vide supra*), which results in the slower back-electron transfer from AcrR$^-$ to the p-xylene radical cation in Scheme 10 [109]. Because the deprotonation of the p-xylene radical cation, which leads to the oxygenated product, competes with the back-electron transfer, the slower back-electron transfer results in the larger product yield. Because the λ values (0.27 and 0.34 eV in $CHCl_3$ and MeCN, respectively) [79] are much smaller than the driving force of the back-electron transfer ($-\Delta G^{\circ}_{BET} = 2.36$ eV) from AcrH$^{\bullet}$ (E^0_{ox} versus SCE $= -0.43$ V) [74] to the p-xylene radical cation ($E^0_{red} = 1.93$ V) [79], the back-electron transfer is deep in the Marcus-inverted region, where the back-electron transfer rate is expected to slow down with the decrease in the λ value, as shown in Fig. 7 [1–4]. Thus, the slower back-electron transfer rate with decreasing the solvent polarity leads to an increase in the product yield, as observed experimentally [109].

   The further improvement of the product yield (100%) was achieved by employing AcrPh$^+$ instead of AcrH$^+$ and this can also be ascribed to the slower back-electron transfer rate for the former than the latter [109]. In the Marcus-inverted region, the back-electron transfer becomes slower with increasing the driving force. Because the E°_{ox} value of AcrPh$^{\bullet}$ (E°_{ox} versus SCE $= -0.55$ V) [110] is more negative than the value of AcrH$^{\bullet}$ (E°_{ox} versus SCE $= -0.43$ V) [74], the driving force of the back-electron transfer from AcrPh$^{\bullet}$ (2.48 eV) is larger than that from AcrH$^{\bullet}$ (2.36 eV). The larger driving force results in the slower back-electron transfer, leading to the improved product yield (Fig. 7) [109]. The enhanced stability of AcrPh$^+$ as a photocatalyst as compared to AcrH$^+$ is ascribed to the steric effect of the phenyl group of AcrPh$^{\bullet}$, which hampers the radical coupling with the deprotonated radicals, that is the deactivation process of the photocatalyst in Scheme 10. Thus, the photocatalytic reactivity in 100% selective photooxygenation of p-xylene to p-toluandehyde as well as highly selective photo-oxygenation of other isomers to the corresponding aromatic aldehydes can be controlled by the redox potentials of the photocatalyst and the solvent polarity [109]. The clean mono-oxygenation of xylenes have also been attained recently via electron transfer from xylenes to periodinane reagents [111].

VI. CONTROL OF PHOTOINDUCED ELECTRON TRANSFER BY COMPLEXATION OF EXCITED STATES WITH METAL IONS

The electron-transfer reactivity of excited states can be controlled by complexation with metal ions, which not only enhance the oxidizing ability of the excited state but also change the spin multiplicity of the excited states [11,12,25]. The lowest excited state of aromatic carbonyl compounds (naphthaldehydes, aceto-naphthones, and 10-methylacridone) is changed from the $n–\pi^*$ triplet to the $\pi–\pi^*$ singlet due to the complexation with metal ions such as $Mg(ClO_4)_2$ and $Sc(OTf)_3$ (OTf = triflate), which act as Lewis acids [112–114]. The change in spin state of the lowest excited state from the triplet to the singlet due to the

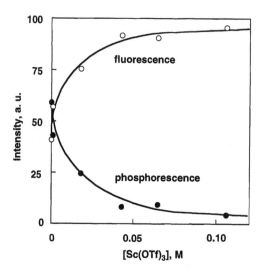

Figure 8 Plots of intensity ratio of fluorescence (○) at λ_{max} = 487 nm at 298 K: phosphorescence (●) at λ_{max} = 548 nm vs [Sc(OTf)₃] for 1-naphthaldehyde (1.5 × 10⁻³ M) in 2-methyltetrahydrofuran at 77 K. (From Ref. 113.)

complexation with Sc(OTf)₃ has been demonstrated clearly in Fig. 8, where the phosphorescence intensity of 1-naphthaldehyde (1-NA) at 548 nm measured at 77 K decreases accompanied by the increased fluorescence intensity at 487 nm measured at 298 K with increasing Sc(OTf)₃ concentration in 2-methyltetrahydrofuran [113]. The change in the lowest excited state may be caused by the complexation of aromatic carbonyl compounds with Sc(OTf)₃. The nonbonding orbitals are more stabilized by the complex formation with Sc(OTf)₃ than π-orbitals due to the stronger interaction between nonbonding electrons and Sc(OTf)₃. Thus, the π,π^* excited state becomes the lowest excited state in the Sc(OTf)₃ complex as compared with the lowest n,π^* triplet excited state in the uncomplexed carbonyl compound. Because the singlet–triplet energy gap is substantially larger in the π,π^* state than the n,π^* state, the singlet π,π^* state being the lowest excited state in the Sc(OTf)₃ complex becomes strongly fluorescent.

The one-electron reduction potentials ($E_{\text{red}}^{\circ}*$) of the fluorescent singlet excited states of metal ion–carbonyl complexes have been determined by adaptation of the free-energy relationship for photoinduced electron transfer from a series of electron donors to the singlet excited states of metal ion–carbonyl complexes as follows [113]. The free-energy change of photoinduced electron transfer from electron donors to the singlet excited states ($\Delta G_{\text{et}}^{\circ}$ in eV) is given by Eq. (6). The dependence of the activation free energy of intermolecular photoinduced electron transfer ($\Delta G_{et}^{\neq}$) on ΔG_{et}° has well been established as given by Eq. (11) [82],

$$\Delta G_{et}^{\neq} = \frac{\Delta G_{et}^{\circ}}{2} + \left[\left(\frac{\Delta G_{et}^{\circ}}{2}\right)^2 + (\Delta G_0^{\neq})^2\right]^{1/2} \qquad [11]$$

where $\Delta G_0^{\neq}$ is the intrinsic barrier that represents the activation free energy when the driving force of electron transfer is zero i.e., $\Delta G_{et}^{\neq} = \Delta G_0^{\neq}$ at $\Delta G_{et}^{\circ} = 0$). The $\Delta G_{et}^{\neq}$ values are obtained from the fluorescence quenching rate constant (k_q) by Eq. (12),

$$\Delta G_{et}^{\neq} = \left(\frac{2.3 k_B T}{e}\right) \log[Z(k_q^{-1} - k_{diff}^{-1})] \qquad [12]$$

where Z is the collision frequency that is taken as 1×10^{11} M^{-1} sec^{-1}, k_B is the Boltzmann constant, and k_{diff} is the diffusion rate constant in MeCN [82]. From Eqs. (6), (11), and (12) is derived a linear relation between $E_{ox}^{\circ} - (\Delta G_{et}^{\neq}/e)$ and $(\Delta G_{et}^{\neq}/e)^{-1}$ as given by Eq. (13):

$$E_{ox}^{0} - (\Delta G_{et}^{\neq}/e) = E_{red}^{0*} - \frac{(\Delta G_0^{\neq}/e)^2}{\Delta G_{et}^{\neq}/e} \qquad [13]$$

Thus, the unknown values of $E_{red}^{\circ}*$ and $\Delta G_0^{\neq}$ can be determined from the intercept and slope of the linear plots of $E_{ox}^{\circ} - (\Delta G_{et}^{\neq}/e)$ versus $(\Delta G_{et}^{\neq}/e)^{-1}$ as shown in Fig. 9. The $E^{\circ}_{red}*$ values of various metal ion–carbonyl complexes thus obtained are summarized in Table 2 [113,114]. The $E^{\circ}_{red}*$ values of the triplet excited states are obtained by adding the triplet excitation energies[115] and are listed in Table 2.

The comparison of the $E_{red}^{\circ}*$ values between the triplet excited states of uncomplexed carbonyl compounds and the singlet excited states of the Mg(ClO$_4$)$_2$ complexes in Table 2 reveals the remarkable positive shifts ($\sim$ 1.2 V) of the $E_{red}^{\circ}*$ values of the singlet excited states of the Mg(ClO$_4$)$_2$–carbonyl complexes as compared to those of the triplet excited states of uncomplexed carbonyl compounds. The comparison of the $E_{red}^{\circ}*$ values in Table 2 reveals the further positive shift (0.14 V) of the $E_{red}^{\circ}*$ value of the singlet excited states of the 1-NA–Sc(OTf)$_3$ complex as compared with the value of the singlet excited state to the 1-NA–Mg(ClO$_4$)$_2$ complex. The overall positive shift from the value of the triplet excited state of 1-NA is as large as 1.3 V, which corresponds to 10^{22} times the acceleration in terms of the rate constant of photoinduced electron transfer in the endergonic region ($\Delta G_{et}^{\circ} > 0$).

In contrast to the case of naphthaldehydes and acetonaphthones, irradiation of the absorption band of 10-methylacridone (AcrCO) results in fluorescence at 413 nm in MeCN [113]. When Sc(OTf)$_3$ is added to an MeCN solution of AcrCO, the absorption bands of 10-methylacridone is red-shifted due to the 1 : 1 complex

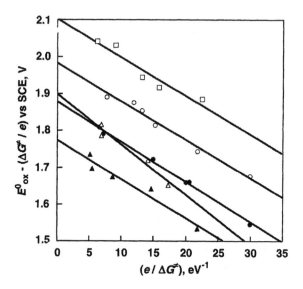

Figure 9 Plots of $E°_{ox} - (\Delta G^{\neq}_{et}/e)$ versus $e/\Delta G^{\neq}_{et}$ for electron transfer from benzene derivatives to the singlet excited states of $Mg(ClO_4)_2$ complexes of 1-naphthaldehyde (1-NA) (○), 2-naphthaldehyde (2-NA) (●), 1-acetonaphthone (△), and 2-acetonaphthone (△), and the singlet excited states of $Sc(OTf)_3$ complexes of 1-naphthaldehyde (□). (From Ref. 113.)

formation between AcrCO and $Sc(OTf)_3$ [113]. The formation constant (K) is determined from the spectral change as 1.2×10^5 M^{-1}, which is much larger than the K values of naphthaldehydes and acetonaphthones (Table 2) [113]. Thus, the Lewis acidity of $Sc(OTf)_3$ is strong enough to form the complex with AcrCO in competition with the coordination of MeCN, which is an abundant weak base. As compared to the $E°_{red}$* value of ^{1}AcrCO* (1.13 V), the $E°_{red}$* value of the ^{1}AcrCO*–$Sc(OTf)_3$ complex is significantly shifted to the positive direction (0.51 V), as shown in Table 2. The $E°_{red}$* value of Me_3SiOTf–AcrCO complex in CH_2Cl_2 is further shifted to the positive direction (0.75 V). This indicates that Me_3SiOTf acts as a stronger Lewis acid than $Sc(OTf)_3$.

Remarkable positive shifts of the $E°_{red}$* values of the singlet excited states of the metal ion–carbonyl complexes as compared to those of the triplet excited states of uncomplexed carbonyl compounds (Table 2) result in a significant increase in the redox reactivity of the Lewis acid complexes versus uncomplexed carbonyl compounds in the photoinduced electron-transfer reactions. For example, photoaddition of benzyltrimethylsilane with naphthaldehydes and acetonaphthones proceeds efficiently in the presence of $Mg(ClO_4)_2$ in MeCN, although

Table 2   Formation Constants (K), Fluorescence Maxima (λ_{max}), Fluorescence Lifetimes (τ), the One-Electron Reduction Potentials ($E^\circ_{red}*$) of the Singlet Excited States of $Mg(ClO_4)_2$, $Sc(OTf)_3$ and Me_3SiOTf Complexes of Aromatic Carbonyl Compounds

Aromatic carbonyl–metal ion salt complex	K^a (M^{-1})	λ_{max} (nm)	τ (nsec)	$E^\circ_{red}*$ versus $SCE^{b,c}$ (V)
1-NA–Mg(ClO$_4$)$_2$	0.17	437	6.7	1.97 (0.83)
1-NA–Sc(OTf)$_3$	2.8	487	10.0	2.11 (0.83)
2-NA–Mg(ClO$_4$)$_2$	0.27	440	10.3	1.87 (0.90)
1-AN–Mg(ClO$_4$)$_2$	—	432	3.3	1.90 (0.60)
2-AN–Mg(ClO$_4$)$_2$	0.51	430	11.8	1.77 (0.65)
AcrCO–Sc(OTf)$_3$	1.2×10^5	474	16.9	1.64 (1.13)^d
AcrCO–Me$_3$SiOTf^e	—	474	20.3	1.88 (1.13)^d
Fl–Mg(ClO$_4$)$_2$	2.2×10^2	504	1.3	2.06 (1.67)
Fl–YbOTf)$_3$	8.8×10^2	500	0.9	2.25 (1.67)
Fl–Sc(OTf)$_3$	3.1×10^4	486	2.4	2.45 (1.67)

^{a} Determined from the spectral change of aromatic carbonyl compounds in the presence of metal ion salts.
^{b} Determined by adaptation of the free-energy relationship for photoinduced electron transfer reactions.
^{c} Values in parentheses are those for the triplet excited states of uncomplexed compounds unless otherwise noted.
^{d} Singlet excited state.
^{e} In CH_2Cl_2.
Source: Data from Refs. 113 and 114.

no photochemical reaction occurs without $Mg(ClO_4)_2$ [112]. The photoaddition reactions proceed via photoinduced electron transfer from $PhCH_2SiMe_3$ to the singlet excited states of the $Mg(ClO_4)_2$–carbonyl complexes (k_{et}), followed by the cleavage of Si—C bond in the radical cation [116,117], and the radical coupling with the carbonyl radical anion (k_p) to yield the adduct in competition with the back-electron transfer to the reactant pair (k_{BET}) as shown in Scheme 11 for the case of the $PhCH_2SiMe_3$/2-NA–$Mg(ClO_4)_2$ system [112].

The use of $Sc(OTf)_3$ instead of $Mg(ClO_4)_2$ as a catalyst makes it possible to undergo the photoaddition of a much weaker electron donor, tetramethyltin (Me_4Sn), with 1-NA via photoinduced electron transfer from Me4Sn to $^1(1$-NA–$Sc^{3+})*$, whereas no photochemical reaction occurs between the corresponding $Mg(ClO_4)_2$ complex and Me_4Sn [113].

Photoaddition of benzyltrimethylsilane with 10-methylacridone (AcrCO) also proceeds via photoinduced electron transfer from $PhCH_2SiMe_3$ to $^1AcrCO*$–$Sc(OTf)_3$ as shown in Scheme 12 [113]. The drastically enhanced

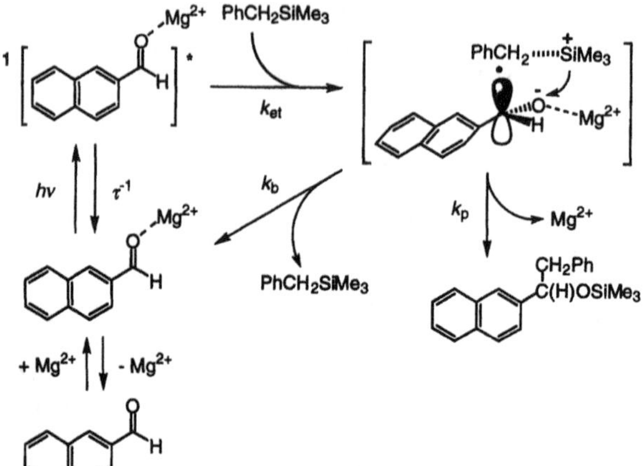

Scheme 11 Photoaddition of PhCH$_2$SiMe$_3$ with 2-naphthaldehyde (2-NA) via photoinduced electron transfer from PhCH$_2$SiMe$_3$ to the singlet excited state of the 2–NA–Mg (ClO$_4$)$_2$ complex. (From Ref. 112.)

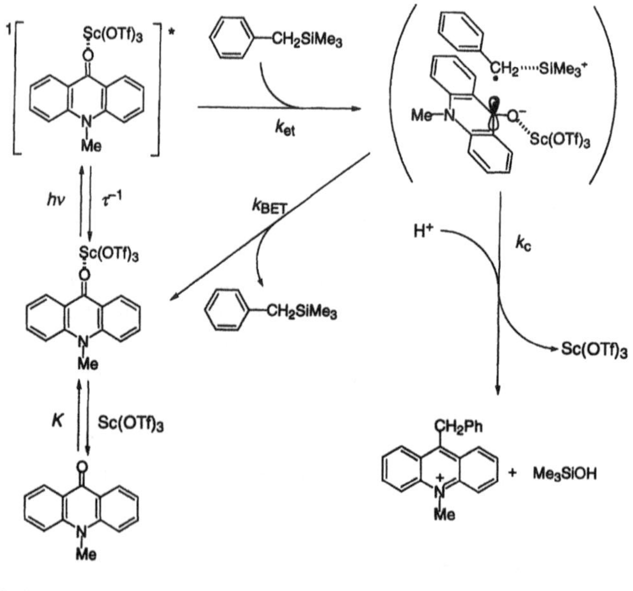

Scheme 12 Photoaddition of PhCH$_2$SiMe$_3$ with AcrCO via photoinduced electron transfer form PhCH$_2$SiMe$_3$ to the singlet excited state of the AcrCO-Sc(OTF)$_3$ complex. (From Ref. 113.)

electron-acceptor ability of the ^{1}AcrCO*–Sc(OTf)$_3$ complex as compared to ^{1}AcrCO* (*vide supra*) makes it possible that electron transfer from PhCH$_2$SiMe$_3$ to ^{1}AcrCO*–Sc(OTf)$_3$ occurs efficiently to produce the radical ion pair (PhCH$_2$SiMe$_3$$^{\bullet+}$ AcrCO$^{\bullet-}$–Sc(OTf)$_3$). The Si—C bond is readily cleaved by the reaction of PhCH$_2$SiMe$_3$$^{\bullet+}$ with AcrCO$^{\bullet-}$–Sc(OTf)$_3$ in the radical ion pair to yield the siloxy adduct (k_c) in competition with the back-electron transfer from AcrCO$^{\bullet-}$–Sc(OTf)$_3$ to PhCH$_2$SiMe$_3$$^{\bullet+}$. The carbon–oxygen bond of the siloxy adduct is readily cleaved by an acid to yield 9-benzyl-10-methylacridinium ion as the final product.

Trimethylsilyl triflate (Me$_3$SiOTf) acts as an even stronger Lewis acid than Sc(OTf)$_3$ in the photoinduced electron-transfer reactions of AcrCO in dichloromethane. In general, such enhancement of the redox reactivity of the Lewis acid complexes leads to the efficient C—C bond formation between organosilanes and aromatic carbonyl compounds via the Lewis-acid-catalyzed photoinduced electron transfer. Formation of the radical ion pair in photoinduced electron transfer from PhCH$_2$SiMe$_3$ to the ^{1}(1-NA)*–Mg(ClO$_4$)$_2$ complex (Scheme 11) and the ^{1}AcrCO*–Sc(OTf)$_3$ complex (Scheme 12) was confirmed by the laser flash experiments [113].

Flavoenzymes are a ubiquitous and structurally and functionally diverse class of redox-active proteins that use the redox-active isoalloxazine ring of flavins to mediate a variety of electron-transfer processes over a wide range of redox potentials [118,119]. The redox reactivity of flavins is controlled by the interaction with apoenzymes, which is affected by noncovalent interactions such as hydrogen-bonding, π–π stacking, and other electrostatic interactions [27]. The redox reactivity of flavins is the most drastically changed by the photoexcitation as compared to the ground state. Thus, photochemistry of flavoenzymes and flavin analogs has been the subject of intense research in photocatalysts for the photobiological redox processes [120–122]. The redox reactivity of the photoexcited states of flavins has been further modulated by complexation with metal ions (*vide infra*) [114,122–124].

Among metal ions, rare earth metal ions have attracted considerable interest as mild and selective Lewis acids in organic synthesis [125–127]. A flavin analog (riboflavin-2′,3′,4′,5′-tetraacetate, Fl) forms the 1:1 and 1:2 complexes with rare earth metal ions [114]. In particular, Sc(OTf)$_3$ has attracted much attention due to its hard character as well as strong affinity to carbonyl oxygen [128–131]. The largest formation constants K_1 and K_2 for the 1:1 and 1:2 complex between Fl and Sc^{3+} are obtained among Fl–metal ion complexes as $K_1 = 3.1 \times 10^4$ M^{-1} and $K_2 = 1.4 \times 10^3$ M^{-1}, respectively [114]. The complexation of Fl with rare earth metal ions results in blue shifts of the fluorescence maximum, shortening of the fluorescence lifetime, and, more importantly, the change in the lowest excited state from the n,π* triplet state of Fl to the π,π* singlet states of Fl–rare earth metal ion complexes, as indicated by the disappearance of the triplet–triplet

(T-T) absorption spectrum of Fl by the complexation with metal ions [114]. The strong complex formation between Fl and rare earth metal ions enhances the oxidizing ability of the excited state of Fl, as indicated by the significant acceleration in the fluorescence quenching rates of Fl–rare earth metal ion complexes via electron transfer from electron donors (e.g., alkylbenzenes) as compared to those of uncomplexed Fl. The one-electron reduction potentials ($E_{red}^{\circ}*$) of the fluorescent singlet excited states of Fl–metal ion complexes have been determined by adaptation of the free-energy relationship for photoinduced electron transfer (*vide supra*) and the $E_{red}^{\circ}*$ values are listed in Table 2 [114]. The order of $E_{red}^{\circ}*$ (versus SCE) values of Fl–metal ion complexes is ^{1}(Fl–2Sc^{3+})* (2.45 V) > ^{1}(Fl–Yb^{3+})* (2.25 V) > ^{1}(Fl–Mg^{2+})* (2.06 V) > ^{1}Fl* (1.67 V), and this order is consistent with that of the formation constants (K_1) of Fl–metal ion complexes [114]. The comparison of the $E_{red}^{\circ}*$ value of ^{1}(Fl–2Sc^{3+})* with that of ^{1}Fl* reveals a remarkable positive shift ($\sim$ 780 mV) of the $E_{red}^{\circ}*$ value of ^{1}(Fl–2Sc^{3+})* as compared to that of ^{1}Fl*. Such a large positive shift of the $E_{red}^{\circ}*$ value results in a significant increase in the reactivity of ^{1}(Fl–2Sc^{3+})* versus uncomplexed Fl in the photoinduced electron-transfer reactions, as shown in in Fig. 10 [114].

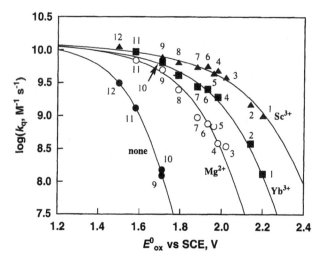

Figure 10 Plots of the logarithm of the quenching rate constant (k_q) for the fluorescence quenching of Fl (1.0×10^{-5} M) in the absence ($\bullet$) and presence of Sc^{3+} ($\blacktriangle$), Yb^{3+} ($\blacksquare$) and Mg^{2+} ($\bigcirc$) (1.0×10^{-2} M) by substituted benzenes: toluene (1), ethylbenzene (2), *m*-xylene (3), *o*-xylene (4), *p*-cymene (5), *p*-xylene (6), 1, 2, 3- trimethylbenzene (7), 1, 2, 4-trimethylbenzene (8), 1, 2, 3, 4-tetramethylbenzene (9), 1, 2, 3, 5-tetramethylbenzene (10), pentamethylbenzene (11) and *m*-dimethoxybenzene (12) versus one-electron oxidation potential (E_{ox}°) of the substituted benzenes in MeCN at 298 K. (From Ref. 114.)

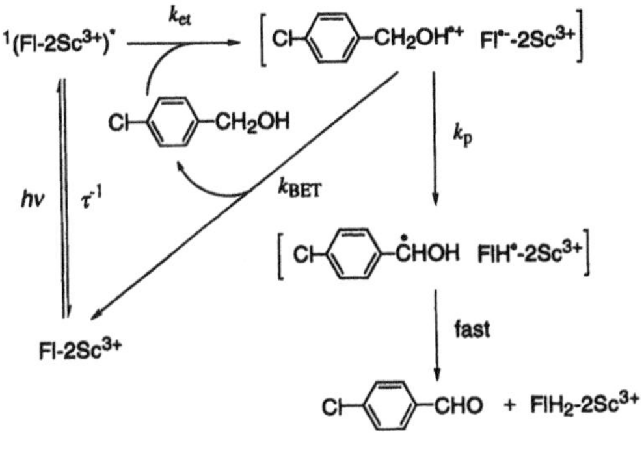

Scheme 13 Photo-oxidation of p-chlorobenzyl alcohol by Fl-2Sc^{3+} via photoinduced electron transfer. (From Ref. 114.)

The remarkable enhancement of the redox reactivity of ^{1}(Fl–2Sc^{3+})* as compared to that of ^{1}Fl* makes it possible to oxidize efficiently p-chlorobenzyl alcohol to p-chlorobenzaldehyde by ^{1}(Fl–2Sc^{3+})*, although no photo-oxidation of p-chlorobenzyl alcohol by Fl occurred in deaerated MeCN [114]. The reaction mechanism of the photo-oxidation of p-chlorobenzyl alcohol by Fl-2Sc^{3+} is summarized as shown in Scheme 13 [114]. The singlet excited state ^{1}(Fl–2Sc^{3+})* is quenched by electron transfer from p-chlorobenzyl alcohol (k_{et}) to give the radical ion pair [p-ClC$_6$H$_4$CH$_2$OH$^{•+}$ Fl$^{•-}$–2Sc^{3+}] in competition with the decay to the ground state. Then, a proton is transferred from the p-chlorobenzyl alcohol radical cation to Fl$^{•-}$–2Sc^{3+} in the radical ion pair (k_p) to give the radical pair [p-ClC$_6$H$_4$CHOH$^•$ FlH$^•$–2Sc^{3+}]. The transient absorption band observed at λ_{max} = 560 nm is assigned due to FlH$^•$–2Sc^{3+} [114]. The hydrogen transfer from p-ClC$_6$H$_4$CHOH$^•$ to FlH$^•$–2Sc^{3+} yields the final products p-ClC$_6$H$_4$CHO and FlH$_2$–2Sc^{3+} [114]. The proton-transfer step (k_p) competes well with the back-electron transfer step to the reactant pair (k_{BET}).

The Φ value for the photo-oxidation of p-chlorobenzyl alcohol (8.0 × 10^{-1} M) in the presence of Fl (2.0 × 10^{-4} M) and metal triflates (1.0 × 10^{-2} M) is largest in the case of Sc(OTf)$_3$ (Φ = 0.17) and decreases in the order Sc^{3+} > La^{3+} > Lu^{3+} > Yb^{3+} > Mg^{2+} [114]. When p-chlorobenzyl alcohol is replaced by p-methoxybenzyl alcohol that is a stronger electron donor, the Fl–Lu^{3+} complex gives the largest Φ value (0.17) [114]. Under atmospheric pressure of oxygen, the photo-oxidation of p-methoxybenzyl alcohol by oxygen proceeds efficiently in the presence of Fl–Lu^{3+}, which acts as an efficient photocatalyst [114]. In contrast with the efficient photocatalysis in the presence of Lu(OTf)$_3$, the photodegradation of photocatalyst (Fl) occurs in the presence of Mg(OTf)$_2$ and the yields of H$_2$O$_2$ as well as p-methoxybenzaldehyde become much lower

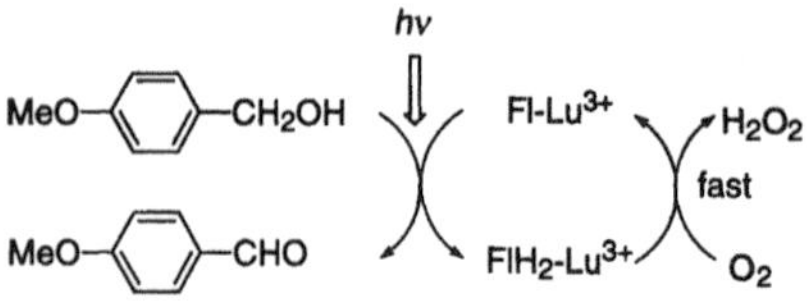

Scheme 14   Photo-oxidation of p-methoxybenzyl alcohol by O_2 catalyzed by $Fl-Lu^{3+}$. (From Ref. 114.)

than those in the presence of $Lu(OTf)_3$ [114]. In this case, $Lu(OTf)_3$ can not only act as an efficient cocatalyst in the photo-oxidation of p-methoxybenzyl alcohol catalyzed by Fl but also prevents Fl from the photodegradation. The Φ value for the photo-oxidation of p-methoxybenzyl alcohol by $Fl-Lu^{3+}$ in oxygen-saturated MeCN increases with increasing the concentration of p-methoxybenzyl alcohol to reach a constant value $\Phi_\infty = 0.17$, which agrees with the corresponding value (0.17) for the photo-oxidation of p-methoxybenzyl alcohol by $Fl-Lu^{3+}$ in deaerated MeCN [114]. Such an agreement indicates that oxidation of the reduced flavin (FlH_2-Lu^{3+}) by oxygen to regenerate $Fl-Lu^{3+}$ is too fast to affect the Φ value, as shown in Scheme 14. The rare earth metal ion accelerates the oxidation of the reduced flavin by oxygen to yield H_2O_2, as reported for the Mg^{2+}-catalyzed oxidation of reduced flavin by oxygen [132]. Although the triplet excited state of Fl is quenched by oxygen that is a well-known triplet quencher [133,134], the Fl–metal ion complexes can act as efficient and stable photocatalysts in the photo-oxidation of benzyl alcohol derivatives by oxygen due to the change in the spin state from the n,π^* triplet to the π,π^* singlet excited state by the complexation with metal ions.

VII.   METAL-ION-PROMOTED PHOTOINDUCED ELECTRON TRANSFER VIA BINDING OF RADICAL ANIONS WITH METAL IONS

As described earlier, the reactivity of photoinduced electron transfer is remarkably enhanced by the complexation of excited states with metal ions. Even if there is no direct interaction between excited states and metal ions, however, metal ions can enhance the reactivity of photoinduced electron transfer when the radical anion produced in photoinduced electron transfer binds with metal ions [11,12,25]. For example, although there is no direct interaction between the triplet excited state of C_{60} ($^3C_{60}^*$) and $Sc(OTf)_3$, an efficient electron transfer occurs from $^3C_{60}^*$ to p-chloranil (Cl_4Q) to produce $C_{60}^{\cdot+}$ and the p-chloranil radical anion $Cl_4Q^{\cdot-}-Sc^{3+}$ complex [135]. In contrast to the facile reduction of C_{60}, the oxidation of C_{60} is rendered more difficult with a highly positive one-electron oxidation potential of 1.26 V versus ferrocene/ferricenium [136,137]. Formation of $C_{60}^{\bullet+}$ in the photoinduced electron transfer from $^3C_{60}^*$ to Cl_4Q is confirmed by appearance of the diagnostic absorption band at 980 nm due to $C_{60}^{\bullet+}$ [137–140] in the transient absorption spectra in Fig. 11 [135]. The decay of $^3C_{60}^*$ at 740 nm is accompanied by the appearance of $C_{60}^{\bullet+}$ at 980 nm, but this species decays at prolonged reaction time (see inset of Fig. 11) [135]. This indicates that photoinduced electron transfer from $^3C_{60}^*$ to Cl_4Q (k_{et}) occurs in the presence of Sc^{3+} to produce $C_{60}^{\bullet+}$ and the $Cl_4Q^{\bullet-}-Sc^{3+}$ complex, both of which decay via a relatively slow back-electron-transfer (k_{BET}) process (Scheme 15) [135]. The strong binding between $Cl_4Q^{\bullet-}$ and Sc^{3+} makes the electron transfer energetically feasible, as reported for the Sc^{3+}-promoted electron-transfer reduction of p-benzoquinone [78].

Formation of the radical intermediate complex $[C_{60}^{\bullet+} \ Cl_4Q^{\bullet-}-S_c^{3+}]$ in Scheme 15 was also confirmed by the ESR spectrum observed in the Sc^{3+}-promoted photoinduced electron transfer from $^3C_{60}^*$ to Cl_4Q in frozen PhCN at

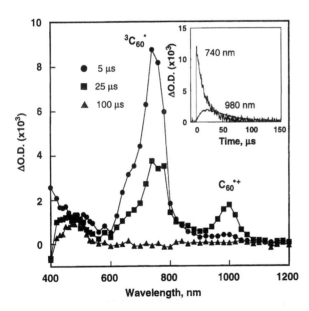

Figure 11 Transient absorption spectra observed in photoinduced electron transfer from C_{60} (1.1×10^{-4} M) to Cl_4Q (4.0×10^{-2} M) in the presence of Sc^{3+} (0.11 M) after laser irradiation at $\lambda = 532$ nm in deaerated PhCN at 298 K (black circles = 5 μsec, black square = 25 μsec and black triangle = 100 μsec). Inset: Time profile of the absorption band at 740 nm and 980 nm due to $^3C_{60}^*$ and $C_{60}^{\bullet+}$. (From Ref. 135.)

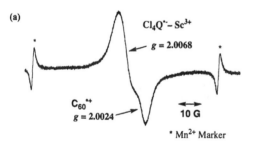

Scheme 15 Sc^{3+} -promoted photoinduced electron transfer from $^3C_{60}^*$ to *p*-chloranil. (From Ref. 135.)

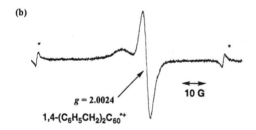

Figure 12 ESR spectra of (a) $C_{60}^{\cdot+}/Cl_4Q^{\cdot-}\text{-}Sc^{3+}$ and (b)1,4-$(C_6H_5CH_2)_2C_{60}^{\cdot+}/$ $Cl_4Q^{\cdot-}\text{-}Sc^{3+}$ generated in photoinduced electron transfer from C_{60} (2.8 × 10^{-4} M) and 1,4-$(C_6H_5CH_2)_2C_{60}$ (2.8 × 10^{-4} M) to Cl_4Q (1.0 × 10^{-2} M) in the presence of Sc^{3+} (5.6 × 10^{-1} M) in deaerated PhCN at 193 K. (From Ref. 135.)

193 K under photoirradiation with a high-pressure mercury lamp [135]. The resulting spectrum is shown in Fig. 12a and consists of two overlapping isotropic ESR signals which have different g values ($g = 2.0024$ and $g = 2.0068$) due to $C_{60}^{\bullet+}$ and $Cl_4Q^{\bullet-}-Sc^{3+}$, respectively [135]. Similarly, the radical cation of a C_{60} derivative, $1,4-(C_6H_5CH_2)_2C_{60}^{\bullet+}$, is formed in the Sc^{3+}-promoted photoinduced electron transfer from $1,4-(C_6H_5CH_2)_2C_{60}$ to Cl_4Q (Fig. 12b) [135].

The rate of Sc^{3+}-promoted photoinduced electron transfer from $^3C_{60}^*$ to Cl_4Q determined from the decay rate of the absorbance due to $^3C_{60}^*$ at 740 nm (inset of Fig. 11) obeys pseudo-first-order kinetics and the pseudo-first-order rate constant increases linearly with increasing the p-chloranil concentration [Cl_4Q] [135]. From the slope of the linear correlation, the second-order rate constant of electron transfer (k_{et}) in Scheme 15 was obtained. The k_{et} value increases linearly with increasing the Sc^{3+} concentration. This indicates that $Cl_4Q^{\bullet-}$ produced in the photoinduced electron transfer forms a 1:1 complex with Sc^{3+} (Scheme 15) [78]. When Cl_4Q is replaced by p-benzoquinone (Q), the k_{et} value for electron transfer from $^3C_{60}^*$ to Q increases with an increase in [Sc^{3+}] to exhibit a first-order dependence on [Sc^{3+}] at low concentrations, changing to a second-order dependence at high concentrations, as shown in Fig. 13 (open circles) [135]. Such a mixture of first-order and second-order dependence on [Sc^{3+}] was also observed in electron transfer from CoTPP (TPP^{2-} = tetraphenylporphyrin dianion) to Q

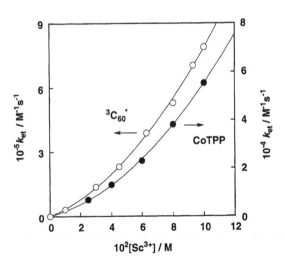

Figure 13 Plots of k_{et} versus [Sc^{3+}] for electron transfer from $^3C_{60}^*$ to Q (open circles) and from CoTPP (1.0×10^{-5} M) to Q (1.0×10^{-2} M) (closed circles) in the presence of Sc^{3+} in deaerated PhCN at 298 K. (From Ref. 135.)

(closed circles in Fig. 13) [135,141]. A mixture of first-order and second-order dependence on [Sc^{3+}] is ascribed to the formation of 1:1 and 1:2 complexes between the semiquinone radical anion ($Q^{\cdot-}$) and Sc^{3+} at the low and high concentrations of Sc^{3+} [Eqs. (14) and (15), respectively],

$$Q^{\cdot-} + Sc^{2+} \xrightleftharpoons{K_1} Q^{\cdot-}-Sc^{3+} \qquad [14]$$

$$Q^{\cdot-}-Sc^{3+} + Sc^{3+} \xrightleftharpoons{K_2} Q^{\cdot-}-(Sc^{3+})_2 \qquad [15]$$

which results in acceleration of the rate of electron transfer [135,141]. The complex formation of $Q^{\cdot-}$ and Sc^{3+} should result in a positive shift of the one-electron reduction potential of Q (E_{red}). The Nernst equation is given by

$$E_{red} = E_{red}^0 + \left(\frac{2.3RT}{F}\right) \log K_1[Sc^{3+}](1 + K_2[Sc^{2+}]) \qquad [16]$$

where E_{red}° is the one-electron reduction potential of Q to $Q^{\cdot-}$ in the absence of Sc^{3+}, and K_1 and K_2 are the formation constants for the 1:1 and 1:2 complexes between $Q^{\cdot-}$ and Sc^{3+}, respectively. Because Sc^{3+} has no effect on the oxidation potential of CoTPP and $^3C_{60}*$, the free-energy change of electron transfer in the presence of Sc^{3+} (ΔG_{et}) can be expressed by

$$\Delta G_{et} = \Delta G_{et}^\circ (2.3RT) \log (K_1[Sc^{3+}] + K_1K_2[Sc^{3+}]^2) \qquad [17]$$

where ΔG_{et}° is the free-energy change in the absence of Sc^{3+}. Thus, electron transfer from CoTPP and $^3C_{60}*$ to Q and Cl_4Q becomes more favorable energetically with an increase in the concentration of Sc^{3+}. Because such a change in the energetics should be directly reflected in the transition state of electron transfer, the dependence of the observed rate constant of electron transfer (k_{et}) on [Sc^{3+}] can be derived from Eq. (17) as given by

$$\frac{k_{et}}{[Sc^{3+}]} = k_0 K_1(1 + K_2[Sc^{3+}]) \qquad [18]$$

where k_0 is the rate constant in the absence of Sc^{3+}. The experimentally observed dependence of k_{et} on [Sc^{3+}] in Fig. 13 is well fitted by Eq. (18) using virtually the same K_2 value (Fig. 13) [135].

The formation of the 1:2 complex between $Q^{\cdot-}$ and Sc^{3+} [Eq. (15)] is confirmed by the ESR spectrum of $Q^{\cdot-}-(Sc^{3+})_2$, which shows superhyperfine structure due to the interaction of $Q^{\cdot-}$ with two Sc nuclei, as shown in Fig. 14 [135]. Because $Q^{\cdot-}-(Sc^{3+})_2$ is unstable because of a facile disproportionation reaction, the $Q^{\cdot-}-Sc^{3+}$ complex was generated in photoinduced electron transfer from dimeric 1-benzyl–1,4-dihydronicotinamide [$(BNA)_2$] [90] to Q at low temperatures [135]. The well-resolved 19 lines of the spectrum clearly indicates the

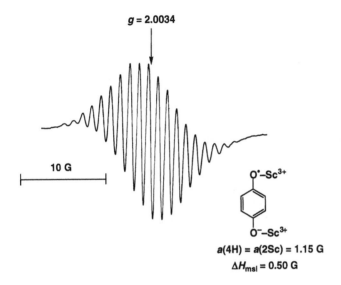

Figure 14   ESR spectrum of the $Q^{\bullet-}$-$(Sc^{3+})_2$ complex observed under photoirradiation of a propionitrile solution containing $(BNA)_2$ (1.6×10^{-2} M), Q (4.9×10^{-2} M), and $Sc(OTf)_3$ (4.4×10^{-2} M) with a high pressure mercury lamp at 203 K. (From Ref. 135.)

hyperfine splitting $[a(4H)]$ due to 4 protons of semiquinone radical anions and superhyperfine splitting $[a(2Sc)]$ by two equivalent Sc^{3+} nuclei ($I = 7/2$). In the complex, Sc^{3+} acts as a strong Lewis acid which can bind strongly with $Q^{\bullet-}$ acting as a base [78]. In the case of Cl_4Q, such a 1:2 complex formation between $Cl_4Q^{\bullet-}$ and Sc^{3+} may be hampered because of the bulky electron-withdrawing substituents, which decrease the basicity of $Cl_4Q^{\bullet-}$ [135].

The Sc^{3+}-promoted photoinduced electron transfer can be generally applied for formation of the radical cations of a variety of fullerene derivatives, which would otherwise be difficult to oxidize [135]. It has been shown that the electron-transfer oxidation reactivities of the triplet excited states of fullerenes are largely determined by the HOMO (highest occupied molecular orbital) energies of the fullerenes, whereas the triplet energies remain virtually the same among the fullerenes [135].

VIII.   CONCLUSIONS AND FUTURE PROSPECTS

As demonstrated in this review, photoinduced electron-transfer reactions are finely controlled by the proper choice of the redox potentials and reorganization energies of electron donors and acceptors, which can be modulated depending on the solvent polarity. In other words, photoinduced electron-transfer reactions can be designed quantitatively based on the redox potentials and the reorganization energies of molecules involved in the electron-transfer reactions based on the Marcus theory of electron transfer [1–4]. Not only simple electron transfers but also redox reactions in which the rate-determining photoinduced electron transfer step is followed by the facile bond-breaking and bond-forming step, leading to the final products, can be well designed by predicting the rate of photoinduced electron transfer and the back-electron transfer to the ground state. The reactivity of photoinduced electron transfer is further controlled by metal ions which can bind strongly with the excited state or/and the radical anions produced in the photoinduced electron transfer. The interactions between metal ions and the products of electron transfer result in the change in the free energy of electron transfer and the reorganization energies associated with the electron transfer. Such promoting effects of metal ions on electron-transfer processes is particularly important to control the redox reactions in which the photoinduced electron-transfer processes are involved as the rate-determining steps followed by facile follow-up steps involving cleavage and formation of chemical bonds. There still remains a wealth of important fundamental questions with regard to fine control of photoinduced electron-transfer reactions, which have been only partially explored in this review, and which certainly deserve a much more detailed attention in future.

ACKNOWLEDGMENTS

The authors gratefully acknowledge the contributions of his collaborators and co-workers mentioned in the references. The authors thank the Ministry of Education, Culture, Sports, Science and Technology, Japan for the financial support.

REFERENCES

1. Marcus, R. A. *Annu. Rev. Phys. Chem.* **1964**, *15*, 155.
2. Marcus, R. A.; Sutin, N. *Biochim. Biophys. Acta* **1985**, *811*, 265.
3. Marcus, R. A. *Angew. Chem. Int. Ed. Engl.* **1993**, *32*, 1111.
4. Bixon, M.; Jortner, J. In *Electron Transfer — From Isolated Molecules to Biomolecules*; Bixon, M., Jortner, J., eds.; J Wiley: New York, 1999; Part 1, pp. 35–202.
5. Newton, M. D. *Chem. Rev.* **1991**, *91*, 767.
6. Balzani, V., ed. *Electron Transfer in Chemistry*; Wiley–VCH: Weinheim, 2001; Vols. 1–5.
7. Eberson, L. *Electron Transfer Reactions in Organic Chemistry*; Springer-Verlag: Heidelberg, 1987.
8. Deisenhofer, J., Norris, J. R., eds. *The Photosynthetic Reaction Center*; Academic Press: San Diego, 1993.

9.   Blankenship, R. E., Madigan, M. T., Bauer, C. E., eds. *Anoxygenic Photosynthetic Bacteria*; Kluwer Academic Publishing: Dordrecht, 1995.

10.   Moser, C. C.; Page, C. C.; Dutton, P. L. In *Electron Transfer in Chemistry*; Balzani, V. ed.; Wiley–VCH: Weinheim, 2001; Vol. 3, pp. 24–38.

11.   Fukuzumi, S. In *Advances in Electron Transfer Chemistry*; Mariano, P. S. ed.; JAI Press: CT, 1992; Vol. 2, pp. 67–175.

12.   Patz, M.; Fukuzumi, S. *J. Phys. Org. Chem.* **1997**, *10*, 129; Fukuzumi, S.; Kochi, J. K. *Bull. Chem. Soc. Japan* **1983**, *56*, 969.

13.   Kochi, J. K. *Organometallic Mechanisms and Catalysis*; Academic Press: New York, 1978.

14.   Kochi, J. K. *Angew. Chem. Int. Ed. Engl.* **1988**, *27*, 1227.

15.   Kochi, J. K. *Acc. Chem. Res.* **1992**, *25*, 39.

16.   Rathore, R.; Kochi, J. K. *Adv. Phys. Org. Chem.* **2000**, *35*, 193.

17.   Kornblum, N. *Angew. Chem. Int. Ed. Engl.* **1975**, *14*, 734.

18.   Bunnett, J. F. *Acc. Chem. Res.* **1978**, *11*, 413.

19.   Savéant, J.-M. *Acc. Chem. Res.* **1993**, *26*, 455.

20.   Chanon, M.; Rajzmann, M.; Chanon, F. *Tetrahedron* **1990**, *46*, 6193.

21.   Julliard, M.; Chanon, M. *Chem. Rev.* **1983**, *83*, 425.

22.   Pross, A. *Acc. Chem. Res.* **1985**, *18*, 212.

23.   Shaik, S. S. *Prog. Phys. Org. Chem.* **1985**, *15*, 264.

24.   Fukuzumi, S. *Bull. Chem. Soc. Japan* **1997**, *70*, 1.

25.   Fukuzumi, S. In *Electron Transfer in Chemistry*; Balzani, V., ed.; Wiley–VCH: Weinheim, 2001; Vol. 4, pp. 3–67.

26.   Fukuzumi, S.; Itoh, S. In *Advances in Photochemistry*; Neckers, D. C., Volman, D. H., von Bünau, G., eds.; Wiley: New York, 1999; Vol. 25, p. 107.

27.   Rotello, V. M. In *Electron Transfer in Chemistry*; Balzani, V. ed.; Wiley–VCH: Weinheim, 2001; Vol. 4, pp. 68–87.

28.   Connolly, J. S.; Bolton, J. R. In *Photoinduced Electron Transfer*; Fox, M. A., Chanon, M., eds.; Elsevier: Amsterdam, 1988; Part D, pp. 303–393.

29.   Wasielewski, M. R. In *Photoinduced Electron Transfer*; Fox, M. A., Chanon, M., eds.; Elsevier: Amsterdam, 1988; Part A, pp. 161–206.

30.   Wasielewski, M. R. *Chem. Rev.* **1992**, *92*, 435.

31.   Gust, D.; Moore, T. A.; Moore, A. L. *Acc. Chem. Res.* **1993**, *26*, 198.

32.   Kurreck, H.; Huber, M. *Angew. Chem. Int. Ed. Engl.* **1995**, *34*, 849.

33.   Gust, D.; Moore, T. A. In *The Porphyrin Handbook*; Kadish, K. M., Smith, K. M., Guilard, R., eds.; Academic Press: San Diego, CA, 2000; Vol. 8, pp. 153–190.

34.   Gust, D.; Moore, T. A.; Moore, A. L. *Acc. Chem. Res.* **2001**, *34*, 40.

35.   Chambron, J.-C.; Chardon-Noblat, S.; Harriman, A.; Heitz, V.; Sauvage, J.-P. *Pure Appl. Chem.* **1993**, *65*, 2343.

36.   Harriman, A.; Sauvage, J.-P. *Chem. Soc. Rev.* **1996**, *26*, 41.

37.   Blanco, M.-J.; Jiménez, M. C.; Chambron, J.-C.; Heitz, V.; Linke, M.; Sauvage, J.-P. *Chem. Soc. Rev.* **1999**, *28*, 293.

38.   Balzani, V.; Juris, A.; Venturi, M.; Campagna, S.; Serroni, S. *Chem. Rev.* **1996**, *96*, 759.

39.   Paddon-Row, M. N. *Acc. Chem. Res.* **1994**, *27*, 18.

40. Verhoeven, J. W. In *Electron Transfer—From Isolated Molecules to Biomolecules*; Jortner, J., Bixon, M., eds.; Wiley: New York, 1999; Part 1, pp. 603–644.
41. Osuka, A.; Mataga, N.; Okada, T. *Pure Appl. Chem.* **1997**, *69*, 797.
42. Sun, L.; Hammarström, L.; Åkermark, B.; Styring, S. *Chem. Soc. Rev.* **2001**, *30*, 36.
43. Imahori, H.; Sakata, Y. *Adv. Mater.* **1997**, *9*, 537.
44. Imahori, H.; Sakata, Y. *Eur. J. Org. Chem.* **1999**, 2445.
45. Guldi, D. M. *Chem. Commun.* **2000**, 321.
46. Guldi, D. M.; Prato, M. *Acc. Chem. Res.* **2000**, *33*, 695.
47. Imahori, H.; Guldi, D. M.; Tamaki, K.; Yoshida, Y.; Luo, C.; Sakata, Y.; Fukuzumi, S. *J. Am. Chem. Soc.* **2001**, *123*, 6617.
48. Fukuzumi, S.; Imahori, H. In *Electron Transfer in Chemistry*; Balzani, V. ed.; Wiley–VCH: Weinheim, 2001; Vol. 2, pp. 270–337.
49. Fukuzumi, S.; Guldi, D. M. In *Electron Transfer in Chemistry*; Balzani, V. ed.; Wiley–VCH: Weinheim, 2001; Vol. 2, pp. 927–975.
50. Guldi, D. M.; Kamat, P. V. In *Fullerenes, Chemistry, Physics, and Technology*; Kadish, K. M., Ruoff, R. S., eds.; Wiley–Interscience: New York, 2000; pp. 225–281.
51. Fuhrhop, J.-H.; Mauzerall, D. *J. Am. Chem. Soc.* **1969**, *91*, 4174.
52. Chosrowjan, H.; Taniguchi, S.; Okada, T.; Takagi, S.; Arai, T.; Tokumaru, K. *Chem. Phys. Lett.* **1995**, *242*, 644.
53. Imahori, H.; Tamaki, K.; Guldi, D. M.; Luo, C.; Fujitsuka, M.; Ito, O.; Sakata, Y.; Fukuzumi, S. *J. Am. Chem. Soc.* **2001**, *123*, 2607.
54. Newton, M. D.; In *Electron Transfer in Chemistry*; Balzoni, V. ed., Wiley-VCH: Weinhein, 2001; Vol. 1, pp. 3–63.
55. Helms, A.; Heiler, D.; McLendon, G. *J. Am. Chem. Soc.* **1992**, *114*, 6227.
56. Imahori, H.; El-Khouly, M. E.; Fujitsuka, M.; Ito, O.; Sakata, Y.; Fukuzumi, S. *J. Phys. Chem. A* **2001**, *105*, 325.
57. Hanson, L. S. In *Chlorophylls*; Scheer, H., ed.; CRC Press: Boca Raton, FL, 1991; pp. 993–1014; Plato, M.; Möbius, K.; Lubitz, W. In *Chlorophylls*; Scheer, H., ed.; CRC Press: Boca Raton, FL, 1991; pp. 1015–1046.
58. Ghosh, A. *J. Phys. Chem. B* **1997**, *101*, 3290.
59. Stolzenberg, A. M.; Strauss, S. H.; Holm, R. H. *J. Am. Chem. Soc.* **1981**, *103*, 4763.
60. Stolzenberg, A. M.; Schussel, L. *J. Inorg. Chem.* **1991**, *30*, 3205.
61. Helaja, J.; Tauber, A. Y.; Abel, Y.; Tkachenko, N. V.; Lemmetyinen, H.; Kilpel-äinen, I.; Hynninen, P. H. *J. Chem. Soc. Perkin Trans. 1* **1999**, 2403.
62. Zheng, G.; Dougherty, T. J.; Pandey, R. K. *Chem. Commun.* **1999**, 2469.
63. Tkachenko, N. V.; Rantala, L.; Tauber, A. Y.; Helaja, J.; Hynninen, P. H.; Lemmetyinen, H. *J. Am. Chem. Soc.* **1999**, *121*, 9378.
64. Tkachenko, N. V.; Vuorimaa, E.; Kesti, T.; Alekseev, A. S.; Tauber, A. Y.; Hynninen, P. H.; Lemmetyinen, H. *J. Phys. Chem. B* **2000**, *104*, 6371.
65. Fukuzumi, S.; Ohkubo, K.; Imahori, H.; Shao, J.; Ou, Z.; Zheng, G.; Chen, Y.; Pandey, R. K.; Fujitsuka, M.; Ito, O.; Kadish, K. M. *J. Am. Chem. Soc.* **2001**, *123*, 10676.

66. Anderson, J. L.; An, Y.-Z.; Rubin, Y.; Foote, C. S. *J. Am. Chem. Soc.* **1994**, *116*, 9763.
67. Dvornikov, S. S.; Knyukshto, V. N.; Solovev, K. N.; Tsvirko, M. P. *Opt. Spectrosc. (USSR)* **1979**, *46*, 385.
68. Stryer, L. *Biochemistry*, 3rd ed.; Freeman: New York, 1988; Chap. 17.
69. Fukuzumi, S.; Tanaka, T. *Photoinduced Electron Transfer*; Fox, M. A., Chanon, M., eds.; Elsevier: Amsterdam; 1988; Part C, pp. 578–635.
70. Fukuzumi, S.; Imahori, H.; Okamoto, K.; Yamada, H.; Fujitsuka, M.; Ito, O.; Guldi, D. M. *J. Phys. Chem. A* **2002**, *106*, 2803.
71. Koper, N. W.; Jonker, S. A.; Verhoeven, J. W.; van Dijk, C. *Rec. Trav. Chim. Pays-Bas* **1985**, *104*, 296.
72. Hapiot, P.; Moiroux, J.; Savéant, J.-M. *J. Am. Chem. Soc.* **1990**, *112*, 1337.
73. Anne, A.; Moiroux, J. *J. Org. Chem.* **1990**, *55*, 4608.
74. Fukuzumi, S.; Koumitsu, S.; Hironaka, K.; Tanaka, T. *J. Am. Chem. Soc.* **1987**, *109*, 305.
75. Koper, N. W.; Verhoeven, J. W. *Proc. K. Ned. Akad. Wet., Ser. B. Phys.* **1983**, *86*, 79.
76. Fukuzumi, S.; Imahori, H.; Yamada, H.; El-Khouly, M. E.; Fujitsuka, M.; Ito, O.; Guldi, D. M. *J. Am. Chem. Soc.* **2001**, *123*, 2571.
77. Ohtsu, H.; Shimazaki, Y.; Odani, A.; Yamauchi, O.; Mori, W.; Itoh, S.; Fukuzumi, S. *J. Am. Chem. Soc.* **2000**, *122*, 5733.
78. Fukuzumi, S.; Ohkubo, K. *Chem. Eur. J.* **2000**, *6*, 4532.
79. Fukuzumi, S.; Ohkubo, K.; Suenobu, T.; Kato, K.; Fujitsuka, M.; Ito, O. *J. Am. Chem. Soc.* **2001**, *123*, 8459.
80. Kavarnos, G. J. *Fundamentals of Photoinduced Electron Transfer*; Wiley–VCH: New York, 1993.
81. Fujita, M.; Ishida, A.; Takamuku, S.; Fukuzumi, S. *J. Am. Chem. Soc.* **1996**, *118*, 8566.
82. Rehm, A.; Weller, A. *Ber. Bunsenges Phys. Chem.* **1969**, *73*, 834; Rehm, A.; Weller, A. *Isr. J. Chem.* **1970**, *8*, 259.
83. Closs, G. L.; Miller, J. R. *Science* **1988**, *240*, 440.
84. Gould, I. R.; Farid, S. *Acc. Chem. Res.* **1996**, *29*, 522.
85. Winkler, J. R.; Gray, H. B. *Chem. Rev.* **1992**, *92*, 369.
86. McLendon, G.; Hake, R. *Chem. Rev.* **1992**, *92*, 481.
87. Mataga, N.; Miyasaka, H. In *Electron Transfer from Isolated Molecules to Biomolecules, Part 2*; Jortner, J., Bixon, M., eds.; Wiley: New York, 1999; p. 431.
88. Chang, R. *J. Chem. Educ.* **1970**, *47*, 563.
89. Cheng, K. S.; Hirota, N. In *Investigation of Rates and Mechanisms of Reactions*; Hammes, G. G., ed.; Wiley–Interscience: New York, 1974; Vol. VI, p. 565.
90. Fukuzumi, S.; Suenobu, T.; Patz, M.; Hirasaka, T.; Itoh, S.; Fujitsuka, M.; Ito, O. *J. Am. Chem. Soc.* **1998**, *120*, 8060.
91. Anne, A.; Moiroux, J.; Savéant, J.-M. *J. Am. Chem. Soc.* **1993**, *115*, 10224.
92. Takada, N.; Itoh, S.; Fukuzumi, S. *Chem. Lett.* **1996**, 1103.
93. Patz, M.; Kuwahara, Y.; Suenobu, T.; Fukuzumi, S. *Chem. Lett.* **1997**, 567.
94. Fukuzumi, S.; Mochida, K.; Kochi, J. K. *J. Am. Chem. Soc.* **1979**, *101*, 5961.

95. Walther, B. W.; Williams, F.; Lau, W.; Kochi, J. K. *Organometallics* **1983**, *2*, 688.
96. Symons, M. C. R. *Chem. Soc. Rev.* **1984**, *13*, 393.
97. Fukuzumi, S.; Kochi, J. K. *J. Org. Chem.* **1980**, *45*, 2654.
98. Franz, G.; Sheldon, R. A. In *Ullmann's Encyclopedia of Industrial Chemistry*; 5th ed.; VCH: Weinheim, 1991.
99. Sheldon, R. A.; Kochi, J. K. In *Metal Catalyzed Oxidation of Organic Compounds*, Academic Press: New York, 1981; Chap. 10.
100. Clarke, R.; Kuhn, A. T.; Okoh, E. *Chem. Br.* **1975**, *11*, 59.
101. Periasamy, M.; Bhatt, M. V. *Tetrahedron Lett.* **1978**, 4561.
102. Ballard, R. E.; McKillop, A. *U.S. Patent 4* **1984**, *482*, 438.
103. Udupa, H. V. K. *Trans. Soc. Adv. Electrochem. Sci. Technol.* **1976**, *11*, 143.
104. Baciocchi, E.; Giacco, T. D.; Roi, C.; Sebastiani, G. V. *Tetrahedron Lett.* **1985**, *28*, 3353.
105. Ho, T.-L. *Synthesis* **1973**, 347.
106. Syper, L. *Tetrahedron Lett.* **1966**, 4493.
107. Barton, D. H. R.; Hui, R. A. H. F.; Lester, D. J.; Ley, S. V. *Tetrahedron Lett.* **1976**, 3331.
108. Bhatt, M. V.; Perumal, P. T. *Tetrahedron Lett.* **1981**, *22*, 2605.
109. Ohkubo, K.; Fukuzumi, S. *Org. Lett.* **2000**, *2*, 3647.
110. Fukuzumi, S.; Ohkubo, K.; Tokuda, Y.; Suenobu, T. *J. Am. Chem. Soc.* **2000**, *122*, 4286.
111. Nicolaou, K. C.; Zhong, Y.-L. *J. Am. Chem. Soc.* **2001**, *123*, 3183.
112. Fukuzumi, S.; Tokuda, Y.; Kitano, T.; Okamoto, T.; Otera, J. *J. Am. Chem. Soc.* **1993**, *115*, 8960.
113. Fukuzumi, S.; Satoh, N.; Okamoto, T.; Yasui, K.; Suenobu, T.; Seko, Y.; Fujitsuka, M.; Ito, O. *J. Am. Chem. Soc.* **2001**, *123*, 7756.
114. Fukuzumi, S.; Yasui, K.; Suenobu, T.; Ohkubo, K.; Fujitsuka, M.; Ito, O. *J. Phys. Chem. A* **2001**, *105*, 10501.
115. Herkstroeter, W. G.; Lamola, A. A.; Hammond, G. S. *J. Am. Chem. Soc.* **1964**, *86*, 4537.
116. Dinnocenzo, J. P.; Farid, S.; Goodman, J. L.; Gould, I. R.; Todd, W. P.; Mattes, S. L. *J. Am. Chem. Soc.* **1989**, *111*, 8973; Cermenati, L.; Freccero, M.; Venturello, P.; Albini, A. *J. Am. Chem. Soc.* **1995**, *117*, 7869.
117. Dockery, K. P.; Dinnocenzo, J. P.; Farid, S.; Goodman, J. L.; Gould, I. R.; Todd, W. P. *J. Am. Chem. Soc.* **1997**, *119*, 1876.
118. Tollin, G. In *Electron Transfer in Chemistry*; Balzani, V., ed.; Wiley–VCH: Weinheim, 2001; Vol. 5, pp. 202–231.
119. Müller, F., ed. *Chemistry and Biochemistry of Flavoenzymes*; CRC Press: Boca Raton, FL, 1991; Vols. 1–3.
120. Cadet, J.; Vigny, P. In *Bioorganic Photochemistry*; Morrison, H., ed.; Wiley: New York, 1990; Vol. 1, pp. 1–272.
121. Heelis, P. F. In *Chemistry and Biochemistry of Flavoenzymes*; Müller, F., ed.; CRC Press: Boca Raton, FL, 1991; Vol. 1, pp. 171–193.
122. Fukuzumi, S.; Tanaka, T. In *Photoinduced Electron Transfer*; Fox, M. A., Chanon, M., eds.; Elsevier: Amsterdam, 1988; Part C, pp. 636–687.

123. Fukuzumi, S.; Kuroda, S.; Tanaka, T. *J. Am. Chem. Soc.* **1985**, *107*, 3020.
124. Fukuzumi, S.; Kuroda, S.; Tanaka, T. *Chem. Lett.* **1984**, 1375.
125. Imamoto, T. *Lanthanides in Organic Synthesis*; Katritzky, A. R., Meth-Cohn, O., Rees, C. W., eds.; Academic Press: London, 1994.
126. Kobayashi, S. *Synlett* **1994**, 689.
127. Molander, G. A.; Harris, C. R. *Chem. Rev.* **1996**, *96*, 307.
128. Ishihara, K.; Kubota, M.; Kurihara, H.; Yamamoto, H. *J. Am. Chem. Soc.* **1995**, *117*, 4413, 6639.
129. Nagayama, S.; Kobayashi, S. *Angew. Chem., Int. Ed.* **2000**, *39*, 567.
130. Kobayashi, S.; Nagayama, S. *J. Am. Chem. Soc.* **1998**, *120*, 2985.
131. Lacôte, E.; Renaud, P. *Angew. Chem., Int. Ed. Engl.* **1998**, *37*, 2259.
132. Fukuzumi, S.; Okamoto, T. *J. Chem. Soc., Chem. Commun.* **1994**, 521.
133. Kearns, D. R. *Chem. Rev.* **1971**, *71*, 395.
134. Foote, C. S. In *Free Radicals in Biology*; Pryor, W. A., ed.; Academic Press: New York, 1976; Vol. 2, p. 85.
135. Fukuzumi, S.; Mori, H.; Imahori, H.; Suenobu, T.; Araki, Y.; Ito, O.; Kadish, K. M. *J. Am. Chem. Soc.* **2001**, *123*, 12418.
136. Xie, Q.; Arias, F.; Echegoyen, L. *J. Am. Chem. Soc.* **1993**, *115*, 9818.
137. Reed, C. A.; Bolskar, R. D. *Chem. Rev.* **2000**, *100*, 1075.
138. Kato, T.; Kodama, T.; Shida, T.; Nakagawa, T.; Matsui, Y.; Suzuki, S.; Shinoharu, H.; Yamaguchi, K.; Achiba, Y. *Chem. Phys. Lett.* **1991**, *180*, 446.
139. Gasyna, Z.; Andrews, L.; Schatz, P. N. *J. Phys. Chem.* **1992**, *96*, 1525; Fulara, J.; Jakobi, M.; Maier, J. P. *Chem. Phys. Lett.* **1993**, *211*, 227.
140. Reed, C. A.; Kim, K.-C.; Bolskar, R. D.; Mueller, L. J. *Science* **2000**, *289*, 101.
141. Fukuzumi, S.; Fujii, Y.; Suenobu, T. *J. Am. Chem. Soc.* **2001**, *123*, 10,191.

6

Molecular Oxygenations in Zeolites

Edward L. Clennan

University of Wyoming, Laramie, Wyoming, U.S.A.

I. INTRODUCTION

Zeolites are crystalline solids consisting of catenated silicon and aluminum tetra-hedra that enclose regular repeating cavities or channels of well-defined size and shape [1]. (Fig. 1) There are over 30 naturally occurring zeolites and over 100 synthetic zeolites without a naturally occurring counterpart. Recent developments in zeolite science have seen the field expand far beyond its initial focus in the 1960s in petroleum processing to include their use in the synthesis of fine chemicals [2]. Properties which make zeolites especially attractive for heterogeneous catalysis of fine chemical synthesis include (1) the availability of zeolites with a variety of pore sizes encompassing a wide range of typical organic chemical diameters, (2) the ability to fine-tune the electrostatic character of the intercavity environment by cation exchange, (3) the ability to synthesize zeolites in which framework aluminum and/or silicon can be substituted with iron, phosphorus, boron, vanadium, and a wide variety of other atoms, and (4) their potential for use in environmentally benign protocols.

In this review we will focus on their use as catalysts and promoters in the introduction of molecular oxygen into organic substrates. Oxidized hydrocarbons serve as important feedstocks for the chemical and pharmaceutical industries. Unfortunately, hydrocarbons are also infamous in their ability to resist oxidation under environmentally benign and easily controlled conditions. The large volume of these materials needed to satisfy the demand of the chemical industry economically precludes all stoichiometric oxidants, with the sole exception of molecular

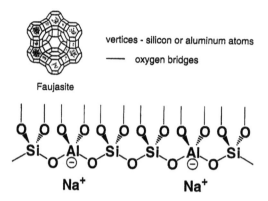

vertices - silicon or aluminum atoms

—— oxygen bridges

Faujasite

Figure 1   The framework composition of a faujasite (e.g., NaY or NaX) depicting the 13-Å-diameter supercage surrounded by tetrahedrally disposed 7.5- Å-diameter windows.

oxygen and possibly hydrogen peroxide. Auto-oxidations, however, are inherently unselective in part as a result of the fact that in many instances the products themselves are more susceptible to oxidation than the substrates. Overoxidation becomes a serious problem as products accumulate, and in a practical sense, it limits conversions of feedstocks to a few percent. In addition, the initially formed products are often peroxides that have a propensity to decompose to produce a wide range of undesirable products.

The use of zeolites can overcome many of these limitations and provide new controlled entries into these oxidized hydrocarbons and new materials. For example, some of the most valuable industrial intermediates are terminally oxidized hydrocarbons, such as n-hexanol or adipic acid, that are not readily available in free-radical chain processes. The ability of zeolites to function as shape-selective catalysts can, in principle, be used to restrict access, by reactant or transition state selectivity, to sites not normally attacked by oxidants [3].

The introduction of molecular oxygen into organic substrates has been accomplished by many different approaches. These molecular oxidations can be conveniently classified using Wagnerova's [4] system, as shown in Fig. 2. In this classification system, molecular oxygenation reactions are placed into 1 of 12 categories depending on (1) the method utilized to overcome the spin barrier encountered between the reactions of ground-state triplet oxygen and singlet-state organic molecules, (2) whether the reaction is initiated either thermally or photochemically, and (3) whether the initiation procedure operates on oxygen or on the organic substrate. The three established routes to overcome the spin barrier are spin inversion (energy transfer), radical formation (one-electron transfer), and coordination to a metal (two-electron transfer). In this review, each of the molecu-

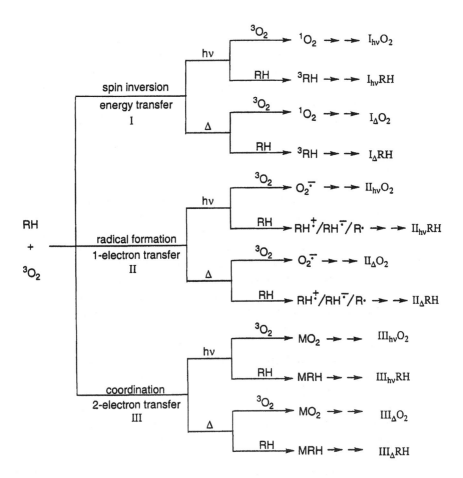

Figure 2 Wagnerova's classification scheme for molecular oxygenations.

lar oxygenation reactions will be referred to by using a symbol composed of three terms: one for each of the branches in the treelike classification system shown in Fig. 2. For example, $I_{hv}O_2$, $II_\Delta RH$, and $III_\Delta O_2$ refer to oxygenation reactions initiated by (1) photochemical spin inversion of oxygen (i.e., a singlet oxygen reaction), (2) thermal conversion of the substrate, RH, to a radical, radical cation, or radical anion, and (3) thermal coordination of oxygen to a metal center, respectively.

In addition, for these 12 Wagnerova mechanistic categories, the possibilities of several hybrid mechanisms also exist. These hybrid mechanisms occur when both reaction partners, the substrate and oxygen, are independently activated by 1 of the 12 mechanistic scenarios depicted in Fig. 2. Several of these hybrid mechanisms will also be discussed. The goal of this review, however, is not to give a comprehensive listing of all zeolite promoted molecular oxygenation reactions. The goal of this review is to give and critically analyze selected examples of zeolite-promoted molecular oxygenations in the hope that this discussion and the systematic classification of these reactions will stimulate further research in this very important area.

II. TYPE I$_{hv}$O$_2$ REACTIONS

The photosensitized formation of the low-lying $^1\Delta_g$ state of oxygen in homogeneous media is one of the oldest and most convenient methods for the generation of this reactive oxidant [5]. However, it was not until 1996 that Li and Ramamurthy [6] reported that irradiation of thionin-, methylene blue-, or methylene green-loaded zeolite X or Y also generated singlet oxygen. This provided the first opportunity for examining intrazeolite singlet oxygen reactions and to take advantage of the ability of zeolites to impart enzymelike organization to reactive complexes. This could be of particular importance in singlet oxygen reactions because despite the ready availability and novel transformations achievable with this powerful oxidant, it suffers from a lack of selectivity [7].

Li and Ramamurthy [6] chose as their first study the examination of the intrazeolite singlet oxygen ene reaction. The mechanistic understanding and scope of the homogeneous $^1\Delta_g O_2$ ene reaction is in a mature stage of development and provides a unique opportunity to probe the influence of the zeolitic environment [8]. The Ramamurthy reactions were run in hexane slurries using zeolite Y doped with approximately 1 molecule of sensitizer per 100 supercages. The doped zeolites were dried under vacuum at ~ 100°C at 10^{-4} torr prior to the photo-oxidation. These conditions promoted the population of the monomeric rather than dimeric dye, suppressed self-quenching, and enhanced singlet oxygen formation. Control reactions established that the reactions occurred in the interior of the zeolite and precluded reactions on the surface or in the hexane solvent. The results for several substrates are shown in Fig. 3.

In order to rationalize the unique regioselectivity in these reactions, the authors tentatively suggested the two different independent models shown in Fig. 4. In model I [9], the alkene is complexed to the cation (sodium in Fig. 4), forcing singlet oxygen to approach the alkene from the opposite direction. The complexation occurs on the least hindered face of the alkene, directing the allylic methyl group to the face approached by oxygen and geometrically precluding the abstraction of H$_a$. In model II [10], the sodium polarizes the alkene to generate a partial positive charge on the most substituted end of the double bond, where it can be effectively dispersed by two alkyl groups.

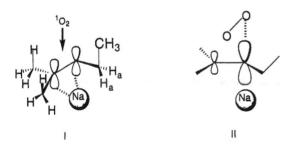

Figure 3 Singlet oxygen ene regiochemistries.

In 1999, Clennan and Sram reported a study of the photo-oxidations of a series of tetrasubstituted alkenes (Fig. 5) in methylene blue-doped zeolite Y [11]. The ene regiochemistries are very sensitive to the size of the allylic substituent, R, in solution. The **A/B** ratio increases from 0.49 to 2.4 as the substituent, R, is changed from methyl to *tert*-butyl. This phenomenon has been attributed [12] to a sterically induced lengthening of the carbon-2 oxygen bond in the perepoxide intermediate **I** and subsequent preferred opening of this long bond (Fig. 5).

The steric effect observed in solution should be amplified in the zeolite if either of the models depicted in Fig. 4 operate because the sodium counterion should force the allylic substituent R to reside to an even greater extent on the face approached by singlet oxygen. This should lead to an increase in allylic hydroperoxy **A** at the expense of **B** in stark contrast to the experimental observation (e.g., the **A/B** ratio does not increase, but actually decreases from 2.4 to 1.58

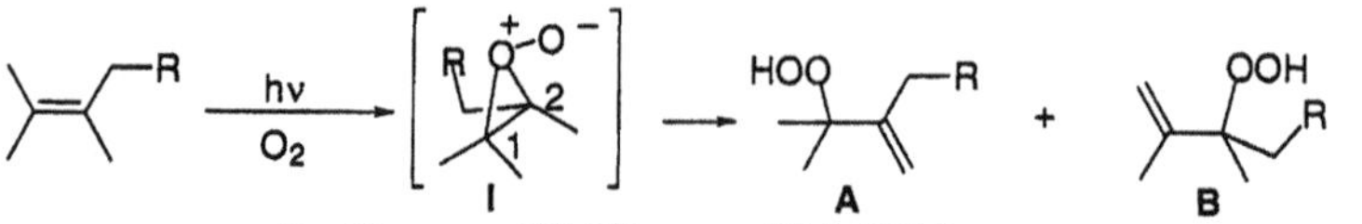

		A	B	A/B
R = Me	CH₃CN	(32.9±3.3%)	(67.1±3.3%)	0.49
	NaMBY	[39.5±3.7%]	[60.5±3.7%]	0.65
R = Et	CH₃CN	(42.0±1.7%)	(58.0±1.7%)	0.72
	NaMBY	[52.7±1.7%]	[47.3±1.7%]	1.11
R =iPr	CH₃CN	(52.6±1.1%)	(47.4±1.1%)	1.1
	NaMBY	[51.8±2.8%]	[48.2±2.8%]	1.07
R = tBu	CH₃CN	(70.6±1.1%)	(29.4±1.1%)	2.4
	NaMBY	[61.2±1.1%]	[38.8±1.1%]	1.58

Figure 5 A comparison of solution (CH₃CN; MB) and zeolite (NaMBY 1%; hexane) photo-oxidants.

for the *t*-butyl allylic substituent; Fig. 5). To rationalize these results, the model depicted in Fig. 6 was suggested. In this model, the important regiochemical-directing element is suggested to be the interaction between the perepoxide oxygen and the sodium counterion. The steric interaction between the sodium and allylic R group generates a preference for perepoxide **P'** at the expense of **P**. As a result, the A/B ratio in every case is closer to 1.0, reflecting equal probability of hydrogen abstraction from the syn methyl groups in perepoxide **P'**. Consequently, the decreased importance of a steric effect that would lengthen the carbon-2 oxygen bond is attributed to preferential formation of **P'** in which the influence of the allylic substituent R would be muted.

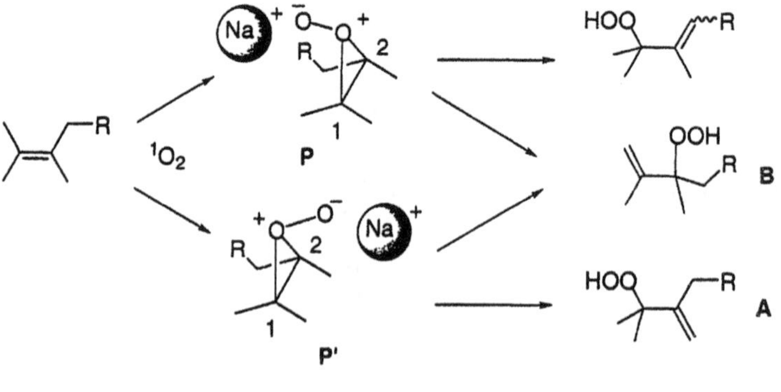

Figure 6 Cation complexation regioselectivity model.

Figure 7 Kinetic isotope effect and open zwitterionic intermediate.

The model in Fig. 6, specifically the intrazeolite presence of a perepoxide, is supported by an isotope effect of 1.04 ± 0.02 for the photo-oxidation of Z-2,3-dimethyl-1,1,1,4,4,4-hexadeutero-2-butene (Fig. 7). This isotope effect is completely inconsistent with an open zwitterion (Fig. 7), which would be expected to collapse to the hydroperoxide with a significant discrimination for hydrogen abstraction ($k_H/k_D > 1$).

The steric aspect of the model in Fig. 6 provides an explanation for the regiochemical change observed for both 2-methyl-2-pentene and 1-methylcyclopentene (Fig. 3), as the reaction is moved from the solution to the interior of the zeolite. In both of these trisubstituted alkenes, the sterically most stable perepoxide with the pendant oxygen residing on the least congested side of the alkene will predominate. As a consequence, hydrogen abstraction is forced to occur from the most substituted end of the alkene. The increased preference for the most substituted end of the alkene could also reflect a contribution from a Markovnikov-directing effect in the interior of the zeolite. This suggestion is supported by two experimental observations: (1) electron-donating substituents, **X** or **Y**, in a series of α, β, β-trimethylstyrenes, **1** (Fig. 8), dramatically favor formation of allylic hydroperoxide **1A** rather than its isomer **1B** during photo-oxidations in the interior of the zeolite but not in solution and (2) the ratio of hydrogen abstraction from the two methyl groups (**2B/2C**) in E-2-methyl-1,1,1-trideuterio-2-butene, **2**, is different in solution and in the zeolite [13]. This ratio would remain the same if a change in the population of **2P** and **2P′** (Fig. 8) were the only thing occurring as the reaction is moved into the zeolite. Allylic hydroperoxide, **2B**, does not decrease as much as **2C**, because the decrease in population of **2P′** in the zeolite is compensated by an increased preference for abstraction from the more highly substituted end of the alkene. (i.e., a Markovnikov effect) The increase in positive charge on the carbon framework that influences the regiochemistry of C—O cleavage in the perepoxide and leads to this Markovnikov effect is a result of the sodium ion acting as an electron sink.

The magnitudes of the increases in the Markovnikov-directing effects, as measured by the change in the **B/C** ratio (Fig. 8), should be similar for all trisubstituted alkenes with inductively similar groups (Fig. 9). That is the case for **2** ([**2B/**

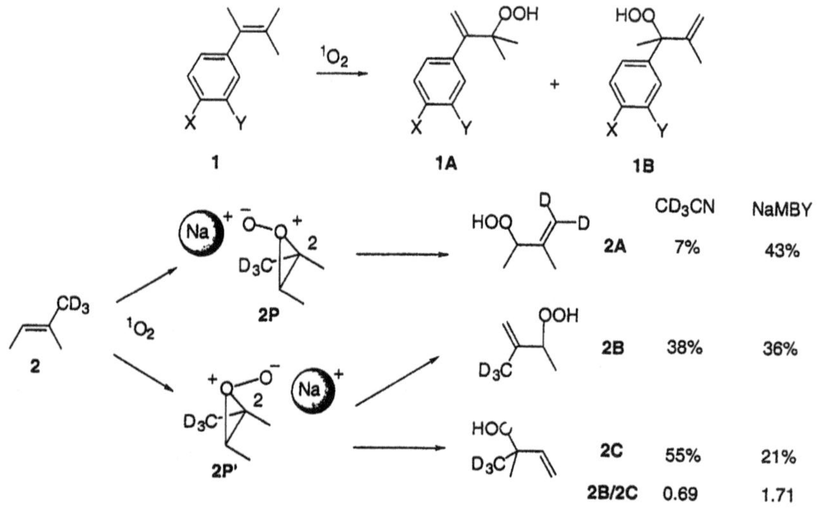

Figure 8 Products from diastereomeric perepoxides formed in intrazeolite photooxygenation of **2**.

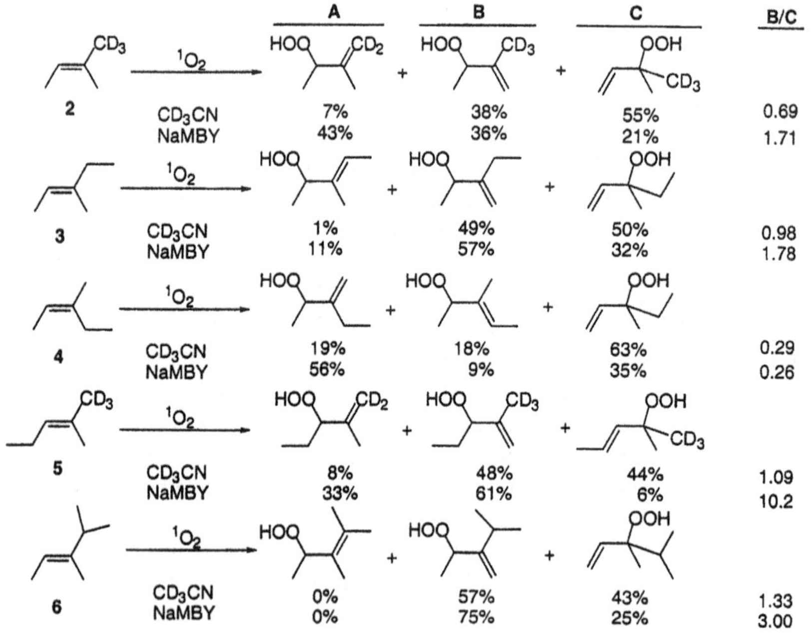

Figure 9 Products from photooxygenation of trisubstituted alkenes.

$2C]_{zeolite}/[2B/2C]_{solution} = 1.71/0.69 = 2.5$), 3 ($[3B/3C]_{zeolite}/[3B/3C]_{solution} = 1.78/0.98 = 1.82$), and 6 ($[6B/6C]_{zeolite}/[6B/6C]_{solution} = 3.00/1.33 = 2.26$) but is not the case for either 4 ($[4B/4C]_{zeolite}/[4B/4C]_{solution} = 0.26/0.29 = 0.90$) or 5 ($[5B/5C]_{zeolite}/[5B/5C]_{solution} = 10.2/1.09 = 9.4$). In both 4 and 5, this apparent discrepancy represents a decreased ability for hydrogen abstraction from an ethyl group to compete with hydrogen abstraction from a methyl group in the zeolite in comparison to solution. This phenomenon cannot be readily explained with the model shown in Fig. 6.

A modification of the model shown in Fig. 6 that can account for all the experimental data is shown in Fig. 10. This model is essentially identical to that presented in Fig. 6 except that it explicitly invokes complexation of the alkene to sodium cations in the zeolite supercage. The complexation of alkali metal cations to alkenes and aromatic rings are well established in both the gas and condensed phases [14,15]. The intrazeolite complexation would force the allylic methyl group in 4 and 5 to occupy the face of the olefin approached by singlet oxygen, thereby preventing hydrogen abstraction from the ethyl group as depicted by model 1 (Fig. 4). As singlet oxygen approaches the alkene, the sodium moves to electrostatically stabilize the perepoxide intermediate to give diastereomeric

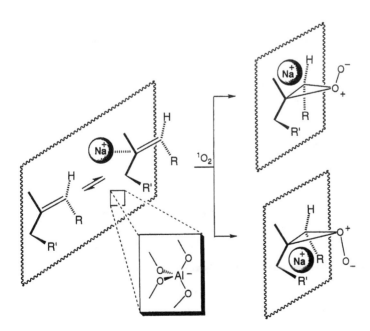

Figure 10 Modified cation complexation regioselectivity model.

perepoxides, as shown in Figs. 6 and 10. The cation may not have to move dramatically because Stratakis and Froudakis [16] have suggested that it may already be displaced toward the least substituted side of the alkene in the alkene–cation complex. Steric interactions in these sodium-complexed perepoxides dictate the population of each diastereomer, and electronic interactions dictate the direction of opening of the perepoxide, as discussed previously during consideration of the model depicted in Fig. 6.

These intrazeolite singlet oxygen ene reactions have synthetic potential because the cis effect observed in solution is suppressed in the zeolite [13]. Consequently, allylic hydroperoxides which are inaccessible by other routes may be available via this new technology. For example, photo-oxidations of aryl-substituted alkenes, **7**, in sensitizer-doped NaY react to generate the allylic hydroperoxides as the major or exclusive product [17]. In contrast, in solution, the hydroperoxides are formed in only 5–20% yields, with 2 + 2 and 4 + 2 adducts dominating the reaction mixtures. In the case of 2-methyl-5-phenyl-2-hexene, **8**, the regioselectivity for **8b** and **8c** improved from 47% to 94% and the diastereoselectivity from 10% to 44% as the reaction is moved from solution into the zeolite [18]:

Ar = phenyl, $pNO_2C_6H_4$-, $pMeOC_6H_4$-, $pCF_3C_6H_4$-, $pPhC_6H_4$-, 1-napthyl, 2-napthyl

	8a	8b 8c
solution	53%	47%(d.e. 10%)
zeolite	6%	94%(d.e. 44%)

These intrazeolite ene reactions are susceptible to many of the problems associated with photo-oxidations in solution. In particular, sensitizer bleaching, competition between type $I_{hv}O_2$, and both type $II_{hv}RH$ and type $II_{hv}O_2$ reactions, and reduced photostability of the products can complicate the synthetic use of these reactions [19]. In addition, many substrates are susceptible to acid-catalyzed rearrangements that can lead to complicated reaction mixtures.

Electron or hydrogen transfer between a substrate and sensitizer is often responsible for initiation of a type $II_{hv}RH$ photo-oxidation. Consequently, the type $II_{hv}RH$ reaction can often be suppressed by physically separating (isolating)

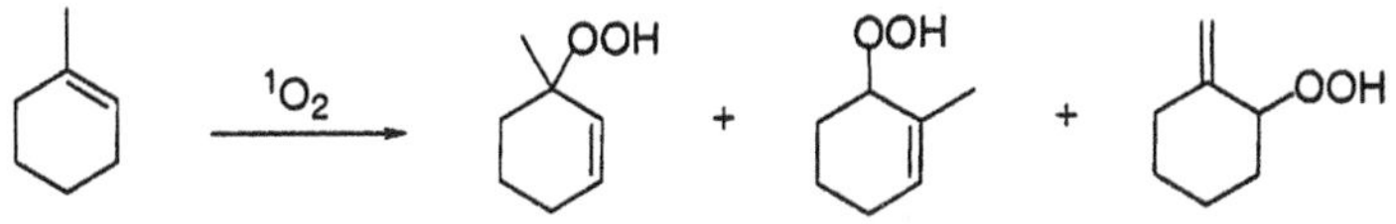

solution	50%	17%	31%
Ru(bpy)$_3^{2+}$/zeolite Y	51%	16%	33%

Figure 11   Photooxygenation of 1-methylcyclohexene.

the sensitizer and substrate. Pettit and Fox [20] isolated the sensitizer in the interior of the zeolite and allowed singlet oxygen to diffuse into solution to react with a substrate. However, this approach, illustrated with Ru(bpy)$_3^{2+}$/NaY (Fig. 11), does not take advantage of the unique environment of the zeolite to influence reaction regio selectivity or stereoselectivity.

   Tung and co-workers [21–23] have taken the opposite approach and have isolated the sensitizer in solution selectively, generating singlet oxygen and allowing it to diffuse into the zeolite to react with the substrate. They have demonstrated the feasibility of this approach by doping ZSM-5 with 1,4-diphenyl-1,3-butadiene, **9**, and *trans*-stilbene, **10**, and by using hypocrellin A (*HA*) and 9,10-dicyanoanthracene (*DCA*) as bulky sensitizers that cannot diffuse into the zeolite. Photo-oxidations of **9** in pentaerythritol trimethyl ether (PTE) using 9,10-dicyanoanthrance as a sensitizer leads to complicated reaction mixtures as a result of competition between the type I$_{hv}$O$_2$ and an electron transfer hybrid type II$_{hv}$O$_2$ – type II$_{hv}$RH (*vide infra*) reaction. (Fig. 12A) Inclusion of **9** into the zeolite, however, completely suppressed the hybrid type II$_{hv}$O$_2$ – type II$_{hv}$RH photooxidation and gave the singlet oxygen product exclusively. Intrazeolite photooxidation of **10** not only completely suppressed the competing electron-transfer route but also suppressed formation of the singlet oxygen product, **11**, observed in solution (Fig. 12B) An alternative singlet oxygen product, benzaldehyde, was formed because diendoperoxide, **11**, was too bulky to form in either the circular sinusoidal (cross section 5.5 Å) or elliptical straight (cross section 5.2 × 5.8 Å) channels of ZSM-5.

   In 1999, Zhou and Clennan [24] reported the first examples of intrazeolite type I$_{hv}$O$_2$ reactions of organosulfides. Photo-oxidation of pentamethylene sulfide, **12**, in NaMBY in hexane slurry resulted in the formation of both the sulfoxide and sulfone. However, the sensitivity of the sulfone, **12SO$_2$**, sulfoxide, **12SO**, ratio to the concentration of **12** was dramatically enhanced in the zeolite in comparison to homogeneous media. The percentage yield of the sulfoxide increased from 46% to nearly 90% in the zeolite and it remained essentially constant in

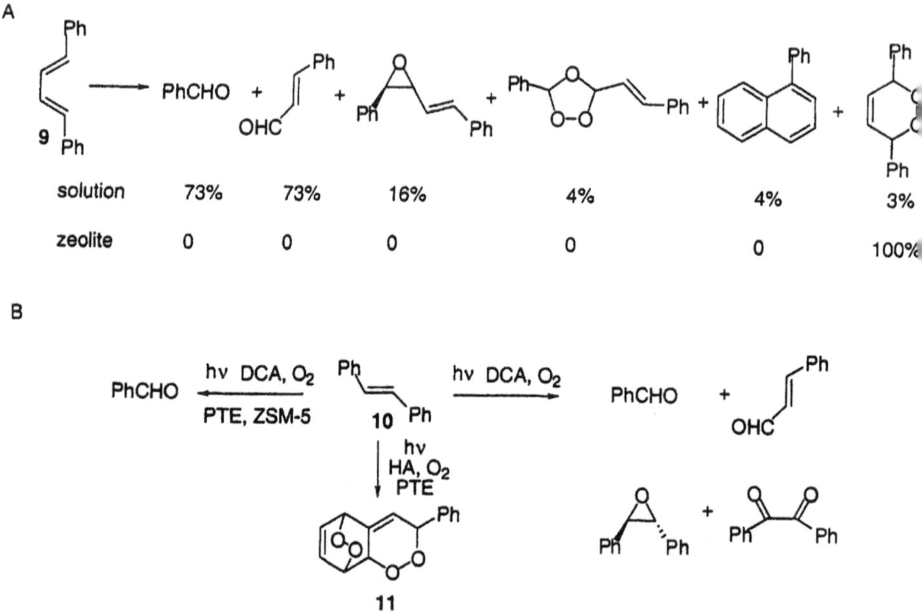

A

| solution | 73% | 73% | 16% | 4% | 4% | 3% |
| zeolite | 0 | 0 | 0 | 0 | 0 | 100% |

B

Figure 12 A comparison of the solution and intrazeolite photooxygenation.

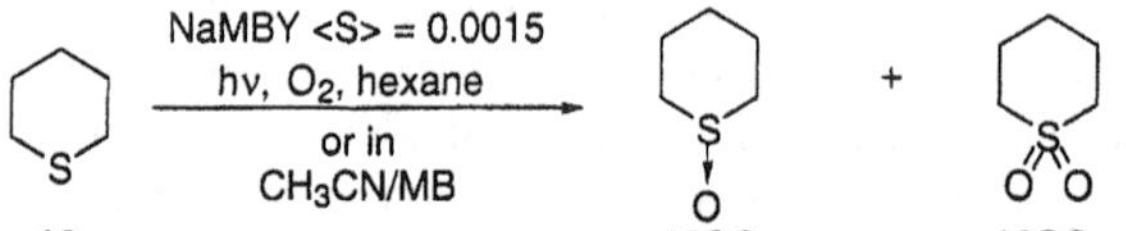

concentration[a]	12SO	12SO₂
.01(0.26)[b]	46.4(89.7)[c]	53.6(10.3)[c]
.02(0.51)[b]	54.0(95.0)[c]	46.0(5.0)[c]
.04(1.04)[b]	68.2(93.7)[c]	31.8(6.3)[c]
.07(1.75)[b]	88.4(92.7)[c]	11.6(7.3)[c]
.10(2.57)[b]	87.1(93.2)[c]	12.8(6.8)[c]

a. concentration Moles/Liter in zeolite NaY and in CH_3CN. b. molecules of substrate per supercage. c. % yield in zeolite(% yield in CH_3CN)

solution as the concentration of **12** was increased from 0.01 M to 0.10 M. This unusual dichotomous behavior is indicative of a change in mechanism as the reaction is moved from solution into the interior of the zeolite.

The established mechanism of sulfide photo-oxidation in solution invokes the novel formation of two intermediates a persulfoxide, **A**, and a hydroperoxy sulfonium ylide, **B**, (Fig. 13A) [25]. In this mechanism the sulfide substrate intercepts the second intermediate, k_{SO}, and does not competitively inhibit the predominant sulfone forming pathway, k_{SO2}. As a consequence, the sulfone/sulfoxide ratio is independent of sulfide concentration. In contrast, the results in the zeolite are inconsistent with this mechanism but are consistent with the trapping of a single intermediate with both sulfide and adventitious sulfoxide (Fig. 13B).

Complexation of sodium to the persulfoxide **A** (Fig. 13B) appears to inhibit intramolecular hydrogen abstraction to form the hydroperoxy sulfonium ylide (**B** in Fig. 13A) and allows a direct reaction of **12** with the sodium-complexed persulfoxide, (**A** in Fig. 13B) to compete. Consistent with this suggestion is the observation that the formation of **13CHO** that emanates from the hydroperoxy sulfonium ylide by Pummerer rearrangement and subsequent cleavage is completely suppressed during photo-oxidations of thiolane, **13**, in NaMBY:

		13SO	13SO₂	13CHO
CH₃CN/MB		10.3%	2.7%	87.0%
NaY/MB		44.2%	55.8%	

The stabilization of the persulfoxide in the electrostatic field of the sodium cations could also promote its formation. Evidence for the enhanced formation of the persulfoxide is provided by an examination of the photo-oxidation of diphenylsulfide, **14**. Photo-oxidation of **14** in CH₃CN for 1 hr resulted in quantitative recovery of the starting material. In contrast, irradiation in NaMBY for the same amount of time resulted in a nearly quantitative conversion to diphenylsulfoxide, **14SO**, and diphenylsulfone, **14SO₂**. The surprising sensitivity of the percentage conversion to the identity of the intrazeolite cation provides corroborating evidence that the strength of the electrostatic field is intimately involved in this phenomenon.

However, an alternative explanation that inhibition of physical quenching (k_q in Fig. 13), rather than enhanced persulfoxide formation, is responsible for the intrazeolite behavior of **14** has not been ruled out. Product formation would

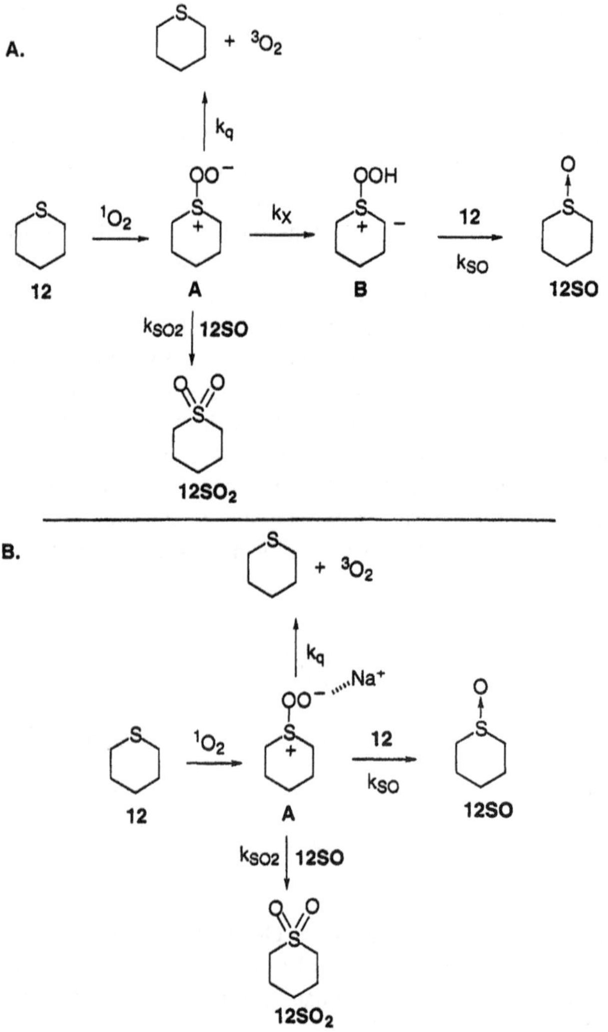

Figure 13 Solution (A) and intrazeolite (B) mechanisms for the photooxygenation of 12.

Ph$_2$S 14	hv, O$_2$	Ph$_2$SO	+	Ph$_2$SO$_2$
% conversion	0.02M 14 zeolite/MB	14SO		14SO$_2$
98.7	NaMBY	89.1%		10.9%
95.2	LiMBY	91.6%		8.4%
38.3	KMBY	95.2%		4.8%
4.8	RbMBY	100%		0%
0.5	CsMBY	100%		0%
50.8	BaMBY	98.6%		1.4%

appear to occur more rapidly if this energy-wasting process were suppressed. Preliminary evidence that supports this possibility has been provided in a study of a series of substrates, **15**, in which the sulfide is tethered to an alkene [26]. The amount of oxidation at sulfur relative to that at the alkene linkage is dramatically

		k$_r$(sulfide)/k$_r$(alkene)
R = H; R' = Ph-	CDCl$_3$	0.14 ± 0.02
	zeolite	2.31 ± 0.22
R = Me; R' =Et-	CDCl$_3$	0.70 ± 0.07
	zeolite	2.26 ± 0.41

enhanced in **15** (R = H; R' = Ph− and in R = Me; R' = Et−) as the reaction is moved from solution into the zeolite. This was argued to be consistent with suppression of physical quenching at sulfur in the interior of the zeolite.

III. TYPE II$_{hv}$RH REACTIONS

In 1999, Sanjuán and co-workers [27] reported a very elegant type II$_{hv}$RH oxidation of cyclohexene to give a mixture of cyclohexane-1, 2-diol, 2-cyclohexenol, and 2-cyclohexenone. The reaction is initiated by excitation of the zeolite-embedded 2,4,6-triphenylpyrylium cation to produce a hydroxy radical (steps 1 and 2

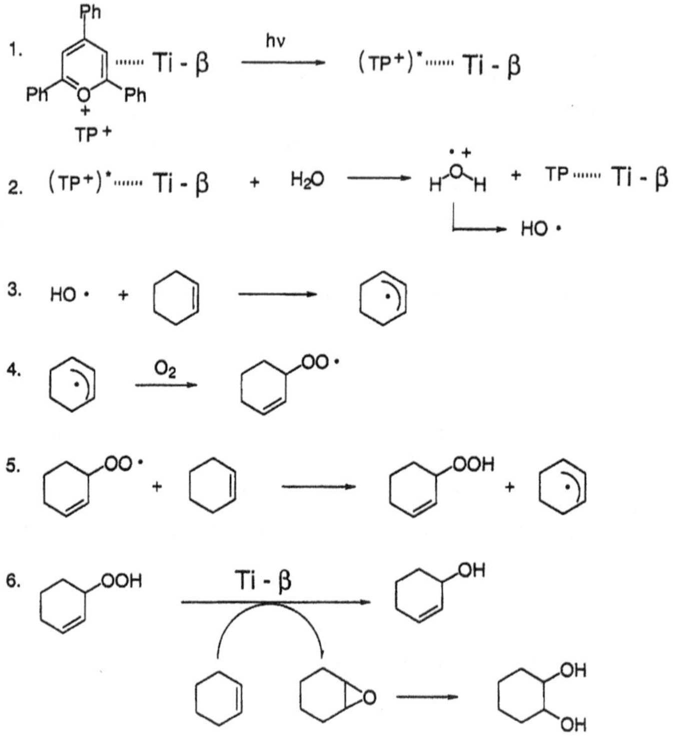

Figure 14 The mechanism for the 2,4,6-triphenylpyrylium induced intrazeolite oxygenation of cyclohexene.

in Fig. 14). In the hydrocarbon activation step the hydroxy radical abstracts an allylic hydrogen from cyclohexene (step 3). The cyclohexenyl radical then reacts with oxygen to generate the allylic hydroperoxide (steps 4 and 5). The in situ generated allylic hydroperoxide then reacts in concert with the titanium sites in the zeolite to epoxidize the cyclohexene, which is subsequently hydrolyzed to the diol (step 6).

IV. HYBRID [TYPE II$_{hv}$O$_2$ – TYPE II$_{hv}$RH] REACTIONS

In 1985, Aronovitch and Mazur [28] reported that irradiation of several organic molecules absorbed on a heterogeneous reaction matrix (silica gel, neutral, acidic, or basic alumina, and Florisil) with 350 nm light in the presence of oxygen re-

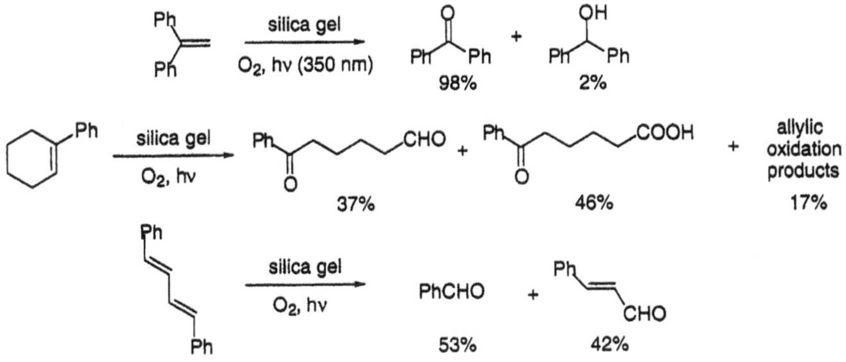

Figure 15 Oxygenation reactions on a silica gel matrix.

sulted in the formation of oxidized products (Fig. 15). The authors postulated that the reaction was initiated by the illumination of oxygen–organic substrate charge-transfer complexes inducing formation of organic substrate radical cation/superoxide ion pairs. Subsequent collapse of these ion pairs produced the observed products. The oxygen charge-transfer complex suggested as an intermediate has ample precedent [29]. Broad tailing bands on the red end of ultraviolet (UV) absorption spectra of oxygen-saturated solutions of organic molecules were first reported in the early 1950s [30,31]. The assignments of these new absorption bands to oxygen charge-transfer complexes (CT complexes) were put on a firm foundation with the observation that the onset of the absorption band was a linear function of the ionization potential of the organic molecule [32]. In addition, the suggestion of charge-transfer complex formation on heterogeneous matrices is supported by the previous spectroscopic observation of oxygen–substrate charge-transfer complexes on both silica gel and in glass matrices [33].

In 1993, Blatter and Frei [34] extended the Aronovitch and Mazur [28] photo-oxidation into zeolitic media, which resulted in several distinctive advantages as described below. Irradiation in the visible region (633 nm) of zeolite NaY loaded with 2,3-dimethyl-2-butene, **16**, and oxygen resulted in formation of allylic hydroperoxide, **17**, and a small amount of acetone. The reaction was followed by in situ Fourier-transform infrared (FTIR) spectroscopy and the products were identified by comparison to authentic samples. The allylic hydroperoxide was stable at $-50°C$ but decomposed when the zeolite sample was warmed to 20°C [35]. In order to rationalize these observations, it was suggested that absorption of light by an alkene/O_2 charge-transfer complex resulted in electron transfer to give an alkene radical cation–superoxide ion pair which collapses

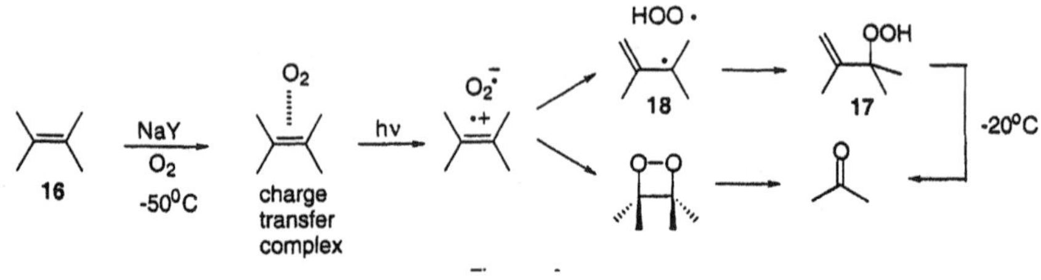

Figure 16 Frei photooxygenation mechanism.

competitively to a hydroperoxy/allylic radical pair, **18**, and a dioxetane (Fig. 16). Subsequent bond formation in the radical pair and cleavage of the dioxetane generates the observed products. A unique feature of this oxidation process and the photo-oxidations of Aronovitch and Mazur [28] are the simultaneous activation of both reaction partners. Consequently, these reactions are really a hybrid of the type $II_{h\nu}O_2$ and type $II_{h\nu}RH$ pathways.

Only a limited number of studies of the chemistry derived by irradiation into oxygen–organic substrate CT bands have been reported. For example, irradiation into the oxygen/1,2,3,4,-tetramethylnapthalene, **19**, charge-transfer band in a variety of solvents resulted in formation of endoperoxide, **20**. In this case, the formation of **20** was attributed to a singlet oxygen reaction because the methanolysis product, **21**, anticipated for a hydrogen abstraction sequence was not observed (Fig. 17). Scurlock and Ogilby [36] directly observed the 1270-nm phosphoresence from singlet oxygen subsequent to the pulsed UV irradiations of the oxygen charge-transfer bands for a variety of organic molecules in solution. Schmidt and Shafii [37] examined the formation of singlet oxygen using a series

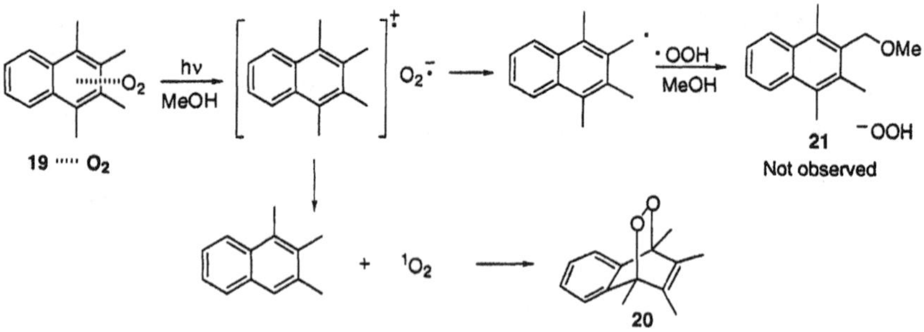

Figure 17 Photooxygenation via an oxygen-charge transfer complex.

of 10 biphenyl sensitizers of varying oxidation potential but nearly invariant triplet energies. Analysis of their results suggests that quenching of the triplet states by oxygen to give singlet oxygen proceeds via two different channels: one without charge-transfer character and one with partial charge-transfer character.

Nevertheless, the formation of allylic hydroperoxide **17** (Fig. 16) was not attributed to a singlet oxygen reaction because replacing NaY with a high silica faujasite resulted in a reversal of the **17**/acetone ratio, a result unprecedented for a singlet oxygen reaction [34]. Further evidence mitigating against singlet oxygen formation by back-electron transfer in the ion pair depicted in Fig. 16 is provided by cyclic voltammetry results, which demonstrate that oxygen reduces the current at the oxidation potential of **16** to approximately one-half the value observed in the absence of oxygen [38]. This is consistent with hydrogen abstraction from the cation radical, followed by trapping with oxygen that inhibits the further oxidation of the allylic radical to the cation [38]. The ability of hydrogen abstraction to compete with singlet oxygen formation is also consistent with the well-established superacidity of cation radicals [39].

Recently, detailed kinetic studies of the hybrid[type $II_{h\nu}O_2$ – type $II_{h\nu}$ RH] photo-oxidations of cyclohexane and cyclohexane-d_{12} in both NaY and BaY have been reported. A kinetic isotope effect k_H/k_D of 5.7 was determined for $\lambda >$ 400 nm in BaY. This substantial isotope effect, which is nearly identical to the isotope effect on the kinetic acidity of cyclohexane, requires that the proton abstraction step, k_{PT}, in the alkane radical cation superoxide ion pair be smaller than the back-electron transfer, k_{ET}, to regenerate the charge-transfer complex (Fig. 18). If k_{PT} were larger than k_{ET}, the rate expression, Eq. (A) in Fig. 18, would be reduced to Eq. (**B**) and only a small isotope effect on k_{ET} would be anticipated.

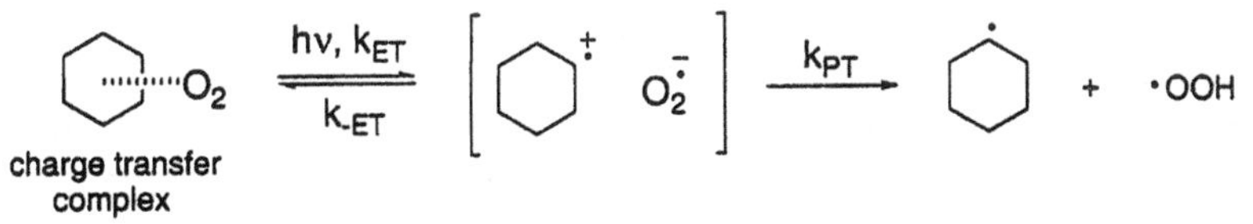

charge transfer
complex

$$\text{rate} = \frac{k_{PT}k_{ET}[\text{CT-complex}]}{k_{PT} + k_{ET}} \quad \textbf{A}$$

if $k_{PT} > k_{ET}$ then:

$$\text{rate} = k_{ET}[\text{CT-complex}] \quad \textbf{B}$$

Figure 18 Rate expressions for reaction via an oxygen-charge complex.

The most remarkable, and potentially the most useful, aspect of the intrazeolite photo-oxidation of **16**, which distinguishes it from the Aronovitch and Mazur results [28], is the fact that NaY stabilizes the excited charge-transfer complex by over $10,000 \, cm^{-1}$, resulting in a shift of the onset of the **16**"'O_2 charge-transfer band from 400 nm in solution to 760 nm in NaY at $-50°C$ [35]. This remarkable stabilization was attributed to the interaction of the highly polar charge-transfer complex with the bulk electrostatic field generated by the extra-framework sodium cations [40]. (An elegant use of time-resolved FTIR has ruled out the possibility that a small number of defect sites in NaY could be responsible for the stabilization of the oxygen–CT complex.) From a practical viewpoint, this phenomenon, which allows generation of the products with much reduced excess internal energy, coupled with the positional restraints of the zeolite pore system provides an unparalleled mild oxidation process. Nevertheless, Frei and co-workers [41] have pointed out three challenges that must be met prior to industrial scale-up. They include (1) increasing efficiency by reduction of scattering to ensure irradiation of all embedded reactants, (2) design of an acceptable method for desorption of products from the zeolite host, and (3) design of a suitable large-scale photochemical reactors.

Frei and co-workers have demonstrated the generality of this new type $\Pi_{h\nu}O_2$ – type $\Pi_{h\nu}RH$ photo-oxidation procedure with *cis*- and *trans*-2-butene [35], propylene [42], and 2-methyl-2-butene [43]. Very large red shifts of the reversibly formed charge-transfer bands were directly observed for these alkenes by diffuse reflectance UV-visible spectroscopy [43]. A relationship between the ionization potentials of the alkenes and the onset of the reaction threshold and/ or charge-transfer bands was noted [42]. For example, 3,3-dimethyl-1-butene, which has a very high ionization potential, was unreactive at all visible wavelengths examined. This relationship, however, is not linear. It was speculated that the lack of linearity could be a result of experimental uncertainties in the estimations of the concentrations of the charge-transfer complexes, or as a result of different quantum efficiencies [43]. The quantum yields for reaction of **16** at 514 nm are 0.18 ± 0.06 and 0.16 ± 0.10 at 633 nm [43]. These remarkably high quantum yields suggest that the deprotonation rates of the alkene radical cations by superoxide (Fig. 16) compete effectively with back-electron transfer to regenerate the ground-state charge-transfer complex (however, see Fig. 18 and associated discussion) [44].

Frei and co-workers also extended this reaction to other zeolites showing that almost identical behavior was observed in BaY, BaX, and in the K^+ and Ba^{2+} forms of zeolite L [45,46]. Xiang et al. [47] have also studied the photo-oxidations of a series of 1-alkenes in the more acidic BaZSM-5 [48] and Ba-β. The extensive polymerization of propylene in these zeolites demonstrates the detrimental effect of Brönsted acid sites on the reaction selectivity. These workers also used ex situ nuclear magnetic resonance (NMR) allowing more detailed analyses of product mixtures than possible by in situ FTIR alone. For example, propionaldehyde was detected by NMR during photo-oxidation of propylene but was not immediately obvious in the FTIR spectrum. Photo-oxidations of longer-chain alkenes were less selective, perhaps as a result of increased sensitivity to Brönsted acid sites. For example, photo-oxidation of 1-butene, **22**,

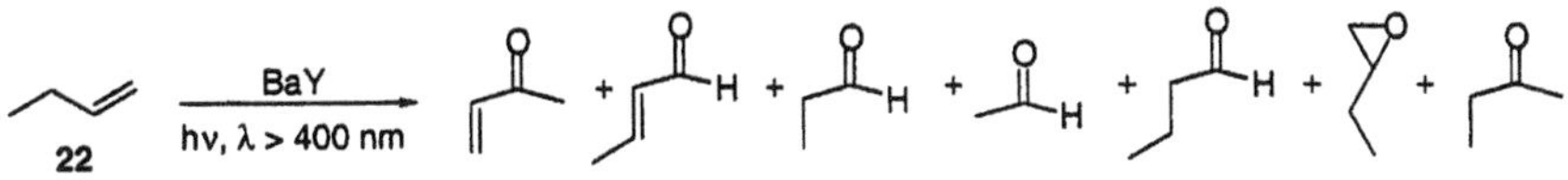

resulted in NMR detection of methyl vinyl ketone, crotonaldehyde, propionalde-hyde, acetaldehyde, butyraldehyde, epoxybutane, and 2-butanone. The formation of acetaldehyde requires a bond migration in 1-butene followed by dioxetane formation and cleavage. Crockett and Roduner [49], reported that the Frei photo-oxidation also occurs in H-mordenite and that 2,5-dimethyl-2,4-hexadiene was converted to either the conjugated allylic hydroperoxide or the endoperoxide. However, no direct evidence for the structures of the products were given.

Most photo-oxidations that occur via the Frei hybrid type $II_{h\nu}O_2$ – type $II_{h\nu}RH$ mechanism have been conducted in the absence of solvent, which can shield the substrate from the intrazeolite electrostatic field. The exceptions include the photo-oxidations of easily ionized substrate [50,51] that will react even in solvent slurries [10]. For example, Ramamurthy and co-workers [52] reported that irradiation into the oxygen–CT bands (≥ 350 nm) for a series of 1,1-diaryleth-ylenes, **23**, in hexane slurries resulted in the formation of oxidized products via the Frei mechanism (Fig. 19).

The Frei photo-oxidation of **23** occurred concurrently with an unanticipated reaction to form a reduced product [52]. This second process (Fig. 20) is favored at the expense of the Frei photo-oxidation at shorter wavelengths. This competing reaction also requires oxygen. To explain this unusual observation, the mechanism depicted in Fig. 20 was proposed. Irradiation directly into the absorption band

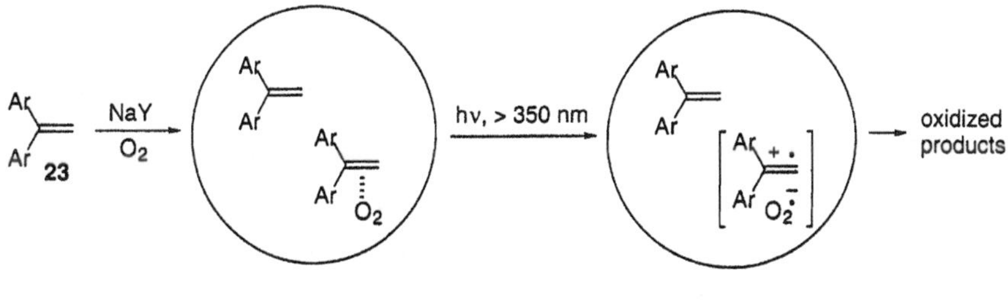

Figure 19   Frei photooxygenation of **23**.

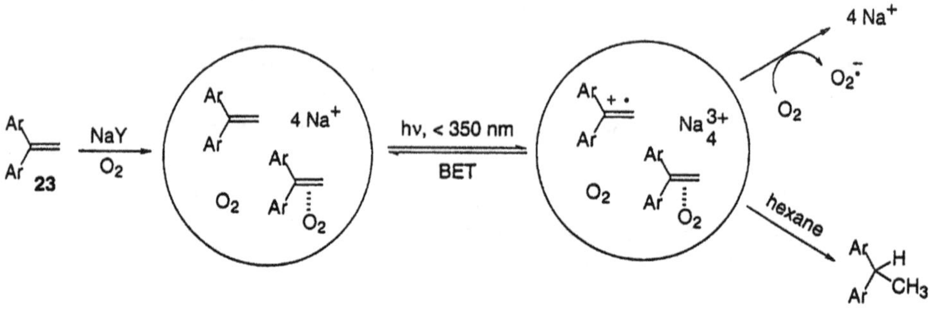

Figure 20   Mechanism of intrazeolite reduction of **23**.

of **23** induces ejection of an electron and formation of the radical cation. The radical cation subsequently abstracts a hydrogen atom from the solvent (hexane or perdeuterohexane) and is converted to the reduced product.

An extremely interesting feature of these mechanisms is the fact that super-oxide and the alkene radical cation are both formed in the reduction (Fig. 20) and also in the Frei oxidation (Fig. 19). In the Frei photo-oxidation, however, they are formed concurrently in a tight ion pair and collapse to product more rapidly than their diffusive separation. In the reduction (Fig. 20), the formation of the radical cation and superoxide occur in independent spatially separated events allowing the unimpeded diffusion of superoxide which precludes back-electron transfer (BET) and formation of oxidized products. The nongeminate formation of these two reactive species provides the time necessary for the radical cation to abstract a hydrogen atom from the solvent on its way to the reduced product.

Frei photo-oxidation also occurs with saturated hydrocarbons. For example, irradiation into the oxygen–CT bands of toluene [53,54] ethane [55], propane [55], isobutane [56], cyclohexane [57], and p-xylene [54] generated with good to excellent selectivity the oxidized products shown in Fig. 21. The success of these oxidations that have no singlet oxygen analogs provides further support for the hydrogen abstraction mechanism suggested by Frei and co-workers (Fig. 16).

Recently Ulagappan and Frei [58] reported a very novel type $II_{h\nu}O_2$ – type $II_{h\nu}RH$ oxidation of 2-propanol in an iron-substituted aluminophosphate molecular sieve, FeAlPO$_4$-5. This redox-active material has the AFI structure characterized by a one-dimensional channel system 7.3 Å in diameter. Loading of 2-propanol and oxygen in the sieve followed by irradiation into the ligand-to-metal charge-transfer band with laser light in the region between 350 and 430 nm resulted in the formation of both acetone and hydrogen peroxide. The authors suggested (Fig. 22) that a ligand-to-metal charge-transfer transition

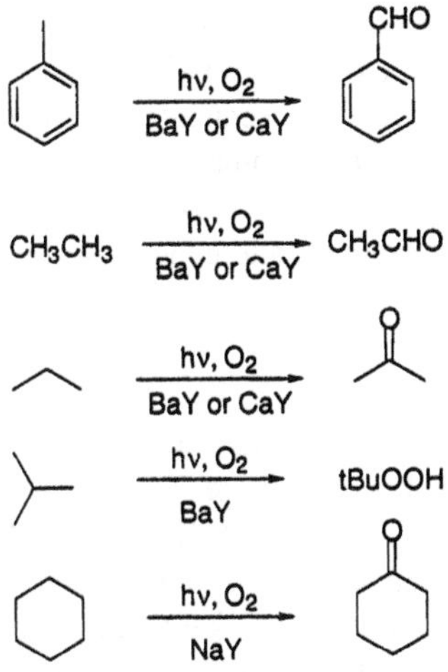

Figure 21 Frei photooxygenations of hydrocarbons.

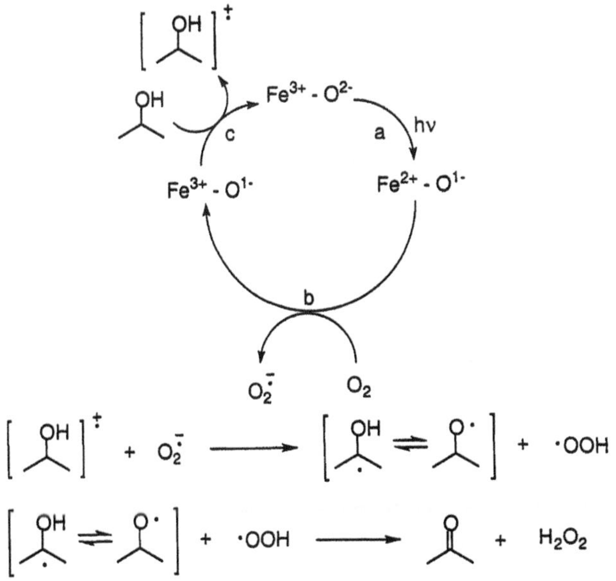

Figure 22 Mechanism of 2-propanol oxygenation in an iron-substituted aluminophosphate molecular sieve.

$(Fe^{3+}-O^{2-} \rightarrow Fe^{2+}-O^{1-}$; process a initiates the oxidation. The transient Fe^{2+} center then serves as a reducing agent to give superoxide and a $Fe^{3+}-O^{1-}$ center (process b). The electron deficient $Fe^{3+}-O^{1-}$ center subsequently oxidizes 2-propanol and in the process regenerates the $Fe^{3+}-O^{2-}$ center (process c). Two hydrogen abstractions, the first to give hydroperoxy radical and the second to give hydrogen peroxide, completes the reaction sequence and generates acetone.

V. TYPE II$_\Delta$O$_2$ REACTIONS

In 1988, Yoon and Kochi [59] reported the thermal formation of superoxide by the intrazeolite reduction of oxygen with methyl viologen, **24**, and diquat, **25**, doped NaY. The reactions were conveniently monitored by the loss of the blue color of **24** and the green color of **25** and by the appearance of the characteristic electron spin resonance (ESR) spectrum of superoxide. Remarkably, the reaction with **25** was reversible. Removal of oxygen resulted in the return of the characteristic green color of **25**, the disappearance of the ESR spectrum of superoxide, and the reappearance of the ESR spectrum of **25**. Repetition of this cycle was accomplished more than a dozen times without degradation of the sample. The reversible behavior was attributed to the near coincidence of the oxidation potential of **25** ($E° = -0.39$ V versus saturated calomel electrode (SCE)) and the intrazeolite reduction potential of oxygen (suggested to lie close to -0.4 V similar to its reduction potential in water). No studies were conducted to utilize the "activated oxygen" for subsequent chemical transformations.

VI. TYPE II$_\Delta$ RH REACTIONS

The use of transition-metal-substituted aluminophosphates (AIPOs) as catalysts for free-radical auto-oxidations of hydrocarbons, a type II$_\Delta$RH reaction, has been explored by several groups [60–62]. In a series of elegant contributions, Thomas and co-workers [3] have taken advantage of the shape selectivity and the oxidizing capability [60,61] of aluminophosphates with 4 at% of their framework Al^{3+} ions substituted with Co(III) or Mn(III) to successfully demonstrate the oxidation of linear hydrocarbons with a high selectivity for the terminal position [63]. These authors have also pointed out that they feel that the problems with solvent leaching have been greatly exaggerated. Many problems can be circumvented by just choosing a benign solvent or by just avoiding the use of any solvent. These reactions were more successful with Co^{III}AlPO-18 and Mn^{III}ALPO-18 than with their AlPO-36 analogs. For example, n-hexane was oxidized at 373 K with a primary selectivity index of 2.1 and 2.5 with Co^{III}AlPO-18 and with Mn^{III}ALPO-18, respectively, but with far lower selectivity with Co^{III}AlPO-36 (0.39) and Mn^{III}ALPO-36 (0; no primary product). The primary selectivity index is defined as the ratio of oxidation at primary relative to secondary and tertiary sites corrected for the number of hydrogens $[1°/(2° + 3°)]$. For the value of 2.5, this corresponds to 9.9% of the primary alcohol, 12.5% of the aldehyde, 43.1% of the carboxylic acid, 8.4% of 2-hexanol, and 23.3% of 2-hexanone with no oxidation observed at C-3. The pore apertures in AlPO-18 and ALPO-36 are 3.8 Å and 6.5 × 7.3 Å in diameter, respectively. The small pore size in AlPO-18 constrains absorption to an end-on mode that restricts access to the catalytically active site to the terminal carbon atoms on the n-alkane. No such constrained approach is enforced in the larger pore system of AlPO-36. Similar regioselectivity was also observed for n-octane, n-decane, and n-dodecane [64].

Oxidation of n-hexane with Co^{III}AlPO-18 with 10% rather than 4% of the framework Al^{3+} ions replaced with Co^{III} resulted in a dramatic enhancement in the formation of adipic acid [65]. It was argued that in these catalysts two Co^{III} ions are ideally separated by 7–8 Å on the inner wall of the zeolite, allowing both methyl groups unfettered access to catalytically active sites. Furthermore, it was demonstrated that 1,6-hexanediol and 1,6-hexanedial served as precursors to the adipic acid. On the other hand, 1-hexanol, hexanoic acid, and hexanal, which were also formed in the reaction, did not serve as precursors for the adipic acid. It is tempting to suggest that the mono-oxidized hexane products were produced in regions of the zeolite where simultaneous access to two catalytically active sites was not possible.

These oxidations suffer from the fact that the high selectivities are only observed at low conversions (<7%). At higher conversions, the carboxylic acid products leach the transition metals out of the zeolite framework into solution where the selectivity index is much lower [63]. As these reactions proceed, the 3+ oxidation states of the metal ions return to their 2+ states, accompanied by their characteristic color change. In the case of MnAlPO-18, the spent catalyst (Mn^{II}) was washed with methanol and reactivated in dry air at 550°C and successfully recycled ($Mn^{II} \rightarrow Mn^{III}$) twice without appreciable loss of activity [64].

The detailed mechanism for these $Co^{III}AlPO$-18- and $Mn^{III}ALPO$-18-cata-lyzed oxidations are unknown, but as previously pointed out (*vide supra*) and by analogy to other metal-mediated oxidations a free-radical chain auto-oxidation (a type II_ARH reaction) is anticipated [63]. This speculation is supported by several experimental observations that include (1) an induction period for product formation in the oxidation of *n*-hexane in CoAlPO-36, (2) the reduction of the induction period by the addition of free-radical initiators, (3) the ability to inhibit the reaction with addition of free-radical scavengers, and (4) the direct observation of cyclohexyl hydroperoxide in the oxidation of cyclohexane [62].

The selective oxidations of the terminal positions of *n*-alkanes are an example of substrate-shape selectivity. Product-shape selectivity has been used to enhance the selectivity of the type II_ARH oxidation of cyclohexane [66–68]. For example, oxidation of cyclohexane at 373 K for 8 hr using FeAlPO-31 (pore aperture 5.4 Å) as a catalyst resulted in 2.5% conversion to a mixture which contained 55.3% of adipic acid and 37.3% of a mixture of cyclohexanol and cyclohexanone [68]. In contrast, oxidation under identical conditions using FeAlPO-5 (pore aperture 7.3 Å) resulted in only 9.2% of adipic acid and 89.5%

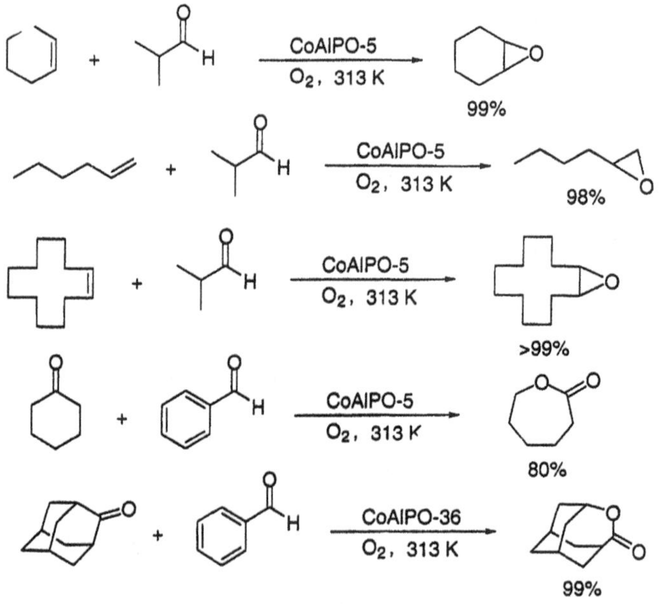

Figure 23 Intrazeolite epoxidations and Baeyer-Villiger reactions in the presence of oxygen.

of a mixture of cyclohexanol and cyclohexanone [68]. In the smaller one-dimensional pore system of FeAlPO-31, sterically bulky intermediates are selectively retained as a result of their reduced propensity to diffuse and are as a consequence susceptible to further oxidation to adipic acid which can more easily adopt a shape to allow its diffusion away from the catalytically active site in the zeolite.

The reactions of aldehydes at 313 K [69] or 323 K [70] in CoAlPO-5 in the presence of oxygen results in formation of an oxidant capable of converting olefins to epoxides and ketones to lactones (Fig. 23). This reaction is a zeolite-catalyzed variant of metal [71–73] and non-metal-catalyzed oxidations [73,74], which utilize a sacrificial aldehyde. Jarboe and Beak [75] have suggested that these reactions proceed via the intermediacy of an acyl radical that is converted either to an acyl peroxy radical or peroxy acid which acts as the oxygen-transfer agent. Although the detailed intrazeolite mechanism has not been elucidated a similar type $II_\Delta RH$ reaction is likely to be operative in the interior of the redox catalysts. The catalytically active sites have been demonstrated to be framework-substituted Co^{II} or Mn^{III} ions [70]. In addition, a sufficient pore size to allow access to these centers by the aldehyde is required for oxidation [70].

In 1999, Zhou and Clennan [76] reported a type $II_\Delta RH$ oxidation of 1,5-dithiacyclooctane, **26**, in CaY. They suggested (Fig. 24) that **26** was activated by an electron transfer to give a radical cation, which subsequently reacted with oxygen to give a peroxysulfonium cation radical intermediate. This scenario is supported by the ease of oxidation of **26** [77] and by the reported propensity of

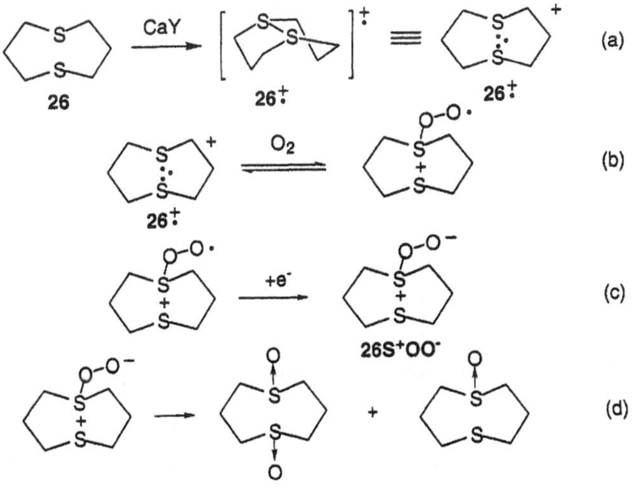

Figure 24 Mechanism of intrazeolite oxygenation of **26**.

zeolites to induced electron transfer from a variety of substrates [70,78–81]. The reaction of oxygen with sulfur radical cations [82], however, is unprecedented and does not occur in solution except at high oxygen pressures [83,84]. Asmus and co-workers [85] have suggested that this lack of reactivity is a result of an orbital mismatch between the σ^*-orbital on the $2\sigma + 1\sigma^*$ radical cation, 26^+, and the lowest unoccupied molecular orbital (LUMO) on oxygen. The successful oxidation in the zeolite appears to be related to the proximity of a reducing equivalent, which converts the peroxysulfonium cation radical to the persulfoxide (step c in Fig. 24) thereby inhibiting the loss of oxygen to regenerate 26^+. (step b in Fig. 24). The decomposition of the persulfoxide, $26S^+OO^-$, to generate the sulfoxide and bissulfoxide is known to occur in homogeneous media [86]. The direct spectroscopic observation of intrazeolite 26^+. by UV-visible ($\lambda_{max} = 420$ nm) and ESR ($g = 2.01123$; $a_H(4H) = 14.50$ G) provides additional support for the mechanism depicted in Fig. 24.

VII. HYBRID [TYPE II$_\Delta$O$_2$ – TYPE II$_\Delta$RH] REACTIONS

Oxidations initiated by thermally induced electron transfer in an oxygen–CT complex represent the thermal analog of the Frei photo-oxidation and are properly classified as hybrid type II$_\Delta$O$_2$–type II$_\Delta$RH oxidations (Fig. 2). Such reactions require either zeolites with high electrostatic fields or substrates with low oxidation potentials. In addition, elevated temperatures are known to promote the thermally initiated electron-transfer step, although the possible intrusion of a classical free-radical initiation chain oxidation at higher temperatures must be considered.

The reaction of propane in CaY appears to be an authentic thermal Frei oxidation [55]. Propane is, itself, inert in CaY but slowly oxidizes at 21°C with complete selectivity for formation of acetone. In contrast, propane oxidation in BaY did not commence until the temperature was raised to 55°C, but even at this elevated temperature, the high regioselectivity for acetone formation argues for a thermal Frei oxidation mechanism.

A thermal oxidation of 2,3-dimethyl-2-butene, **16**, occurs in NaY when the temperature of the oxygen-loaded zeolite in raised above -20°C [35]. Similar thermally initiated oxidations were not observed for the less electron rich *trans*- or *cis*-2-butene. Remarkably, pinacolone was conclusively identified as one of the products of the reaction of **16**. This ketone is not a product of the photochemical Frei oxidation (*vide supra*) and underscores the very different character of these two reactions and the complexity of the oxygen/**16** potential energy surface. A rationale for the different behavior could lie in the different electronic states of the reactive oxygen–CT complex in the thermal and photochemical reactions. Irradiation could produce an excited triplet-state CT complex ($^3[\mathbf{16}\cdots\mathbf{O_2}]^*$) and/or ion pair ($^3[\mathbf{16}^+\cdot\mathbf{O_2}^-\cdot]^*$) with different accessible reaction channels than those available to a vibrationally excited ground-state triplet complex ($^3[\mathbf{16}\cdots\mathbf{O_2}]$) and/

or ion pair ($^3[16^{+\cdot}O_2^{-\cdot}]$). Clearly, more work is needed to delineate the details of both the thermal and photochemical pathways.

Stratakis and Stravroulakis [87] reported a hybrid type $II_\Delta O_2$–type $II_\Delta RH$ oxidation of limonene in methyl viologen, 24^{2+}, doped NaY. These workers suggested that substrate activation was achieved by spontaneous electron transfer from the zeolite framework to 24^{2+} followed by electron transfer from limonene. Oxygen activation by superoxide formation, on the other hand, occurred by the well-established reduction of oxygen by $24^{+\cdot}$ (*vide supra*) [59]. Subsequent reaction of the activated reaction partners generated the endoperoxide, ascaridole, in 3–15% yields. The yield was suppressed by a competitive dehydrogenation of limonene radical cation to form *p*-cymene (Fig. 25).

Vanoppen et al. [88] have reported the gas-phase oxidation of zeolite-adsorbed cyclohexane to form cyclohexanone. The reaction rate was observed to increase in the order NaY < BaY < SrY < CaY. This was attributed to a Frei-type thermal oxidation process. The possibility that a free-radical chain process initiated by the intrazeolite formation of a peroxy radical, however, could not be completely excluded. On the other hand, liquid-phase auto-oxidation of cyclohexane, although still exhibiting the same rate effect (i.e., NaY < BaY < SrY < CaY), has been attributed to a homolytic peroxide decomposition mechanism [89]. Evidence for the homolytic peroxide decomposition mechanism was provided in part by the observation that the addition of cyclohexyl hydroperoxide dramatically enhanced the intrazeolite oxidation. In addition, decomposition of cyclohexyl hydroperoxide followed the same reactivity pattern (i.e., NaY < BaY

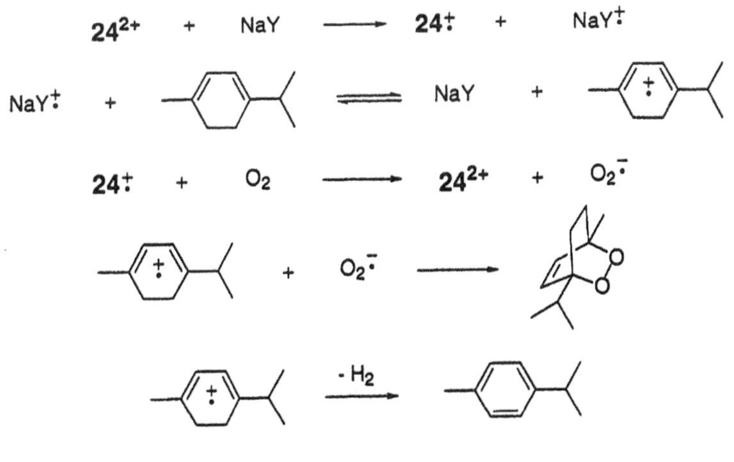

Figure 25 Oxygenation of limonene in methyl viologen, **24**, doped NaY.

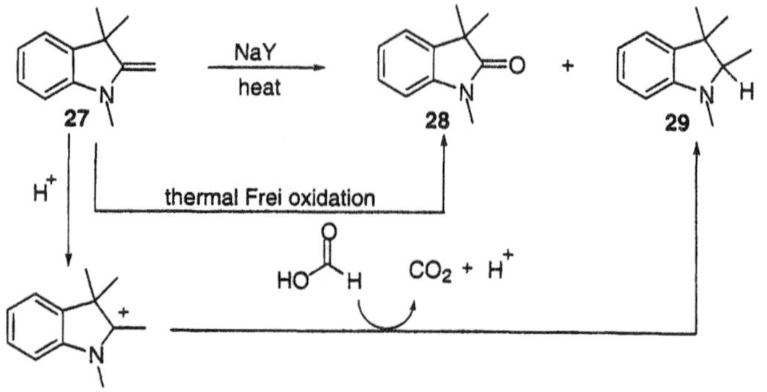

Figure 26 Mechanism for intrazeolite oxygenation and reduction of **27**.

$<$ SrY $<$ CaY) as observed in the "uninitiated" cyclohexane auto-oxidation. These workers believe that in both the gas- and liquid-phase photo-oxidations that electrostatically promoted electron transfer to generate a cyclohexane radical cation–superoxide ion pair occurs. However, only under liquid-phase conditions is there a continuous medium in which radical reactions can propagate themselves.

Garcia and co-workers [90] have reported a novel thermally induced concurrent oxidation and reduction of 1,3,3-trimethyl-2-methyleneindoline, **27**. This reaction is reminiscent of the concurrent photoinduced oxidation and reduction of 1,1-diarylethylenes, **23**. (Figs. 19 and 20). A thermal Frei oxidation (hybrid type $II_\Delta O_2$ - type $II_\Delta RH$) was utilized to rationalize the formation of indolinone, **28**. However, a carbocation intermediate generated by protonation of **27** was suggested to be on the reaction surface for the formation of the reduced product, **29** (Fig. 26). Formaldehyde formed from decomposition of the dioxetane on the Frei oxidation surface or a derived product (e.g., formic acid) was suggested to act as the reducing agent. It is possible that a thermal one-electron oxidation, similar to the photoejection mechanism suggested for the uncomplexed 1,1-diarylethylene (Figs. 19 and 20), is not available thermally. However, the similarity of these two reactions suggests that re-examination of both systems is warranted.

VIII. CONCLUSION

We have provided examples of both type I and type II intrazeolite photo-oxidations. Intrazeolite type III oxidations can be identified by analogy to the many published solution examples [91]. However, definite experimental evidence supporting intrazeolite type III oxidations are more difficult to establish than for the metal mediated type II oxidations. Fundamentally, one can describe the difference between the metal-mediated type II and III processes as outer sphere versus inner sphere electron transfers [92]. An alternative way of thinking about these processes is that the metal complexes formed in the type II processes have short lifetimes and dissociate into the open-shell species ($O_2^{-\bullet}$, $RH^{+\bullet}$, $RH^{-\bullet}$, or $R^{\bullet}$) which then react with the second reaction component. In the type III processes, the metal complexes have much longer lifetimes and react prior to dissociation directly with the second reaction component. Clearly, more work needs to be done to establish the mechanisms these fundamentally important oxidation processes.

REFERENCES

1. Dyer, A. *An Introduction to Zeolite Molecular Sieves*; Wiley: New York, 1988.
2. Rabo, J. A.; Schoonover, M. W. *Appl. Catal. A: Gen.* **2001**, *222*, 261–275.
3. Thomas, J. M.; Raja, R.; Sankar, G.; Bell, R. G. *Acc. Chem. Res.* **2001**, *34*, 191–200.
4. Wagnerova, D. M. *Z. Phys. Chem. Int.* **2001**, *215*, 133–138.
5. Kasha, M.; Brabham, D. E. *Singlet Oxygen Electronic Structure and Photosensitization*; Wasserman, H. H. and Murray, R. W., eds.; Academic Press: New York, 1979; Vol. 40, pp. 1–33.
6. Li, X.; Ramamurthy, V. *J. Am. Chem. Soc.* **1996**, *118*, 10,666–10,667.
7. Clennan, E. L. *Tetrahedron* **2000**, *56*, 9151–9179.
8. Frimer, A. A. *Singlet O_2*; CRC Press: Boca Raton, FL, 1985.
9. Robbins, R. J.; Ramamurthy, V. *J. Chem. Soc., Chem. Commun.* **1997**, 1071–1072.
10. Ramamurthy, V.; Lakshminarasimhan, P.; Grey, C. P.; Johnston, L. J. *J. Chem. Soc., Chem. Commun.* **1998**, 2411–2424.
11. Clennan, E. L.; Sram, J. P. *Tetrahedron Lett.* **1999**, *40*, 5275–5278.
12. Clennan, E. L.; Chen, X.; Koola, J. J. *J. Am. Chem. Soc.* **1990**, *112*, 5193–5199.
13. Clennan, E. L.; Sram, J. P. *Tetrahedron* **2000**, *56*, 6945–6950.
14. Ma, J. C.; Dougherty, D. A. *Chem. Rev.* **1997**, *97*, 1303–1324.
15. Barich, D. H.; Xu, T.; Zhang, J.; Haw, J. F. *Angew. Chem. Int. Ed. Engl.* **1998**, *37*, 2530–2531.
16. Stratakis, M.; Froudakis, G. *Org. Lett.* **2000**, *2*, 1369–1372.
17. Stratakis, M.; Kosmas, G. *Tetrahedron Lett.* **2001**, *42*, 6007–6009.
18. Stratakis, M.; Rabalakos, C. *Tetrahedron Lett.* **2001**, *42*, 4545–4547.
19. Shailaja, J.; Sivaguru, J.; Robbins, R. J.; Ramamurthy, V.; Sunoj, R. B.; Chandrasekhar, J. *Tetrahedron* **2000**, *56*, 6927–6943.
20. Pettit, T. L.; Fox, M. A. *J. Phys. Chem.* **1986**, *90*, 1353–1354.
21. Tung, C.-H.; Wang, H.; Ying, Y.-M. *J. Am. Chem. Soc.* **1998**, *120*, 5179–5186.
22. Tung, C.-H.; Wu, L.-Z.; Zhang, L.-P.; Li, H.-R.; Yi, X.-Y.; Song, K.; Xu, M.; Yuan, Z.-Y.; Guan, J.-Q.; Wang, H.-W.; Ying, Y.-M.; Xu, X.-H. *Pure Appl. Chem.* **2000**, *72*, 2289–2298.
23. Tung, C.-H.; Wu, L.-Z.; Yuan, Z.-Y.; Guan, J.-Q.; Wang, H.-W.; Zhang, L.-P. *J. Photosci.* **1999**, *6*, 85–90.
24. Zhou, W.; Clennan, E. L. *J. Am. Chem. Soc.* **1999**, *121*, 2915–2916.
25. Toutchkine, A.; Clennan, E. L. *J. Org. Chem.* **1999**, *64*, 5620–5625.

26. Zhou, W.; Clennan, E. L. *Org. Lett.* **2000**, *2*, 437–440.
27. Sanjuán, A.; Alvaro, M.; Corma, A.; García, H. *J. Chem. Soc., Chem. Commun.* **1999**, 1641–1642.
28. Aronovitch, C.; Mazur, Y. *J. Org. Chem.* **1985**, *50*, 149–150.
29. Zhu, J.; Liu, Z.; Cao, W.; Feng, X. *Chin. Sci. Bull.* **1998**, *43*, 1798–1802.
30. Evans, D. F. *J. Chem. Soc.* **1953**, 345–347.
31. Munck, A. U.; Scott, J. F. *Nature* **1956**, *177*, 587.
32. Tsubomura, H.; Mulliken, R. S. *J. Am. Chem. Soc.* **1960**, *82*, 5966–5974.
33. Ishida, H.; Takahashi, H.; Sato, H.; Tsubomura, H. *J. Am. Chem. Soc.* **1970**, *92*, 275–280.
34. Blatter, F.; Frei, H. *J. Am. Chem. Soc.* **1993**, *115*, 7501–7502.
35. Blatter, F.; Frei, H. *J. Am. Chem. Soc.* **1994**, *116*, 1812–1820.
36. Scurlock, R. D.; Ogilby, P. R. *J. Phys. Chem.* **1989**, *93*, 5493–5500.
37. Schmidt, R.; Shafii, F. *J. Phys. Chem. A* **2001**, *105*, 8871–8877.
38. Onodera, K.; Furusawa, C.-I.; Kojima, M.; Tsuchiya, M.; Aihara, S.; Akaba, R.; Sakuragi, H.; Tokumaru, K. *Tetrahedron* **1985**, *41*, 2215–2220.
39. Dinnocenzo, J. P.; Banach, T. E. *J. Am. Chem. Soc.* **1989**, *111*, 8646–8653.
40. Vasenkov, S.; Frei, H. *J. Phys. Chem. B* **1998**, *102*, 8177–8182.
41. Frei, H.; Blatter, F.; Sun, H. *Chemtech* **1996**, *26*, 24–30.
42. Blatter, F.; Sun, H.; Frei, H. *Catal. Lett.* **1995**, *35*, 1–12.
43. Blatter, F.; Moreau, F.; Frei, H. *J. Phys. Chem.* **1994**, *98*, 13,403–13,407.
44. Vasenkov, S.; Frei, H. *J. Phys. Chem. B* **1997**, *101*, 4539–4543.
45. Blatter, F.; Sun, H.; Vasenkov, S.; Frei, H. *Catal. Today* **1998**, *41*, 297–309.
46. Vasenkov, S.; Frei, H. *Photo-Oxidation in Zeolites*: Ramamurthy, V. and Schanze, K. S., eds.; Marcel Dekker: New York, 2000; Vol. 5, pp. 295–323.
47. Xiang, Y.; Larsen, S. C.; Grassian, V. H. *J. Am. Chem. Soc.* **1999**, *121*, 5063–5072.
48. Myli, K. B.; Larsen, S. C.; Grassian, V. H. *Catal. Lett.* **1997**, *48*, 199–202.
49. Crockett, R.; Roduner, E. *Electron Transfer Reactions in H-Mordenite*: Weitkamp, J., Karge, H. G., Pfeifer, H. and Hölderich, W., eds.; Elsevier Science B. V.: Amsterdam, 1994; Vol. 84, pp. 527–534.
50. Takeya, H.; Kuriyama, Y.; Kojima, M. *Tetrahedron Lett.* **1998**, *39*, 5967–5970.
51. Matsubara, C.; Kojima, M. *Tetrahedron Lett.* **1999**, *40*, 3439–3442.
52. Lakshminarasimhan, P.; Thomas, K. J.; Johnston, L. J.; Ramamurthy, V. *Langmuir* **2000**, *16*, 9360–9367.
53. Sun, H.; Blatter, F.; Frei, H. *J. Am. Chem. Soc.* **1994**, *116*, 7951–7952.
54. Panov, G. I.; Uriarte, A. K.; Rodkin, M. A.; Sobolev, V. I. *Catal. Today* **1998**, *41*, 365–385.
55. Sun, H.; Blatter, F.; Frei, H. *Catal. Lett.* **1997**, *44*, 247–253.
56. Blatter, F.; Sun, H.; Frei, H. *Chem. Eur. J.* **1996**, *2*, 385–389.
57. Sun, H.; Blatter, F.; Frei, H. *J. Am. Chem. Soc.* **1996**, *118*, 6873–6879.
58. Ulagappan, N.; Frei, H. *J. Phys. Chem. A* **2000**, *104*, 490–496.
59. Yoon, K. B.; Kochi, J. K. *J. Am. Chem. Soc.* **1988**, *110*, 6586–6588.
60. Lin, S. S.; Weng, H. S. *Appl. Catal.* **1994**, *118*, 21–31.
61. Krauschaar-Czarnetzki, B.; Hoogervorst, W. G. M.; Stork, W. H. *Oxidation of Saturated Hydrocarbons Involving CoAPO Molecular Sieves as Oxidants and as Cata-

lysts; Weitkamp, J., Karge, H. G., Pfeifer, H. and Hölderich, W., eds.; Elsevier Science: Amsterdam, 1994; Vol. 84C, pp. 1869–1876.

62. Vanoppen, D. L.; De Vos, D. E.; Genet, M. J.; Rouxhet, P. G.; Jacobs, P. A. *Angew. Chem. Int. Ed. Engl.* **1995**, *34*, 560–563.

63. Thomas, J. M.; Raja, R.; Sankar, G.; Bell, R. G. *Nature* **1999**, *398*, 227–230.

64. Raja, R.; Thomas, J. M. *J. Chem. Soc., Chem. Commun.* **1998**, 1841–1842.

65. Raja, R.; Sankar, G.; Thomas, J. M. *Angew. Chem. Int. Ed. Engl.* **2000**, *39*, 2313–2316.

66. Sankar, G.; Raja, R.; Thomas, J. M. *Catal. Lett.* **1998**, *55*, 15–23.

67. Raja, R.; Sankar, G.; Thomas, J. M. *J. Am. Chem. Soc.* **1999**, *121*, 11926–11927.

68. Dugal, M.; Sankar, G.; Raja, R.; Thomas, J. M. *Angew. Chem. Int. Ed. Eng.* **2000**, *39*, 2310–2313.

69. van Breukelen, H. F. W. J.; Gerritsen, M. E.; Ummels, V. M.; Broens, J. S.; van Hooff, J. H. C. *Application of CoAPO-5 Molecular Sieves as Heterogeneous Catalysts in Liquid Phase Oxidation of Alkenes with Dioxygen*; Chon, H., Ihm, S.-K. and Uh, Y. S., eds.; Elsevier Science: Amsterdam, 1997; Vol. 105, pp 1029–1035.

70. Raja, R.; Thomas, J. M.; Sankar, G. *J. Chem. Soc. Chem. Commun.* **1999**, 525–526.

71. Yamada, T.; Takai, T.; Rhode, O.; Mukiyama, T. *Chem. Lett.* **1991**, 1–4.

72. Yamada, T.; Rhode, O.; Takai, T.; Mukaiyama, T. *Chem. Lett.* **1991**, 5–8.

73. Nam, W.; Kim, H. J.; Kim, S. H.; Ho, R. Y. N.; Valentine, J. S. *Inorg. Chem.* **1996**, *35*, 1045–1049.

74. Kaneda, K.; Haruna, S.; Imanaka, T.; Hamamoto, M.; Nishiyama, Y.; Ishii, Y. *Tetrahedron Lett.* **1992**, *33*, 6827–6830.

75. Jarboe, S. G.; Beak, P. *Org. Lett.* **2000**, *2*, 357–360.

76. Zhou, W.; Clennan, E. L. *J. Chem. Soc., Chem. Commun.* **1999**, 2261–2262.

77. Musker, W. K. *Acc. Chem. Res.* **1980**, *13*, 200–206.

78. Adam, W.; Corma, A.; Miranda, M. A.; Sabater-Picot, M.-J.; Sahin, C. *J. Am. Chem. Soc.* **1996**, *118*, 2380–2386.

79. Corma, A.; Fornés, V.; García, H.; Miranda, M. A.; Primo, J.; Sabater, M.-J. *J. Am. Chem. Soc.* **1994**, *116*, 2276–2280.

80. Li, X.; Ramamurthy, V. *Tetrahedron Lett.* **1996**, *37*, 5235–5238.

81. Pitchumani, K.; Corbin, D. R.; Ramamurthy, V. *J. Am. Chem. Soc.* **1996**, *118*, 8152–8153.

82. Glass, R. S. *Sulfur Radical Cations*; Page, P. C. B., ed.; Springer-Verlag: Berlin, 1999; Vol. 205, pp. 1–87.

83. Riley, D. P.; Correa, P. E. *J. Chem. Soc., Chem. Commun.* **1986**, 1097–1098.

84. Riley, D. P.; Smith, M. R.; Correa, P. E. *J. Am. Chem. Soc.* **1988**, *110*, 177–180.

85. Schäfer, K.; Bonifacic, M.; Bahnemann, D.; Asmus, K.-D. *J. Phys. Chem.* **1978**, *82*, 2777–2780.

86. Clennan, E. L.; Wang, D.-X.; Yang, K.; Hodgson, D. J.; Oki, A. R. *J. Am. Chem. Soc.* **1992**, *114*, 3021–3027.

87. Stratakis, M.; Stavroulakis, M. *Tetrahedron Lett.* **2001**, *42*, 6409–6411.

88. Vanoppen, D. L.; De Vos, D. E.; Jacobs, P. A. *Cation Effects in the Oxidation of Adsorbed Cyclohexane in Y Zeolite: An In Situ IR Study*; Chon, H., Ihm, S.-K. and Uh, Y. S., eds.; Elsevier Science: Amsterdam, 1997; Vol. 105, pp. 1045–1051.

89. Vanoppen, D. L.; De Vos, D. E.; Jacobs, P. A. *J. Catal.* **1998**, *177*, 22–28.
90. Casades, I.; Alvaro, M.; García, H.; Esplá, M. *J. Chem. Soc., Chem. Commun.* **2001**, 982–983.
91. Jäger, E.-G.; Knaudt, J.; Schuhmann, K.; Guba, A. *Binding and Activation of Dioxygen by Biomimetic Metal Complexes*; Adam, W., ed.; Wiley–VCH: Weinheim, 2000, pp. 249–280.
92. Eberson, L. *Electron Transfer Reactions in Organic Chemistry*; Springer-Verlag: Heidelberg, 1987; Vol. 25.

7

Organic–Inorganic Composites as Photonic Antenna

Huub Maas, Stefan Huber, Abderrahim Khatyr, Michel Pfenniger, Marc Meyer, and Gion Calzaferri
University of Berne, Berne, Switzerland

I. INTRODUCTION

A photonic antenna is an organized multicomponent arrangement in which several chromophores absorb the incident light and channel the excitation energy to a common acceptor component. Imaginative attempts to build an artificial antenna different from ours have been presented in the literature. Multinuclear luminescent metal complexes, multichromophore cyclodextrins, Langmuir–Blodgett films, dyes in polymer matrices, and dendrimers have been investigated. Sensitization processes in silver halide photographic materials and also the spectral sensitization of polycrystalline titanium dioxide films bear, in some cases, aspects of artificial antenna systems (for references see Ref. 1). The materials which have been reported by us so far are of a bidirectional type. They are based on zeolite L as a host and able to collect and transport excitation energy [1–5]. The transport is enabled by specifically organized dye molecules which mimic the natural function of chlorophyll. The zeolite L crystals consist of a continuous one-dimensional tube system. We have filled each tube with chains of joined but electronically noninteracting dye molecules. Light shining on the cylinder is first absorbed and the energy is then transported by the dye molecules inside the tubes to a desired region.

a)   b)

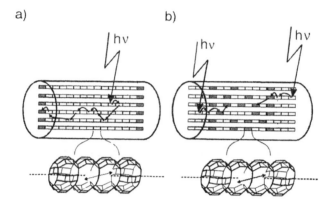

Figure 1 Representation of a cylindrical nanocrystal consisting of organized dye molecules acting as donors (light gray rectangles) and acceptors (dark gray rectangles). (a) The
donors are in the middle part of the crystal and the acceptors at the front and the back of
each channel. (b) The donors are at the front and the back of each channel and the acceptors
are in the middle part. The enlargements show details of a channel with a dye and its
electronic transition moment (arrow), which is parallel with respect to the channel axis
for long molecules and bent for shorter ones. The diameter of the channel windows is
0.71 nm and the largest free diameter is 1.26 nm. The center-to-center distance between
two channels is 1.84 nm.

The principle of two types of photonic antenna materials is illustrated in
Fig. 1. The rectangles represent sites for dye molecules. Light gray rectangles
are sites which contain donors and the dark ones contain acceptors. Both donors
and acceptors are assumed to be strongly luminescent dyes. The dye molecule
in Fig. 1a, which has been excited by absorbing an incident photon, transfers its
electronic excitation to neighboring dye molecules. After a series of energy-
migration steps, the electronic excitation reaches an acceptor. The energy migration is in competition with spontaneous emission, radiationless decay, and photochemically induced degradation. Very fast energy migration is therefore crucial
if an acceptor is to be reached before other processes take place. The energy
quantum can be guided to a reaction center once it has been captured by the
acceptor. These conditions impose not only spectroscopic but also decisive geometrical constraints on the system. Recently, we have been able to reverse the
scheme in Fig. 1a having acceptor dyes in the center and the donors at both ends,
as illustrated in Fig. 1b. The material was investigated by stationary energy-
migration experiments on an ensemble and space- and time-resolved measurements on single crystals [6].

The general synthesis concept of these inorganic–organic composites is based on the specific geometrical constraints imposed by the parallel arrangement of one-dimensional channels of the host [4,5]. Various synthesis strategies for the preparation of chromophores in porous silica and minerals have been reviewed recently [7]. The structure of zeolite L is illustrated in Fig. 2. The number of parallel channels which coincide with the c axis of the hexagonal framework is equal to 1.07 r^2_{cyl}, where r_{cyl} is the radius of the crystal (in nm). The length of a unit cell along the c axis is 0.75 nm. This means that a crystal of, for example, 600 nm diameter and 300 nm length, gives rise to about 100,000 parallel channels, each consisting of 400 unit cells (ucs) [8–11].

Control of the shape and size of the crystals is a necessary prerequisite for particular applications. Large crystals of a few hundred to a few thousand nanometers are very useful for studying the optical and photophysical properties of dye–zeolite composites on single crystals by means of optical microscopy methods. Crystals in the range of a few tenths to a few hundred nanometers are needed for high-efficiency photonic antenna materials, useful as fluorescent microprobes in cell biology and analytical chemistry [12], for developing a new generation of dye-sensitized solid-state solar cells [13], or for preparing a new generation of light-emitting diodes. We have recently shown how fine-tuning of the size of zeolite L crystals in the size range of 30 nm up to 3000 nm can be realized by changing the composition of the starting gel for otherwise constant reaction conditions. It was thus possible to extend the investigations on energy migration in Py$^-$-loaded zeolite L crystals, modified with Ox$^-$ as a luminescent acceptor at the ends of the crystals [5].

Some dye molecules which have been inserted into zeolite L are given in Table 1, where we also give abbreviations used in this chapter. It is important to distinguish among three types of dye molecules: (i) Molecules small enough to fit into a single unit cell; some examples are biphenyl (BP), hydoxy-TEMPO, fluorenone, and naphtalene; (ii) molecules the size of which makes it hard to guess if they align along the c-axis or if they find a way to fit into a single unit cell; Ox$^-$, Py$^-$, and Th$^+$ are molecules of this type [15]; (iii) molecules which are so large that they have no other choice but to align along the c axis; many examples fit into this category: POPOP, DMPOPOP, MBOXE, pTP, and DPH are some of them. It is important to know more precisely which of the type ii and iii molecules can arrange in a way so that they can interact via their π-system and which are only in physical contact (i.e., have negligible electronic interaction).

II. INSERTION EQUILIBRIA

In this chapter, we describe equilibria between dyes inside the zeolite L channels and dyes outside either in gas phase or in solution. We consider dye molecules

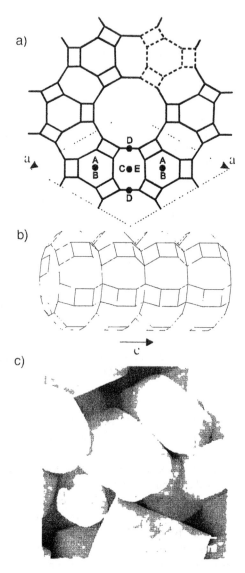

Figure 2   (a) Zeolite L framework with the different cation positions A to E and the ε-cage (dashed) viewed perpendicularly to the *c* axis. (b) Side view of the 12-ring channel along the *c* axis. (c) SEM image of zeolite L.

Table 1 Dye Molecules and Abbreviations

BP	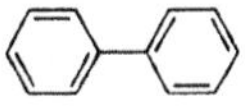	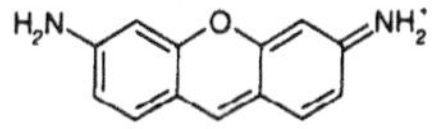	Py^{+}
pTP	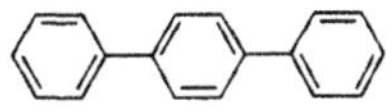	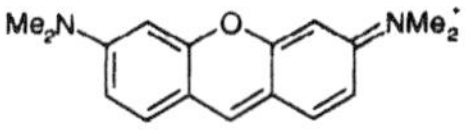	PyGy^{+}
DPH	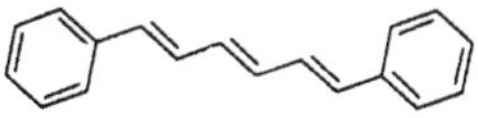	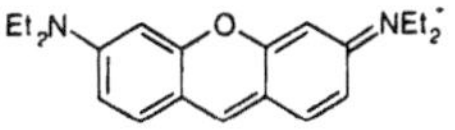	PyB^{+}
PBOX	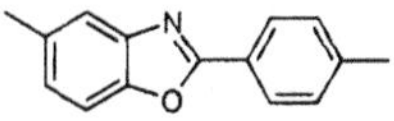	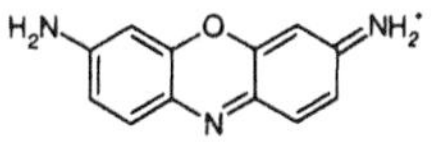	Ox^{+}
MBOXE	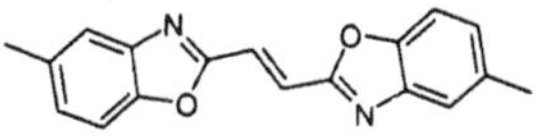	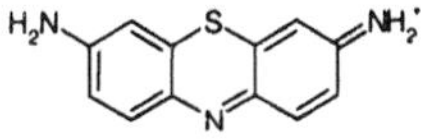	Th^{+}
POPOP	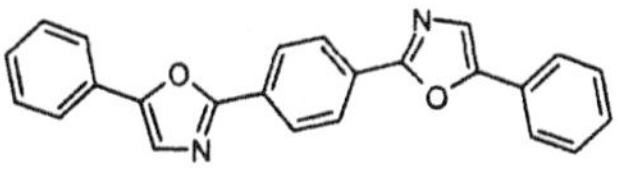	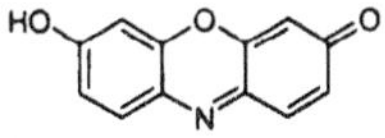	ResH
DMPOPOP	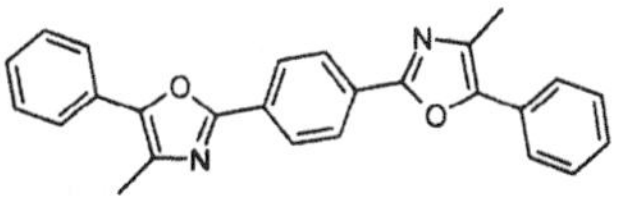	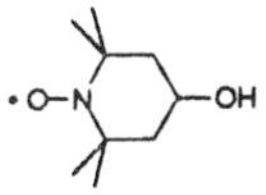	Hydroxy-TEMPO
DSC	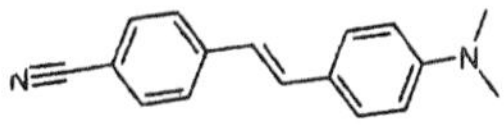	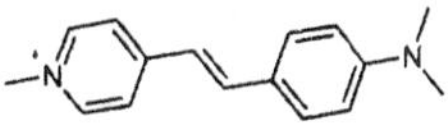	DMSI^{+}
Fluorenone	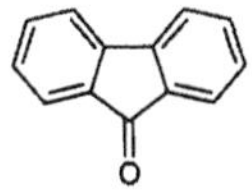	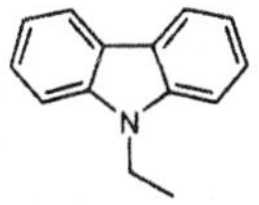	N-Ethylcarbazole[a]
Naphtalene[a]	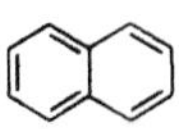	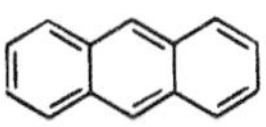	Anthracene[a]
MV^{2+}	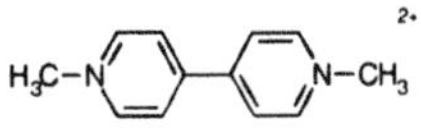		

[a] From Ref. 14.

which only interact with the zeolite framework, including the cations and small molecules like water present in the channels, but that they do not interact with themselves for geometrical reasons. We rely on a recently published study of noninteracting particles in microporous materials [16]. Some consequences of results reported there are discussed with respect to the following three cases:

> *Solid–gas equilibrium*: Dye molecules D in the gas phase (g) are in equilibrium with dye molecules in the channels of the zeolite Z.

$$ZD_{r-1} + D(g) \rightleftarrows ZD_r \qquad\qquad [1]$$

> The parameter r counts the number of sites occupied by dye molecules. Its values range from 0 to n_{box}. n_{box} is equal to the number of sites in one channel. In case of a 300-nm-long zeolite and a 1.5-nm-long dye which occupies two unit cells, n_{box} is equal to 200.
>
> *Displacement equilibrium*: Neutral dye molecules D in the zeolite ZD_rX_{p-x} can be displaced by x molecules of X. The states of X(out) and D(out) have to be specified.

$$ZD_{r-1} X_p + D(\text{out}) \rightleftarrows ZD_r X_{p-x} + xX(\text{out}) \qquad\qquad [2]$$

> *Ion-exchange equilibrium*: In most experiments described here, monovalent cationic dyes have been used. D_s^+ and M_s^+ denote the dye cation and the alkali metal cation in solution. Z stands for zeolite and Y describes the cation concentration inside the zeolite. For monovalent cations and dyes which occupy two unit cells in zeolite L (e.g., Py$^+$ or Ox$^+$), we must use $Y_{n_{box}-r} = [(M_{18}^+)_{n_{box}-r} (M_{17}^+)_r]$ to describe the state of a given channel. An empty site contains 18 cations M$^+$. Only one of them can be exchanged by a singly charged dye D$^+$ cation. By the exchange of rD^+ molecules, the number of sites containing 18 cations is reduced by r, and r sites containing only 17 alkali cations are formed:

$$ZY_{n_{box}-(r-1)} D_{r-1} + D_S^+ \rightleftarrows ZY_{n_{box}-r} D_r + M_S^+ \qquad\qquad [3]$$

Using the abbreviation SG for solid–gas, DI for displacement, and IE for ion exchange, the equilibrium constants for these three cases can be expressed as follows:

$$K_r^{SG} = \frac{[ZD_r]}{[ZD_{r-1}][D(g)]} \qquad\qquad [4]$$

$$K_r^{DI} = \frac{[ZD_r X_{p-x}]}{[ZD_{r-1} X_p][D(\text{out})]} [X(\text{out})]^x \qquad\qquad [5]$$

$$K_r^{IE} = \frac{[Z_{n_{box}-r} D_r]}{[ZY_{n_{box}-(r-1)} D_{r-1}][D_S^+]} [M_S^+] \qquad\qquad [6]$$

These equations show that the equilibrium constants depend on r. All three cases correspond to the situation expressed in Eqs. (26)–(29) of our study on particle distribution in microporous materials [16]. This means that the r dependence of the equilibrium constants K_r^{SG}, K_r^{DI}, and K_r^{IF} can be described by the same formula:

$$K_{r-1} = K_r\left(\frac{r}{r+1}\frac{n_{box}-r}{n_{box}-r+1}\right) \qquad [7]$$

It is sufficient to know, for example, K_1 from which all other K_r can be calculated. The decrease of K_r with increasing r is due to the fact that the entropy of the system decreases with increasing loading. It is most pronounced for very low and for very high loading. This is illustrated in Fig. 3, where the dependence of the equilibrium constant K_r as a function of the occupation probability is shown for $K_1 = 7.75 \times 10^8$. From this, it follows that the dye insertion is complete for low loading but that the situation changes for higher loadings. This fact must be taken into account when doing experiments of the type described in this chapter.

For a better understanding of the consequences of this relation, it is useful to discuss an example more explicitly. We choose the ion-exchange equilibrium [Eq. (3)]. The total concentration of dye molecules inside an ensemble of zeolite nanocrystals dispersed in a solvent $[D_Z]_{tot}$, expressed with respect to the total volume under consideration, is

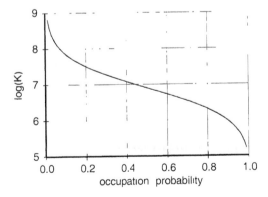

Figure 3 Dependence of the equilibrium constant K for insertion of a dye in zeolite L as a function of the occupation probability p_r, calculated for $K_1 = 7.75 \times 10^8$.

$$[D_Z]_{tot} = \sum_{r=1}^{n_{bo\lambda}} r\,[ZY_{n_{bo\lambda}-r}\,D_r] \qquad [8]$$

The total number of channels A_0, expressed in terms of the total number of zeolite L unit cells n_{uc} is given by

$$A_0 = \frac{n_{uc}}{n_S\,n_{box}} \qquad [9]$$

where n, is the number of unit cells required by one dye molecule (n, $= 2$ for, e.g., Py^+ or Ox^+). Using Eq. (16) of Ref. 16, the individual concentrations $[ZY_{n_{bo\lambda}-r}D_r]$ can be expressed as

$$[ZY_{n_{bo\lambda}-r}\,D_r] = \frac{([D_S^+][M_S^+])^r\,\prod_{j=0}^r K_j}{\sum_{t=0}^{n_{box}}\left\{([D_S^+][M_S^+])^t\,\prod_{j=0}^t K_j\right\}}\,A_0 \qquad [10]$$

In this equation, K_0 is equal to 1, by definition. Using the abbreviation

$$f_r = \begin{cases} 1, & r = 0 \\ \dfrac{r}{r+1}\cdot\dfrac{n_{bo\lambda}-r}{n_{bo\lambda}-r+1}, & r = 1, 2, ..., n_{\,bo\lambda} \end{cases} \qquad [11]$$

and therefore $K_{r+1} = K_r f$, if $r \geq 1$, we find after some rearrangement for $r = 1, 2, \ldots, t_{box}$ that

$$[ZY_{n_{bo\lambda}-r}\,D_r] = \frac{([D_S^+][M_S^+]\,K_1)^r\,\prod_{j=0}^{r-1} f_j^{r-j}}{\sum_{t=0}^{n_{box}}\left\{([D_S^+][M_S^+]K_1)^t\,\prod_{j=0}^t f_j^{t-j}\right\}}\,A_0 \qquad [12]$$

By using this equation, it is possible to calculate the distribution of channels containing a certain amount of dye molecules as a function of the dye concentration in the solvent (see Fig. 4). Equation (12) can be inserted into Eq. (8), which leads to the following expression:

$$[D_Z]_{tot} = \frac{A_0}{1 + \sum_{t=1}^{n_{box}}\left\{([D_S^+][M_S^+]K_1)^t\,\prod_{j=0}^t f_j^{t-j}\right\}}$$

$$\sum_{r=1}^{n_{box}} r([D_S^+][M_S^+]\,K_1)^r\,\prod_{j=0}^{r-1} f_j^{r-j} \qquad [13]$$

This equation can be used to determine the equilibrium constant K_1 because the dye concentration in solution $[D_S^+]$, the cation concentration in solution

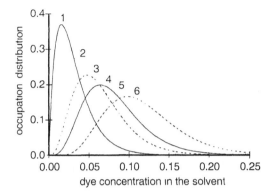

Figure 4 Distribution of channels containing one, two, three, four, five, or six dye molecules in equilibrium as a function of the dye concentration in the solvent, calculated by means of Eq. (12) for $K_1 = 7.75 \times 10^8$, $[M_s^+] = 1.9 \times 10^{-3} M$, $n_{box} = 100$, and $A_0 = 4.4 \times 10^{-7} M$. The free dye concentration is expressed in units of the total number of available sites $\frac{1}{2}$uc.

$[M_s^+]$, the total number of channels available A_0, the number of sites per channel n_{box}, and the total dye concentration $[D_z]_{tot}$ can be measured. A numerical solution of Eq. (13) is easy to obtain.

Figure 5 shows the total concentration of dye molecules in the channels of zeolite L $[D_z]_{tot}$ expressed as occupation probability p versus the dye concentration in solution in units of the total number of available sites, $\frac{1}{2}$uc. From the results illustrated, it follows that it is easy to prepare materials with low loading, but that sophisticated techniques are needed for high loading.

Only few data are available for this kind of analysis. In addition to the above-presented data, exchange isotherms have been measured for Th$^+$ and for MV^{2+} –zeolite L[17]. An experimental result for potassium zeolite L suspended in water and exchanged at room temperature with MV^{2+} is shown in Fig. 6. For details, see Ref. 18.

For dyes which only reluctantly enter the zeolite channels or which are insufficiently soluble or stable in aqueous solution, the exchange equilibrium can be influenced by means of an ionophore:

$$K^+ + \text{crypt} \rightleftarrows K\text{crypt}^+ \qquad [14]$$

For potassium zeolites, cryptofix 222 and cryptofix 222BB, for example, can be used. The structures together with the stability constants K, of the complexes (cryptates) of cryptofix 222 and cryptofix 222BB with potassium are shown in

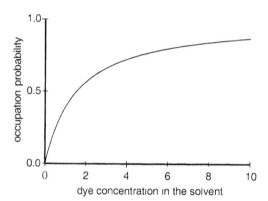

Figure 5 Total concentration $[D_z]_{tot}$ of dyes in the channels expressed as occupation probability p versus the concentration of free dyes in solution expressed in units of the total number of available sites ½uc. calculated with the same parameters as used in Fig. 4.

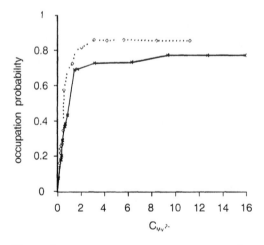

Figure 6 Occupation probability p of the MV^{2+}-loaded zeolite L at room temperature as a function of C_{MV2+}, which is the number of moles of MV^{2+} per mole of zeolite L unit cells present in the aqueous suspension. Experiments with self-synthesized zeolite L (dotted curve) and with commercial zeolite (Linde-type ELZ-L, Union Carbide) (solid curve). (From Ref. 18.)

Table 2 Structures of Potassium Selective Cryptands Cryptofix 222 and Cryptofix 222BB and the Stability Constants of the Matching Cryptates

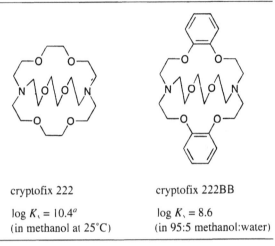

cryptofix 222

$\log K_s = 10.4^a$
(in methanol at 25°C)

cryptofix 222BB

$\log K_s = 8.6$
(in 95:5 methanol:water)

[a] From Ref. 19.
[b] From Ref. 20.

Table 2. For example, these cryptands allow to insert $PyGY^+$ and PyB^- from an *n*-butanol solution, in which, otherwise, the potassium ions would not be sufficiently soluble.

III. ENERGY MIGRATION

The two cationic dyes, Py^+ as a donor and Ox^+ as an acceptor, were found to be very versatile for demonstrating photonic antenna functionalities for light harvesting, transport, and capturing, as illustrated in Fig. 7. They can be incorporated into zeolite L by means of ion exchange, where they are present as monomers because of the restricted space. In this form they have a high fluorescence quantum yield and favourable spectral properties. The insertion of the dyes can be visualised by means of fluorescence microscopy. The fluorescence anisotropy of Ox^+-loaded zeolite L has recently been investigated in detail by conventional and by confocal microscopy techniques [15].

The occupation probability p of the sites with a dye is equal to the number of occupied sites divided by the number of sites available. Using this, the Förster-

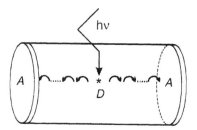

Figure 7 Simplified view of a bidirectional photonic antenna. The middle part shown in light gray contains donor molecules D. After excitation of D to D*, the excitation energy migrates with equal probability to the left and to the right until it reaches an acceptor A (dark gray) which captures the excitation energy and emits it as red-shifted light.

type energy transfer [21,22] and the energy migration rate constant k_{ij} from a molecule i to a molecule j in this material can be expressed by the following [2]:

$$k_{ij} = \frac{9(\ln 10)}{128\pi^5 \, N_A \, n^4} \frac{\phi_i}{\tau_i} G_{ij} J_{ij} p_i p_j \qquad [15]$$

where ϕ_i and τ_i (sec^{-1}) are the fluorescence quantum yield and the intrinsic fluorescence lifetime of the donor, N_A is the Avogadro number, n is the refractive index of the medium, G_{ij} (Å^{-6}) represents the geometrical constraints of the sites in the crystal and the relative orientation of the transition moments, p_i and p_j are the occupation probabilities of the sites with excited donors i and acceptors j in the ground state. The spectral overlap J_{ij} (cm^3/M) is equal to the integral of the corrected and normalized fluorescence intensity $f_i(\tilde{\lambda})$ of the donor multiplied by the extinction coefficient $\varepsilon_j(\tilde{\lambda})$ of the acceptor as a function of the wave number $\tilde{\lambda}$:

$$J_{ij} = \int_0^\infty \varepsilon_j(\tilde{\lambda}) f_i(\tilde{\lambda}) \frac{d\tilde{\lambda}}{\tilde{\lambda}^4} \qquad [16]$$

We have investigated the temperature dependence of the spectral overlap of the following donor/acceptor pairs in the channels of zeolite L: Py$^+$/Py$^+$, Ox$^+$/Ox$^+$, and Py$^+$/Ox$^+$. Dye-loaded zeolite L layers were prepared on circular quartz plates (16 mm in diameter) by depositing a calculated volume of an aqueous suspension of dye-loaded zeolite L (occupation probability $p = 0.01$) and drying overnight. The resulting layers were of about 3000 nm average thickness. Figures

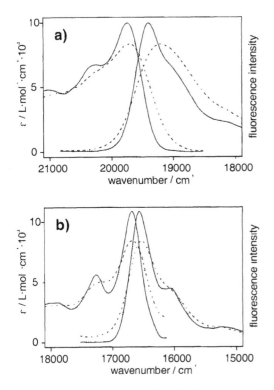

Figure 8 Fluorescence and excitation spectra of Py$^-$–zeolite L (a) and Ox$^-$–zeolite L (b) at 80 K (solid curve), 193 K (dotted curve), and 293 K (dashed curve). The fluorescence spectra have been scaled to the same height as the corresponding excitation spectra.

8 and 9 show the excitation and fluorescence spectra for the investigated donor/acceptor pairs at three different temperatures. For a specific dye molecule, the maximum of the excitation spectrum measured at room temperature was set equal to the extinction coefficient at the absorption maximum in aqueous solution. The integrals of the excitation spectra were then normalized to the integral of the corresponding spectrum at room temperature. This is reasonable because the oscillator strength f of a transition $n \leftarrow m$ does not depend on the temperature [3,23].

The spectral overlap of the investigated donor/acceptor pairs does not change significantly in the temperature range from 80 K to 300 K (see Fig. 10). The large difference between the absolute values of the overlap integrals of Ox$^+$/Ox$^+$ and Py$^+$/Py$^+$ is due to a different Stokes shift (140 cm^{-1} for Ox$^+$, 560 cm^{-1} for Py$^+$). Note that there is a nice mirror symmetry between excitation

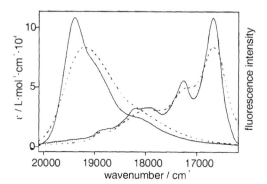

Figure 9 Fluorescence spectra of Py⁺–zeolite L (left) and excitation spectra of Ox⁺–zeolite L (right) at 80 K (solid curve), 193 K (dotted curve), and 293 K (dashed curve). The fluorescence spectra have been scaled to the same height as the corresponding excitation spectra.

and fluorescence spectra for these two cases. This is not given in the case of the Py⁺/Ox⁺ overlap integral, which concerns the opposite side of the spectra (low-energy side of the fluorescence spectra and high-energy side of the excitation spectrum).

   The results shown so far suggest that the spectral overlap between the absorption and fluorescence spectrum of a molecule does not change significantly

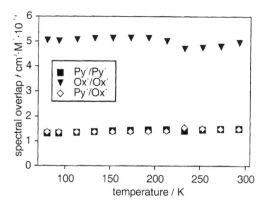

Figure 10 Temperature dependence of the spectral overlap of the investigated donor/acceptor pairs in the channels of zeolite L.

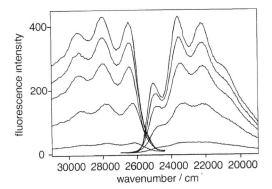

Figure 11 Fluorescence and excitation spectra of DPH-loaded zeolite L at 100, 180, 200, 240, and 293 K (top to bottom). The fluorescence spectra have been scaled to the same height as the corresponding excitation spectra. The excitation spectra have not been normalized to the same integral in order to demonstrate the decreasing fluorescence intensity with increasing temperatures.

with temperature. However, there are cases where this is not true. For DPH in zeolite L ($p = 0.5$), we found an increasing spectral overlap with increasing temperature. In Fig. 11, we chose a different way of presenting the spectra, because the fluorescence quantum yield of DPH strongly depends on the temperature. The excitation spectra were not normalized to the same integral for graphical presentation. Figure 12 compares the values of the spectral overlap integral at different temperatures.

We describe properties of Py$^+$-loaded zeolite L of which the channel ends are modified with Ox$^+$. Ox$^+$ acts as an acceptor which becomes excited via radiationless energy transfer from an excited Py$^+$ moiety. If radiationless relaxation is not considered, Ox$^+$ can lose its excitation energy only by luminescence, as it cannot transfer the energy back to Py$^+$ because of its lower energy. Fluorescence of excited Py$^+$, internal conversion, and intersystem crossing compete with the energy migration and energy-transfer processes. The efficiency of energy migration among the Py$^+$ molecules along the crystal can be investigated by measuring the front–back trapping efficiency $T_{FB,\infty}$. The front–back trapping efficiency is equal to the ratio of fluorescence intensity emitted by the acceptor Ox$^+$ (I_A) divided by the total fluorescence intensity ($I_A + I_D$) [2,24]:

$$T_{FB,\infty} = \frac{I_A}{I_D + I_A} \qquad [17]$$

Two important parameters that we can vary independently in order to test the energy migration efficiency are the loading with donor molecules Py$^+$, which

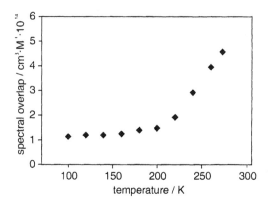

Figure 12 Temperature dependence of the spectral overlap of DPH-loaded zeolite L. Note that the values are 10 times smaller compared with the overlap integrals in Fig. 10.

we express as p_{py}, and the length l_{cyl} of the crystals. The front–back trapping efficiency $T_{FB,\infty}$ can be tested by exciting the crystals at a wavelength where the acceptor absorption is negligible. We expect that $T_{FB,\infty}$ decreases with increasing crystal length l_{cyl} for otherwise constant parameters, specifically the constant p_{py}, because the excitation energy has to migrate over an increasingly large distance to reach an acceptor. This can be tested by using materials with different average crystal lengths.

We show in Fig. 13a experimental results obtained with crystals of the following average length: (1) 300 nm; (2) 500 nm; (3) 850 nm; (4) 1400 nm; (5) 2400 nm. These crystals were loaded with Py$^+$ in such a way that the occupation probability was always the same, namely $p_{py} = 0.11$. They were then modified with two Ox$^+$ molecules on average at both ends of each channel. The fluorescence of a thin layer on quartz was measured at room temperature after specific excitation of Py$^+$ at 460 nm. It shows a strong increase of the Ox$^+$ emission with decreasing crystal length l_{cyl}. The front–back trapping efficiency increases from 0.33 up to 0.91. This means that in the 300-nm crystals, 91% of the emitted light is due to energy migration among the Py$^+$ molecules and, finally, transfer to the luminescent acceptor Ox$^+$. Experiments with crystals in the size range of 50 nm show a similar behavior.

We also prepared a wide range of Py$^+$ loading ranging from 0.007 to 0.182 for zeolite crystals with a length of approximately 650 nm. The crystals were modified with Ox$^+$ acceptors as before. Figure 13b shows the emission spectra upon excitation at 470 nm. The spectra are scaled to the same height at the Py$^+$ emission maximum, which allows better comparison. The Ox$^+$ emission has its

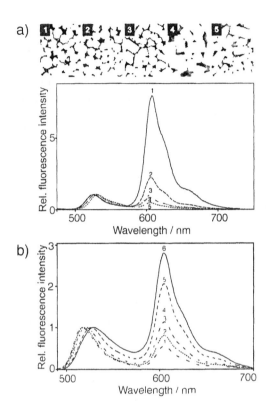

Figure 13 (a) Electron microscopy pictures of zeolite L samples with different mean crystal length l_{cyl}: (1) 300 nm; (2) 500 nm; (3) 850 nm; (4) 1400 nm; (5) 2400 nm (the bars in *1–4* correspond to 1 μm, in 5 to 3 μm). Fluorescence spectra of Py⁺-loaded and Ox⁺-modified zeolite L crystals with constant Py⁺ loading ($p_{py} = 0.11$) as a function of crystal length after specific excitation of only Py⁻ at 460 nm. The Ox⁺ modification was, on average, two molecules at both ends of each channel. (From Ref. 5.) (b) Fluorescence intensity after specific excitation of only Py⁻ at 470 nm of Py⁺-loaded and Ox⁺-modified zeolite L crystals (650 nm length and 600 nm diameter) as a function of Py⁺-loading p_{py}: (1) 0.007, (2) 0.012, (3) 0.031, (4) 0.058, (5) 0.106, (6) 0.182. The Ox⁺ modification was, on average one molecule at both ends of the channel. All spectra are scaled to the same height at the maximum of the Py⁺ emission. (From Ref. 5.)

maximum at 605 nm and its intensity increases strongly with increasing p_{py} from samples 1–6. In fact, the trapping efficiency increases from 0.36 up to 0.74.

Some shift of the Py$^+$ maximum is observed in both cases of Fig. 13. This wavelength shift may be due to self-absorption and re-emission. In the case of Fig. 13a, the absorption depth increases with increasing crystal size despite of the constant p_{py}. In Fig. 13b, self-absorption and re-emission increases with increasing Py$^+$ loading, which is easily comprehensible. This phenomenon has been discussed more quantitatively in Ref. 3.

The time evolution of the excitation distribution along the cylinder axis of this material is of great interest [1,2,6]. We report calculated results based on the theory described in Ref. 2, obtained for a crystal consisting of 90 slabs, each of which corresponds to the thickness of a site. The photophysical data of Py$^+$ as donor and Ox$^+$ as acceptor were used to calculate the excitation distribution of the donors. Immediately after excitation, all slabs have the same excitation probability. The excitation distribution is shown after 5, 10, 50, and 100 psec. We observe in Fig. 14a that the slabs close to the acceptor layers lose their excitation energy very quickly. The trapping rate is proportional to the gradient of the excitation distribution at the position of the acceptors. Hence, it depends not only on the remaining excitation probability but also on the excitation distribution. This is in contrast to the donor fluorescence rate, which depends only on the excitation distribution of the donors. The fluorescence rate of the acceptors is proportional to their excitation probability. We, therefore, expect a fluorescence decay behavior, as illustrated in Fig. 14b. The fluorescence decay of the donors becomes much faster in the presence of acceptors, because of the depopulation due to the irreversible energy transfer to the acceptors. The fluorescence intensity stemming from the acceptors initially increases because excited states must first be populated via energy transfer from the donors. It therefore reaches a maximum before it decays. The experimental decay curves in Fig. 14c for Py$^+$ donors and Ox$^+$ acceptors measured by frequency domain time-resolved spectroscopy show the predicted behavior. The rise of the acceptor fluorescence intensity due to the "pumping" of the donors is impressive. Investigations on different donor loadings show that it depends on this parameter, as expected.

A sophisticated bidirectional antenna material with three dyes is illustrated in Fig. 6 of Ref. 25. After selective excitation of dye1, located in the middle part, the light energy is carried spectrally from the blue to green (dye2) to red (dye3) and spatially from the crystal center to its left and right ends. The stacking of the dyes in the crystal can be seen in microscopy images taken on relatively large crystals. We have shown this for a 2000 nm crystal with POPOP in the middle, followed by Py$^+$ and then by Ox$^+$. The different color regions which can be observed are impressive. After selective excitation of the POPOP, the middle of the crystal fluoresces shows the blue fluorescence of POPOP and the red emission of Ox$^+$ appears at both ends. Between these two zones, the superposition of the

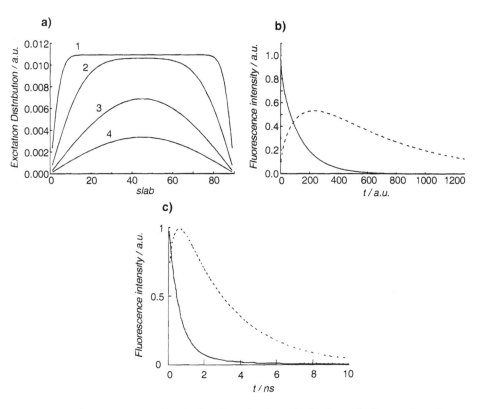

Figure 14 (a) Excitation distribution along the channel axis of a zeolite L crystal consist-ing of 90 slabs (occupation probability $p = 0.3$) under the condition of equal excitation probability at $t = 0$ calculated for front–back trapping. Fluorescence of the donors is taken into account. (*1*) $t = 5$ psec. (*2*) $t = 10$ psec. (*3*) $t = 50$ psec. and (*4*) $t = 100$ psec after irradiation. (b) Predicted fluorescence decay of the donors in absence of acceptors (dotted curve), in the presence of acceptors at both ends (solid curve), and fluorescence decay of the acceptors (dashed curve). (c) Measured fluorescence decay of Py$^+$-loaded zeolite L ($p_{py} = 0.08$) (dotted curve), Py$^+$-loaded zeolite L ($p_{py} = 0.08$) with, on average, one Ox$^+$ acceptor at both ends of each channel (solid curve), and fluorescence decay of the Ox$^+$ acceptors (dashed curve), scaled to 1 at the maximum intensity. The experiments were conducted on solid samples of a monolayer of zeolite L crystals with a length of 750 nm on a quartz plate.

fluorescence of all three dyes results in emission of white light. The microscopy images show very nicely the sequence of the inserted dyes in the channels.

After this overview of photonic antenna properties, we now turn to some intriguing optical properties we have recently observed in different dye–zeolite guest–host composites.

IV. OPTICAL PROPERTIES

We describe observations made on dye-loaded zeolite L crystals of dimensions which are about equal or a few times larger than the wavelength of visible light. The refractive index of zeolites is in the same range as that of quartz. It is expected to vary to some extent depending on the cations, the water content, and the dye loading. Thus optical effects observed in tiny glass fibers are expected to appear in zeolite L crystals as well. The luminescent dyes inserted into the channels cause some new phenomena, also because of the pronounced anisotropy of their light absorption and emission properties and because the refractive index changes in regions of strong absorption. The orientation of the transition moments of Ox^+ in zeolite L is 72° with respect to the c axis, whereas that of POPOP, DMPOPOP, and similar molecules is parallel to this axis [1,15,25].

A. *Fata Morgana* Effects in Dye-Loaded Zeolite L Crystals

Optical effects due to refraction and total internal reflection have been observed in dye-loaded zeolite L crystals of 2.5 μm length and 1.4 μm diameter by means of an optical microscope equipped with polarizers, a narrow band, and cutoff filters [23]. An astonishing effect taking place in an Ox^+-loaded crystal is seen in Fig. 15. Looking at the polarized red emission, a homogeneous intensity distri-

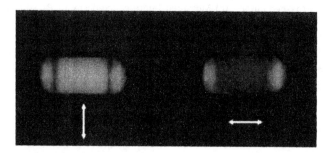

Figure 15   Polarized fluorescence microscopy images of a 2.5-μm-long Ox^+-loaded zeolite L crystal after excitation at 545–580 nm (cutoff 605 nm). The arrows indicate the transmission direction of the emission polarizer.

bution is observed over the whole crystal, with the exception of two perpendicular dark lines which separate the luminescence at both ends. The fluorescence in the middle part of the crystal decreases when turning the polarizer by 90°, but the two wings at both sides appear with about the same intensity as earlier. Obviously, the light observed at both ends of the crystal is not polarized, whereas that observed in the middle part is strongly polarized. Ox$^+$ molecules located at the outer surface of the crystals can be quantitatively destroyed by treating the material with a hypochlorite solution [3,5]. Such a treatment does not alter the optical properties of the material. The effect (dark lines and wings) disappears, however, upon refractive index matching (e.g., when applying an immersion oil). This means that the optical phenomenon is not due to some molecules present at the outer surface. We have to seek another explanation.

The luminous nonpolarized wings are due to part of the emission built up in the middle of the crystal. They can be understood when taking total internal reflection into account. Refraction of the emission from the object O at the zeolite L–air interface observed at an angle α_2 is explained in Fig. 16a. The emission can appear between 0° and the critical angle for total reflection $\alpha_{1,max}$, which is 42° for a refractive index of 1.49. The used objective of the microscope collects

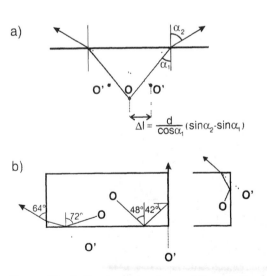

Figure 16 Refraction and pathway of the emission from the object O hitting the zeolite L–air interface. (a) Refraction of the emission from the object O at the zeolite L–air interface observed at an angle α_2. The emission appears in the range between 0° and the critical angle for total reflection $\alpha_{1,max}$ which is 42° (d is the distance between O and the interface). (b) Pathway of the emission of a molecule O.

all light that is emitted under an angle $<64°$ and, therefore, the object appears at the positions O' in a circle with radius Δl around the object O.

An incident photon hitting the wall at an angle α_1 larger than 42° is totally reflected and can only leave the cylinder at its ends (see Fig. 16b). In the case shown on the right side of Fig. 16b, it appears as luminous nonpolarized wings. Hence, the emission observed in an optical microscope appears not at the origin of the molecule but outside of the crystal at both ends. This means that we "see" the emitting molecules at another place in space than they are, similar to a mirage, which shows an image that may be hundreds of kilometers away but can appear to be closer than it is. This phenomenon is called *fata morgana*. We are not sure if we understand the second part of the observation, namely the two dark regions which separate the polarized emission of the bulk from the nonpolarized wings. It is perhaps due to interference phenomena.

Another nice *fata morgana* can be observed in DMPOPOP–zeolite L microcrystals, the luminescence of which is strongly polarized along the c axis. Looking at standing crystals, a nonpolarized blue ring with a dark center is observed; see Fig. 17a. The dark part disappears immediately, the whole luminescent spot shrinks, and the luminescence intensity decreases when adding a drop of a solvent with the same refractive index as the zeolite. This shows that the blue ring is caused by refraction at the crystal–air interface.

Two kinds of refraction are responsible for the appearance of the luminescent ring with a dark spot in the middle. Emission at O, which hits the wall at an angle between 17° and 42°, as shown in Fig. 18a, is refracted but can still be detected by the microscope. It appears to originate further away from the center

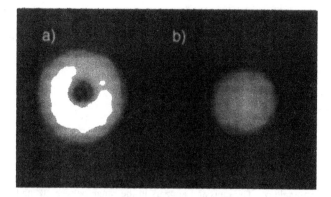

Figure 17 Fluorescence microscopy images of a standing 2.5-μm-long DMPOPOP-loaded zeolite L crystal upon excitation at 330–385 nm and observed with a cutoff filter (410 nm); (a) in air and (b) in a refractive-index-matching solvent.

a)   b)

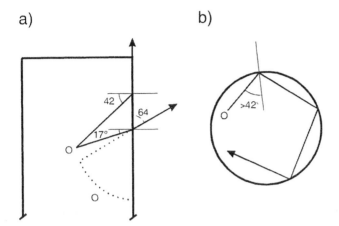

Figure 18 Refraction and pathway of the emission from the object O hitting the zeolite L–air interface. (a) Emission at O which hits the wall at an angle between 17° to 42° can be detected by the microscope. Its refraction makes it appear to originate from the region O′ (i.e., closer to the wall). (b) Emission at O which hits the wall at an angle greater than 42° is totally reflected and can travel around the crystal until it reaches one of the two ends.

than it is in reality. As a consequence, the center appears darker. Total reflection as shown in Fig. 18b causes a photon to travel on a helical pathway until it reaches the top or the bottom of the cylinder. If the photon encounters another excited molecule on its way, stimulated emission can occur. The information from where the emission originates is nearly totally lost. An angle of total reflection can only be realized by excited molecules which are near the side walls. This again makes the center appear darker. The phenomenon seems to reveal the characteristics of a very tiny ring resonator which perhaps resembles the ring resonator reported recently for much larger crystals [26].

B. Polarization of the Fluorescence

Dye molecules of interest for energy transport in our system have an oblong form and a strong $\pi–\pi^*$ transition with a transition moment parallel to the molecules' long axis. The absorption and emission of light from these molecules is therefore strongly polarized. In an ensemble of many molecules, this polarization can only be observed when the molecules are ordered. Their width allows them to penetrate the one-dimensional channels of zeolite L. The geometrical constraints lead to an anisotropic organisation of the dyes and result in a net polarization anisotropy.

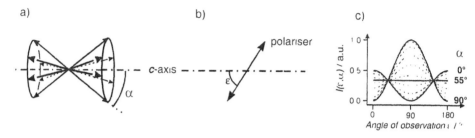

Figure 19 (a) Distribution of the transition moments on a double cone with a half-opening angle α. (b) Polarization direction observed, when a single crystal is examined by means of a polarizer. (c) Relative intensity of the observed fluorescence as a function of the observation angle ε with respect to the crystal c axis, for different half-cone angles α.

How exactly the molecules are oriented inside the channels depends on their specific shape and on the adsorption interaction between the dyes and the channel walls or charge compensating cations. Because of the dye's oblongness, a double-cone-like distribution in the channels is a reasonable model. This distribution is illustrated in Fig. 19a. The arrows represent the transition moments of the dyes and α describes the half-opening angle of the double cone. The hexagonal structure of the zeolite L crystal hence allows six equivalent positions of the transition moments on this double cone with respect to the channel axis.

Figure 19b shows how the fluorescence polarization of a single zeolite L crystal can be analyzed by means of a polarizer, which only transmits the indicated polarization direction. In Fig. 19c, the theoretical relative intensity of the observed fluorescence is plotted versus the angle ε between the observed polarisation and the crystal c axis for different values of α. If α is equal to 0°, the molecules' transition moment coincides with the crystal c axis and maximum fluorescence can be observed at ε equal to 0° or 180°. Such a crystal emits parallel to the crystal axis and remains dark if the polarizer is set perpendicular to it. If α is equal to 90°, then the maximum fluorescence is observed perpendicular to the c axis and the crystal emits no light parallel to it. For all α values between 0° and 90°, there is a gradual change in behavior from one extreme to the other. The difference between the maximum and minimum fluorescence intensity is reduced, and at the magic angle of 54.7°, no fluorescence anisotropy can be observed although the transition moments are not randomly oriented in the crystal.

In the case of Ox^+ in zeolite L, a half-cone angle α of 72° was obtained from quantitative measurements on single crystals [15]. The orientation of the transition moments with respect to the zeolite channels can be determined directly form the fluorescence polarization, whereas the orientation of the transition moments with respect to the molecular axis cannot be determined from these experiments. Strong electric fields can change the orientation of the dipole moments (i.e. the "Stark effect") [27]. Geometrical estimates of the maximum angle of the double cone in the case of Ox $^+$ led to the conclusion that the angle α for the molecules cannot be larger than 40°. The observation of an $\alpha = 72°$ for the transition moments was interpreted by the existence of a remarkable Stark effect in these materials [15]. The arguments do not apply for molecules oriented along the channel axis.

In Fig. 12 in Ref. 25, fluorescence microscopy images of different dye-loaded zeolite L single crystals are shown. Each line consists of three pictures of the same sample, but with different polarizations of the fluorescence observed. In the first one, the total fluorescence of the crystals is shown, and in the others, the fluorescence with the polarization direction indicated by the arrows is displayed. The zeolite was loaded with the following dyes: (A) Py $^+$, (B) PyGY $^+$, (C) PyB $^+$, (D) POPOP (see Table 1). Most crystals show a typical sandwich structure with fluorescent dyes at the crystal ends and a dark zone in the middle. This situation can be observed when the diffusion of the dyes in the channels has not yet reached its equilibrium situation. It illustrates nicely how the molecules penetrate the crystals via the two openings on each side of the one-dimensional channels.

Py $^+$ is a molecule with similar dimensions as Ox $^+$ and, therefore, the behavior of the fluorescence polarization is the same as already described for the Ox $^+$–zeolite L sample: maximum fluorescence perpendicular to the crystal axis and minimum fluorescence parallel to it. PyGY $^+$ is slightly larger than Py $^+$, because of the methyl groups. The observed fluorescence polarization in this sample changes significantly compared to the Py $^+$. The maximum fluorescence intensity is now measured parallel to the c axis and the difference between maximum and minimum intensity is rather small. According to Fig. 19c, this is an indication that the transition moments have changed to an angle α below 55°. In the case of PyB $^+$ (with four ethyl groups), the intensity difference between minimum and maximum is even more pronounced, indicating an angle α near 0°. Finally, the fluorescence of POPOP-zeolite L crystals is analyzed. POPOP has an estimated molecular length of 19 Å and therefore occupies approximately 3 ucs. The fluorescence polarisation shows very nicely that the transition moments of the POPOP molecules are arranged parallel to the c axis, which means that α is equal to 0°.

V. FURTHER DEVELOPMENTS

In natural photosynthesis, complex arrays of antennas collect the solar energy and convert it into chemical potential energy that drives the chemistry of the photosynthetic machinery. The processes involved are very fast and highly efficient [28–30].

The dye–zeolite composites reported so far show fascinating photonic antenna properties which are perhaps comparable to some extent to those of natural systems. Tuning their chemical and photochemical behavior, organizing information exchange between their inside and the external world, but also organizing individual crystals on a surface in order to realize, for example, monodirectional functionalities remain a challenge which we address in this section.

A. Closure and Stopcock Molecules

The synthesis principle we are using is based on the fact that molecules can diffuse into individual channels [3,4,31]. This means that, under appropriate conditions, they can also leave the zeolite by the same manner. In some cases, however, it is desirable to block their way out, so as to stabilize the structure. This can be done by adding a closure molecule after the synthesis has been completed. A variety of "closure" molecules which seal the channels completely or only partially can be used, depending on the requirement. An example already used is the addition of fluorenone which enters readily but leaves the structure reluctantly. If more complete sealing is needed, molecules bearing appropriate reactive groups can be used. Functionalization of the closure molecules is an option for tuning wettability, refractive index, chemical reactivity, and other properties.

External trapping and injection of quanta is more demanding. The general approach we are using to solve this problem is to add "stopcock" molecules, as illustrated in Fig. 20 [1]. A stopcock molecule generally consists of three components: a head, a spacer, and a label. The tail moiety (spacer + label) has a longitudinal extension of more than one u.c. along the c-axis. The head moiety has a lateral extension that is larger than the channel width and prevents the head from penetrating into the channel. The channels are therefore terminated in generally plug like manner.

The main function of the stopcocks is to connect the antenna function of the crystal to its surroundings. They act as bridges between the dye molecules inside the channels and the outside world by either trapping excitation energy coming from the inside or injecting excitation energy into the dye-loaded zeolite crystal. Therefore, stopcock traps have a sufficiently large spectral overlap integral with donor molecules inside the channels, whereas stopcock donors have a large spectral overlap integral with acceptor molecules inside the channels.

Strongly fluorescent molecules are suitable for these energy-transport functions. Because fluorescence is usually quenched by dimerisation [32,33], the design must be such that the chromophores do not electronically interact with each other. Both heads and labels can be fluorescent, depending on the needs. Fluorescent labels have the advantage of being protected by the zeolite frame-

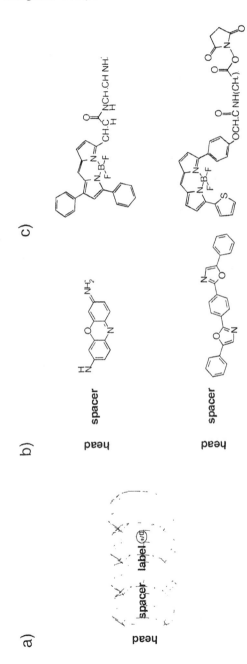

Figure 20 (a) Typical shape of a stopcock molecule located at the end of a zeolite channel. (b) Fluorescent molecules which have already been inserted in zeolite L are modified with an inert head in order to build stopcock molecules with fluorescent tails. (c) Examples of molecules which can be used as stopcocks with a fluorescent head.

work. They are stabilized chemically and cannot form quenching dimers due to lack of space. Fluorescent heads, on the other hand, are closer to the external surface, which is desirable for coupling the antenna to a device.

The labels are based on organic and silicon–organic backbones. The simplest form of a label is nonreactive. Its only function is to enter and to hook to the channel. Nonreactive labels can be both neutral or cationic. Reactive labels enter the zeolite channels and then undergo a chemical reaction under the influence of irradiation, heat, or a sufficiently small reactant. In this way, they hook themselves inside the zeolite channels or, perhaps, bind to molecules already present.

The heads of the stopcock molecules are too large to be able to enter the channels. Typical functionalized groups such as bucky balls [34], chelating centers [35], and others can be used. Some examples are given in Fig. 21. Similarly to the labels, heads can be reactive or nonreactive. Reactive heads may have "arms", which can interact with each other to form a "monolayer polymer" or bind to the zeolite external surface.

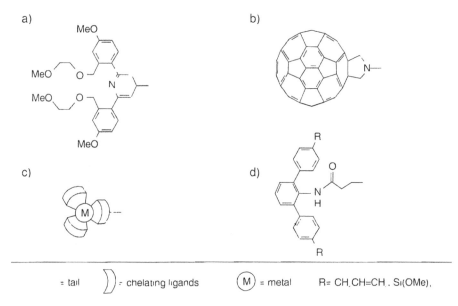

Figure 21 Schematic representation of several head moieties: (a) fluorescent chemosensor [36]; (b) C_{60}-based heads [34]; (c) ligands containing suitable binding sites in a correct arrangement for metal ion complexation [35]; (d) head moieties functionalized with reactive sites [37].

In some cases, it is desirable to add a spacer which elongates the stopcock molecule so that the length of the tail can be controlled. This can be a tool to improve the solubility of the whole molecule. Polar groups might help to bind the molecules more strongly inside of the zeolite channels. Spacers which are sufficiently flexible so that they can bend the tail into the zeolite channels include, for example, aliphatic chains, polyethers, or amides.

Positively charged stopcocks can be plugged in the zeolite channels by ion exchange, whereas neutral stopcocks can be added by dehydration of the zeolite channels and adsorption from a nonaqueous solution or from the gas phase. The zeolite's external surface consists of a coat and a base. These two surfaces differ in a number of properties so that the interactions can be tuned. For MFI- and FAU-type zeolites, as an example, it was reported that guest molecules bind to the holes on the external surface much more strongly than on the framework between the holes [38,39].

The specific coupling of the photonic antenna via the stopcocks to a target depends on the functionality envisaged. Direct linkage to a semiconductor surface, embedding into a semiconducting polymer, organization of crystals bearing ionic stopcocks by means of charged polymers [40,41], or linkage to sites of biological interest are possibilities.

First experimental results on dye-loaded zeolite L systems modified with commercial stopcock molecules on the external surface show that electronic excitation energy can be transferred from molecules inside the channels to the stopcocks and vice versa and that the stopcocks prefer to adsorb on the cylinder base instead of the coat [42].

B. Monodirectional Antenna Materials

In monodirectional antenna systems, the energy is transported in one direction only. We discuss the possibilities sketched in Fig. 22 to create asymmetrical

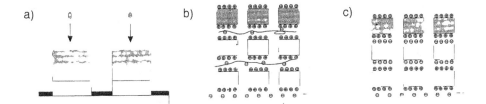

Figure 22 Schematic view of three strategies for the preparation of monodirectional antenna materials. (a) Organisation of zeolite L crystals on a substrate. The channels are oriented perpendicular to the surface. Dye molecules can only be inserted from one side. (b) Multilayer organization of charged polymers and zeolite L crystals loaded with different dyes. (c) Multilayer organization of oppositely charged zeolite L crystals.

antenna materials. Either the insertion of the dyes into the zeolite channels is controlled such that the molecules can penetrate the channels only from one side, or crystals containing different dyes are organized appropriately.

Asymmetric loading of zeolite L crystals is possible if the channel openings are selectively closed on one side so that the dyes can be inserted from the other side in sequence by the same procedure as used for the preparation of the bidirectional antenna materials [4]. Organizing the crystals on a substrate, maybe by linking them covalently [37], in order to inhibit the penetration of dye molecules from the substrate side, is a way to realize this. Microstructured substrates have been used to organize relatively large zeolite ZSM-5 crystals in a plane [43]. Other forces used for this purpose are adhesion [44] and electric fields [45]. The morphology and the size of the crystals are important in all cases. ZSM-5 nanoscale seed crystals were organized with their largest area surface on positively charged polymers on a gold surface. Calcination and growth of the silicalite then led to a densely packed and oriented polycrystalline layer [46]. Zeolite L crystals can be organized by modifying them with charged stopcock molecules. This leads to an anisotropic distribution of the zeolite surface charges and gives rise to attractive electrostatic interactions with an oppositely charged substrate. The crystal growth of AlPO-5 in the presence of floating anodized alumina can be controlled to obtain vertically aligned hexagonal crystals [47].

To further suppress the insertion of dyes into the channels from the substrate side, the channels can be sealed on one side by adsorption of a polymer onto the substrate [48,49]. Alternatively, stopcocks can be functionalised specifically with electroreactive constituents like ethynylic [50] vinylic [51], or pyrrolic groups [52,53]. The substrate serves as electrode under the influence of an electrochemical potential and only the adjacent heads of the stopcocks react to form a covalently cross-linked structure. The stopcocks are then removed from the other side of the crystals to unblock the channels for the dye insertion.

The formation of monodirectional antenna materials by organizing different dye-loaded crystals is illustrated in Figs. 22b and 22c. The assembly of layered organic–inorganic composites is currently a major research area [54,55]. An increasing number of articles concerning the assembly of zeolite crystals on substrates via ionic or covalent bonding, some by the use of appropriate organic additives, have been published [56–60]. This knowledge can be used for alternate adsorption of charged polymers and zeolite L crystals modified with stopcocks bearing a charged head, as illustrated in Fig. 22b. Polymeric materials ranging from biopolymers such as proteins [61,62] to inorganic macromolecules and clays [63,64] have been assembled by means of related techniques. Luminescent polymers can be used as connectors to transport the energy from one crystal to another one [65,66]. A simpler concept consists in arranging differently charged crystals in the manner shown in Fig. 22c.

VI. APPLICATIONS

The tunability in size and properties of the materials described in the previous sections is so large that we expect them to find applications in different fields such as optoelectronics, pigments, molecular probes, and educational tools. We focus on five applications among these. Some of them are already feasible, whereas others require further development.

A. Educational Tools

There are not many demonstration experiments which show the process of energy transfer and energy migration well. We found that the energy transfer between dye molecules inside the channels of zeolite L works so well that it can be used as a nice educational tool to show both energy transfer and energy migration. The reason for this is that the dye molecules can be brought close together without forming dimers. We can vary the mean distance between the molecules by varying the occupation probability of the dyes, which is directly proportional to the concentration of the dyes in the zeolite channels. The volume of a zeolite crystal is given by

$$V_Z = \pi r_{cyl}^2 l_{cyl} \qquad [18]$$

where r_{cyl} is the radius of the zeolite crystal and l_{cyl} is the length of the zeolite crystal. The number of unit cells of one crystal, n_{uc}, is given by

$$n_{uc} = \frac{2\pi}{\sqrt{3}} l_{cyl} \, r_{cyl}^2 \, \frac{l}{|c||a|^2} \qquad [19]$$

where c and a are the primitive vectors of the zeolite L framework. Now, the number of channels in a zeolite crystal can be written as

$$n_{ch} = \frac{n_{uc}|c|}{l_{cyl}} \qquad [20]$$

If n_s is the number of unit cells that forms a site and p_{dye} is occupation probability of dyes, then the number of dyes in the zeolite crystal is.

$$n_{dye} = \frac{l_{cyl} n_{ch}}{n_s |c|} p_{dye} \qquad [21]$$

Now, we can calculate a mean volume element that a dye occupies in the zeolite crystal, and if we consider this volume element as a cube, we can calculate the length of the cube. If we use a 1:1 mixture of donor D and acceptor A, this length is approximately the mean distance between D and A:

$$R_{DA} \simeq \left(\frac{V_Z}{n_{dye}}\right)^{1/3} = \left(\frac{1}{2} \sqrt{3} \, |c||a|^2 \frac{n_s}{p_{dye}}\right)^{1/3} \qquad [22]$$

This equation shows that the mean distance between two molecules is only dependent on some geometrical factors and on the occupation probability of the dyes. Thus, the occupation probability can be regarded as a tool to vary the mean distance between the dyes.

We concentrate on the two cationic dyes Py^+ and Ox^+, which can be incorporated into zeolite L from an aqueous solution with about equal rates. Both dyes have a large fluorescence quantum efficiency and, therefore, radiationless relaxation is not considered. If a 1:1 mixture of the two dyes is used, we can look at the mechanism of energy transfer from Py^+ (donor D) to Ox^+ (acceptor A). We can also modify Py^+-loaded zeolite L crystals with one molecule of Ox^+ on average at all channel endings. By varying the occupation probability of Py^+, energy migration to the crystal endings can be tuned from poor to efficient.

The energy transfer in these materials can easily be observed when a series of samples is put in a black box, as shown in Fig. 23. The samples are excited using a Mini Mag-Lite® AA flashlight and a Schott DAD 8–1 interference filter at 486.7 ± 5 nm. The light beam has to be perpendicular to the interference filter and to the sample. The emission light is observed through a Schott OG 515 cutoff filter. The sample and the cutoff filter are not perpendicular to the observation angle in order to minimize reflection effects. The transmission spectra of the two filters and the absorption and the emission spectra of Py^+ and Ox^+ are plotted in Fig. 24.

With a 1:1 mixture of the two dyes, we can show the energy transfer from the Py^+ donors to the Ox^+ acceptors very clearly. The seven luminescent samples

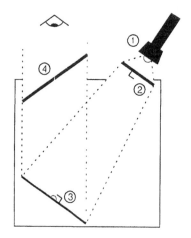

Figure 23 Black box for observation of the energy transfer and energy migration demonstration experiments: (1) Mag-Lite® AA flashlight, (2) Schott DAD 8–1 interference filter at 486.7 ± 5 nm, (3) sample, (4) Schott OG 515 cutoff filter.

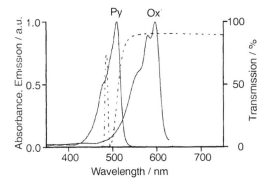

Figure 24 Absorption (solid curve) and emission spectra (dotted curve) of Py⁺ and Ox⁺ in zeolite L. The dashed lines show the transmission of the Schott DAD 8–1 interference filter and the Schott OG 515 cutoff filter.

shown in Fig. 18 of Ref. 25 consist of zeolite L crystals of 300 nm average length containing different amounts of Py⁺ and Ox⁺. In all cases, Py⁺ was specifically excited at 486.7 ± 5 nm.

Reference samples 1 and 7 contain only Py⁺ or Ox⁺, respectively. The other samples are 1:1 mixtures of the two dyes with increasing occupation probability. We calculate the following mean donor–acceptor distances: (2) 187 Å; (3) 109 Å; (4) 87 Å; (5) 69 Å; (6) 55 Å. The Förster radius for Py⁺ to Ox⁺ energy transfer in a medium of refractive index of 1.4 is about 70 Å, based on the Py⁺/Ox⁺ spectral overlap, which is 1.5×10^{-13} cm³/M. In sample 2, we observe mainly the green luminescence of Py⁺, which means that the dyes are too far apart for Förster-type energy transfer. The yellow color of 3 is due to a mixture of green and red luminescence, which means that energy transfer is significant. The energy transfer becomes more and more efficient, and by sample 6, the red luminescence of the Ox⁺ is dominant.

One would expect that the color of the samples 2–6 in diffuse reflection mode is the same because there is a 1:1 mixture of the two dyes in all cases. This, however, is not observed because of the high concentration of dye molecules inside the zeolite. If the concentration of dye molecules inside the zeolite channels is high enough, a saturation effect is observed, (i.e., light of specific wavelengths is totally absorbed and this changes the absorption spectrum as is shown in Fig. 25). Therefore, the visual color of the different samples observed (as diffuse reflection) changes from yellow to red with increasing loading.

A similar experiment can be done to show energy migration. One can modify Py⁺-loaded zeolite L crystals with, on average, one Ox⁺ molecule at

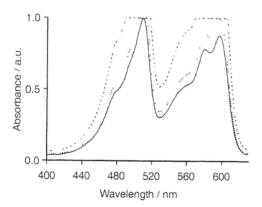

Figure 25 Calculated saturation effect for a 1:1 mixture of Py⁻ and Ox⁺. This effect changes the visual color of the material.

both ends of each channel. By varying the occupation probability of Py⁺, and thus the mean distance between the Py⁺ molecules, the energy-migration efficiency can be varied (see Fig. 13). If energy migration is efficient, the electronic excitation energy will be trapped at the crystal endings by the Ox⁺ molecules and the red emission of Ox⁺ molecules will be visible. If, however, the energy migration is not very efficient, the green emission of the Py⁺ molecules will dominate.

B. Pigments

A promising field with potential for industrial application is the use of dye-loaded zeolites as pigments [67–69]. The encapsulation of dyes in zeolite L leads in general to a substantially increased stability of the organic molecules. The matrix protects them from oxidative agents and other reactive species and the spatial constraints can eliminate photoisomerization reactions. Additionally, the processing properties of such pigments only depends on the zeolite surface, which can be tuned to a large extent, and not on the specific dye characteristics.

The use of highly fluorescing dyes leads to much brighter colors. In particular, the xanthene dyes (Ox⁺, Py⁺, PyGY⁺, PyB⁺) act as optical brighteners because the fluorescence in the visible can be initiated by absorption in the UV region of the spectrum. By the use of only two different dyes, a large palette of different colors can already be generated. New interesting color effects can be controlled via the energy transfer, saturation, and reabsorption phenomena in these materials. Applying the closure and stopcock molecules discussed in Section V.A. allows to fine-tune the specific properties of the pigments.

C. Color-Changing Medium

A light-emitting device (LED) display consists of a matrix of contacts made to the bottom and top surfaces of each light-emitting element, or pixel. To generate a full-color image, it is necessary to vary the relative intensities of three closely spaced, independently addressable pixels, each emitting one of the primary colors red, yellow, or blue. Several techniques have been proposed for producing the three colors in each pixel. One method is the use of a single blue or ultraviolet LED to pump organic fluorescent wavelength converters, also known as color-changing media (CCM), as illustrated in Fig. 26 [70].

The dye–zeolite L antenna systems bear the ideal properties to act as color-changing media in such an application. Due to the high concentration of dye monomers in the zeolite matrix, almost all incoming light can be absorbed within the dimensions of one monolayer of such crystals [3]. Through the high-energy-transfer efficiency reported in Fig. 13, about 90% of the absorbed light can then be converted into light of a longer wavelength. The anisotropic arrangement of the dye molecules leads to spatially well-defined emission of partially polarized light. The scattering of light at the crystal surface can be minimized by the use of nanocrystals or can be eliminated by embedding the material into a medium with the same refractive index.

D. Nanoscaled Lasers

As we reported in Section IV.A, refraction and total internal reflection can occur in dye-loaded zeolite L microcrystals. Therefore, a bundle of light rays in, for example, a POPOP-loaded zeolite composite can circulate inside the hexagon. If the emission can circulate often enough in the same volume and the loss of the zeolite L ring cavity is small enough, lasing should be possible.

Laser activity was, indeed, recently reported in a pyridine-2-loaded nano-porous AIPO-5 molecular sieve [26]. Like zeolite L, this compound consists of hexagonally arranged one-dimensional channels along the crystal long axis which can act as a matrix to arrange geometrically suitable dyes along the c axis in a supramolecular manner. The thickness of the used AIPO-5 crystals ranged from

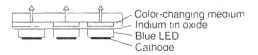

Figure 26   Scheme of a pixel for generating full color. Different colors are generated by transforming the emitted blue or UV light from the LEDs with the help of color-changing media.

4.5 to 22 μm and the corresponding length from more than 10 to 100 μm. The authors assumed that the resonator is built up by total internal reflection leading to a ring cavity. However, using a resonator with a size of the wavelength or less, classical optics arguments might not be sufficient.

To overcome the lasing threshold, the light amplification has to be larger than unity. For this reason, the loss from the cavity has to be small. Whereas the length of one round-trip in the largest crystals we are using is 3 μm, it is about 13.5 μm for the smallest described AlPO-5 laser cavity. The consequences of this are not well understood yet.

To obtain the spectrum of modes in a first approximation, one could fit an integral number of wavelengths into the beam pathway to get constructive interference after a round trip. Using the same arguments as in Ref. 26, the wavelength interval $\Delta\lambda$ can be calculated between two constructive modes of a POPOP-loaded zeolite L laser with a diameter of ~1 μm, giving a round-trip length of $L \simeq 3$ μm at an emission wavelength of 440 nm. From the relation:

$$\Delta\lambda = \lambda^2 \frac{1}{nL} \qquad [23]$$

a value $\Delta\lambda = 43$ nm is obtained, allowing only single-line emission for dyes with a narrow emission band. This has already been reported for an AlPO-5 pyridine-2-loaded sample with a diameter of ~4.5 μm and a free spectral range for the sample of ~24 nm.

In contrast to the AlPO-5 system, we are able to fine-tune the dye concentration in a broad range. Furthermore, the morphology of the material does not change with different dye loadings, because the dyes are inserted into the previously synthesized zeolite L. For neutral molecules, it is possible to vary the loading from $p \simeq 0.001$ molecules per site up to a nearly filled zeolite L with a loading of $p \simeq 1$, whereas for ionic dyes, a maximal loading of $p \simeq 0.4$ molecules per site has been realized until now. Therefore, a very high optical density of monomeric aligned dye molecules can be reached, shortening the minimum optical pathway for laser action.

Depending on the size of an incorporated dye, the angle of the transition dipole moment to the c axis lies between 0° for long molecules and 72° for smaller ones. Therefore, if a small molecule is inserted into the channels of zeolite L, part of the emission will be parallel to the c axis. Due to the flat and parallel ends of appropriately prepared zeolite crystals, one can envisage to arrange crystals between two mirrors or to add a reflecting layer on individual crystals. This might lead to a microlaser with a plane-parallel resonator. Apart from experimental difficulties, the realization of a dye-loaded zeolite L microlaser appears to be feasible.

E. Dye-Sensitized Solar Cells and LEDs

The currently available solar cells are an attractive source of renewable energy. They are, however, still expensive for large-scale applications. During the last decades, some effort has been made to find an inexpensive and efficient alternative for the crystalline silicon p/n junctions. Remarkable work has been done on developing p/n junctions with organic semiconductor materials [71–73]. More groups, however, have focused on dye sensitization of metal oxide semiconductors [74–77]. In this type of dye-sensitized solar cells, the dyes have the function of absorbing light and, subsequently, injecting an electron in the metal oxide semiconductor. After electron injection, the dyes have to be regenerated, usually by means of a redox couple. A different kind of dye-sensitized solar cell was proposed by Dexter [78,79] in 1979. He described sensitization of the semiconductor by energy transfer instead of electron injection, followed by the production of an electron–hole pair in the semiconductor (Scheme 1). Although Dexter published this idea in 1979, only a few groups have tried to observe energy transfer from a dye to a semiconductor [80–82].

The dyes do not have to be regenerated in this case because they do not exchange electrons. Energy transfer to the semiconductor works well if the distance between the donor and the semiconductor is on the order of the critical distance for Förster energy transfer. Only a very thin semiconductor layer is needed, because the electron–hole pairs form near the surface. The flexibility in tuning the energy of the donors is large and only limited by the energy gap of the semiconductor, which must be equal or smaller than the excitation energy of the donor. Electron transfer is prevented by introducing a insulating layer.

The antenna effect as it is found in natural photosynthetic systems is an attractive tool for increasing light absorption of solar cells. Some of the work done on dye sensitization of polycrystalline titanium dioxide shows aspects of antenna behavior [76,83–87]. Most of the problems in the systems where an electron is injected into the semiconductor arise in the regeneration process of

---

Two Ideas on Energy Transfer Phenomena: ... and Sensitization of Photovoltaic Cells
" . Energy transfer can also produce e-h pairs in a semiconductor, although this effect seems not to have been considered heretofore. .. "

Scheme 1 Citation of D. L. Dexter. (From Ref. 79.)

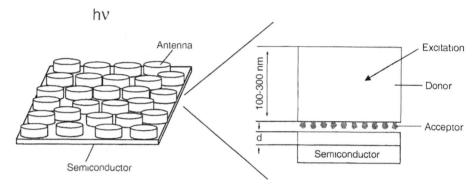

Figure 27 Sensitized solar cell based on dye-loaded zeolite L antenna systems. The antenna systems absorb light and transport their energy mainly along the c axis of the crystals to the semiconductor surface. Electron–hole pairs are formed in the semiconductor by energy transfer from the antenna system to the conduction band of the semiconductor.

the dyes. If the principle of energy transfer is used instead of electron injection, this regeneration process can be avoided. The photonic antenna material developed in our group is based on energy migration and energy transfer and it is challenging to combine its properties with the ideas of Dexter to form an antenna-sensitized solar cell. It appears to be feasible to put small crystals with a relatively large diameter with their c axis perpendicular to the surface of a semiconductor (see Fig. 27). In this case, of course, a one-directional antenna as described in Section V.B is needed. The excitation energy is transported to the edge of the

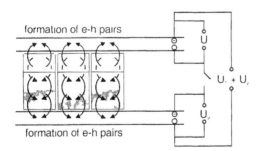

Figure 28 Tandem solar cell where monodirectional antenna systems with three dyes are put between two n-type semiconductors with different band gaps.

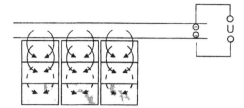

Figure 29 Light-emitting nanocrystals. The electrical source supplies excitation energy that can be transferred from the semiconductor to the nanocrystals. The dyes inside the crystals emit their energy by fluorescence because of their high-fluorescence quantum yield.

crystal by energy migration. At the edge of the crystal, energy transfer takes place from a stopcock molecule to the semiconductor over a certain distance. Tandem solar cells are more advanced high-efficiency systems [88]. In this concept, one-directional antenna materials are put between two n-type semiconductors with different band gaps, as shown in Fig. 28. Both n-type semiconductors generate electrons, but the band gap of the first one is tuned to the red dye, whereas the band gap of the second one is tuned to the blue.

The system discussed so far can be reversed. If we reverse the current and put a voltage over the semiconductor, it is possible to transfer energy from the semiconductor to the acceptor antenna, as shown in Fig. 29. This results in light-emitting nanocrystals. The color of the emitted light can be tuned by adapting the ratio of the blue, the yellow, and the red emitting dyes. Conventional LEDs are excited electrically. In this case, the excitons are statistically distributed over singlet and triplet states. According to literature, one-quarter of the formed excitons are singlet excitons, which are responsible for the electroluminescence, whereas three-quarters are triplet excitons, which mainly decay nonradiatively [89–92]. In the case of direct energy transfer from a semiconductor to dye-loaded nanocrystals, mostly the excited singlet states are filled, leading to higher luminescence efficiency.

VII. CONCLUSIONS

We have shown that the basic concept of filling hexagonal crystals, which consist of one-dimensional channels with molecular diameter and tuneable length, with luminescent dyes leads to new inorganic–organic host–guest materials with fascinating properties and challenging options for chemical modifications. From this result, applications in different fields are of great practical interest. The concepts described are not limited to zeolite L as a host. A nice feature of this zeolite is, however, that neutral as well as cationic dyes can be inserted. For many dyes, insufficient space is available for electronic overlap, so that they keep the properties of monomers. Additionally, the base of the cylindrical crystals is chemically sufficiently different with respect to the coat, so that fine-tuning of stopcock molecules is feasible. Only few experiments have been made with the 30–100 nm material so far [42]. However, this will change soon, because it offers challenging opportunities for sensor applications, as molecular probes, or for incorporation into photoconductive polymers.

ACKNOWLEDGMENTS

This work was supported by the Swiss National Science Foundation Project NFP47 (4047-057481) and by the Bundesamt für Energiewirtschaft Project 10441. We also thank René Bühler for his contribution to realize the experiments reported in Section V.A and for the synthesis of the zeolite L materials.

REFERENCES

1. Calzaferri, G.; Maas, H.; Pauchard, M.; Pfenniger, M.; Megelski, S.; Devaux, A. In *Advances in Photochemistry*; Neckers, D. C.; Wolff, T.; Jenks, W. S. ed.; *2002*; Vol. 27, in press.
2. Gfeller N.; Calzaferri, G. *J. Phys. Chem. B* **1997**, *101*, 1396.
3. Calzaferri, G.; Brühwiler, D.; Megelski, S.; Pfenniger, M.; Pauchard, M.; Hennessy, B.; Maas, H.; Devaux, A.; Graf, U. *Solid State Sci.* **2000**, *2*, 421.
4. Pauchard, M.; Devaux, A.; Calzaferri, G. *Chem. Eur. J.* **2000**, *6*, 3456.
5. Megelski, S.; Calzaferri, G. *Adv. Funct. Mater.* **2001**, *11*, 277.
6. Pauchard, M.; Huber, S.; Méallet-Renault, R.; Maas, H.; Pansu, R.; Calzaferri, G. *Angew. Chemie, Int. Ed.* **2001**, *40*, 2839; *Angew. Chemie* **2001**, *113*, 2921.
7. Schulz-Ekloff, G.; Wöhrle, D.; Van Duffel, B.; Schoonheydt, R. *Microporous Mesoporous Mater.* **2002**, *52*, 91.
8. Breck, D. W. *Zeolite Molecular Sieves*; Wiley: New York, *1974*.
9. Bärlocher, C.; Meier, W. M.; Olsen, D. H. *Atlas of Zeolite Framework Types*; Elsevier: Amsterdam, *2001*.
10. Ohsuna, T.; Horikawa, Y.; Hiraga, K. *Chem. Mater.* **1998**, *10*, 688.
11. Anderson, P. A.; Armstrong, A. R.; Porch, A.; Edwards, P. P.; Woodall, L. J. *J. Phys. Chem. B* **1997**, *101*, 9892.
12. Lakowicz, J. R. *Principles of Fluorescence Spectroscopy*, Kluwer, Academic/Plenum, New York, *1999*.
13. Calzaferri, G. *Chimia* **1998**, *52*, 525.
14. Hashimoto, S.; Hagiri, M.; Matsubara, N.; Tobita, S. *Phys. Chem. Chem. Phys.* **2001**, *3*, 5043.

15. Megelski, S.; Lieb, A.; Pauchard, M.; Drechsler, A.; Glaus, S.; Debus, C.; Meixner, A. J.; Calzaferri, G. *J. Phys. Chem. B* **2001**, *105*, 25.
16. Kunzmann, A.; Seifert, R.; Calzaferri, G. *J. Phys. Chem. B* **1999**, *103*, 18.
17. Calzaferri, G.; Gfeller, N. *J. Phys. Chem.* **1992**, *96*, 3428.
18. Hennessy, B.; Megelski, S.; Marcolli, C.; Shklover, V.; Bärlocher, C.; Calzaferri, G. *J. Phys. Chem. B* **1999**, *103*, 3340.
19. Cox, B. G.; Schneider, H.; Stroka, J. *J. Am. Chem. Soc.* **1978**, *100*, 4746.
20. Dietrich, B.; Lehn, J. M.; Sauvage, J. P. *J. Chem. Soc., Chem. Commun.* **1973**, *1*, 15.
21. Förster, T. *Ann. Phys. (Leipzig)* **1948**, *2*, 55.
22. Förster, T. In *Fluoreszenz Organischer Verbindungen*; Vandenhoeck & Ruprecht: Göttingen, **1951**.
23. Calzaferri, G.; Rytz, R. *J. Phys. Chem.* **1995**, *99*, 12141.
24. Gfeller, N.; Megelski, S.; Calzaferri, G. *J. Phys. Chem. B* **1999**, *103*, 1250.
25. Calzaferri, G.; Pauchard, M.; Maas, H.; Huber, S.; Kathyr, A.; Schaafsma, T. *J. Mater. Chem.* **2002**, *12*, 1.
26. Braun, I.; Ihlein, G.; Laeri, F.; Nöckel, J. U.; Schulz-Ekloff, G.; Schüth, F.; Vietze, U.; Weiss, Ö.; Wöhrle, D. *Appl. Phys. B: Lasers Opt.* **2000**, *70*, 335.
27. Renn, A.; Bucher, S. E.; Meixner, A. J.; Meister, E. C.; Wild, U. P. *J. Lumin.* **1988**, *39*, 181.
28. Fetisova, Z. G.; Freiberg, A. M.; Timpmann, K. E. *Nature* **1988**, *334*, 633.
29. Hu, X.; Schulten, K. *Phys. Today* **1997**, *50*, 28.
30. Scholes, G. D.; Fleming, G. R. *J. Phys. Chem. B* **2000**, *104*, 1854.
31. Pfenniger, M.; Calzaferri, G. *ChemPhysChem* **2000**, *4*, 211.
32. Kasha, M.; Rawls, H. R.; El-Bayoumi, M. A. *Pure Appl. Chem.* **1965**, *11*, 371.
33. McRae, E. G.; Kasha, M. *J. Chem. Phys.* **1958**, *28*, 721.
34. Guldi, D. M.; Maggini, M.; Menna, E.; Scorrano, G.; Ceroni, P.; Marcaccio, M.; Paolucci, F.; Roffia, S. *Chem. Eur. J.* **2001**, *7*, 1597.
35. Balzani, V.; Juris, A.; Venturi, M.; Campagna, S.; Serroni, S. *Chem. Rev.* **1996**, *96*, 759.
36. Mello, J. V.; Finney, N. S. *Angew. Chem. Int. Ed.* **2001**, *40*, 1536; *Angew. Chem.* **2001**, *113*, 1584.
37. Ha, K.; Lee, Y.-J.; Lee, H. J.; Yoon, K. B. *Adv. Mater.* **2000**, *12*, 1114.
38. Turro, N. J.; Lei, X.-G.; Li, W.; Liu, Z.; McDermott, A.; Ottaviani, M. F.; Abrams, L. *J. Am. Chem. Soc.* **2000**, *122*, 11649.
39. Turro, N. J.; Lei, X.-G.; Li, W.; Liu, Z.; Ottaviani, M. F. *J. Am. Chem. Soc.* **2000**, *122*, 12571.
40. Ladam, G.; Schaad, P.; Voegel, J. C.; Schaaf, P.; Decher, G.; Cuisinier, F. *Langmuir* **2000**, *16*, 1249.
41. Cassier, T.; Lowack, K.; Decher, G. *Supramol. Sci.* **1998**, *5*, 309.
42. Maas, H.; Calzaferri, G. *Angew. Chem. Int. Ed.* **2002**, *41*, 2284; *Angew. Chem.* **2002**, *114*, 2389.
43. Scandella, L.; Binder, G.; Gobrecht, J.; Jansen, J. C. *Adv. Mater.* **1996**, *8*, 137.
44. Lainé, P.; Seifert, R.; Giovanoli, R.; Calzaferri, G. *New J. Chem.* **1997**, *21*, 453.
45. Caro, J.; Finger, G.; Kornatowski, J.; Richter-Mendau, J.; Werner, L.; Zibrowius, B. *Adv. Mater.* **1992**, *4*, 273.

46. Mintova, S.; Schoeman, B.; Valtchev, V.; Sterte, J.; Mo, S.; Bein, T. *Adv. Mater.* **1997**, *9*, 585.

47. Tsai, T.-G.; Chao, K.-J.; Guo, X.-J.; Sung, S.-L.; Wu, C.-N.; Wang, Y.-L.; Shih, H.-C. *Adv. Mater.* **1997**, *9*, 1154.

48. Caro, J.; Zibrowius, B.; Finger, G.; Buelow, M.; Kornatowski, J.; Huebner, W. *German Offen.* DE 41 09 038 A1, *1992*.

49. Caro, J.; Zibrowius, B.; Finger, G.; Richter-Mendau, J. *German Offen.* DE 41 09 037 A1, *1992*.

50. Shiga, K.; Inoguchi, T.; Mori, K.; Kondo, K.; Kamada, K.; Tawa, K.; Ohta, K.; Maruo, T.; Mochizuki, E.; Kai, Y. *Macromol. Chem. Phys.* **2001**, *202*, 257.

51. Calvert, J. M.; Schmehl, R. H.; Sullivan, B. P.; Facci, J. S.; Meyer, T. J.; Murray, R. W. *Inorg. Chem.* **1983**, *22*, 2151.

52. Deronzier, A.; Moutet, J. C. *Coord. Chem. Rev.* **1996**, *147*, 339.

53. Diaz, A. F.; Kanazawa, K. K.; Gardini, G. P. *J. Chem. Soc., Chem. Commun.* **1979**, *14*, 635.

54. Zanetti, M.; Lomakin, S.; Camino, G. *Macromol. Mater. Eng.* **2000**, *279*, 1.

55. Decher, G. *Science* **1997**, *277*, 1232.

56. Yan, Y.; Bein, T. *J. Am. Chem. Soc.* **1995**, *117*, 9990.

57. Yan, Y.; Bein, T. *J. Phys. Chem.* **1992**, *96*, 9387.

58. Bein, T. *Chem. Mater.* **1996**, *8*, 1636.

59. Li, Z.; Lai, C.; Mallouk, T. E. *Inorg. Chem.* **1989**, *28*, 178.

60. Boudreau, L. C.; Tsapatsis, M. *Chem. Mater.* **1997**, *9*, 1705.

61. Lvov, Y.; Ariga, K.; Ichinose, I.; Kunitake, T. *J. Am. Chem. Soc.* **1995**, *117*, 6117.

62. Kong, W.; Zang, X.; Gao, M. L.; Zhou, H.; Li, W.; Shen, J. C. *J. Macromol. Rapid Commun.* **1994**, *15*, 405.

63. Kotov, N. A.; Dékány, I.; Fendler, J. H. *J. Phys. Chem.* **1995**, *99*, 13065.

64. Lvov, Y.; Ariga, K.; Ichinose, I.; Kunitake, T. *Langmuir* **1996**, *12*, 3038.

65. Binder, F.; Calzaferri, G.; Gfeller, N. *Solar Energy Mater. Solar Cells* **1995**, *38*, 175.

66. Nguyen, T.-Q.; Wu, J.; Tolbert, S. H.; Schwartz, B. J. *Adv. Mater.* **2001**, *13*, 609.

67. Hölderlich, W.; Lauth, G.; Wagenblast, G.; Albert, B.; Lamm, G.; Reichelt, H.; Grund, C.; Gruettner-Merten, S. German Patent DE 4207745 (assigned to W. Hölderlich), *1993*.

68. Hölderlich, W.; Lauth, G.; Müller, U.; Brode S. German Patent DE 4131447 (assigned to BASF), *1993*.

69. Hölderlich, W.; Lauth, G.; Wagenblast, G.; Schefczik, E. German Patent DE 4207339 (assigned to W. Hölderlich), *1993*.

70. Forrest, S.; Burrows, P.; Thompson, M. *IEEE Spectrum* **2000**, *38*, 29.

71. Wöhrle, D.; Meissner, D. *Adv. Mater.* **1991**, *3*, 129.

72. Rostalski, J.; Meissner, D. *Solar Energy Mater. Solar Cells* **2000**, *63*, 37.

73. Wienke, J.; Schaafsma, T. J.; Goossens, A. *J. Phys. Chem. B* **1999**, *103*, 2702.

74. Nazeeruddin, M. K.; Kay, A.; Rodicio, I.; Humphry-Baker, R.; Müller, E.; Liska, P.; Vlachopoulos, N.; Grätzel, M. *J. Am. Chem. Soc.* **1993**, *115*, 6382.

75. Cahen, D.; Hodes, G.; Grätzel, M.; Guillemoles, J. F.; Riess, I. *J. Phys. Chem. B* **2000**, *104*, 2053.

76. Koehorst, R. B. M.; Boschloo, G. K.; Savenije, T. J.; Goossens, A.; Schaafsma, T. J. *J. Phys. Chem. B.* **2000**, *104*, 2371.
77. Van der Zanden, B.; Goossens, A. *J. Phys. Chem. B* **2000**, *104*, 7171.
78. Dexter, D. L. *J. Chem. Phys.* **1953**, *21*, 836.
79. Dexter, D. L. *J. Lumin.* **1979**, *18/19*, 779.
80. Farzad, F.; Thompson, D. W.; Kelly, C. A.; Meyer, G. J. *J. Am. Chem. Soc.* **1999**, *121*, 5577.
81. Hayashi, T.; Castner, T. G.; Boyd, R. W. *Chem. Phys. Lett.* **1983**, *94*, 461.
82. Gole, J. L.; DeVincentis, J. A.; Seals, L. *J. Phys. Chem. B.* **1999**, *103*, 979.
83. Bignozzi, C. A.; Argazzi, R.; Schoonover, J. R.; Meyer, G. J.; Scandola, F. *Solar Energy Mater. Solar Cells* **1995**, *38*, 187.
84. Bignozzi, C. A.; Schoonover, J. R.; Scandola, F. In *Progress in Inorganic Chemistry*; Karlin, K. D., ed.; Wiley: *1997*; Vol. 44, p. 1–95.
85. Tributsch, H. In *Proceedings IPS-10*, Tian, Z. W.; Cao Y., eds.; International Academic Publishers: Beijing, *1993*, pp. 235–247.
86. Memming, R.; Tributsch, H. *J. Phys. Chem.* **1971**, *75*, 562.
87. Kerp, H. R.; Donker, H.; Koehorst, R. B. M.; Schaafsma, T. J.; Van Faasen, E. E. *Chem. Phys. Lett.* **1998**, *298*, 302.
88. He, J.; Lindström, H.; Hagfeldt, A.; Lindquist, S.-E. *Solar Energy Mater. Solar Cells* **2000**, *62*, 265.
89. Samuel, I. D. W.; Beeby, A. *Nature* **2000**, *403*, 710.
90. Cleave, V.; Yahioglu, G.; Le Barny, P.; Friend, R. H.; Tessler, N. *Adv. Mater.* **1999**, *11*, 285.
91. Brown, A. R.; Pichler, K.; Greenham, N. C.; Bradley, D. D. C.; Friend, R. H. *Chem. Phys. Lett.* **1993**, *210*, 61.
92. Burroughes, J. H.; Bradley, D. D. C.; Brown, A. R.; Marks, R. N.; Mackay, K.; Friend, R. H.; Burns, P. L.; Holmes, A. B. *Nature* **1990**, *347*, 539.

8

Photo-Switching Spiropyrans and Related Compounds

**Jonathan Hobley, Martin J. Lear, and
Hiroshi Fukumura**
Tohoku University, Sendai Miyagi, Japan

I. INTRODUCTION

Spiropyrans (NIPS and BIPS), naphthopyrans or chromenes (CHR1) and spirooxazines (NOSI) in Scheme 1 have been extensively researched over the past five
decades because of their properties of photochemical and thermal optical switch-
ability. Their photochemical and thermal transformations from an ultraviolet
(UV)-light-absorbing state to a visible-light-absorbing state can be harnessed for
many practical applications [1–5]. Figure 1 shows a typical spectrum of a spiropyran in each of its two forms. These spectral changes result from a change in the
extent of conjugation upon breaking or forming the sp^3 carbon–oxygen bond in
the ring-closed form of the molecule to switch to or from the ring-open mero-
form isomers as shown in Scheme 2. In the ring-closed form, the two ring systems
are orthogonal and electronically independent, whereas in the mero-forms, the
ring system's π-clouds are connected. Scheme 2 shows the photochemically and
thermally inducible transformations of a typical example of each type of molecule.

The isomers that are shown in Scheme 2 are only those that have been
proven to exist in thermal equilibrium, although, in principle, eight isomers cis
and trans about the three methine bridge bonds are possible, as we shall discuss
in depth throughout this chapter. The nomenclature used for these different isomers in this chapter is the same as that used by the majority of authors. For

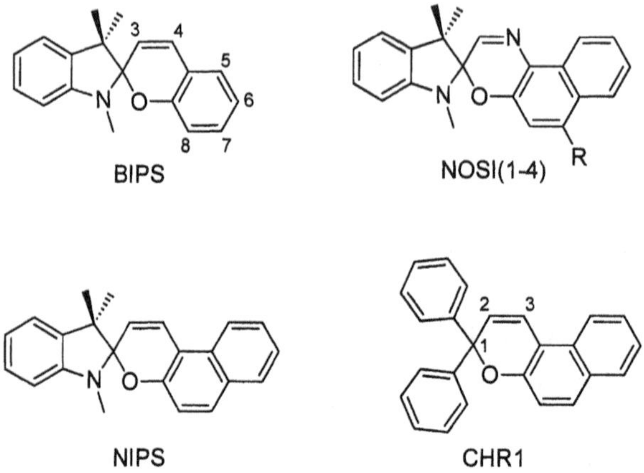

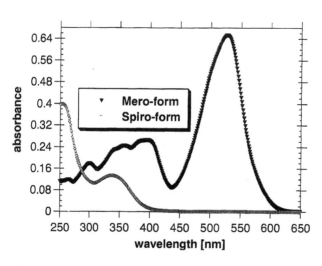

BIPS   NOSI(1-4)

NIPS   CHR1

Scheme 1

Figure 1   Absorption spectra of the ring spiro- and mero-form of 6,8-dinitro-BIPS.

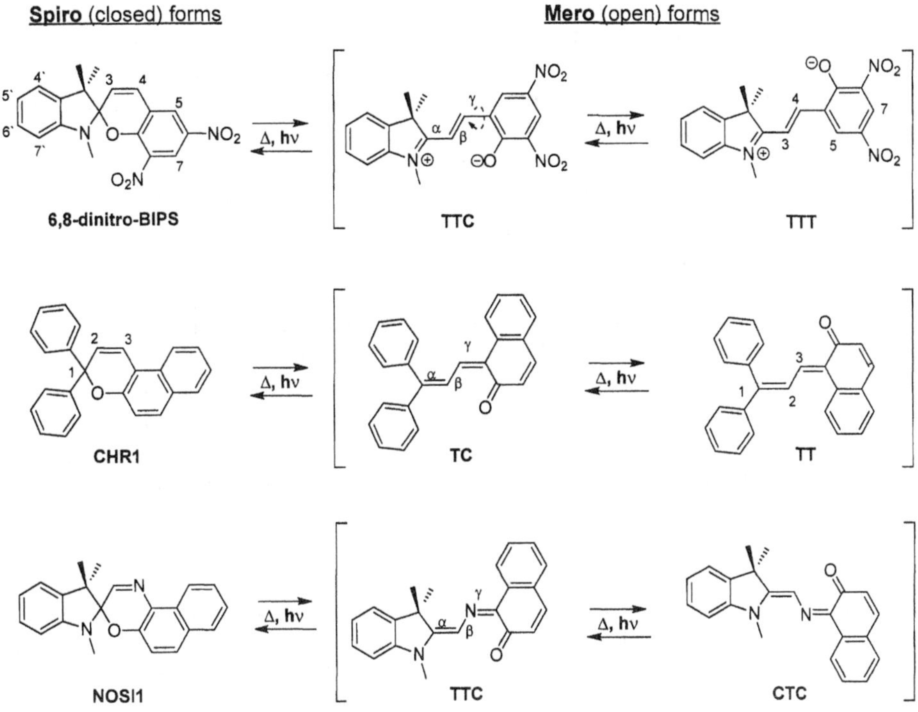

Spiro (closed) forms   **Mero (open) forms**

6,8-dinitro-BIPS   TTC   TTT

CHR1   TC   TT

NOSI1   TTC   CTC

Scheme 2

example, CCC has three cis methine bonds, whereas TTC has the α-bond trans, the β-bond trans, the γ-bond cis, and so forth. The numbering of carbon sites is included where it will simplify the discussion throughout the chapter. As much as possible, the abbreviations are consistent with those used by referenced works. CHR is short for chromene, which is a common name for naphthopyrans and benzopyrans, BIPS has become the usual name for the benzopyran-based spiropyrans and, by analogy, NIPS is adopted for the naphthopyran-based spiropyrans. NOSI is an abbreviation for naphthoxazine spiroindoline, which is simply a longer name for the naphthoxazine-based spiro-oxazines.

In each case the photo-coloration reaction begins with the photo-induced scission of an sp^3 C—O bond in the ring-closed form followed by a series of geometrical changes resulting in the more planar and more conjugated mero-form isomers [6–10]. The nature of the mero-form is different for each of the three classes of compound that we will describe.

Of these three classes of compound, the spiropyrans have been the main focus of research and, more recently, spiro-oxazine and naphthopyran have been receiving more attention due to their high fatigue resistance and their resulting commercial viability. These compounds have already found uses in many photo-switching applications such as photo-switchable holographic gratings [11,12], light-modulated metal ion recognition, [13–16], in photo-controllable bioactivity [15,17–21], in data recording and processing [3,22–24], and most successfully in photochromic opthalmic lenses [2,3]. However the spiropyran has often been considered too unstable with respect to continued photo-switching to be used in commercial products. Spirooxazine has a quite remarkable stability to repeated photo-switching and is extensively used in commercial applications. Four books have already been partly dedicated to these compounds [1,3,4,25] and the aim of this chapter is to bring the reader up to date on the new developments in our understanding of the photochemistry of these molecules and to give a general critical description of the previously reported work based on the advances in our knowledge. One surprising aspect of these systems is the often apparently conflicting conclusions concerning the reaction mechanisms of these molecules; however, this can be rationalized to some extent by the range of solvents and substituents that have been used. Also, there have been many advances in our knowledge and technology over the 50 years during which these compounds have been studied. Different experimental protocols inevitably produce their own variations in results. The first photochemical studies on these compounds were carried out using continuous-wave (CW) radiation from conventional light sources [6–8], whereas the more modern work uses femtosecond laser excitation [26–31].

As an illustration of the amount of research literature relating to these compounds, upon entering the topic spiropyran in our literature search we found nearly 500 hits, spiro-oxazine produced nearly 80 hits and naphthopyran produced 60 hits. Of course, no literature search will find all of the relevant articles and we can only assume that we have merely found the tip of the iceberg. Now is a good opportunity to help to consolidate the hard work of many authors by reviewing their painstaking work. We will try to logically follow the full photochemical and thermal reaction cycle that these compounds follow, starting from their initial ring-opening reaction to form merocyanine isomers. In turn, we will describe the nature of the merocyanine isomers and then describe how these can subsequently reform the ring-closed molecule.

II. RING-OPENING REACTIONS

A. Thermal Ring Opening

In their closed forms, spiropyran, spiro-oxazine, and some naphthopyrans are chiral about the sp^3 carbon atom that joins their orthogonal or out-of-plane ring systems or substituents [32–37]. Thermally, the spiro-oxazine spiropyran and naphthopyrans can all undergo spontaneous and repetitive cleavage of the sp^3 carbon–oxygen bond resulting in the inversion of one enantiomer to form the other mirror image. In the case of spiropyran, this process can be observed using ^{1}H-NMR (nuclear magnetic resonance) by observing spectral line broadening on the nonequivalent geminal methyl groups on the indoline ring system. This was first reported by Toppet et al. [32] and later by other workers [33–37]. It is even possible to observe the inversion of these compounds after they have just been eluted from a chiral column [37]. In fact, by making the geminal substituents on the indoline ring asymmetric, it is possible to achieve photo-switchable optical activity because the two enantiomers will have different energies [35]. The activation energy for racemization is between 46 and 103 kJ/mol for various spiropyrans in various solvents. Scheme 3 shows the proposed mechanism for the racemization reaction, which should proceed via either a CCT or CCC intermediate, as shown.

This can be reliably proposed as the mechanism, because the rate of racemization is calculated to be 18 sec^{-1} for 6,8-dinitro-BIPS in dimethyl sulfoxide (DMSO) at room temperature [36], whereas the rate of ring opening to form the merocyanine isomers trans about the central β-methine bond is 3×10^{-5} sec^{-1} for the same compound in aqueous mixtures [18] and was reported to be a slow process taking days to complete in DMSO [36]. Therefore, merocyanine isomers trans about the central methine β-bond cannot be implicated as the relevant intermediates in the racemization reaction. This has been the conclusions of most workers in this field since the first observation of this inversion process by Toppet et al. [32]. The charge separation of the CCC or CCT intermediate form is suggested because the activation energy is lower in polar solvents and when nitro groups are present on the benzopyran ring to delocalize any negative charge from the C9 oxygen. [37] $\Delta G^{\ddagger}$ of racemization for BIPS is 97 kJmol^{-1}, for 6-nitro-BIPS it is 86 kJmol^{-1} and for 6,8-dinitro-BIPS it is 60 kJmol^{-1} in DMF and 46 kJmol^{-1} in DMSO.

Thermal ring opening to form the merocyanine form is less dependent upon the type of substituent keeping a $\Delta G\ddagger$ of near to 101 kJmol^{-1} for BIPS, 6-nitro-BIPS and 6,8-dinitro-BIPS. [37] The ring opening reaction's rate-determining step is probably a rotation cis to trans about the central β-methine bond and this is also consistent with the polar CCC or CCT transition state which have a rather double β-methine bond.

B. Photochemical Ring Opening

1. Spiropyran Ring Opening

   a. Overview. A description of the ring-opening reaction for spiropyrans should be split into three categories. One category is for nitro-substituted compounds in polar media and a second for the nitro-substituted compounds in nonpo-

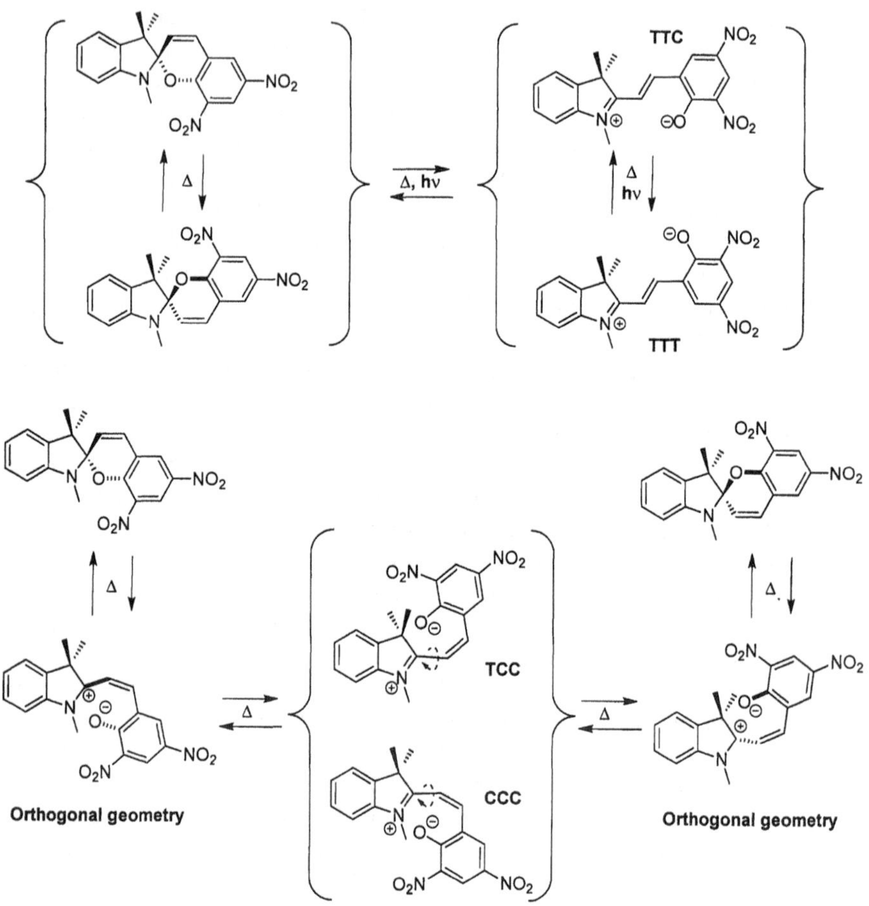

Scheme 3

lar media. The third category is for the ring-opening reaction for unsubstituted spiropyrans. Briefly, nitro substitution enhances the quantum yield of merocyanine formation [25] and this has often been attributed to the opening of a triplet-state manifold for the reaction. On the other hand, the unsubstituted compounds ring open exclusively in a singlet-state manifold [27–29]. The different behavior in polar and nonpolar media is easily understood when considering the large dipole moment of the zwitterionic spiropyran merocyanine [25,36], which will be stable and soluble only when formed in a polar medium, but spontaneously aggregates when formed in a nonpolar solvent such as hexane [24,38–45].

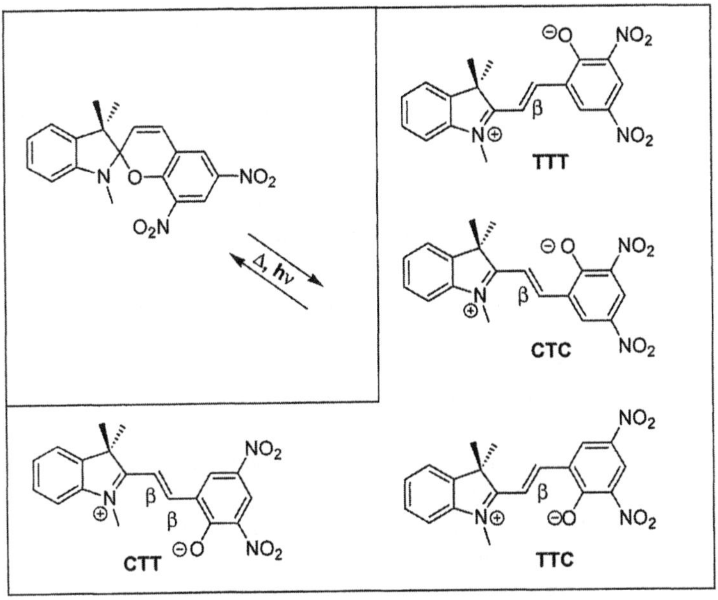

Scheme 4

Much discussion about the photo-formation of the merocyanines is based on the possibility of four stable merocyanine isomers that are trans about the β-bond on the methine bridge. Commonly, these are referred to as TTC, CTT, CTC, and TTT [6–8,28,36] and they are shown in Scheme 4. Note that these are "possible" or hypothesized structures. Gorner et al. even suggest that isomers cis about the central β-methine bond may have stability and equilibrate with the more planar isomers [14,46–51]. Certainly, the four isomers cis about the cental β-methine bond could have transient stability.

b. Unsubstitued Spiropyrans. In the case of the spiropyrans, the reaction proceeds via a singlet pathway for unsubstituted compounds in nonpolar and polar solvents. Aramaki et al. [52] studied an *N*-methoxyethyl NIPS using picosecond transient absorption and picosecond time-resolved resonance Raman spectroscopy. They reported that after 287 nm excitation of their compound in polar and nonpolar solvents, the full photochromic reaction occurred within 50 psec, producing a merocyanine species absorbing at the probe wavelength of 574 nm. Subsequently, no further probe beam absorption changes occurred up to 1.5 nsec after excitation. No evolution in time-resolved resonance Raman spectra occurred after the merocyanine had formed in the 50-psec pulse convolution, and the spectrum after this time was the same as the thermally equilibrated merocyanine form. Notably and only in polar solvents, a second transient absorption (between 400 and 460 nm) was observed both before the merocyanine isomer(s) was formed and when the merocyanine isomer(s) was re-excited at 574 nm. The intermediate absorbing at 400–460 nm had a lifetime of less than the 50-psec excitation pulse and these authors estimated the actual lifetime to be around 11 psec. No back reaction from the merocyanine isomer(s) to the spiro-form was noted when the sample was irradiated at either 574 or 425 nm, where the merocyanine absorbs, with delays of 100 psec to 1 nsec after 287-nm excitation. Further information was revealed by double pumping the sample solution and determining resonance Raman scattering by the photo-produced merocyanine form as a function of the 574-nm probe beam pulse energy. The higher the probe fluence, the more the resonance Raman peaks due to the merocyanine form becoming depleted, suggesting that pumping at 574 nm reduces the merocyanine ground-state absorption. The transient absorption observed by pumping with 287 nm and by double pumping with 287 nm and then 574 nm was therefore assigned to the merocyanine $S_n \leftarrow S_1$ transition. The overall mechanism that was proposed to account for all of the data that were obtained for this molecule is given in Scheme 5. The only criticism of this mechanism is that in the same article, the authors noted that the transient that they assigned as S_1 was not observed in cyclohexane either as a transient bleach or absorption at the probe wavelengths. This is possible if there were solvent shifts in the S_1 absorption spectra; however, they chose three probe wavelengths (458, 436, and 416 nm) and one would imagine that they would observe some absorption change due to S_1 formation. It is therefore conceivable that their assigned S_1 state is actually another transient that does not form as a metastable state in cyclohexane.

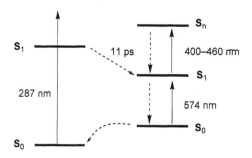

Scheme 5

Ernsting and Arthen-Engeland studied the photochemical spiropyran ring-opening reaction in the nonpolar n-pentane for BIPS, NIPS, and 6-nitro-BIPS, using transient absorption spectroscopy [27–29]. For BIPS and NIPS, within 1 psec of photoexcitation, they noted a broad absorption from 380 to 680 nm. This absorption was assigned to the spiro compound $Sn \leftarrow S_1$ transitions. In the case of BIPS and NIPS, the merocyanine absorption bands rose with respective time constants of 0.9 and 1.4 psec. These transient spectra were the same as spectra taken several microseconds after excitation. These authors suggested that a distribution of two or more isomers, that are trans about the β-bond, is already established within 10 psec of ultraviolet (UV) excitation of the spiropyran and that no further significant changes occur after this time. The findings of these authors relating to 6-nitro-BIPS will be dealt with later. The data for Ernsting et al. relating to NIPS [28,29] should be compared with the data of Aramki and Atkinson [52] that was measured in cyclohexane and not with the data from the polar solvents in which Aramaki saw the 11-psec transient. Considering this point, the two descriptions of the reaction are very similar.

Takahashi et al. [53] measured resonance Raman spectra of BIPS equilibrated with its merocyanine and also time-resolved resonance Raman spectra during the ring-opening reaction in various polar and nonpolar solvents. Their conclusion was that in the case of BIPS, no spectral changes occur in the resonance Raman spectra after 200 ns (20 ns in other related work [54]). This is consistent with the work of Ernsting et al. [27–29] and Aramaki et al. [52]. Takahashi et al. further noted that there are clear differences between the resonance Raman spectra of the merocyanines in different solvents. The simplest spectrum was obtained in cyclohexane, which had just one strong band at 1496 cm^{-1}, which the authors assigned to one merocyanine isomeric form. Other bands become prominent in more polar solvents, namely at 1534 and 1234 cm^{-1} and at 1446 and 1305 cm^{-1}. These authors pair these bands and assign them to two different merocyanine species. A further species is assigned to a spectral band at 1350 cm^{-1}. This gives a total of four merocyanine forms speculated to exist in these different solvents. Further, some of these bands in acetonitrile showed a temperature dependence, with the 1487 cm^{-1} band decreasing in relative intensity. In summary, these authors give assignments presented in Table 1.

Another explanation for their resonance Raman results could be a change in the zwitterionic nature of the merocyanine isomers in the different solvents which may result in changes in the Raman transition probabilities, or the spectral changes could be due to solvent shifts of the absorption spectrum, resulting in a change in the relative contribution of the different vibrational modes to each resonance Raman spectrum. We note that in the same article, the authors report the transient absorption spectra of the merocyanine forms, which clearly show that the BIPS spectrum in cyclohexane has more discrete vibrational modes than are observed in the polar solvents, which show more spectral broadening. Al-

Table 1 Band Assignments for Resonance
Raman Bands of BIPS in Different Solvents

Band (cm^{-1})	Assignment
1496	TTC
1534	CTT
1446	CTC
1350	TTT
1234	CTT
1305	CTC

though isomeric variance has been found for the spiropyran merocyanines [36,55], the rich variety suggested by these authors has not been effectively demonstrated for any spiropyran system so far.

Later, Chibisov and Gorner studied variously substituted BIPS compounds [14,46–51]. In the case of BIPS and NIPS and their non-nitro-substituted analogs, a singlet manifold was again proposed for the ring opening. They proposed that this ring-opening pathway should pass through a CCC isomer, but stated that it may not correspond to a potential energy minimum [50]. This is in accordance to previous conclusions. We should take care that no direct proof is ever given for the involvement of the CCC isomeric form over CCT in the spiropyran photochemical ring-opening reaction. Its involvement is merely suggested by calculation [28] or speculation and it may just be an intuitive choice.

To summarize the data obtained for BIPS and NIPS in polar solvents, Scheme 5 describes the photochemical ring-opening reaction, where the reaction forms the merocyanine excited singlet state in around 11 psec or less and the stable ground-state merocyanine distribution also forms on the picosecond time scale. The reaction is similarly rapid in nonpolar solvents except that the assigned merocyanine S_1 state is not observed as a transient form. The state assigned to the S_1 of the merocyanine, which is observed in polar solvents, may equally be assignable to another transient. The exact identity is not proven beyond doubt.

c. Nitro-BIPS and Nitro-NIPS in Polar Solvents. The simple arguments and proposed mechanisms which have been put forward to explain the behavior of the spiropyran compounds that are not nitro-substituted may have lulled the reader into a false sense of security because the proposed mechanism is relatively straightforward, as are the discussions. However, when we begin to consider the nitro-substituted analogs, suddenly the mechanisms become rather messy and the discussions become rather convoluted. This is true in polar solvents and slightly more so in the nonpolar solvents.

Lenoble and Becker studied 6-nitro-BIPS in acetonitrile [56]. In nitrogen-purged solutions, they reported an initial fast absorption rise (within the pulse) when monitoring at around 440–570 nm. The absorption at 440 nm decayed in a biexponential manner with a 32-nsec and a longer microsecond lifetime. The biexponential lifetimes of this species (440 nm) were quenched by oxygen for 11 and 70 nsec and it should, therefore, be a triplet state. In the deoxygenated solutions monitored over longer time scales, a transient having peaks at 400 and 560 nm grew with a lifetime of 24 μsec and transients at 620 nm and 440 nm decayed with a similar time scale. Bubbling oxygen quenched the formation of both the 620-nm shoulder and the 440-nm peak. From this information, the 440-nm peak and the 620-nm shoulder were associated with a triplet state and the 400-nm and 560-nm peaks were assigned to the merocyanine form. All rise times were less than 400 μsec in the presence of oxygen.

Further, Takahashi et al. [57] reported no spectral change in an air atmosphere between 200 nsec and 2 msec for 6-nitro-BIPS in methanol and acetonitrile, a result that is consistent with Lenoble and Becker's findings [56]. Kaliskey and co-workers [58] collected data for a nitro-substituted BIPS in acetonitrile with many similarities to that of Lenoble and Becker and further noted that the buildup of the 560 nm absorption is concentration dependent, which implies a bimolecular reaction such as aggregation leads to the absorption changes in this spectral region.

Gorner and Chibisov [46–49, 51] also worked on the ring-opening reaction of 6-nitro-BIPS derivatives in polar solvents. Forming rapidly within the laser pulse, they also reported a transient with absorption maxima at around 440 nm and 600 nm for a variety of nitro-substituted compounds in various polar solvents. This transient is again quenched by oxygen and has a short lifetime of less than 10 μsec. These authors postulate that this species could be the merocyanine triplet state. A longer-lived species again assigned to the merocyanine had peaks at 400 nm and around 560 nm (depending on both the compound and solvent). This description is very similar to previously reported findings except for the assignment of the triplet state to that of the merocyanine. Further, Gorner and Chibisov report a blue shift in the merocyanine absorption envelope occurring on the longer microsecond time scale. Again, this blue shift has previously been reported for 6-nitro-BIPS in acetonitrile by both Takahashi et al. [57] and Kaliskey et al. [58]. Lenoble and Becker suggested that the shift was due to the breakup of merocyanine dimers that had formed by reactions occurring with excited triplet states. On the other hand, Gorner and Chibisov preferred invoking an equilibration with a second "cis" merocyanine isomer having a red shifted spectrum compared to the more planar trans form. Against this argument is that no stable isomers cis about the methine β-bond have been reported at room temperature in any NMR investigation and such isomers are not usually predicted to be stable [28,36,55]. Further, the fully cis isomer CCC or even CCT should be far less planar than a trans merocyanine and thus one would expect that such an isomer would have a spectrum that was blue-shifted with respect to the trans form. It may be possible that the blue spectral shift Gorner and Chibisov and other authors are observing is an isomeric redistribution of the stable planar isomers that are trans about the β-bond. Chibisov and Gorner's general scheme (for polar and nonpolar solvents) for nitro-substituted spiropyran in all solvents is shown in Scheme 6. In this case, the triplet state (440 nm and 600 nm) is an equilibrium between a perpendicular and a trans merocyanine form.

An alternative scheme (Scheme 7) proposed by Lenoble and Becker [56] will be developed and shown in the next section when we discuss the nonpolar solvent case. However, for an immediate comparison, in Scheme 7 the triplet state that was oxygen quenched is assigned to a nonplanar transient historically named ³X, which has an orthogonal spiropyran type of conformation, as opposed to the proposed triplet merocyanine of Gorner and Chibisov.

d. 6-Nitro-BIPS in Nonpolar Solvents. Ernsting et al. examined the fast picosecond photochemistry of 6-nitro-BIPS in *n*-pentane [29]. They noted

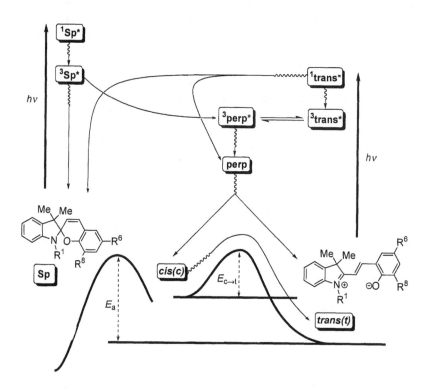

Scheme 6

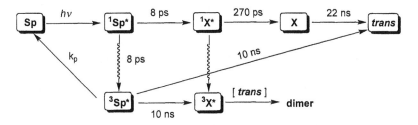

Scheme 7

that, after picosecond laser excitation, two bands grew, having peak maxima at 430 and 575 nm. These bands had the same rise time of ~20.5 psec. In the microsecond spectrum of the same system, the 430-nm peak had disappeared and the 575-nm peak had broadened. This behavior appears similar to the behavior in polar solvents and the initial 430- and 575-nm peaks are triplet states because they can be quenched by oxygen, as we will describe. As we shall see, the case of the nonpolar solvents is also complicated by aggregation phenomena.

Lenoble and Becker [56] carried out transient absorption on 6-nitro-BIPS in N_2-purged n-hexane solution. After an 11-nsec pulse (instrument response 400 nsec), they observed three new bands at 370, 430, and 570 nm and a shoulder at 630 nm. The absorption rise kinetics were second order with a rate constant of 2.6×10^6 mol^{-1} sec^{-1}. The 430-nm peak decayed with the same rate constant. Oxygen bubbling quenched both the species absorbing at 430 nm and the shoulder at 630 nm. In nitrogen-purged hexane, exciting with a 200 psec pulse and measuring transient absorption from times of 1–800 nsec monitoring at 370, 430, 530, and 570 nm, the absorption at 370, 530, and 570 nm had a 10-nsec first-order rise time. The 430-nm absorption had a 10-nsec rise and a several microsecond decay (beyond their detection time range). After bubbling, the solution with oxygen, the rise times at 370, 530, and 570 nm increased from 10 to 22 nsec, the 430-nm rise time decreased to 4 nsec, and its decay decreased to 33 nsec. Importantly, from these data, the kinetics of the bands at 370, 530, and 570 nm are assigned to merocyanine, whereas that of the band at 430 nm is assigned to a triplet intermediate with a short lifetime from 33 nsec to many microseconds, depending on the oxygen concentration. The absorption at 630 nm was ascribed to dimers that were suggested to form from the reaction between the merocyanine and the triplet species absorbing at 430 nm. In all cases, the dimerization or aggregation is characterized by the formation of a shoulder on the merocyanine spectrum at around 630 nm. Lenoble and Becker noted that this shoulder did not appear in an oxygen-purged solution, hence their involvement of the triplet state in the dimerization reaction. Lenoble and Becker's general reaction scheme [56] is shown in Scheme 7. This has been based not only on their work but also that of others. In this scheme, "Sp" is spiropyran, "trans" is a merocyanine isomer, and "X" is an intermediate retaining the parent orthogonal geometry, but with the spiro C—O bond broken. The triplet state ^3X is responsible for the oxygen-quenchable absorption at 430 nm. ^1X had been proposed by Alfimov et al. in picosecond studies for a transient forming in 8 psec [59]. Kaliskey and Williams had found a transient forming within 270 psec, which they assigned to a precursor to the trans merocyanine that was assigned to X ground state [58]. When oxygen is introduced, the triplet pathway through ^3X is quenched and "trans" merocyanine forms in 22 nsec exclusively via the X ground state. The oxygen prevents the formation of ^3X and then dimer bands are unable to form via "trans" and ^3X bimolecular reaction. In the absence of oxygen, ^3Sp forms the "trans" form directly in 10 nsec and also ^3X forms in 10 nsec. The transient absorption assigned to ^3X appeared to decay with the same rate constant as the shoulder due to aggregates forming in microseconds.

Takahashi et al. [57] also reported time-resolved resonance Raman spectra for 6-nitro-BIPS in cyclohexane. After 200-nsec, peaks were observed at 1600, 1550, 1523, 1480, 1265, 1234, 1212, and 1110 cm^{-1}. By 10 μsec, new bands at 1594, 1537, and 1358 cm^{-1} had formed while the other bands remained constant. Up to 25 μsec, the 1594, 1537, and 1358-cm^{-1} bands continued to increase in intensity. After 25 μsec, the bands at 1480, 1265, 1234, and 1212 cm^{-1} began to decrease until 2 msec, by which time they had disappeared. Note that for the same compound in acetonitrile and methanol, no time dependence was observed between 200 ns and 2 ms. These authors assigned the observed spectral changes to five transient species, again including four different merocyanine isomers. Again, we caution that such isomeric variance has not been found in NMR studies [36,55] and that the resonance Raman spectra will be variable with the varying absorption spectra that were reported by other authors [56,58]. Even aggregation that is known to occur in these alkane solutions could affect their spectra.

Photo-aggregation of the nitro-BIPS compounds continues to fascinate researchers and articles are being produced still regarding this phenomenon [24,38–45,60–62]. It is, of course, noteworthy before we continue with this discussion that spiropyran merocyanines are zwitterions possessing a significant dipole moment and that their stability in a nonpolar solvent will therefore be minimal. Spontaneous precipitation upon formation may be expected as the final outcome of the photo-formation of the merocyanine.

Krongrauz was the first to report on the formation of aggregates upon the photo-formation of 6-nitro-BIPS merocyanine [38–40]. He suggested that the aggregates were formed by the interaction between the spiropyran form and the merocyanine, possibly forming a CT complex. Since then, many mechanisms have been proposed for the dimerization and aggregation during the photoconversion of the spiropyran merocananines. Mostly these suggest the formation of aggregates involving spiropyan and the merocyanine; however, the final aggregate has been shown to be made exclusively of the merocyanine by Fourier-transform infrared (FTIR) and Raman analysis of the precipitates that formed after photolysis [42,44]. In this case, the lack of involvement of the spiropyran is apparent because of the lack of any band at around 1650 cm^{-1} attributable to the alkene C$=$C stretch of the closed-form pyran ring [44]. The spectrum of 6-nitro-BIPS in its spiropyran form compared to the spectrum for the same compound in its photo-aggregated form is shown in Fig. 2. These two spectra can be directly compared with the spectrum of the pure merocyanine of 6,8-dinitro-BIPS crystallized from acetone. From these spectra, we can see the C$=$C stretch of the closed spiropyran form is absent in the spectrum of the pure merocyanine and also in that of the photo-aggregates. This suggests that the photo-aggregates contain pure merocyanine. However, we cannot say that under certain circumstances, mixed aggregates of spiropyran and merocyanine may form, as suggested by Kaliskey et al. and Krongrauz et al. [38–40]. The carbonyl region of the aggregate FTIR spectrum is notably more complex than the pure merocyanine, which may just reflect the different intermolecular interactions in the aggregated form. Further, the position of the carbonyl peak is characteristic of a zwitterionic form, which has a relatively low bond order for a carbonyl group.

Sato et al. [42] measured resonance Raman spectra of 6-nitro-BIPS solid aggregates. They assigned the aggregates that they collected from cyclohexane as being the same as Takahashi groups "mixture" of isomers in acetonitrile [57] (not cyclohexane) because their spectra were similar. Sato et al. concluded that the aggregates were pure merocyanines with no ring-closed form present.

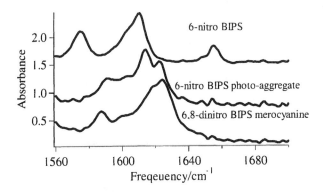

Figure 2 FTIR spectra of 6-nitro-BIPS spiropyran closed form, 6-nitro-BIPS photo-aggregate, and 6,8-dinitro-BIPS merocyanine.

2. Spiro-oxazine Ring-Opening Reaction

Spiro-oxazine is widely held to undergo its photochemical ring-opening reaction via the excited singlet state [4]. When studied using femtosecond and picosecond pulsed laser photolysis, the reaction is, in general, very fast and over in times of a few tens of picoseconds to a few nanoseconds, depending on the particular compound and solvent [26,63–65]. We should point out that this is not always the conclusion from nanosecond studies [66–68]. Nanosecond and picosecond pulsed laser experiments are not similar in that a nanosecond pulse can excite products that are produced over the entire duration of the laser pulse. There are clearly going to be fewer photo-products to re-excite within a shorter laser pulse. The femtosecond pulse, however, is notorious for its ultrahigh intensity and, as such, can produce its own artifacts that have to be fully considered.

Early picosecond studies were carried out by Schneider et al. [63] on the parent spiro-oxazine (NOSI1 in Scheme 8) and similar derivatives. In a back-to-back work, they also described a complimentary CARS (coherent anti-Stokes Raman spectroscopy) investigation [69]. Simply put, these authors found that the closed spiro-oxazine ring opened in 2–12 psec after laser excitation. The reaction was slower in more viscous solvents. An intermediate state formed within the excitation pulse and preceded the formation of merocyanine forms. This transient was named X in deference to the X transient named by Heiligman-Rim et al. for the spiropyran primary photoproduct [8]. (See also the previous section.) The name "X" has since been adopted by other workers for the spiro-oxazines [26,65].

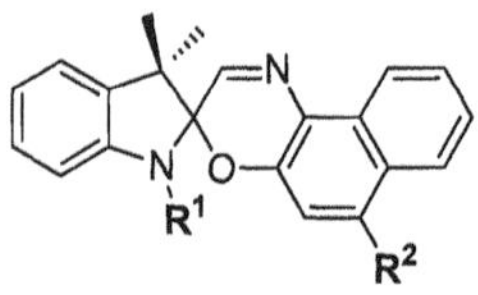

	R¹	R²
NOSI1	CH_3	H
NOSI2	CH_3	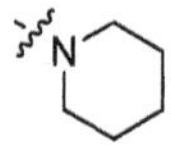 (piperidine)
NOSI3	CH_3	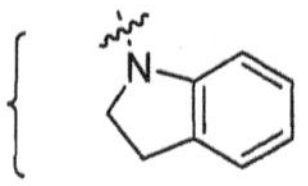 (indoline)
N-^{i}Bu-NOSI3	^{i}Bu	(indoline)
NOSI4	CH_3	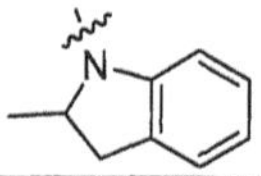 (indoline)

Scheme 8

Based on fluorescence lifetime measurements, the merocyanine form appeared to be a mixture of two or three isomers. The fluorescence had two or three component decays. Their CARS investigation backed up this conclusion of isomeric variance.

Aramaki and Atkinson were also active in work on the spiro-oxazines [65]. They noted that for NOSI1 in many polar and nonpolar solvents the picosecond time-resolved resonance Raman spectra simply built up over 50 psec with no shape evolution. The same finding was concluded from transient absorption measurements over the same time scale. The spectra/absorbances were then constant for 1.5 nsec. These authors suggest that only two isomers can be expected to contribute to the merocyanine spectra because those trans about the γ-methene bridge bond attached to the naphthalene ring are sterically crowded due to short interproton distances. There was no evidence for the X transient in their study; however, the 50-psec convoluted pulse profile may be expected to mask this short-lifetime species even if it were present.

Tamai and Masuhara [26] also worked on NOSI1, but in 1-butanol. They could examine femtosecond dynamics for the C—O bond breaking and formation of a primary photo-product X, which formed within 1 psec and had a broad absorption with peaks at 450 and 700 nm. The spectrum of X then evolved, forming a broad merocyanine-type spectrum, which itself evolved with time to form the usual merocyanine spectrum in that solvent after less than 400 psec. The spectral broadening was said to be either due to the formation of a vibrationally hot ground state or to an equilibration between isomeric forms because the spectrum that formed at early times was similar to the spectrum usually obtained in cyclohexane. Tamai's spectra are shown in Fig. 3.

Wilkinson et al. [64] worked in a range of solvents on a range of compounds using the same transient absorption experimental protocol. The compounds that they studied are shown in Scheme 8. While confirming the results of the previous authors [26] for the NOSI1, wherever experimental similarity existed, these authors also found several other transient phenomena for other compounds.

In the case of NOS13 and N-isobutyl NOS13 flashed in alcohol solutions and monitored at 436 nm, a rapid rise within 15 psec was followed by a rather slow decay over 4 nsec. After 4 nsec, the spectrum was the same as the merocyanine spectrum in that solvent. The spectral changes for N-isobutyl NOS13 are shown in Fig. 4. This 4-nsec decay was not observed in cyclohexane. Increasing the viscosity of the alcohol increased the lifetime of the longer transient decay, indicating that it was a bond-rotation step. However, the addition of the N-isobutyl group did not affect the rate of this long-lifetime decay. The longer-lifetime transient was assigned to a TCC isomer because if the CCC isomer was responsible for the absorption, then the presence of the N-isobutyl group would be expected to slow the decay rate, as the naphthalene ring would have to pass close to this bulky alkyl group during the conformational change to form the merocyanine

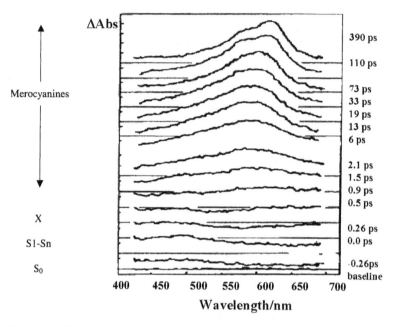

Figure 3   Subpicosecond time-resolved absorption spectra for the ring opening of NOSI1 in 1-butanol. (Data courtesy of N. Tamai [26].)

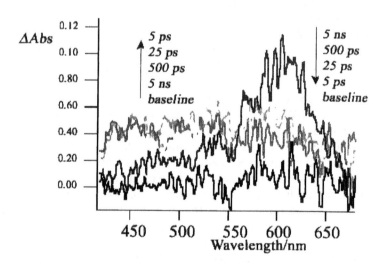

Figure 4   Time-resolved absorbance changes for *N*-isobutyl NOSI3 in 1-butanol from 5 psec to 5 nsec.

isomers trans about the central methine bond. This rate retardation is observed in the thermal ring-closure reaction, which, by the same logic, we therefore suggest occurs via the CCC isomer [3,70].

The spectrum of the assigned TCC transient is broad and featureless and, in this way, similar to transients assigned to the orthogonal "X" transient. It could be that in the present case, the observed transient is also similar to X, but even if this is the case, then the subsequent isomerization should occur through the TCC isomer. In Fig. 5, the rise and decay profiles for NOSI3 and N-isobutyl NOSI3 and the other compounds studied are shown.

The addition of the CH_3 group on NOSI4 lengthens the lifetime of the assigned TCC transient in 1-butanol, and this must be the result of twisting the 6'-indoline out of plane, reducing electron donation from the nitrogen group. This

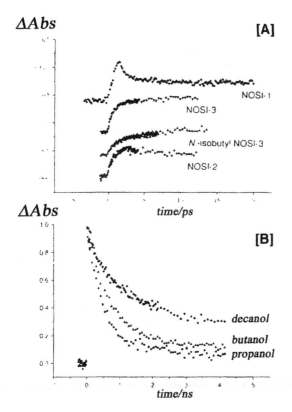

Figure 5 [A] Transient absorption rises for various NOSI compounds in 1-butanol. [B] Transient absorption rises and decays for N-isobutyl NOSI3 in different alcohols with different viscosities.

may suggest that the long-lifetime transient is bipolar with a negative charge on the oxygen. This polarity for the transient assigned as TCC is also indicated by the fact that it only appears to be stabilized in polar media.

The transient assigned to TCC formed more slowly in 1-butanol for *N*-isobutyl NOSl3 (12.3 psec) than for NOSl3 itself (7.6 psec); however, no solvent viscosity effect was observed among the formation times in 1-propanol (12.6 psec), 1-butanol (12.3 psec), and 1-decanol (13.0 psec) for the same *N*-isobutyl NOSl3.

For *N*-isobutyl NOSl3, the situation in cyclohexane was simple and the ring-opening reaction appeared to be over, with a lifetime of 5 psec. Note that only one probe wavelength (436 nm) was used, giving no spectral resolution. Hence, some transient states could not manifest even if they were there.

In acetonitrile *N*-isobutyl, NOSl3 had slightly more complex kinetics probing at 436 nm, exhibiting a fast rise within the excitation pulse followed by a decay with a lifetime of 54 psec. This was followed by a longer decay of 1.7 nsec. Similar transient species could be suggested in this case.

In the case of NOSl2, different kinetics were observed at 436 nm. In cyclohexane, there was a rapid rise, with a lifetime of 6.6 psec followed by a decay with a 100-psec lifetime. In 1-butanol, there was a rapid rise (lifetime = 4.3 psec), a decay (43-psec lifetime), and a second longer decay within a 1.4-nsec lifetime. These findings were confirmed by picosecond time-resolved resonance Raman spectroscopy. In these Raman studies in cyclohexane, a single rate constant was observed, whereas in 1-butanol, three spectral components grew with different time constants. The data were said to be consistent with the photo-formation of two or three isomers trans about the central methine bond; however, other transient species could be responsible for the observed kinetics because the absorption envelope obviously shifts and this would affect the resonance Raman bands.

When exciting with nanosecond pulsed lasers, Kellmann et al and Chibisov and Gorner independently noted a microsecond component to the formation of the merocyanine of NOSl1, NOSl3, and other compounds. There was no spectral shape change that occurred during this slower rise that could be associated with a normal trans merocyanine isomerization reaction; however, Chibisov and Gorner tentatively assigned this longer coloration component to an isomeric equilibration step. Bohne et al. [71,72] showed data for NOSl1 formation in cyclohexane and acetonitrile that did not show any longer component rise occurring in microsecond scales or less. Also, Hobley et al. [73] did not observe any microsecond order component of ring opening in toluene for NOSl3 or any of the compounds that they studied, including NOSl1. We can alternatively suggest that the reason for seeing this minor microsecond rise in the case of nanosecond excitation is that when the laser fluence is sufficiently high and conversion to the colored merocyanine form was significant, then the merocyanine itself may become excited within the laser pulse. As we will describe later, when merocyanines are irradiated, they can sometimes form a transient state that absorbs at around 300 nm [71,72], but not in the visible, and this state recovers to the usual merocyanine isomer. This is especially found to be so in nonpolar solvents. Gorner and Chibisov themselves report this transient bleaching phenomenon [74].

To finalize this section, we will comment on the isoindoline-derived spiro-oxazine NOSl5. This compound also photo-converts efficiently and very rapidly into their open merocyanine. In this case, the reaction is over within 30 psec and has a biexponetial rise with 2- and 15-psec lifetimes. The merocyanine spectrum does not evolve in time after this time. A broad absorption occurs around time zero, which is either the excited singlet state or the often hypothesized transient "X." The data for NOSl5 are shown in Fig. 6 The biexponential absorption rise

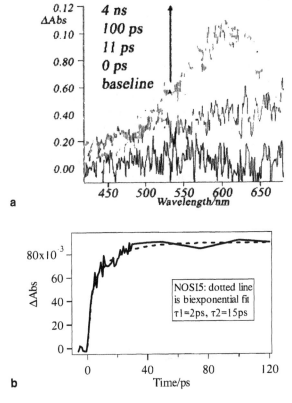

a

b

Figure 6 (a) Transient absorption spectra for NOSl5 irradiated with 388-nm 180 fsec light in 1-butanol. (b) Transient rise kinetics for NOSl5 in 1-butanol.

rather suggests that the broad transient is not due to the S_1 state absorption spectrum and should be a distinct transient formed outside of the excitation pulse.

3.   Naphthopyran Ring Opening

The naphthopyran ring-opening reactions have not been as well studied as they have for the spiropyrans and spiro-oxazines. Aubard et al. [75] recently reported that in acetonitrile and hexane, irradiation of CHR1 resulted in a broad transient spectrum after 0.8 psec, having three maxima at 360–370 nm, 500 nm, and 650 nm. At 1.8 psec, a well-defined band forms at 425 nm. From 10 to 100 psec, there is little further evolution except for a continued growth in the peak at 425 nm of about 15%. There is also a decrease in the overall bandwidth. The mechanism in Scheme 9 has been proposed, where B_1 B_2 and B_3, are isomers of the mero-form. Three isobestic points were identified in the transient spectra at different times, suggesting four transient states. Forming between 0.8 and 1.6 psec, the B_1 state was assigned as the cis isomer. This had a spectrum similar to that obtained for Tamai's "X" transient of the spiro-oxazine NOSI1, which was obtained at subpicosecond time scales [26].

The authors attribute the B_2 and B_3 states to the isomers TC and TT and support this claim with the statement that the thermal ring-closure reaction for this compound that lasts for tens of seconds shows two components. This fact alone means that their assignment is unlikely because any species equilibrating in 21 psec would not show two decay components over tens of seconds. Also, the isomerization between TT and TC forms of typical napthopyrans has been shown to result in a spectral shift [76,77], which is not observed in this case. Further, for the same and similar compounds, the two mero-forms have clearly resolved noncoalesced ^{1}H-NMR peaks [78,79] and the rate constant for

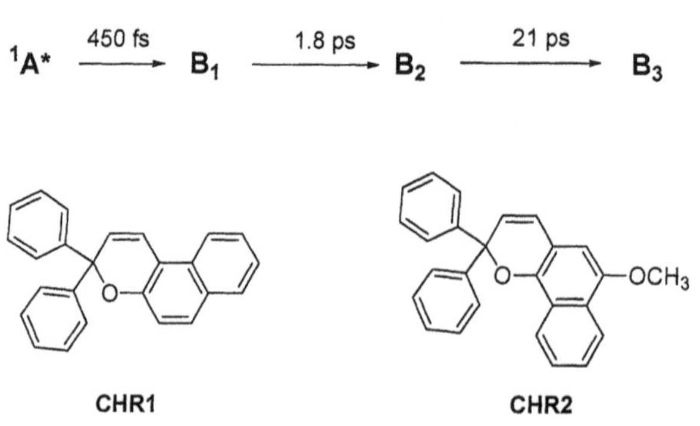

$$^1A^* \xrightarrow{\text{450 fs}} B_1 \xrightarrow{\text{1.8 ps}} B_2 \xrightarrow{\text{21 ps}} B_3$$

CHR1   CHR2

Scheme 9

the transformation of TT to TC is also reported to be 4 sec^{-1} [76]. It is likely that pumping the naphthopyran with 266-nm light results in a photo-product with excess vibrational energy, which dissipates with a lifetime of 21 psec, resulting in spectral narrowing as hot bands are cooled.

III. THE PHOTO-MERO-FORMS

A. Spectra

Spiropyran merocyanine spectra shift markedly to the blue as the solvent polarity increases [4,25], as shown in Fig. 7a for 6,8-dinitro-BIPS merocyanine. This is generally accepted to imply that they have a zwitterionic character caused by the donation of electron density from the indoline nitrogen to the phenolic C9 oxygen [4,25]. The rational behind this assignment to the zwitterion is based simply on the fact that if the zwitterion is in the highest occupied molecular orbital (HOMO) state, then the corresponding quinoidal resonant form is in the lowest unoccupied molecular orbital (LUMO) state and changing the solvent to a more polar one

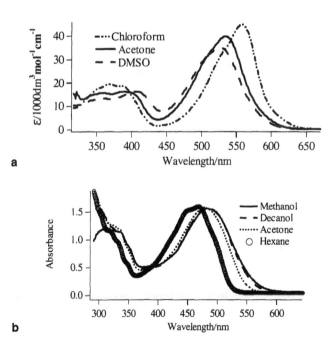

Figure 7 (a) Solvent shifts for 6,8-dinitro-BIPS merocyanine. (b) Solvent shifts for the CHR2 TT mero-form.

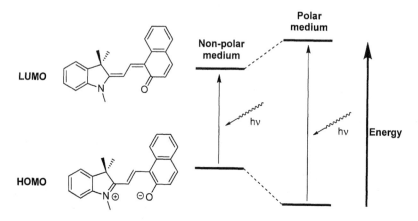

Scheme 10

relatively stabilizes the HOMO more than the LUMO, resulting in an increase in the $S_1 \leftarrow S_0$ energy gap. This is shown in Scheme 10.

The opposite direction of spectral shift with solvent polarity is observed for the spiro-oxazine [4] and naphthopyran mero-forms and this is generally accepted to infer a quinoidal HOMO and a zwitterionic LUMO state in their cases. This shift is shown also in Fig. 7b for a naphthopyran TT isomer (CHR2). Of course, H-bonding and other specific interactions will also affect the position of the mero-form spectral maximum. In the case of the substituted BIPS, spiro-oxazine, and CHR2, there is evidence to support their respective assignments to a zwitterionic and quinoidal forms, including x-ray [36,80–85] and NMR [36,55,86–88] data.

For the napthopyran CHR2, it can be easily shown that the two distinct merocyanine isomeric forms (TT and TC) have markedly different spectra [30]. This is because these merocyanines have a long lifetime and can be separated from each other by high-performance liquid crystallography (HPLC) [77] or even by crystallization [30]. The spectra of the different forms are shown in Fig. 8.

Clearly, the TC isomer has its spectrum red-shifted compared to the TT isomeric form. Although spectra for individual isomeric forms of spiro-oxazine and spiropyran merocyanines are not available, it has been shown that the different isomers have very different spectra. In fact, Abe and co-workers [89] have shown that the merocyanine of spiro-oxazine can be converted photochemically between two states having different absorption spectra. Even earlier work carried out at low temperature and or with visible-light irradiation suggested that the spiropyran merocyanine isomers also exhibit significant differences between their absorption spectra [6–8].

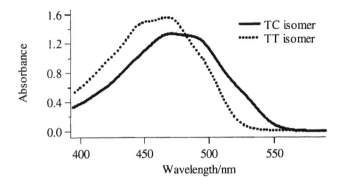

Figure 8 Mero-form spectra for CHR2.

We have already described how and why the mero-form spectra shift in solvents of varying polarity and now we will describe the specific properties of each type of mero-form.

B. Spiropyran Merocyanines

These spiropyran-derived merocyanines have large dipole moments and are essentially zwitterions. The dipole moment of the spiropyran zwitterion has been estimated to be 12.7 D (Debye) [90], 14.2 D, [91] 11.3 D [92], and 15–20 D [93] in different studies. This dipolar character is the key to understanding the behavior of these compounds. The excited-state dipole moment is reported to be lower than that in the ground state and this supports the conclusions based on observed spectral shifts. The central β-methine bond is the most double-bonded form and the two terminal methine bonds (α and γ) are of low bond order. The nitrogen carries a positive charge, whereas the phenolic oxygen carries a negative charge. Only the TTC and TTT isomeric forms have ever been identified using crystallographic techniques [36,80–85]. Usually, the TTC isomer crystallizes [80–85], but the TTT form has also been isolated once [36]. From their crystallographic work, Aldoshin et al. [80–84] proposed that the negative charge on the phenyl oxygen and the positive charge on the indoline nitrogen were delocalized along the path of conjugation with C3 and C4 (numbering as in Scheme 2), taking a partial negative and partial positive charge, respectively. These authors have also suggested that the phenyl oxygen is involved in H-bonding with C3—H. However, this is based on calculated C—H bond lengths. The crystal structures of the TTT and TTC forms are given in Fig. 9 for 6,8-dinitro-BIPS [36] and 6-nitro-8-bromo-BIPS [80], respectively. Selected bond length are given in Table 2 for the TTT isomer.

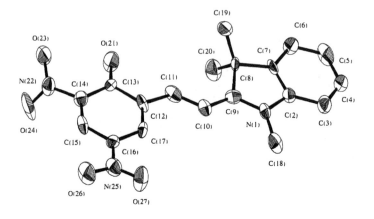

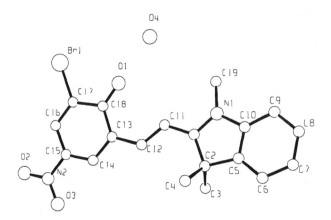

Figure 9 X-ray structures for TTT and TTC merocyanines of spiropyrans.

Recent studies have been carried out using ^{1}H-NMR and ^{13}C-NMR spectroscopy [36,55,94], the results of which lend overall support to the conclusions made from crystallographic data. The carbonyl carbon (C9) resonance at 182 ppm is characteristic of a partial double C9—O bond, and the C3 resonance at around 111 ppm and C4 resonance at 152 ppm illustrates that a charge separation exists between these two carbons. The C2' resonance at 169 ppm indicates a positive charge on the indoline nitrogen. Further, from ^{1}H-NMR, we note the

Table 2 Selected Bond Lengths for the 6,8-Dinitro BIPS TTT Merocyanine Isomer

Bond	Length (Å)	Bond	Length (Å)
N(1)—C(2)	1.410(4)	N(1)—C(9)	1.326(5)
N(1)—C(18)	1.452(4)	C(9)—C(10)	1.406(4)
C(10)—C(11)	1.349(5)	C(11)—C(12)	1.424(4)
C(12)—C(13)	1.455(6)	C(13)—O(21)	1.249(4)
C(14)—N(22)	1.446(5)	N(22)—O(23)	1.195(3)
N(22)—O(24)	1.215(4)	C(16)—N(25)	1.433(4)
N(25)—O(26)	1.226(3)	N(25)—O(27)	1.213(4)

position of the *N*-methyl group resonance at 3.94 ppm, which shows up the positive charge on the indoline nitrogen. The position of the C3—H is rather downfield, which is a probable result of its proximity to the phenyl oxygen on C9 in the more stable TTC isomeric form. ^{13}C-NMR data are summarized in Scheme 11 [94].

Nuclear Overhauser enhancement spectroscopy (^{1}H-^{1}H NOESY and NOE) experiments show only the TTC isomer in equilibrium with the TTT isomer for 6,8-dinitro-BIPS [36,55] and 6-nitro,8-bromo-BIPS. The TTC form dominates the equilibrium. Spectral broadening for several proton resonances in the spectra of 6,8-dinitro-BIPS and 6-nitro,8-bromo-BIPS indicate a rapid exchange between these two isomeric forms. The activation energy for this isomerisation is reported to be 43.6 kJ mol and the energy difference between the the TTC and TTT forms is 4.6 kJ mol [55].

The C3—H is labile and can be replaced by deuterium from D_2O or from MeOD. [36,55,94,95]. The fact that the C3 proton can undergo facile isotopic exchange in nonacidic conditions is further evidence for the negative charge on C3. Protonation of the spiropyran merocyanine forms appears to occur at C3

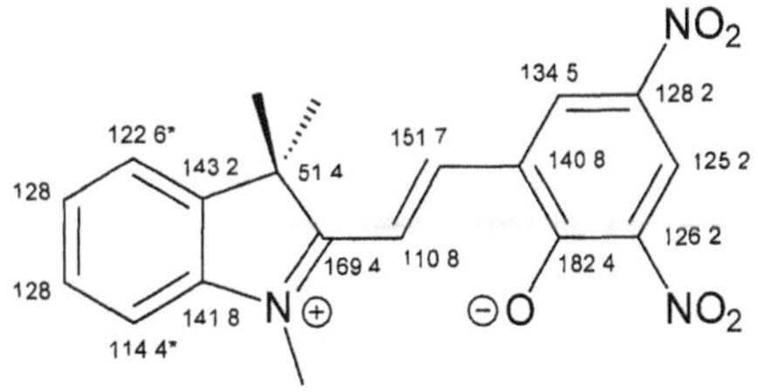

Scheme 11

when the reaction is conducted on the initially open form and this is confirmed by the reversible spectral change in the merocyanine visible absorption spectrum upon acidification [36]. In this case, the spectrum blue-shifts upon acidification to give a yellow colored species in line with a loss of conjugation. However, in the case where protonation is initiated via the ring-closed form, the reaction yields a product that is protonated at the phenolic oxygen and the open form retains a strong absorption in the red [96]. In this latter case, the protonation of the phenyl appears to protect the C3 proton from further attack. Capping the phenyl oxygen may prevent electron donation from the oxygen to C3 and this indicates the resonance involvement of the phenyl oxygen with the isotopic exchange reaction. Protonation of the phenyl oxygen also alters the ^{1}H-NMR spectrum significantly. The N-methyl protons shift to around 4.2 ppm, indicating that conjugation from the phenolic oxygen and the indoline nitrogen is blocked, leaving the nitrogen with a nearly full positive charge.

Facile deuterium isotopic exchange has facilitated a series of isotopomer studies using ^{13}C-NMR and solid-state deuterium-magic angle spin (D-MAS) NMR. [94,95,97]. The ^{13}C-NMR isotopomer study reveals long-range and large deuterium-isotope-induced line shifts at C9, which indicate an interaction between the C3 proton or deuteron with the C9 site, probably mediated through the C9—O bond. This supports Aldoshin's suggestion of a C—H bond to the C9—O [80–84]. Smaller isotope shifts were observed at the C9 site in the case of a C4-deuterated isotopomer even though the site of deuteration is closer to C9 through bond in this case. However, a C4—H bond is, in fact, further away from the C9 site if a through space dipole interaction is considered. The isotope shift map of the molecule 6,8-dinitro-BIPS is shown in Scheme 12.

Solid-state D-MAS NMR (solid-state NMR) on crystalline samples of C3-deuterated 6,8-dinitro-BIPS merocyanine, 6-nitro-8-bromo-BIPS merocyanine, and the 6-nitro-BIPS closed spiropyran form also indicates an asymmetric electric field gradient for the C3 deuteron [95]. This further supports the above suggestion that the negative charge on the phenolic oxygen significantly influences the chemical environment of the C3 hydrogen. Table 3 lists the parameters obtained from this study. The size, or strength, of the quadrupolar interaction on the deuterium is given by the quadrupole coupling constant (QCC):

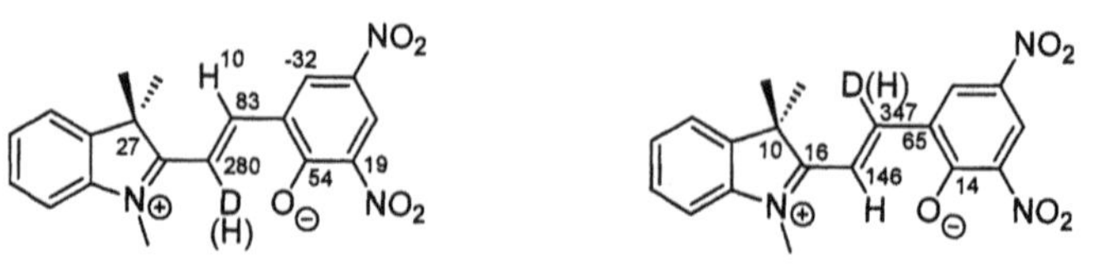

Scheme 12

Table 3 Isotropic ^{2}H-NMR Chemical Shifts (δ_{iso}) and Quadrupole Coupling Parameters (QCC, η) Determined Experimentally for Selected Molecules

Sample	δ_{iso} (ppm)	QCC (kHz)	η	Method[a]
NaDCO$_3$	12 (5)	149 (10)	0.182 (5)	MAS
NaDCO$_3$	13.5 (5)	149 (2)	0.20 (3)	MAS
3D-6-nitro-BIPS	4.7 (4)	174 (2)	0.00 (3)	MAS
3D-6,8-dinitro-BIPS	6.9 (5)	178 (2)	0.10 (3)	MAS
3D-6-nitro-8-bromo-BIPS	7.5 (5)	177 (2)	0.10 (3)	MAS
2D-1,3-diphenyl-1-hydroxypropene-3-one	nd	188.5 (1)	0.118 (2)	ADLF
benezene–D$_6$ (at 87 K)	nd	183 (1)		QE
			0.040 (5)	
1,4-dibromobenzene–D$_4$	nd	176 (5)	0.09 (2)	MAS
HDC=O	nd	170 (2)	< 0.15	μwave
α-(COOH)$_2$ ·2D$_2$O (at 203 K)	8.2	230 (2)	0.062 (6)	QE
		219 (2)	0.16 (3)	

Note: Error margin ($\pm$) of last decimal given in parentheses; nd = not determined.
[a] MAS = ^{2}H-MAS NMR spectroscopy at 298 K. ADLF = nuclear quadrupole resonance by level crossing (at 77 K). QE = quadrupole echo ^{2}H (static) NMR spectroscopy at low temperature of the solid state. μwave = microwave spectroscopy of the gaseous state.

$$\frac{e^2 \, q_{zz} \, Q}{h} \qquad [1]$$

Its magnitude is governed by the amount of electronic and nuclear nuclear charge that lies along the z axis along the C—D bond. The shape of the interaction (electric field gradient tensor) is described by the asymmetry parameter (η), which ranges from 0 to 1:

$$\eta = \frac{q_{xx} - q_{yy}}{q_{zz}} \qquad [2]$$

where q_{xx}, q_{yy}, and q_{zz} are the electric field gradient along the x, y, and z axis, respectively Q is the nuclear electric quadrupole moment.

From Table 3, we can see the that QCC does not increase substantially in going from the spiropyran to the merocyanines, suggesting that the C—D bond polarization is similar in each case or that any effect on the QCC due to C—D bond polarization changes is itself canceled by multiple effects such as bond-length changes or H-bonding. More significant is the nonzero value of η, which suggests that the electric field gradient experienced by the requisite deuteron is asymmetric. This can easily be imagined, as in the more stable TTC form, the deuteron is directed toward the C9—O bond, which is carrying a negative charge.

Theoretical studies have been carried out for the 6-nitro-BIPS merocyanine form [92]. The results of this work again show the probable delocalization of charge across the line of conjugation through the methine bridge between the C9—O and the indoline nitrogen. The relative charges on C3 and C4 are similar at around -0.2 in the closed spiropyran from; however, in the open form, C3 takes a negative charge of more than 0.5 and C4 takes a positive charge of around 0.1. The phenolic oxygen and the indoline nitrogen themselves have relative charges of around -0.6 and 0, respectively.

All studies show the zwitterionic nature of the spiropyran merocyanine beyond any doubt and we can further see how the acid–base characteristics of the C9 oxygen and indoline nitrogen are delocalized as alternating charges on the methine bridge, leading to the labile properties of the bridge itself.

C. Spiro-oxazine Merocyanines

The ^{1}H-NMR spectra of photo-produced spiro-oxazine merocyanines show a relatively low zwitterionic character. These have a typical N-methyl resonance at 3.59–3.68 ppm [85,87], as compared to 3.94 ppm for the spiropyran merocyanines [36,55] and around 2.74–2.94 ppm for the ring-closed forms [87,88]. Rys et al. found that protonation reactions of the merocyanine do not occur easily on the methine bridge, even when they are carried out under mildly acidic conditions (pH 3) [87]. Instead, they are restricted to the naphthalene oxygen and this was found even though the authors of this work suggested that the merocyanine is in thermal equilibrium with its protonated form. In this case, protonation of the naphthalene oxygen resulted in a shift of the N-methyl resonance to 4.47 ppm. Interestingly, the excited-state deuteration of the methine bridge carbon may be suggested by the fact that under prolonged pulsed laser irradiation of the merocyanine in a photo-stationary state in the presence of MeOD, some deuterium incorporation occurred [72]. This may be caused by the reaction of a zwitterionic merocyanine excited state with the solvent.

Two permanent merocyanines have been reported for the spiro-oxazines [85]. These were NOSI1 heteroanellated by imidazo [1,2-a]pyridine and imidazo [1,2-a]pyrimadine. Several tests have been conducted to determine the nature of these species. ^{1}H -NMR data show that the indoline nitrogen is not highly charged and the crystal structure indicates that the ground state is essentially the quinoidal form. The most stable form was found to be the TTC isomeric form by x-ray analysis. The dipole moment of these permanent spiro-oxazine merocyanines was around 3.84 D, which is much lower than the values reported for spiropyran merocyanines.

Two independent NOE experiments on NOSI1 merocyanine [88,98] show that the TTC isomer is the most stable form in solution, but one of these studies also reported that the CTC isomer is also present in equilibrium [88]. These two isomeric forms are in rapid exchange on the ^{1}H-NMR time scale, as we see only one resonance for the two types however, upon saturating the methine proton, significant NOE occurs at both the germinal methyl groups and the *N*-methyl groups. The *N*-methyl resonance at 3.68 ppm still indicates a partial positive charge on the indoline nitrogen, because in the closed form, the *N*-methyl resonance is at 2.94 ppm. This necessarily implies a reduced bond order for the terminal (α and γ) double bonds on the methine bridge and this would allow a ready rotation about these bonds for isomer exchange to occur.

D. Naphthopyran Mero-forms

Naphthopyran mero-forms have been studied using ¹H-NMR and NOE studies [78,79]. It has been concluded that there are two isomeric forms that are possible for the mero-form when the terminal methine carbon atom (C1 in Scheme 1 and C42 in Fig. 10) is symmetrically substituted, as shown in Scheme 1. A fourth species has been identified in photo-stationary states of naphthopyran and its mero-forms; however, this has not yet been clearly identified. The two main mero-isomers are not in rapid exchange, as two peaks can be clearly identified in the ¹H-NMR spectrum.

Ultraviolet spectroscopic studies also show that the interchange between the isomeric forms is slow [76]. In one reported case (CHR2), the thermal interchange between the TT and TC isomers is, in fact, blocked and the TT isomer can be crystallized once it has been formed [30,77]. Crystallographic studies on this permanent TT isomer show that the ground state is quinoidal. A quinoidal structure is also the intuitive choice because we expect electron donation from the C1

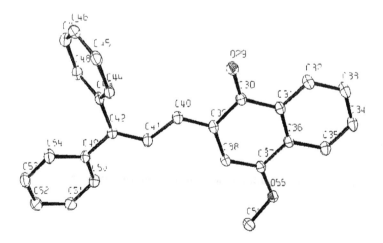

Figure 10 X-ray structure for the TT isomer of CHR2.

Table 4   Selected Bond Lengths for the
CHR2 TT Mero-form Crystal

Bond	Length (Å)
O(29)—C(30)	1.231
C(30)—C(39)	1.484
C(39)—C(40)	1.372
C(40)—C(41)	1.429
C(41)—C(42)	1.362
C(37)—O(55)	1.363
O(55)—C(56)	1.430

to the naphthalene oxygen to be limited. This accounts for why we observe separate ^{1}H-NMR resonances for the different naphthopyran mero-forms at room temperature, but not for spiro-oxazine and spiropyran merocyanines in which the terminal methine bond orders are reduced. The crystal structure for the CHR2 permanent TT mero-isomer is shown in Fig. 10, and the selected bond lengths of importance are given in Table 4.

IV.   RING CLOSURE

A.   Thermal Ring Closure

1.   Spiropyran Ring Closure

Spiropyran thermal ring closure is faster in nonpolar solvents than in polar solvents. This would be expected due to the polarity of this zwitterionic form, which will be relatively stabilized in a polar solvent, because a transition state would be expected to be less charge separated if it were cis about the central β-methine bond. The activation energy for ring closure is typically between 75 and 105 kJ mol depending on the solvent polarity [51]. The reaction is generally reported to be first order [25,51].

The thermal reverse reaction's value of ΔG^* increased from 77 to 92 to 111 kJ mol respectively for BIPS, 6-nitro-BIPS, and 6,8-dinitro-BIPS probably due to stabilization of the merocyanine form by delocalization of the negative charge on the C9 oxygen [37].

2.   Spiro-oxazine Ring Closure

Spiro-oxazine ring closure takes the opposite trend with respect to solvent polarity compared to the spiropyrans [4,25,68]. Again, we can rationalize this with the fact that the HOMO is quinoidal and not zwitterionic. The rate is slowed down significantly by the addition of a bulky group on the indoline nitrogen [3,70]. This observation suggests that a CCC intermediate is involved in ring closure because only this intermediate would be sterically hindred by the presence of a bulky group in this position. In fact, for a homologous series of spiro-oxazines, it was found that ΔH was rather constant, whereas the ΔS decreased as the alkyl chain length increased [70]. This suggests that the alkyl chain has to adopt a specific orientation prior to ring closure, moving out of the way of the advancing naphthalene ring. These data are shown in Table 5.

Abe and co-workers [89] worked on the spiro-oxazine system NOSI6 shown in Scheme 13 and found markedly different ring-closure kinetics. Note that in the case of Abe and co-workers' compound NOSI6, there are no unfavorable interactions between the proton on the naphthalene and the methine proton in either the TTT and CTT forms. In fact, TTT and CTT should be the most planar forms. Compare that to the case of the NOSI1 merocyanine, which was studied using ^{1}H-NOE and found to be an equilibrium mixture of TTC and CTC [88]. Ring closure would clearly be a more favorable series of rotations to form the closed-form geometry for CTC and TTC than it would be for CTT and TTT.

In the latter two cases, a single bond rotation would not lead to the oxygen coming close to the ring-closure geometry, whereas for TTC and CTC, it would. This may account for Abe and co-workers' finding that the NOSI6 merocyanine has a relatively long lifetime. There are analogous arguments for naphthopyran mero-forms, as we shall see [99].

Again, many groups have reported first-order kinetics for the ring-closure reaction [100–102]; however, Pottier et al. [103] and Chibisov and Gorner [68] report a second minor component to the fade kinetics occurring over tens of milliseconds. In the report by Chibisov and Gorner, it appears that there is no spectral shape evolution accompanying the fast component of the ring-closure reaction over the millisecond time scale. These authors propose a ground-state potential energy surface, with a minimum corresponding to a transient state. They

Table 5 Thermodynamic Properties for a Homologous Series of N-Substituted NOSI3 in Toulene

Compound	ΔH (kJ/mol)	ΔS (JK/mol)	E_a (kJ/mol)	k (294 K) (sec^{-1})
NOSI3	70.6	−16.6	73.2	0.077
N-Propyl NOSI3	69.3	−32.8	71.8	0.059
N-Isobutyl NOSI3	69.0	−37.6	71.6	0.040

Source: Ref. 70.

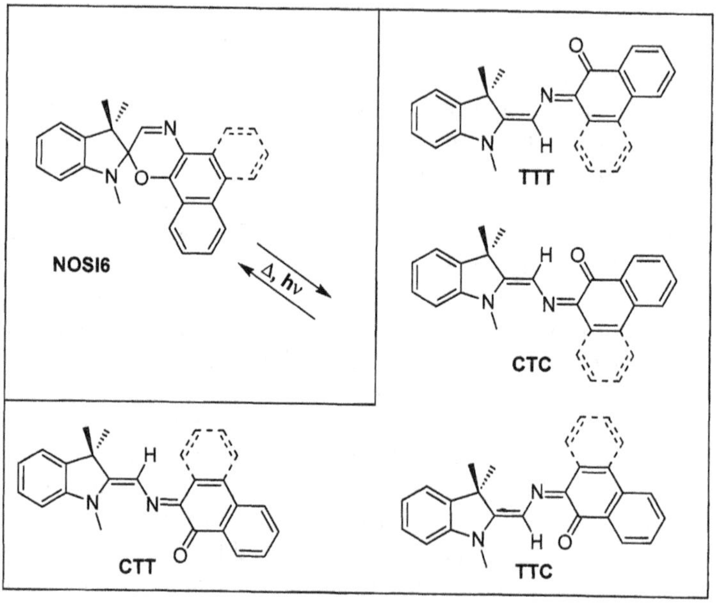

Scheme 13

propose that the thermal ring closure passes through a CCC isomer, which is in agreement with the proposed thermal fade reaction of other authors [64,70]. In this case the rapid component to their fade must correspond to an equilibration between a CCC or twisted isomer and the isomers that are trans about the β-bond on the methine bridge. The intermediate state should then form the closed form rather slowly. This sounds plausible because the central methine bond has a low bond order for the quinoidal merocyanines. The activation energies of both the fast and slow components were found to be similar [68]. Racemization of spiro-linked closed forms is sometimes possible to observe after elution from a chiral column and this would need to occur in seconds or even minutes [37] and may pass through a similar CCC or TCC intermediate. Chibisov and Gorner note that the minor component to the merocyanine fade rate is faster in nonpolar solvents, whereas the major component is slower in nonpolar solvents.

This finding would support a quinoidal structure (Scheme 14) in the thermal ring-closure reaction because if the nonpolar species were stabilized, it would increase its rate of formation and decrease its rate of ring closure by raising the overall activation barrier. Also the double β-bond in the zwitterionic form would be hard to twist.

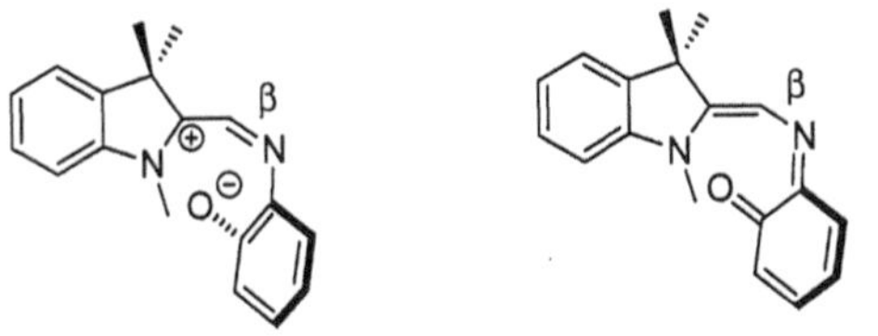

Scheme 14

3. Naphthopyran Ring Closure

In comparison to spiro-oxazine and spiropyran, the naphthopyran thermal ring-closure reaction is a rather slow process, taking tens of seconds at room temperature and often more. The closure reaction kinetics exhibit two clearly distinguishable components [76,78,79]. A third photo-product also exists in photo-stationary states [78,79]. This third photo-product is a minor form that has not been clearly assigned, but, logically, it may be a cis isomeric form due to the likely facile rotation about the central β-methine bond. This bond is close to being single bonded, so it would be a low activation process to form any mero-form cis about the β-bond. The main point regarding the naphthopyran mero-form fade kinetics is that the longer-lived mero-form is the TT isomer [30,76–79]. For both CHR1 and CHR2, the TT isomer cannot be expected to ring close in a single isomerization step about the β-bond, and the high bond order of the terminal α- and γ-methine bonds must hinder their rotation. We can expect that the TT isomer therefore becomes trapped. In the case of CHR1, there is an unfavorable interaction between one of the hydrogens on the naphthalene ring and the methine hydrogen on C2 (C41 in crystal structure) in the TT form [99]. This destabilizes the TT form and the ring-closure rate is increased. Probably, the ring closure from TT is as follows:

$$TT \rightarrow TC \rightarrow CHR1 \qquad [3]$$

In the case of CHR2, no steric crowding of protons occurs for the TT isomer, which can therefore adopt a planar geometry. In this case, the TT isomer is extremely stable and does not thermally fade [30,77].

V. PHOTOCHEMICAL RING CLOSURE AND MERO-FORM PHOTOCHEMISTRY

A. Spiropyran Photochemical Ring Closure and Merocyanine Photochemistry

The ring-closure reaction of 6,8-dinitro-BIPS and analoges substituted at the indoline nitrogen has been studied by two groups: one using femtosecond irradiation [31] (388 nm and 180 fsec) and the other using nanosecond 532-nm laser photolysis [104]. The femtosecond work is not in agreement with the nanosecond studies.

In the femtosecond work [31] in acetonitrile, there was a rapid bleaching of the merocyanine within the excitation pulse, followed by a fast kinetic component of around 0.5 psec assigned to relaxation of the S_1 level from vibrationally hot states. Next, there was biexponential partial ground-state recovery with lifetimes of around 70 and 350 psec. The absorption changes were matched by changes in the observed stimulated emissions from the merocyanine S_1 state, due to the probe beam arrival. This indicates that the S_1 state was the species being probed. Further, the two ground-state recovery lifetimes from the transient absorption studies were confirmed by fluorescence lifetime measurements. These data can be explained by the two possible mechanisms given in Schemes 15 and 16.

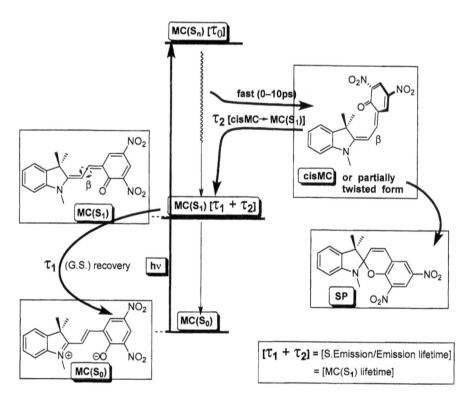

Scheme 15

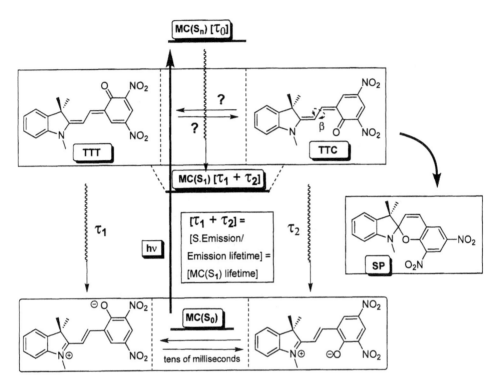

Scheme 16

The mechanism in Scheme 15 suggests the rapid formation of an intermediate that is cis about the central β-methine bond, which leads eventually to the ring-closed form but which can also re-form the S_1 state as the equilibrium S_1 concentration depletes. Scheme 16 is based on the fact that two possible conformers for the merocyanine have been identified (TTC and TTT) [36] and these may have different ground-state recovery times, but TTC and TTT generate the closed form with equal efficiency. Both of these schemes have merit and both have possible flaws; however, what is clear is that there is no triplet-state involvement and that the bleaching reaction generates the spiropyran closed form essentially within 1.5 nsec.

Uncertainties in the mechanisms are as follows. The cis intermediate in Scheme 15 must form very fast (possibly from hot S_1 states) but live in a large potential energy minimum in order to gradually recover to the S_1 state to give the longer-lived component that forms the ground state. This is possible but not proven. In favor of this scheme is that the 388-nm laser pulse is certainly forming a hot S_1 state because the merocyanine absorption maximum is in the visible region. In the alternative Scheme 16, there are apparently two isomers that have different ground-state recoveries. However, these two chemically distinct species do not seem to change their populations over time, even as they form the closed form and recover to the ground state. This is because we do not observe any spectral changes that we would expect if the two forms having different conformations and different electronic properties changed their relative concentration. In fact, the 1-nsec bleached-state spectrum was almost the exact inverse of the merocyanine room-temperature spectrum. This means that both TTC and TTT would either have identical or very similar spectra, which is hard to believe, or that they could happen to regenerate ring-closed forms with identical quantum efficiencies even though they have different S_1 lifetimes. Although possible, this is not proven. Further, in favor of Scheme 15 is that the amplitude of the longer component has an opposite sign to the faster component in the merocyanine visible absorption region and this is not likely if the two lifetimes are both due to similar merocyanine isomers, but it is possible if the longer component $\tau 2$ is caused by the repopulation of the merocyanine excited state from another species that does not absorb as strongly in the 400–600-nm region. These decay amplitudes are shown in Fig. 11.

Recall the Aramaki and Atkinsons result [52] that pumping the N-methoxy-ethyl NIPS closed form with 287-nm light and the merocyanine with 574-nm light in polar solvents both resulted in the formation of what they assigned as

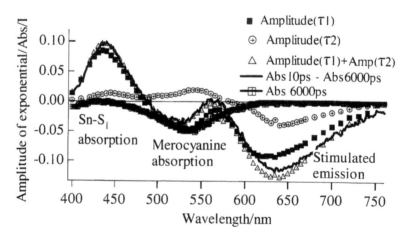

Figure 11   Transient absorption spectra and amplitudes of kinetic components for 6,8-dinitro-BIPS ring-closure reaction in acetonitrile.

the merocyanine S_1 level (Scheme 5), but this visible-light irradiation did not result in any bleaching. This highlights the fact that all systems differ upon changing the substituents, and facile ring closure or opening that occurs in one system cannot necessarily be applied to every system.

The nanosecond studies that have been performed on similar systens [48–51,104] are apparently at odds with the previously mentioned femtosecond study. However, 532-nm nanosecond laser pulses have different characteristics compared to a 388-nm femtosecond pulse and this may explain the discrepancies. Again, different substituents can also affect the outcome, and certainly the 6,8-dinitro-BIPS system is not a standard to gauge all other spiropyran systems. In fact, no typical molecule or molecular system exists which can be used as an example for describing the behavior of all other nominally similar systems.

Takeda et al. [104] found that the steady-state irradiation of N-octadecyl 6,8-dinitro-BIPS resulted in the formation of the closed form. This is in agreement with the femtosecond studies. However, the fade reaction showed three components when monitored at 500 nm in 1,2-dichloroethane. The first component was manifest as a bleach that occurred within the instrument response time of 25 nsec. This was followed by two longer components with 240-nsec and 3.4 µsec lifetimes. The spiropyran ring-closed form evolved with two exponential components with lifetimes of 620 nsec and 4 µsec. The N-octadecyl group may affect the outcome compared to the N-methylated form.

Gorner and Chibisov studied the photochemical ring-closure reaction of several spiropyran merocyanine forms [48–51]. Their results are generally in contrast to the case for the 6,8-dinitro-substituted compounds already described, which had essentially bleached to form the ring-closed form. The general conclusion for Gorner and Chibisov's experiments was that the major pathway was a cis to trans isomerization in the merocyanine form. For 6-nitro-BIPS, bleaching to the closed form did occur in the more polar solvents, especially in alcohols. They assigned two transient states: one to a triplet form of the merocyanine that could be quenched by oxygen and the other to a "cis" transient. In this case, they allude that the cis transient observed for 6-nitro, 8-bromo-BIPS may be associated with the isomerization observed for the same compound using ^{1}H-NMR that is assigned to the TTC to TTT transformation [55]. It seems from Gorner and Chibisov's discussion that they also consider the possible formation of an "essentially cis isomer structure," which appears to mean one cis component about the central β-methine bond. The proof for a stable merocyanine isomer that is cis about the central β-bond is not available and usually these isomers are not considered to be stable contributors to the spiropyran merocyanine absorption envelope or population. It may be that Gorner and Chibisov observed an "essentially cis" isomer or simply a population change from the equilibrium distribution essentially trans merocyanine isomers. Arguably, if a cis isomer is plausible for the quinoidal spiro-oxazine merocyanines, then it is not so logical to assign the same cis isomer in the case of the zwitterionic spiropyran merocyanines, because in the former case, β-bond rotation will be a low-energy process, whereas in the latter case, it will not. Also, examination of the spectra presented by Gorner and Chibisov revealed that the "essentially cis isomer" spectrum for the spiropyran merocyanine is apparently red-shifted [51], whereas that of the spiro-oxazine cis isomer that they propose to explain the fast component in the thermal fade reaction must be blue-shifted [68]. This means that any "essentially cis" isomer must have some through-space CT character from the C9—O to the indoline nitrogen. In view of all the facts, it seems most likely that Gorner and Chibisov observed an isomeric redistribution of the four stable planar isomers that are trans about the β-bond; however, in favor of their assignment is that the methine bond orders in the excited state would plausibly favor twisting about the central bond and not the α or γ bonds.

B. Spiro-oxazine Photochemical Ring Closure and Merocyanine Photochemistry

Schneider et al. [63] investigated the photochemistry of the spiro-oxazine merocyanines pumping and probing at 570 nm in acetonitrile. The found that the solution bleached within the <5-psec pulse duration. The bleached state recovered with at least a biexponential behavior, and from their fluorescence decay measurements, three exponentials were required to fit the decay. They attribute these findings to the possibility of three merocyanine isomers that are in equilibrium. Their compounds feature geminal ethyl groups on the indoline moieties and this may influence the system as compared to NOSI1.

Monti et al. [105] also used fluorescence lifetimes to monitor the merocyanine forms of NOSI1. In ethanol, they identified two components to the decay having lifetimes of 15 and 700 psec. The longer-lifetime decay is a very minor component. In the case of NOS12 in ethanol, 85% of the decay was attributed to the fastest component with a lifetime of 20 psec, but to fit the decay, it was necessary to use a further two components. This is in agreement with Wilkinson et al.[64], who suggested three components to the merocyanine formation from the NOS12 closed form based on picosecond time-resolved resonance Raman which they attributed to equilibration of three merocyanines trans about the β-bond. Monti et al. further found that in acetonitrile, NOSI1 had one component decay with a lifetime of 20 psec. Clearly, solvent and substituents are important factors.

Spiro-oxazine (NOSI1) photo-induced ring closure reactions were first described by Bohne et al., who used two-laser two-color excitation in the UV and visible regions [71,72]. In this work, they found that photoexciting the merocyanine in cyclohexane leads to a bleached product which recovers quantitatively to the merocyanine form over 30 μsec. The transient bleach state had an absorbance maximum at 300 nm but did not absorb in the visible. The recovery lifetime was 4 µsec in saturated hydrocarbons such as cyclohexane, but shorter (around 800 ns) in benzene and toluene. Bohne et al. did not assign a definite structure to their bleached state. In acetonitrile, this bleached state showed very little or no recovery and the final bleached state was assigned to the ring-closed form.

Chibisov and Gorner [68] also later reported that excitation of the spirooxazine merocyanines in the visible region (532 nm) resulted in the bleaching of the merocyanine absorption. This bleaching did not alter the merocyanine's spectral shape, but the absorption grew at the short-wavelength region below 400 nm; that is, the transient's spectrum was significantly blue-shifted. In nonpolar solvents, this bleached state recovered with first-order kinetics, but in nonpolar solvents, there was again nearly no recovery. Chibisov and Gorner assigned this transient to what they call a cis intermediate, which may mean a cis isomer about the central β-methine bond. This assignment would be reasonable in view of the blue shift in the transient spectrum that would be expected upon loss of planarity and conjugation. However, the methine bond orders in the excited state would plausibly favor twisting about the α or γ bonds and not the central methine bond.

Abe and co-workers [89] noted that in the case of his compound NOSI6 (Scheme 13), it was possible to transform the merocyanine absorption envelope by irradiation in the visible region of the spectrum. Hence, the different isomers must have different absorption spectra as we expect.

In summary, the merocyanine S_1 lifetime depends on its particular isomeric form and this, itself, is affected by the solvent and substituents. The longest merocyanine isomer singlet-state lifetimes reported are 4 nsec and the shorter ones are tens of picoseconds. After excitation of the merocyanine, a bleached state can form, which leads to the closed form in polar solvents, but the full ring closure is blocked in nonpolar media and the bleached state recovers to its original state. The bleached state is possibly an isomer cis about the central β-methine bond. Scheme 17 describes the transient bleaching of the merocyanine state.

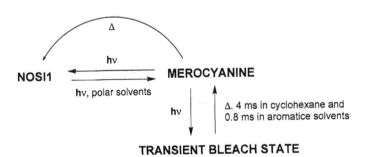

Scheme 17

Overall, the quantum yield of ring closure is reported to be low when measured in toluene [100] and this is in accordance with the transient studies. Further, in the transient study by Bohne et al. [71,72], the dye laser used to bleach the merocyanine form was rather powerful at 100–300 mJ/pulse. Even using this strong laser pulse on a merocyanine solution with an absorbance of ~0.25, they only permanently bleached the mero-form absorbance by one-third in acetonitrile and one can, therefore, imagine that the bleaching yield is also rather low in acetonitrile.

C. Naphthopyran Photochemical Ring Closure and Mero-form Photochemistry

The naphthopyran mero-forms have been the least studied in terms of their photochemistry. Recently, one naphthopyran TT merocyanine isomer, that of CHR2, was studied using steady-state, nanosecond, and femtosecond laser photolysis in several solvents pumping at 388 nm with a 200-fsec laser pulse. The photochemistry of the TT isomer was dominated by TT to TC isomerization, as shown in steady-state and nanosecond laser studies. The yield of the TT to TC photoreaction was inherently low in all solvents studied due to highly efficient recovery to the ground state, which occurred in competition with transient formation perhaps involving intersystem crossing. The yield was lowest of all in polar solvents. Several intermediate states were identified in the first 6 nsec of the reaction, which were tentatively assigned. These included a possible twisted MC T_1 state. Other distinct states were assigned to being TT-S_1, TT-T_1 and a further state to TC-T_1. The full photochemical conversion from TT to TC was a slower process than 6 nsec. The kinetics are summarized in Table 6. The overall scheme that was put forward to account for the data is shown in Scheme 18.

Clearly more data are needed to characterize the naphthopyran mero-form photochemistry, as it is not even unambiguously proven that single-photon excitation can bring about the full photo-transformation from the TT isomer to the TC

Table 6 Lifetimes of Transient States of the CHR2 TT Merocyanine in Different Solvents

Solvent	Dielectric constant	Viscosity (Pa s)	$\tau 1$ (psec)	$\tau 2$ (psec)	$\tau 3/\tau 4$ (psec)
Cyclohexane	2.0	0.9	14.7	—	Nonzero baseline (nzb)
Decanol	8.1	10.9	1.3	17.3	~4000 psec + nzb
Methanol	32.66	0.5929	2.1	12.5	~1500 psec + nzb
Cyclohexane	2.0	0.9	1.9	15.4	~300 psec + nzb

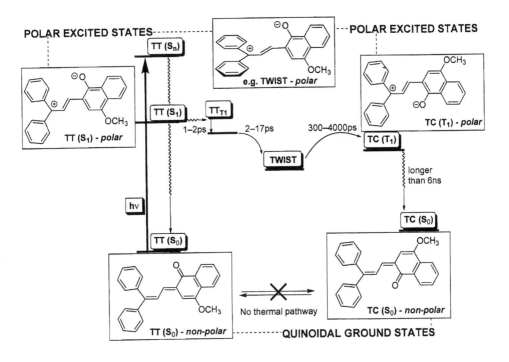

Scheme 18

isomer. It does, however, seem logical that the TT mero-form would tend to transform to the TC form in the excited state because the bond orders are reduced on the α and β bonds.

VI.   PHOTO-TRANSFORMATIONS IN THE CRYSTALLINE STATE

Spiropyran and the related compounds all undergo interesting photo-transformation in the crystalline state. There are examples of all these compounds undergoing both multi-photon- and single-photon-induced transformations in the crystalline state [77,106–110].

In the case of the single-photon transformation, this is not a general phenomenon and few examples exist. Recently, the crystalline-state photo-transformation of spiropyran [107], naphthopyran [77], and spiro-oxazine [106] chemical systems have been reported. None of these studies, however, give information about transient states except that the study on naphthopyran system described a longer-lived mero-form of CHR2 to be the TT isomer, as has been confirmed in solution. Further, there is no evidence that these are not simply surface reactions.

The work by Asahi and co-workers [108–111] on multiple-photon femtosecond excitation is perhaps more exciting and innovative. They reported that when crystals of NOSII are irradiated with a femtosecond laser with a fluence of 0.4 mJ/cm, they do not form merocyanines with long lifetimes, even after prolonged irradiaition. However, by using femtosecond time-resolved diffuse reflectance spectroscopy, an absorption at 495 nm due to the S_1 state is observed, which rapidly evolves (within a 3-psec lifetime) into a broad spectrum with two peaks at 460 and 760 nm. This spectrum is similar to the transient found by Tamai et al. [26] in their early studies on femtosecond photolysis in solution. In both the present case and in Tamai's work, the transient was assigned to a photoproduct "X" with the spiro C—O bond broken. This form has a lifetime of 2 nsec in the crystalline state, compared with 470 fsec in solution, and does not evolve into the open merocyanine form at low laser fluence in the solid state. Instead, this transient reforms the ground-state spiro-oxazine ring-closed form. These authors convincingly argue that this result is similar to the case when flashing the 2′-methylated form of NOSII [72], which forms a photo-product with a broken spiro C—O bond that cannot rotate to its ring open form due to steric hindrance. In this case, the photo-product also experiences a long lifetime and has a similar spectrum.

In the case of photoexcitation by an intense femtosecond laser pulse the irradiated powder becomes more permanently colored, gaining an absorption spectrum comparable to a usual merocyanine form. This photo-coloration is threshold dependent and does not occur at fluences below 0.5 mJ/cm. Thermally, this colored form bleaches with complex kinetic components that exist for between minutes and hours.

Observation of the spectral evolution with time was revealing. A species with an absorption maximum around 650 nm formed in times of 10 nsec. This transient was similar to that which formed in 1-butanol within 2 psec after femtosecond excitation [26], but thereafter rapidly evolved to its final colored form. In the crystalline state, however, this transient clearly had a longer lifetime. This transient absorption change was assigned to a nonplanar photo-product transforming to a metastable merocyanine in both the solution study and the crystalline-state study. The main difference between photolysis in the crystal as compared to that in solution is that the time scales for photo-transformation are drastically increased.

Upon absorption of an intense laser pulse by a solid sample, temperature is known to rise [112]; however, the cooling time should be a microsecond-order event. In the present study, the authors eliminated the possibility of thermal effects contributing to the photo-coloration by simply splitting the laser pulse and delaying the two pulses by up to a few nanoseconds. In this case, the overall two-

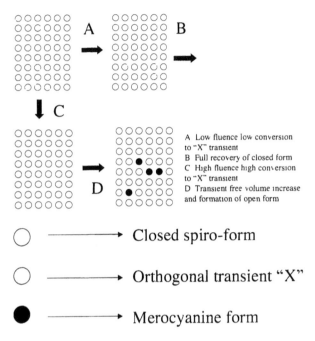

A Low fluence low conversion to "X" transient
B Full recovery of closed form
C High fluence high conversion to "X" transient
D Transient free volume increase and formation of open form

○ ──────→ Closed spiro-form

○ ──────→ Orthogonal transient "X"

● ──────→ Merocyanine form

Figure 12 Schematic for the mechanism of cooperative crystalline-state isomerization.

pulse coloration decreased as the time delay increased to 5 nsec, which is a similar time scale to the decay of the transients. This means that the effect is photochemical because no decay would occur if it were a thermal effect.

A cooperative model for isomerization was proposed to explain the effect of intense laser irradiation. In this model, the formation of excited or transient states in close proximity can transiently provide enough free volume for isomerization to occur. This work with NOSI1 has been complemented by studies using 6-nitro-BIPS and NIPS [109,110]. This cooperative model is shown in Figure 12.

VII. SENSITIZED RING OPENING

As a separate issue, the ring-opening reaction of the spirooxazine and the spiropyran can be sensitized by triplet energy donors[73,113–115]. A typical absorption rise for a benzophenone-sensitized ring-opening reaction is compared to that of the unsensitized fast absorption rise for NOSI3 in Fig. 13 [73,113]. In this case, both solutions were optically matched at the excitation wavelength.

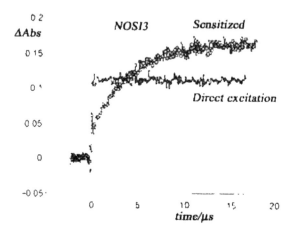

Figure 13   Sensitized ring opening compared to that following direct excitation.

Table 7   Quantum Efficiencies of Ring Opening by Direct Excitation and by Sensitization

Compound	Solvent	ε (dm³/mol/cm)	Φ
NOSI1 [73]	Cyclohexane (degas)	24.000	0.41
NOSI1 [73]	Cyclohexane	24.000	0.43
NOSI1 [73]	Ethanol	51.000	0.24
NOSI1 [73]	Toluene	31.000	0.33
NOSI2 [73]	Toluene	32.000	0.74
NOSI3 [73]	Toluene	48.000	0.64
NOSI3 [73]	Toluene (degas)	48.000	0.60
NOSI3 [73]	Cyclohexane	40.000	0.72
N-Isobutyl NOSI3 [73]	Toluene	56.000	0.72
NOSI4 [73]	Toluene	43.000	0.58
NOSI1/2′acetonaphthone (Et = 246 kJ/mol) [73]	Toluene	31.000	1.07
NOSI2/2′acetonaphthone [73]	Toluene	32.000	0.98
NOSI3/2′acetonaphthone [73]	Toluene	48.000	0.98
NOSI3/p-MeOacetonaphthone (Et = 301 kJ/mol) [73]	Toluene	48.000	0.97
NOSI1/camphorquinone [113]	Methylcyclohexane		1
NOSI1/fluoreneone [113]	Methylcyclohexane		1
NOSI2/camphorquinone (Et = 213 kJ/mol) [113]	Methylcyclohexane		0.54
NOSI2/fluorenone (Et = 222 kJ/mol) [113]	Methylcyclohexane		0.7

It is clear from this that the sensitized reaction has a higher quantum efficiency than the the reaction following direct photo-activation. A comparison of quantum efficiencies for direct and sensitized photo-ring opening is given in Table 7, showing that the sensitized process has an efficiency of unity.

The triplet energy for NOSI1 and NOSI3 has been determined to be around 200–212 kJ mol in two separate studies [73,114] using sensitizers of different energy to donate triplet energy to the ring-closed form. Note in Table 7 that the sensitized quantum efficiency tends to drop with lower triplet-energy donors. This triplet energy is probably dominated by the naphthoxazine part of the molecule, which is electronically isolated from the indoline ring system. The indoline ring would itself be expected to have higher triplet energy like that of indole (301 kJ/mol) [116]. Although a triplet state is available for the unsubstituted BIPS, NIPS, and NOSI compounds, there is no evidence that direct excitation leads to the formation of any triplet state and probably the ring-opening reaction via the excited singlet-state manifold is much faster than the intersystem crossing.

VIII. SUMMARY

In this chapter, we have attempted to summarize some 50 years of research on spiropyrans and related compounds. Although there are often conflicting conclusions regarding the photochemistry of these species, this must be expected as our knowledge evolves. Classifying compounds into definite categories is in itself a risky business for a photochemist, as changing solvents, substituents, and such like, can change everything. We can, however, offer a few general guidelines in this field.

Spiro-oxazines ring open via an excited singlet state very rapidly and often in less than a few picoseconds. In polar solvents, we often see a transient that can be assigned to a nonplanar ring-opened form and this is either the transient "X," which retains the orthogonal spiro-oxazine conformation or the CCT merocyanine isomer (arguably CCC). The reaction can proceed via a triplet manifold if it is sensitized. The efficiency of photoring closure is low, and in any case, this only occurs in polar solvents. In nonpolar solvents, a transient bleached state forms that recovers to the open merocyanine form. Thermal ring closure probably occurs via the CCC isomer, which appears to have some stability. The spiro-oxazine merocyanines are quinoidal and their isomeric composition is dominated by steric interactions. Solvent and substituents seem to affect the distribution. Only TTC and CTC isomers have been definitely proven to exist so far.

Unsubstituted spiropyrans similarly exhibit rapid ring-opening photochemistry to the spiro-oxazines also via a singlet state. One transient state has been identified also in polar solvents. This was assigned as S_1, but by analogy with the spiro-oxazines, this may be similar to "X" that was proposed by many workers. Photochemical ring closure appears to be inefficient and we often observe non-ring-closed intermediate states that are either cis or, more likely, trans about the β-bond. This photo-induced redistribution of isomers recovers to its original equilibrium. The excitation of the merocyanine form leads to the formation of some triplet state. Thermal ring closure again probably occurs via the CCC non-planar isomer.

Nitro-substitution especially at the 6-position of BIPS opens up a triplet pathway for photo-isomerization. This pathway runs in parallel to the singlet manifold. This increases the yield and, in turn, may lead to photo-aggregation that is observed for these compounds. Photochemical ring closure to the spiropy-ran form is more efficient for these 6-nitro-substituted compounds. The photo-chemistry of 6-nitro-BIPS merocyanine is similar to that of unsubstituted BIPS(s); however, the 6,8-dinitro compound efficiently cyclizes upon excitation to form the spiropyran closed form via a singlet manifold.

Spiropyran meocyanines are highly zwitterionic and exist in two observable merocyanine states. Only the TTC and TTT isomers have been proven to exist so far.

Naphthopyran rings open rapidly in a few picoseconds to form their mero-form isomers. The interchange between the isomers which form is very slow and sometimes does not even occur unless photo-activated. Photochemical ring clo-sure of the TT mero-isomer does not occur. Instead, it isomerizes to the TC mero-form. Thermal ring closure is rather a slow process showing two clear components due to the two isomeric forms.

REFERENCES

1. Crano, J. C.; Guglielmetti, R. J. eds. *Organic Photochromic and Thermochromic Compounds*; Plenum: New York, *1998*. Vol. 1.
2. Van Gemert, B. *Mol. Cryst. Liq. Cryst.* **2000**, *344*, 57–62.
3. McArdle, C. B. ed., *Applied Photochromic Polymer Systems*; Blackie: London, *1992*.
4. Durr, H.; Bouas-Laurent, H. eds. *Photochromism, Molecules and Systems*; Elsevier: Amsterdam, *1990*.
5. Berkovic, G.; Krongauz, V.; Weiss, V. *Chem. Rev.* **2000**, *100*, 1741–1753.
6. Fischer, E.; Hirshberg, Y. *J. Chem. Soc.* **1952**, 4522.
7. Heiligman-Rim, R.; Hirshberg, Y.; Fischer, E. *J. Phys. Chem.* **1962**, 66, 2465–2470.
8. Heiligman-Rim, R.; Hirshberg, Y.; Fischer, E. *J. Phys. Chem.* **1962**, 66, 2470–2477.
9. Chu, N. Y. C. *Can. J. Chem.* **1982**, *61*, 300.
10. Becker, R. S.; Michl, J. *J. Am. Chem. Soc.* **1966**, *88*, 5931.
11. Hobley, J.; Fukumura, H.; Goto, M. *Appl. Phys. A* **1999**, *69*, S945–S948.
12. Hobley, J.; Goto, M.; Kishimoto, M.; Fukumura, H.; Uji-i, H.; Irie, M. *Mol. Cryst. Liq. Cryst.* **2000**, *345*, 299–304.
13. Tamaki, T.; Ichimura, K. *Chem. Commun.* **1989**, 1477–1479.
14. Gorner, H.; Chibisov, A. *J. Chem. Soc. Faraday Trans.* **1998**, 94(17), 2557–2564.

15. Inouye, M. *Mol. Cryst. Liq. Cryst.* **1994**, *246*, 169–172.
16. Teranishi, T.; Yokoyama, M.; Sakamoto, H.; Kimura, K. *Mol. Cryst. Liq. Cryst.* **2000**, *344*, 271–276.
17. Pfiefer, U.; Fukumura, H.; Misawa, H.; Kitamura, N.; Masuhara, H. *J. Am. Chem. Soc.* **1992**, *114*, 4417–4418.
18. Pfiefer-Fukumura, U. *J. Photochem. Photobiol. A; Chem.* **1997**, *111*, 145–156.
19. Willner, I.; Willner, B. *Mol. Cryst. Liq. Cryst.* **2000**, *344*, 15–22.
20. Willner, I.; Rubin, S. *Angew. Chem. Int. Ed.* **1996**, *35*, 367.
21. Willner, I. *Acc. Chem. Res.* **1997**, *30*, 347.
22. Kawata, S.; Amistoso, J. O. *Mol. Cryst. Liq. Cryst.* **2000**, *344*, 23–30.
23. Raymo, F. M.; Giordani, S. *J. Am. Chem. Soc.* **2001**, *123*, 4651–4652.
24. Tamaoki, N.; Van Keuren, E.; Matsuda, H.; Hasegawa, K.; Yamaoka, T. *Appl. Phys. Lett.* **1996**, *69*(9), 1188–1190.
25. Bertelson, R. C. *Techniques of Chemistry III*; G. H. Brown, ed.; Wiley–Interscience: New York, *1971*.
26. Tamai, N.; Masuhara, H. *Chem. Phys. Lett.* **1992**, *191*(1,2), 189–194.
27. Ernsting, N. P. *Chem. Phys. Lett.* **1989**, *159*(5,6), 526–531.
28. Ernsting, N. P.; Dick, B.; Arthen-Engeland, T. *Pure Appl. Chem.* **1990**, *62*(8), 1483–1488.
29. Ernsting, N. P.; Arthen-Engeland, T. *J. Phys. Chem.* **1991**, *95*, 5502–5509.
30. Hobley, J.; Malatesta, V.; Hatanaka, K.; Kajimoto, S.; Williams, S. L.; Fukumura, H. *Phys. Chem. Chem. Phys.* **2002**, *4*, 180–184.
31. Hobley, J.; Pfiefer-Fukumura, U.; Bletz, M.; Asahi, T.; Masuhara, H.; Fukumura, H. *J. Phys. Chem.* **2002**, *106*, 2265–2270.
32. Toppet, S.; Quintens, W.; Smets, G. *Tetrahedron* **1975**, *31*, 1957–1958.
33. Marevtsev, W. S.; Zaichenko, N. L.; Ermakova, V. D.; Beshenko, S. I.; Linskii, V. A.; Gradyushko, A. T.; Cherkashin, M. I. *Izv. Akad. Nauk SSSR, Ser. Khim.* **1981**, 1591–1596 (English translation).
34. Zaichenko, N. L.; Lyubimov, A. V.; Marevtsev, V. S.; Cherkashin, M. I. *Izv. Akad. Nauk SSSR, Ser. Khim.* **1988**, 1543–1544 (English translation).
35. Eggers, L.; Buss, V. *Angew. Chem. Int. Ed.* **1997**, *36*(8), 881–883.
36. Hobley, J.; Malatesta, V.; Millini, R.; Montanari, L.; Neil Parker, Jr. W. O. *Phys. Chem. Chem. Phys.* **1999**, *1*, 3259–3267.
37. Kiesswetter, R.; Pustet, N.; Brandl, F.; Mannschreck, A. *Tetrahedron Asymm.* **1999**, *10*, 4677–4687.
38. Krongauz, V. A.; Parshutkin, A. A. *Photochem. Photobiol.* **1972**, *15*, 503–507.
39. Krongauz, V. A.; Fishman, S. N.; Goldburt, E. S. *J. Phys. Chem.* **1978**, *82*(23), 2469–2474.
40. Krongauz, V. A.; Goldburt, E. S. *Nature* **1978**, *271* (5 Jan), 43–45.
41. Onai, Y.; Kasatani, K.; Kobayashi, M.; Shinohara, H.; Sato, H. *Chem. Lett.* **1990**, 1809–1812.
42. Sato, H.; Shinohara, H.; Kobayashi, M.; Kiyokawa, T. *Chem. Lett.* **1991**, 1205–1208.
43. Li, Y.; Zhou, J.; Wang, Y.; Zhang, F.; Song, X. *J. Photochem. Photobiol. A: Chem.* **1998**, *113*, 65–72.

44. Uznanski, P. *Syn. Met.* **2000**, *109*, 281–285.
45. Tachibana, H.; Yamanaka, Y.; Sakai, H.; Abe, M.; Matsumoto, M. *J. Lumine.* **2000**, *87–89*, 800–802.
46. Gorner, H.; Atabekyan, L. S.; Chibisov, A. K. *Chem. Phys. Lett.* **1996**, *260*, 59–64.
47. Chibisov, A. K.; Gorner, H. *J. Photochem. Photobiol. A: Chem.* **1997**, *105*, 261–267.
48. Gorner, H. *Chem. Phys.* **1997**, *222*, 315–329.
49. Chibisov, A. K.; Gorner, H. *J. Phys. Chem. A* **1997**, *101*, 4305–4312.
50. Chibisov, A. K.; Gorner, H. *Phys. Chem. Chem. Phys.* **2001**, *3*, 424–431.
51. Gorner, H. *Phys. Chem. Chem. Phys.* **2001**, *3*, 416–423.
52. Aramaki, S.; Atkinson, G. H. *J. Am. Chem. Soc.* **1992**, *114*, 438–444.
53. Takahashi, H.; Yoda, K.; Isaka, H.; Ohzeki, T.; Sakaino, Y. *Chem. Phys. Lett.* **1987**, *140*(1), 90–94.
54. Yuzawa, T.; Ebihara, K.; Hiura, H.; Ohzeki, T.; Takahashi, H. *Spectrochimio. Acta* **1994**, *50A*(8,9), 1487–1498.
55. Hobley, J.; Malatesta, V. *Phys. Chem. Chem. Phys.* **2000**, *2*, 57–59.
56. Lenoble, C.; Becker, R. S. *J. Phys. Chem.* **1986**, *90*, 62–65.
57. Takahashi, H.; Murakawa, H.; Sakaino, Y.; Ohzeki, T.; Abe, J.; Yamada, O. *J. Photochem. Photobiol. A: Chem.* **1988**, *45*, 233–241.
58. Kaliskey, Y.; Orlowski, T. E.; Williams, D. J. *J. Phys. Chem.* **1983**, *87*, 5333–5338.
59. Krysanov, S. A.; Alfimov, M. V. *Chem. Phys. Lett.* **1982**, *91*(1), 77–80.
60. Kaliskey, Y.; Williams, D. J. *Chem. Phys. Lett.* **1982**, *86*(1), 100–104.
61. Kaliskey, Y.; Williams, D. J. *Macromolecules* **1984**, *17*, 292–296.
62. Tachibana, H.; Yamanaka, Y.; Sakai, H.; Abe, M.; Matsumoto, M. *Mol. Cryst. Liq. Cryst.* **2000**, *345*, 149–154.
63. Schneider, S.; Mindl, A.; Elfinger, G.; Melzig, M. *Ber. Bunsenges Phys. Chem.* **1987**, *91*, 1222–1224.
64. Wilkinson, F.; Worrall, D. R.; Hobley, J.; Jansen, L.; Williams, S. L.; Langley, A. J.; Matousek, P. *J. Chem. Soc. Faraday Trans.* **1996**, *92*(8), 1331–1336.
65. Aramaki, S.; Atkinson, G. H. *Chem. Phys. Lett.* **1990**, *170*(2,3), 181–186.
66. Kellmann, A.; Tfibel, F.; Dubest, R.; Levoir, P.; Aubard, J.; Pottier, E.; Guglielmetti, R. J. *J. Photochem. Photobiol. A: Chem.* **1989**, *49*, 63.
67. Kellmann, A.; Tfibel, F.; Guglielmetti, R. J. *J. Photochem. Photobiol. A: Chem.* **1995**, *91*, 131.
68. Chibisov, A. K.; Gorner, H. *J. Phys. Chem. A* **1999**, *103*, 5211–5216.
69. Schneider, S.; Mindl, A.; Elfinger, G.; Melzig, M. *Ber. Bunsenges. Phys. Chem.* **1987**, *91*, 1225–1228.
70. Hobley, J. Ph.D. thesis, Loughborough University of Technology, *1994*.
71. Bohne, C.; Fan, M. G.; Li, Z. J.; Lusztyk, J.; Scaiano, J. C. *J. Chem. Soc., Chem. Commun.* **1990**, 571.
72. Bohne, C.; Fan, M. G.; Li, Z. J.; Liang, Y. C.; Lusztyk, J.; Scaiano, J. C. *J. Photochem. Photobiol. A: Chem.* **1992**, *66*, 79–90.
73. Hobley, J.; Wilkinson, F. *J. Chem. Soc. Faraday Trans.* **1996**, *92*(8), 1323–1330.
74. Chibisov, A. K.; Gorner, H. *J. Phys. Chem.* **1999**, *103*, 5211–5216.
75. Aubard, J.; Maurel, F.; Buntinx, G.; Poizat, O.; Levi, G.; Guglielmetti, R.; Samat, A. *Mol. Cryst. Liq. Cryst.* **2000**, *345*, 215–220.

76. Ottavi, G.; Favaro, G.; Malatesta, V. *J. Photochem. Photobiol. A: Chem.* **1998**, *115*, 123–128.
77. Hobley, J.; Malatesta, V.; Millini, R.; Giroldini, W.; Wis, L.; Goto, M.; Kishimoto, M.; Fukumura, H. *Chem. Commun.* **2000**, 1339–1340.
78. Delbaere, S.; Luccioni-Houze, B.; Bochu, C.; Teral, Y.; Campredon, M.; Vermeersch, G. *J. Chem. Soc. Perkin Trans. 2* **1998**, 1153–1157.
79. Delbaere, S.; Micheau, J. C.; Teral, Y.; Bochu, C.; Campredon, M.; Vermeersch, G. *Photochem. Photobiol.* **2001**, *74*(5), 694–699.
80. Aldoshin, S. M.; Atovmyan, L. O.; D'yachenko, O. A.; Gal'bershtam, A. *Izv. Akad. Nauk SSSR, Ser. Khim.* **1982**, *12*, 2262–2270 (English translation).
81. Aldoshin, S. M.; Atovmyan, L. O. *Izv. Akad. Nauk SSSR, Ser. Khim.* **1986**, *9*, 1859–1865 (English translation).
82. Aldoshin, S. M.; Atovmyan, L. O.; Kozina, O. A. *Izv. Akad. Nauk SSSR, Ser. Khim.* **1988**, *5*, 914–917 (English translation).
83. Aldoshin, S. M. *Mol. Cryst. Liq. Cryst.* **1994**, *246*, 207–214.
84. Nakatsu, K.; Yoshioka, H. *Kwansei Gakuin Univ. Nat. Sci. Rev.* **1996**, *1*, 21–38.
85. Lareginie, P.; Lockshin, V.; Samat, A.; Guglielmetti, R.; Pepe, G. *J. Chem. Soc. Perkin Trans. 2.* **1996**, 107–111.
86. Zaichenko, N. L.; Lyubimov, A. V.; Marevtsev, V. S.; Cherkashin, M. I. *Izv. Akad. Nauk SSSR, Ser. Khim.* **1988**, *5*, 941–945 (English translation).
87. Rys, P.; Weber, R.; Wu, Q. *Can. J. Chem.* **1993**, *71*, 1828–1833.
88. Delbaere, S.; Bochu, C.; Azaroual, N.; Buntinx, G.; Vermeersch, G. *J. Chem. Soc. Perkin Trans. 2* **1997**, 1499–1501.
89. Horii, T.; Miyake, Y.; Nakao, R.; Abe, Y. *Chem. Lett.* **1997**, 655–656.
90. Lapienis-Grchowska, D.; Kryszewsky, M.; Nadolski, B. *J. Chem. Soc. Faraday Trans.* **1979**, 2(75), 312.
91. Levitus, M.; Glasser, G.; Neher, D.; Aramendia, P. F. *Chem. Phys. Lett.* **1997**, *277*, 118–124.
92. Cottone, G.; Noto, R.; La Manna, G.; Fornili, S. L. *Chem. Phys. Lett.* **2000**, *319*, 51–59.
93. Bletz, M.; Pfeifer-Fukumura, U.; Kolb, U.; Bauman, W. *J. Phys. Chem.* **2002**, *106*, 2232–2236.
94. Hobley, J.; Malatesta, V.; Giroldini, W.; Stringo, W. *Phys. Chem. Chem. Phys.* **2000**, *2*, 53–56.
95. Neil Parker, W. O., Jr.; Hobley, J.; Malatesta, V. *J. Phys. Chem.*, *106*, 4028.
96. Gabbutt, C. D.; Hepworth, J. D.; Heron, B. M. *Dyes Pigments* **1999**, *42*, 35–43.
97. Hobley, J.; Malatesta, V.; Millini, R.; Niel Parker, W. O., Jr. *Mol. Cryst. Liq. Cryst.* **2000**, *345*, 329–334.
98. Nakamura, S.; Uchida, K.; Murakami, A.; Irie, M. *J. Org. Chem.* **1993**, *58*, 5543–5545.
99. Kumar, A.; Van Gemert, B.; Knowles, D. B. *Mol. Cryst. Liq. Cryst.* **2000**, *344*, 217–222.
100. Wilkinson, F.; Hobley, J.; Naftaly, M. *J. Chem. Soc. Farady Trans.* **1992**, *88*, 1511.
101. Favoro, G.; Masetti, F.; Mazzucato, U.; Ottavi, G.; Allegrini, P.; Malatesta, V. *J. Chem. Soc. Farady Trans.* **1994**, *90*, 333.

102. Favoro, G.; Malatesta, V.; Mazzucato, U.; Ottavi, G.; Romani, A. *J. Photochem. Photobiol. A: Chem.* **1995**, *87*, 235.

103. Pottier, E.; Dubest, R.; Guglielmetti, R.; Tardieu, P.; Kellmann, A.; Tfibel, F.; Levoir, P.; Aubard, J. *Helv. Chim. Acta* **1990**, *73*, 303.

104. Takeda, J.; Ikeda, Y.; Mihara, D.; Kurita, S.; Sawada, A.; Yokoyama, Y. *Mol. Cryst. Liq. Cryst.* **2000**, *345*, 191–196.

105. Monti, S.; Malatesta, V.; Bortolus, P.; Magde, D. *Photochem. Photobiol.* **1996**, *64*(1), 87–91.

106. Benard, S.; Yu, P. *Chem. Commun.* **2000**, 65–66.

107. Godsi, O.; Peskin, U.; Kapon, M.; Natan, E.; Eichen, Y. *Chem. Commun.* **2001**, 2132–2133.

108. Asahi, T.; Masuhara, H. *Chem. Lett.* **1997**, 1165–1166.

109. Suzuki, M.; Asahi, T.; Masuhara, H. *Mol. Cryst. Liq. Cryst.* **2000**, *345*, 51–56.

110. Suzuki, M.; Asahi, T.; Masuhara, H. *Phys. Chem. Chem. Phys.* **2002**, *4*, 185–192.

111. Asahi, T.; Suzuki, M.; Masuhara, H. *J. Phys. Chem.* **2002**, *106*, 2335–2340.

112. Fukumura, H.; Hatanaka, K.; Hobley, J. *J. Photochem. Photobiol. C. Rev.* **2001**, *2*, 153–167.

113. Favaro, G.; Malatesta, V.; Mazzucato, U.; Ottavi, G.; Romani, A. *Mol. Crys. Liq. Cryst.* **1994**, *246*, 299–302.

114. Eloy, D.; Escaffre, P.; Gautron, R.; Jardon, P. *J. Chem. Phys.* **1992**, *89*, 897.

115. Reeves, D. A.; Wilkinson, F. *J. Chem. Soc. Faraday Trans.* **1973**, *69*, 1381–1390.

116. Scaiano, J. C. *Handbook of Photochemistry*; CRC Press: Cleveland, OH, *1989*. p. 382.

Index

Alkene-oxygen complex, 291
Allenes, thioselenation, 199
α,β-epoxy ketone, 191
Aromatics-oxygen complex, 292
Aza-di-π-methane (ADPM), 5
  electron transfer, 22–27
  mechanism, 21
  reactive state, 21
  sensitized by electron acceptors,
    22–30
  sensitized by electron donors, 30–33

Benzopyrans, photoreactions, 355
β,γ-unsaturated aldehyde
  competition between ODPM and
    DPM, 14
  ODPM, 8–20
    examples, 10
    limits, 19
    reactive state, 13, 17
β,γ-unsaturated imine, 6
β,γ-unsaturated oxime, 7
Biscarbenes, 1,2-phenylene linked, 172,
  174
Biscarbenes, 1,3-phenylene linked, 159
  electronic states, 160

Biscarbenes, 1,4-phenylene linked, 142
  computations, 145
  spin state, 143
Bischlorocarbene, 1,4-phenylene linked,
  155
Bisfluorocarbene, 1,4-phenylene linked,
  157
Bisnitrenes, 1,4-phenylene
  computations, 151
  IR studies, 150

Carbenes and nitrites, computational
  studies, 138
Charge separation, 230
Charge shift reaction, 186
Chromenes, photoreactions, 355
Closure molecules, 334

Diphenylhexatriene-zeolite, 323
Distyrylbenzenes, photoanimation, 204
Donor-acceptor linked assemblies, 229
Dye
  distribution within zeolites, 317
  occupational probability within
    zeolites, 318
Dyes in zeolites, 278

Dyes, structures, 313
Dye-zeolite composites, 311

Electron transfer, 185
  cosensitization, 189
  dimerization reactions, 192
  donor acceptor complexes, 209
  energetics, 186
  fragmentation reactions, 190
  metal ion effect, 253
  Paterno-Buchi reaction, 217
  photoanimation, 197, 203
  photocarbonylation, 200
  photocyclization, 201, 204
  photocycloaddition, 214
  photo-NOCAS reaction, 213
  photooxygenation, 199, 251
  photothioselenation, 199
  radical addition, 195
  radical anion chemistry, 206
  sensitized, 188
  sensitizers, 189
  silyl enol ethers, 201
  solvent effect, 242
  vinylcyclopropane, 205
Electrostatic field within zeolites, 287
Energy migration, 310, 320
Epoxidation, 300
Excited state ordering, metal ion effect,
  254

Faujasite, 276
Flavones, dimerization, 194
Fluorescence anisotropy, 319
Forster type energy transfer, 320
Free rotor effect, 13
Frei photooxidation, 290–298
  mechanism, 292

Geraniol, photocyclization, 204

Halocarbenonitrenes, 165–171
Hydrocarbon-oxygen complex, 293
Hydroxy radical, 290

Inorganic-organic composites, 311
Intramolecular electron transfer, 231

Lasers, dyes within zeolites, 343
Light emitting diodes, dyes within
  zeolites, 343
Limonene, oxygenation, 303

Marcus
  equation, 233
  inverted region, 244
  plot, 234
Markovnikov effect, 281
Merocynanine
  crystal structures, 378
  from naphthopyran, 383
  from spiro oxazine, 382
  ring closure
    photochemical, 387
    thermal, 384
    transient studies, 388–392
  structure
    dipole moment, 377
    NMR studies, 379–380
    solvent effect, 375
  triplet, 363
Metal ion-carbonyl complex, 256
Meta-phenylenecarbenonitrene, 163

Naphthaldehyde, metal ion effect, 253
Naphthopyran, ring opening, transient
  studies, 374
Naphthopyrans, photoreaction, 355
Norrish type I process, 17

1-Aza, 1,4 dienes, examples of aza-di-
  π-methane, 27–30
Ortho-phenylenecarbenonitrene, 173
Oxa-di-π-methane (ODPM), 3, 4
Oxazine-zeolite, 322–328
Oxidation of dienes in zeolites, 285
Oxidation of olefins in zeolites
  diastereoselectivity, 284
  mechanistic model, 283
Oxidation of organosulfides in zeolites,
  285–289

Oxidation via activation of
     hydrocarbons, 299
Oxygenation, classification, 277
Oxygenation via thermal electron
     transfer, 302
Oxygen-organic complex, stabilization
     by zeolites, 294

Para-phenylenecarbenonitrene, 150
Perepoxide intermediate, 281
Phenylcarbene, 136
     ground spin state, 136
Phenylnitrene, 136
Photoalkylation, electron transfer, 247
Photocatalysis, 238
Photodecarbonylation, 16
Photo-Fries
     amides, 75–85
     carbamates, 85
     carbonates, 85
     formates, 87
     isotope effect, 60
     magnetic field effect, 60
     mechanism, 45
     naphthol esters, 69, 71
     oxalates, 87
     phenolic esters, 67
     polymers, 109
     reactive state, 47
     selenoesters, 93
     solvent effect, 49
     substituent effect, 51
     sulfonates, 88
     synthetic applications, 97–106
     theoretical studies, 66
     thioesters, 92
     time resolved studies, 61
     transient studies, 54–66
     within cyclodextrins, 115
     within micelles, 116
     within nafions, 118
     within polyethylenes films, 119
     within supercritical CO_2, 120
     within zeolites, 116–118
Photonic antenna, 319

Photooxygenation on silicasurface, 291
Photoreduction within zeolites, 296
Photosynthetic reaction center, 228
Pigments, dyes within zeolites, 342
Polarization anisotropy, 331
Polarized
     fluorescence microscopy, 328
     red emission, 328
Porphyrin-fullerene linked molecules,
     233
Pthalimide, photocyclization, 211
Pyronine-zeolite, 321–328

Radical anions, stabilization by metal
     ions, 262
Radical cations, nucleophilic capture,
     193
Radical ion pair, 230, 245
Redox reactions, 238
Rehm-Weller equation, 186
Retropinacol reaction, 190

Silylamino enones, photocyclization,
     202
Silylenol ethers, photocyclization, 201
Singlet oxygen in zeolites, 278
Singlet oxygen reactions in zeolites,
     278–289
     regioselectivity, 280
Singlet-triplet splittings in carbenes,
     137
6-nitrobenzospiropyrans
     photoaggregation, 366
     ring opening, 363
     solvent effect, 364
Solar cells, dyes within zeolites, 345
Solvent reorganization energy, 248
Spirooxazine
     picosecond studies, 368
     subpicosecond studies, 370
Spirooxazines, photoreaction, 355
Spiropyran
     ring opening
          triplet sensitization, 397
          solid state photochemistry, 395

Spiropyrans
  photoreaction, 355
  ring opening, 357
    medium effect, 358
    picosecond studies, 359
    resonance Raman study, 359
    time resolved resonance
      Raman, 361
Stopcock molecules, 334
Superoxide, 241, 298

Tri-π-methane, 34
  2-Aza, 1,4 dienes, examples of aza-
    di-π-methane, 21

Vinylcyclopropanes, photocyclization,
  206

Zeolite L, 309
Zeolite, structure, 276

9 780367 395513